Springer Theses

Recognizing Outstanding Ph.D. Research

Aims and Scope

The series "Springer Theses" brings together a selection of the very best Ph.D. theses from around the world and across the physical sciences. Nominated and endorsed by two recognized specialists, each published volume has been selected for its scientific excellence and the high impact of its contents for the pertinent field of research. For greater accessibility to non-specialists, the published versions include an extended introduction, as well as a foreword by the student's supervisor explaining the special relevance of the work for the field. As a whole, the series will provide a valuable resource both for newcomers to the research fields described, and for other scientists seeking detailed background information on special questions. Finally, it provides an accredited documentation of the valuable contributions made by today's younger generation of scientists.

Theses are accepted into the series by invited nomination only and must fulfill all of the following criteria

- They must be written in good English.
- The topic should fall within the confines of Chemistry, Physics, Earth Sciences, Engineering and related interdisciplinary fields such as Materials, Nanoscience, Chemical Engineering, Complex Systems and Biophysics.
- The work reported in the thesis must represent a significant scientific advance.
- If the thesis includes previously published material, permission to reproduce this must be gained from the respective copyright holder.
- They must have been examined and passed during the 12 months prior to nomination.
- Each thesis should include a foreword by the supervisor outlining the significance of its content.
- The theses should have a clearly defined structure including an introduction accessible to scientists not expert in that particular field.

More information about this series at http://www.springer.com/series/8790

David A. Petrone

Stereoselective Heterocycle Synthesis via Alkene Difunctionalization

Bulky Phosphine Ligands Enable Pd-Catalyzed Arylhalogenation, Arylcyanation and Diarylation

Doctoral Thesis accepted by
the University of Toronto, Canada

Springer

Author
Dr. David A. Petrone
Laboratory for Organic Chemistry
ETH Zürich
Zürich
Switzerland

Supervisor
Prof. Mark Lautens
Department of Chemistry
University of Toronto
Toronto
Canada

ISSN 2190-5053 ISSN 2190-5061 (electronic)
Springer Theses
ISBN 978-3-030-08472-1 ISBN 978-3-319-77507-4 (eBook)
https://doi.org/10.1007/978-3-319-77507-4

Printed on acid-free paper

This Springer imprint is published by the registered company Springer International Publishing AG part of Springer Nature
The registered company address is: Gewerbestrasse 11, 6330 Cham, Switzerland

Supervisor's Foreword

The synthesis of medicinally interesting molecules and novel materials has been dramatically impacted by the discovery of catalytic methods to make carbon–carbon and carbon–heteroatom bonds. The vast majority of these reactions follow closely related mechanistic pathways. By far, the most common and fundamental processes involve oxidative addition and reductive elimination as key steps in the catalytic cycle. These two reactions are characterized by a change in oxidation state of the metal and the cleavage (oxidative addition) or formation (reductive elimination) of a bond of interest.

The most common process, as it concerns carbon–halogen (C–X) bonds, is oxidative addition, which is then followed by reaction of the metal–halogen bond and subsequent reductive elimination to form a C–C or C–Y bond. The inverse reaction, namely reductive elimination to form a C–X bond, is far less thoroughly studied and is generally not the favored pathway. Early work by Hartwig and Milstein focused on stoichiometric reactions and uncovered key requirements for successful C–X reductive bond formation. In particular, sterically bulky ligands and substrates were found to be essential to form C–Br bonds.

In 2010 and 2011, Stephan Newman from our group demonstrated catalytic reversible oxidative addition as a key step in the formation of Csp^2–Br and Csp^3–I bonds using palladium catalysts coordinated to bulky ligands such as ${}^tBu_3P$. In addition, Newman discovered that a C–C and C–I bond could be formed and called this reaction a carboiodination. His work set the stage for the thesis work of David Petrone.

Chapter 1 of the thesis investigated the diasteroselectivity in carbohalogentation reactions to make novel heterocyclic products. Among the scaffolds prepared by this methodology were isochromans, chromans, and dihydroisoquinolines. Petrone then used the carboiodination reaction in the formal synthesis of a natural product. His target was an alkaloid known as (+)-corynoline and he was able to prepare it in enantiomerically and diasteromerically enriched form using aryliodination as a key step.

Chapter 2 set out to address a particular challenge arising from the synthesis described above, namely, the displacement of a neopentyl C–I bond by cyanide. An alternative strategy was developed involving a carbocyanation rather than a carboiodination. An improved yield and efficiency of the synthesis was achieved by adding cyanide as a nucleophile to the reaction. With the goal of carbocyanation achieved, Petrone extended his study to addition of C–CN across aromatic systems such as indoles, resulting in a dearomative carbocyanation. Dearomatic processes have become of increasing interest, but less was known about difunctionalization of the C=C of the indole. A particular challenge was the easy epimerization at the benzylic cyanide that results from the carbocyanation. Fortunately, careful choice of solvent and catalyst overcame this hurdle.

Chapter 3 set out to generalize the carbofunctionalization of the indole, and a 1,2-arylation was achieved by combining carbopalladation with a Suzuki-type coupling to form a diarylation of the indole alkene. Aryl boroxines proved to be the ideal nucleophiles so that carbopalladation of the alkene proceeds faster than direct Suzuki coupling.

Chapter 4 highlights a particularly important finding when considering reversible oxidative addition of C–I bonds. Petrone prepared substrates with two aryl C–I bonds and, through a careful synthetic and mechanistic study, showed that each undergoes insertion by bulky Pd(0) complexes at a similar rate. By linking an acceptor in proximity to one of the aryl C–I bonds, it was possible to do a carboiodination while leaving the other aryl C–I bond "untouched." Of course that bond had also undergone reaction, but in a reversible fashion. In fact by adding a second acceptor (external), both C–I bonds could be orthogonally reacted. One underwent carboiodination, and the other a Heck reaction. This study demonstrated the power of reversible oxidative addition in polyhalogenation compounds.

Overall Petrone thesis built on a recent finding in the group and took the chemistry into completely new and unexpected ways. The potential of carbofunctionalization was revealed through synthetic and mechanistic work, and the reactions he discovered were applied to target molecules with biological activity.

Toronto, Canada
February 2018

Mark Lautens

Preface

My Ph.D. studies in the research group of Prof. Mark Lautens at the University of Toronto focused on using palladium catalysts modified by bulky phosphine ligands to promote the highly stereoselective formation of several classes of heterocycles. My initial venture into this research topic was initiated during study aimed to examine the stereoselective synthesis of chromans and isochromans via the cyclization strategy which has been designated "arylohalogenation" throughout this thesis. This project was initiated by a talented post-doctoral fellow named Dr. Hasnain Malik who mentored me during my studies. The knowledge gained throughout this project led to our group's more general interest in the stereoselective syntheses of other classes of halogenated heterocycles.

From here, in collaboration with Hyung Yoon and Dr. Harald Weinstabl, a highly scalable and asymmetric route to dihydroisoquinolinone scaffolds was developed using the aryliodination methodology. This reaction was later utilized in key step in our synthesis of the bioactive alkaloid natural product (+)-corynoline. The difficulties associated with a particular cyanation step within our validated synthetic route led to the development of an efficient and highly stereoselective arylcyanation reaction as a solution, which minimized the use of costly ligands and excess amounts of toxic cyanating reagents. Since the arylcyanation reaction forms two carbon–carbon bonds, all while incorporating a useful nitrile functional group handle, we aimed to expand the scope of this transformation to the preparation of fused indolines. With the assistance of Andy Yen and Nicolas Zeidan, an efficient and highly stereoselective preparation of fused indolines was accomplished using simple indole substrates in a dearomative process. This efficient dearomative difunctionalization concept was expanded to the stereoselective indole diarylation in a joint effort between myself and a visiting Japanese scholar named Dr. Masaru Kondo.

Substrates that are functionalized with a carbon–halogen bond allow for a high degree of regiocontrol in cross-coupling reactions since the metal catalyst has a high propensity to activate the molecule via oxidative addition at this site. However, because of the potential for over-coupling, non-site elective coupling, and catalyst deactivation, synthetic planning can become challenging when a molecule contains multiple carbon–halogen bonds. A visiting German student Matthias Lischka and I developed a protocol to utilize diiodinated aromatic compounds in both a site-selective Pd-catalyzed intramolecular aryliodination and an intramolecular aryliodination/intermolecular Mizoroki-Heck reaction. These two processes relied on the unique ability of the judiciously chosen palladium catalyst to undergo reversible oxidative addition to aromatic carbon–halogen bonds, and this rare characteristic is what allows this reaction to proceed to high levels of conversion.

In general, this thesis brings to light the true power of sterically hindering ligands beyond their commonplace in challenging cross-couplings. Their unique characteristics have opened up several new avenues regarding carbon–carbon and carbon–halogen bond formation, and are surely to have continued and expanding impact on many fields which rely on catalysis to prepare molecules via the formation of bonds that would otherwise be difficult.

Zürich, Switzerland David A. Petrone
January 2018

Parts of this thesis have been published in the following journal articles:

1. *"Modern Transition Metal-Catalyzed Carbon−Halogen Bond Formation,"* **Petrone, D.A.**; Ye, J.; Lautens, M. *Chem. Rev.* **2016**, *116*, 8003–8104.
2. *"Pd-Catalyzed Dearomative Diarylation of Indoles,"* **Petrone, D.A.**; Kondo, M.; Zeidan, N.; Lautens, M. *Chem. Eur. J.* **2016**, *22*, 5684.
3. *"Dearomative Indole Bisfunctionalization via a Diastereoselective Palladium-Catalyzed Arylcyanation,"* **Petrone, D.A.**; Yen, A.; Zeidan, N.; Lautens, M. *Org. Lett.* **2015**, *17*, 4838–4841.
4. "*Pd-Catalyzed Carboiodination: Early Developments to Recent Advancements*," **Petrone, D.A.***, Le, C.M.*; Newman, S.G.; Lautens, M., in *New Trends in Cross-Coupling: Theory and Application*, T. Colacot, Ed.; The Royal Society of Chemistry: Cambridge, **2015**, chapter 7. *Equal contribution.
5. *"Synergistic Steric Effects in the Pd-Catalyzed Alkyne Carbohalogenation: Synthesis of Tetrasubstituted Vinyl Halides,"* Le, C.M.; Menzies, P.; **Petrone, D. A.**; Lautens, M.: *Angew. Chem. Int. Ed.* **2015**, *54*, 254–257.
6. *"Diastereoselective Palladium-Catalyzed Arylcyanation/Heteroarylcyanation of Enantioenriched N-Allyl Carboxamides,"* Yoon, H.*; **Petrone, D.A.*;** Lautens, M.: *Org. Lett.* **2014**, *16*, 6420–6423. *Equal contribution.
7. *"Additive Effects in the Pd-Catalyzed Carboiodination of Chiral N-Allyl Carboxamides,"* **Petrone, D.A.**; Yoon, H.; Weinstabl, H.; Lautens, M.: *Angew. Chem. Int. Ed.* **2014**, *53*, 7908–7912.
8. *"Harnessing Reversible Oxidative Addition: Application of Diiodinated Aromatic Compounds in the Carboiodination Process,"* **Petrone, D.A.**; Lischka, M.; Lautens, M.: *Angew. Chem. Int. Ed.* **2013**, 52, 10635–10638.
9. *"Functionalized Chromans and Isochromans via a Diastereoselective Pd-Catalyzed Carboiodination."* **Petrone, D.A.**; Malik, H.A.; Clemenceau, A.; Lautens, M.: *Org. Lett.* **2012**, *14*, 4806–4809.
10. *"A Conjunctive Carboiodination: Indenes by a Double Carbopalladation–Reductive Elimination Domino Process."* Jia, X.; **Petrone, D.A.**; Lautens, M.: *Angew. Chem. Int. Ed.* **2012**, *51*, 9870–9872.

Acknowledgements

First and foremost, I thank Prof. Mark Lautens for his exceptional guidance and support throughout my Ph.D. studies. The experiences that I have been afforded while in his research group have been second to none, and I have always been given the opportunity to test my own hypotheses and to be creative. I sincerely thank you for accepting me into your research group during a time where you were by no means obligated to.

I wish to extend my deepest gratitude to the exceptional friends and colleagues whom I have had the privilege to work alongside over the past 5 years. The Lautens group is a revolving door of scientists where I have met over one hundred amazing people. I wish all of you the best. To Dr. Hasnain Malik, thank you for being the first to show me the ropes in the Lautens group and for being an excellent mentor. To Dr. David Candito, I am deeply indebted to you for your patience and assistance. Your passion and enthusiasm for research is contagious, and you taught me that the path to success is never straight. To Zafar Quereshi, thanks for always being a voice of logic and reason. To Drs. (Prof.) Xiaodong Jia, Patrick Liu, Harald Weinstabl, Juliane Kelitz, Marcel Sickert, Steffen Kress, Charles Loh, and Ivan Franzoni, I thank you for each bringing a unique and valuable outlook on life with you here to Canada. Together you have imparted a library of knowledge on me. To Dr. Juntao Ye, you have been a fantastic mentor and I am fortunate to have worked alongside you for so long.

To all of the lab three mates especially Dr. Christine Le, Hyung Yoon, Perry Menzies, Richard Huang, and Jordan Evans, thank you for the great times and conversations. To Drs. Lei Zhang and Jenn Tsoung as well as Mo El-salfiti, Adam Friedman, Thomas Johnson, Andy Yen, Nicolas Zeidan, and Alvin (Young-Jin) Jang, it was a pleasure to work with you, and I wish you all the best in the future.

During my time in the Lautens group, I have had the honor to work with many visiting students. Much of their work has gone into the success of the projects either mentioned or not mentioned in this thesis. Thank you to Dr. Antonin Clemenceau, Geoffrey Barral, Martin Mitura, Matthias Lischka, and Dr. Masaru Kondo. Your hard work and dedication never went unnoticed.

My foray into chemistry began with conversions with Dr. Robert Reed from the University of Guelph. Thank you for your enthusiasm and refreshingly honest viewpoint about science. My first research experience was in the research group of Prof. Kathryn Preuss. Thank you for your guidance, as well fostering a fantastic group to work in. Professor William Tam was the one who inspired me to become an organic chemist. I am extremely thankful for all the help that you have given me over the years. I have been fortunate to have great committee members during my time here. I wish to thank Profs. Mark Taylor, Andrei Yudin, and Datong Song for their guidance, as well as their time spent reading this thesis. Professor Bruce Arndtsen is gratefully acknowledged for taking the time to read this thesis and providing helpful feedback.

We are very fortunate to have world-class technical scientists that support our daily research in the Department of Chemistry at the University of Toronto. Without your hard work and dedication, we would be stuck at a standstill. I would specifically like to thank Drs. Darcy Burns and Timothy Burrows as well as Dmitry Pichugin for the immense number of hours they have invested, providing us with a top-notch NMR facility. Dr. Alan Lough and Dr. Matthew Forbes have also been instrumental in our X-ray and mass spectrometry endeavors, respectively.

To Michelle, progressing through graduate school has always been easier because of your undying support and understanding.

Lastly, I would like to thank my parents. None of this would have been possible without you. Thank you for putting your children first, and always planning for the future. My appreciation stretches the length of this manuscript.

Contents

Abbreviations

Å	Angstrom
aq.	Aqueous
BINAP (*R*) or (*S*)	2,2′-bis(diphenylphosphino)-1,1′-binaphthyl
Bu	Butyl
Bz	Benzoyl
cat.	Catalyst or catalytic
Cat	Catechol
crotyl	But-2-en-1-yl
Cy	Cyclohexyl
DABCO	1,4-diazabicyclo[2.2.2]octane
DCM	Dichloromethane (CH_2Cl_2)
dba	Dibenzylideneacetone
DIBAL	Diisobutylaluminum hydride
DIOP	2,3-*o*-isopropylidene-2,3-dihydroxy-1,4-bis(diphenylphosphino)butane
dippb	Diisopropylphosphinobutane
DME	1,2-dimethoxyethane
DMF	*N,N*-dimethlyformamide
DMA	*N,N*-dimethlyacetamide
DMSO	Dimethylsulfoxide
dppb	Bis(diphenylphosphino)butane
dppf	1,1′-Bis(diphenylphosphino)ferrocene
dppp	Bis(diphenylphosphino)pentane
EDG	Electron-donating group
EWG	Electron-withdrawing group
ee	Enantiomeric excess
EI	Electron impact
ESI	Electrospray ionization
equiv.	Equivalent(s)
Et	Ethyl

EtOAc	Ethyl acetate
FMO	Frontier molecular orbitals
g	Gram(s)
h	Hour(s)
HRMS	High-resolution mass spectrometry
IR	Infrared
J	Coupling constant (NMR spectroscopy)
L	Generic ligand
LG	Leaving group
m	*Meta*
MeCN	Acetonitrile
Me	Methyl
MHz	Megahertz
Min	Minute(s)
mL	Milliliter(s)
mol	Mole(s)
mp	Melting point
Ms	Methanesulfonyl
MS	Molecular sieves or mass spectrometry
m/z	Mass over charge
NBS	*N*-bromosuccinimide
NF	Nonaflate (nonafluorobutanesulfonate)
NMR	Nuclear magnetic resonance
NMP	*N*-methyl-2-pyrrolidone
o	*Ortho*
p	*Para*
Ph	Phenyl
PMB	Para-methoxybenzyl
PMP	1,2,2,6,6-pentamethylpiperidine
ppm	Parts per million
Pr	Propyl
QPhos	1,2,3,4,5-Pentaphenyl-1′-(di-*tert*-butylphosphino)ferrocene
quant.	Quantitative
R	Generic chemical group
RT	Room temperature
s	Second(s)
S_N2	Bimolecular nucleophilic substitution
S_NAr	Nucleophilic aromatic substitution
T	Temperature
TBAB	Tetrabutylammonium bromide
TBAI	Tetrabutylammonium iodide
^{t}Bu	*Tert*-butyl
TEMPO	2,2,6,6-tetramethylpiperidin-1-yl)oxyl
TFA	Trifluoroacetate
THF	Tetrahydrofuran

TLC	Thin-layer chromatography
TMEDA	Tetramethylethylenediamine
TMS	Tetramethylsilyl
Tol	Tolyl
μW	Microwave
X	Generic halide/heteroatom
Y	Generic halide/heteroatom
Z	Generic halide/heteroatom

List of Figures

List of Tables

List of Schemes

Chapter 1
Diastereoselective Synthesis of Heterocycles via Intramolecular Pd-Catalyzed Alkene Aryliodination

1.1 Introduction

1.1.1 Carbon–Carbon Bond Formation via Pd-Catalyzed Cross-coupling

The construction of the carbon–carbon single bond is one of the most important focus in organic synthesis. Due to the ubiquity of this bond in organic molecules, as well as the infinite number of both substitutional and stereochemical combinations, the development of diverse methods which forge this bond has remained a strong focus. As a testament to the value of these methods, the 2010 Nobel Prize in chemistry was awarded to Richard F. Heck, Ei-ichi Negishi and Akira Suzuki for "*palladium-catalyzed cross-coupling in organic synthesis*".[1,2] This important recognition prompted a rapid increase in the world-wide effort to invent new, and elaborate on existing classes of Pd-catalyzed cross-couplings.

Barring some variations, cross-couplings reactions generally occur between an organo(pseudo)halide electrophile **1.1** and an organometallic nucleophile **1.2**. When these two partners react in the presence of a catalyst (i.e. Pd, Ni) a product **1.3** containing a new carbon–carbon or carbon–heteroatom bond is generated with concomitant formation of a metal halide byproduct **1.4** (Scheme 1.1). The electrophilic component can typically be an alkyl, vinyl or aryl (pseudo)halide, whereas the nucleophilic component is typically an alkyl, vinyl or aryl boron (Suzuki-Miyaura),

Portions of this chapter have appeared in print. See Refs. [1–4].

[1]"The Nobel Prize in Chemistry". Nobelprize.org, 6 April 2016 http://www.nobelprize.org/nobel_prizes/chemistry/laureates/2010/.
[2]Reference [5].

D. A. Petrone, *Stereoselective Heterocycle Synthesis via Alkene Difunctionalization*, Springer Theses, https://doi.org/10.1007/978-3-319-77507-4_1

Scheme 1.1 General representation of the cross-coupling reaction

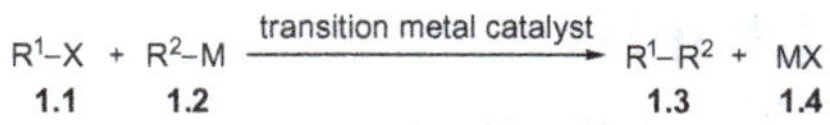

Scheme 1.2 General representation of the Mizoroki-Heck reaction

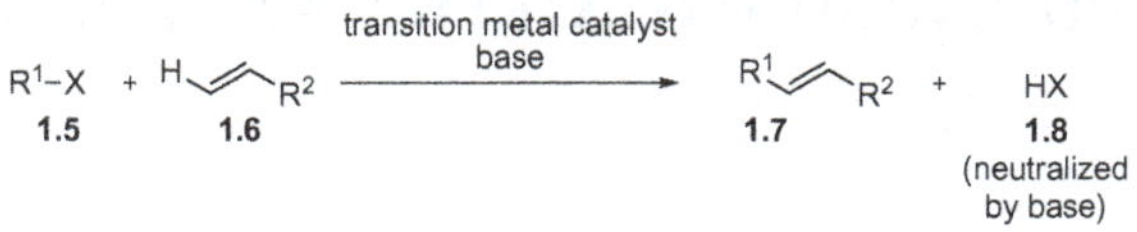

tin (Stille), zinc (Negishi), magnesium (Kumada-Corriu), silicon (Hiyama-Denmark) or metalated amide (Buchwald-Hartwig) reagent.[3]

However, the above classification of cross-coupling reactions does not encompass the Mizoroki-Heck reaction. This formal olefin C–H functionalization typically utilizes a palladium catalyst to couple an organo(pseudo)halide **1.5** with an olefin **1.6** containing at least one vinylic C–H bond. This reaction produces a new olefin-containing product **1.7** with concomitant formation of HX **1.8** (Scheme 1.2). The latter component is typically neutralized by a stoichiometric amount of base which allows for catalyst turnover.

Furthermore, there are numerous other so-called cross-coupling reactions which do not precisely fit the above categories. The Fujiwara-Moritani reaction accomplishes the vinylation of an aromatic C–H bond under oxidative conditions, and negates the need to utilize pre-functionalized organohalide components (i.e. **1.5**). The Sonogashira reaction couples terminal alkynes with aryl halides, yet does not rely on the use a stoichiometric organometallic nucleophile (i.e. **1.2**). Instead, a copper acetylide is generated from the terminal alkyne in catalytic amounts in the presence of base and a Cu(I) salt, and effectively transmetallates its acetylene ligand to palladium (vide infra). The traditional Tsuji-Trost allylation reaction uses allylic halides, acetates and carbonates as the electrophile in the presence of pre-formed metal enolates (e.g. Na salt of dimethyl malonate) to forge allylic C–C bonds.

1.1.1.1 General Mechanisms for Cross-couplings and the Mizoroki-Heck Reaction

Generally, the mechanisms of palladium-catalyzed cross coupling and the Mizoroki-Heck reaction overlap in regards to only the activation of the organohalide electrophile, which occurs by oxidative addition of a Pd(0) complex to the carbon–halogen bond. The former reaction class proceeds by initial oxidative addition of the Pd(0) pre-catalyst to carbon–halogen bond of the organohalide electrophile to generate $R^1Pd(II)X$ species **1.9**. The nucleophile can transmetallate

[3]Reference [6].

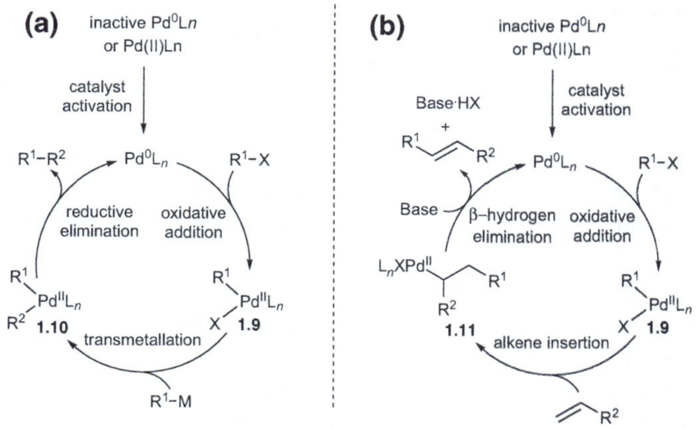

Scheme 1.3 General mechanisms for **a** classical cross-coupling reactions; and **b** the Mizoroki-Heck reaction

its carbon-based fragment to species **1.9**, in turn generating $R^1Pd(II)R^2$ complex **1.10** which undergoes carbon–carbon, carbon–oxygen, carbon–nitrogen or carbon-sulfur bond-forming reductive elimination (Scheme 1.3a). The Mizoroki-Heck reaction mechanism involves the production of **1.9** in a similar fashion, yet instead of undergoing a transmetallation with an organometallic reagent, it undergoes alkene migratory insertion in the presence of an olefin partner to generate 2° alkyl Pd(II) complex **1.11**. Since species **1.11** is in the presence of suitable β-hydrogen atoms, it may undergo β-hydride elimination to generate the olefin-containing product.

Oxidative Addition of Metal Complexes to Carbon–Halogen Bonds[4]

The elementary step of oxidative addition typically involves an active low valent Pd (0) complex undergoing heterolytic cleavage of a polarized carbon–halogen bond via a concerted, three-centered transition state (Scheme 1.4).[5,6] Overall, palladium donates two non-bonding electrons to form two new metal–ligand bonds which are oriented *cis* to one another (i.e. *cis*-**1.12**). Complex **1.12** can undergo rearrangement to the thermodynamically favored *trans*-**1.13**.[7] The classical order of reactivity with

[4]See Ref. [3].

[5]Reference [7].

[6]Reference [8].

[7]Reference [9].

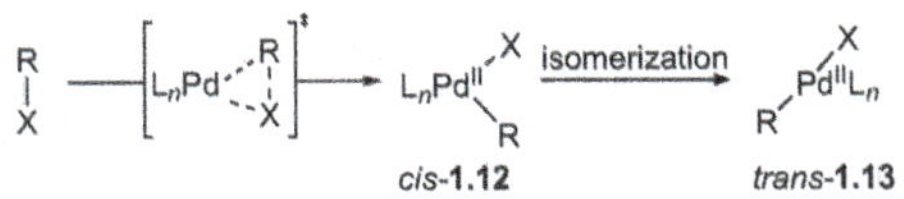

Scheme 1.4 General mechanism for oxidative addition of a Pd(0) complex to an organohalide

respect to aryl halides is I > Br > Cl ≫ F, where C–F bonds are highly unreactive. However, recent efforts in the area are actively shifting this longstanding paradigm.[8,9,10,11]

Oxidative addition is generally promoted by strongly σ-donating ligands which increase electron density at palladium. In these cases, the Pd center becomes more susceptible to donate an electron pair to facilitate carbon–halogen bond breakage.[12,13] Conversely, π-accepting ligands counteract this process by rendering palladium more electron poor. Sterics also play an important (albeit less straightforward) role in oxidative additions. Although oxidative addition results in increased steric bulk around the metal center by directly increasing the number of ligands by two, large and sterically hindered ligands are typically used to favor challenging oxidative additions. Although somewhat counterintuitive, this behavior is rationalized by the fact that extremely bulky ligands drive the formation of highly active PdL_1 or PdL_2 species in situ, which have more open coordination sites overall to permit rapid oxidative addition.[14]

Transmetallation

Transmetallation refers to the replacement of the pseudo(halide) ligand in the Pd(II) complex with the organic component of the organometallic reagent.[15] However, this step can be generally characterized by the transfer of ligands between two metal species. Furthermore, due to the fact that a number of organometallic reagents can undergo transmetallation with ArPd(II)X complexes, there is a continuum of mechanistic scenarios which are possible for this crucial step. In many cases, transmetallation can act as a driving force in the reaction due to the formation of a MX species. Numerous studies have been conducted to gain insight into the

[8] Reference [10].
[9] Reference [11].
[10] Reference [12].
[11] Reference [13].
[12] Reference [14].
[13] Reference [15].
[14] Reference [16].
[15] Reference [17].

transmetallation of various organometallic reagents including organosilicon,[16,17] organoboron,[18,19,20,21] and organotin reagents.[22,23,24,25]

Reductive Elimination from Transition Metal Complexes[26]

Reductive elimination results in the formation of a new bond between two suitable *cis*-oriented ligands on a metal center **1.14**. Generally, this process is thought to proceed via a three-member transition state (i.e. **1.15**) whereby the oxidation state of the metal is reduced by two (Scheme 1.5). This key elementary transformation can be promoted by various factors, including steric and electronic perturbations of the ancillary ligand(s) and substrate. The factors which affect the ability of a suitable transition metal complex to undergo reductive elimination are generally opposite to those seen for the corresponding oxidative addition processes. In this regard, electron-poor metal centers are better candidates to undergo reductive elimination when compared with electron-rich metal centers with similar steric properties. This is because as this step proceeds through the transition state, there is an increase of electron density at the metal center. Additionally, a metal complex which possesses more sterically encumbering ligands will generally lead to faster rates of reductive elimination when compared to metal complexes possessing less bulky ligands with similar electronic properties.[27]

Olefin Insertion into Metal–Carbon Bonds

The main factor separating the Mizoroki-Heck cross-coupling reaction and other cross-couplings reactions is the insertion of an olefin into a palladium–carbon bond. This results in the formation of a new carbon–carbon bond as well as a new carbon–palladium bond. Generally, coordination of the olefin to the $R^1Pd(II)X$ complex

---

[16]Reference [18].

[17]Reference [19].

[18]References [20, 21].

[19]Reference [22].

[20]Reference [23].

[21]Reference [24].

[22]Reference [25].

[23]Reference [26].

[24]Reference [27].

[25]Reference [28].

[26]See Ref. [3].

[27]Reference [29].

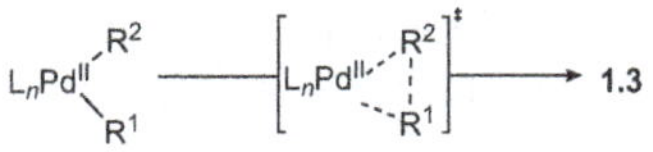

1.14 1.15

Scheme 1.5 General mechanism for reductive elimination from a $R^1Pd(II)R^2$ complex

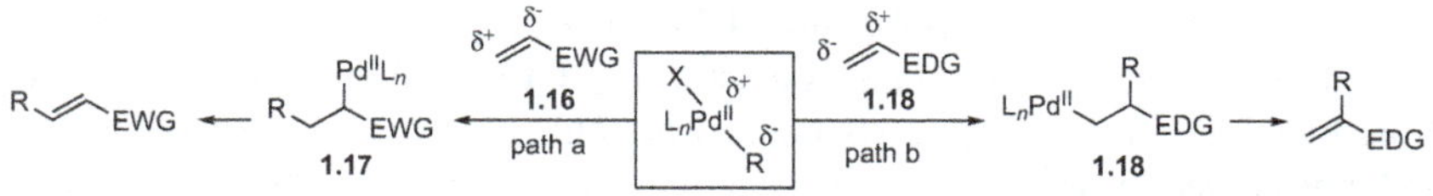

Scheme 1.6 Regioselectivity outcomes for the insertion of olefins containing electron-withdrawing (path a); and electron-donating substituents (path b)

occurs prior to this step, and may either occur via an associative displacement of a monodentate ligand from the metal center, or it may proceed by replacement of the halide ligand leading to a cationic Pd(II) species.[28] This step of the reaction mechanism determines the regiochemical outcome of the reaction. Furthermore, activated olefins are generally employed due to their greater ability to coordinate and ultimately undergo insertion. Olefins possessing electron withdrawing substituents **1.16** (i.e. acrylates, α,β-unsaturated ketones, and nitroalkenes) will result in the new carbon–carbon bond being formed at the β-position (**1.17**, Scheme 1.6a), whereas olefins possessing electron-donating substituents **1.18** (i.e. enol ethers) will result in the formation of the new carbon–carbon bond at the α-position adjacent the heteroatom (**1.19**, Scheme 1.6b). This is due to the partial positive (δ^+) and negative (δ^+) charges carried by both the metal–carbon bond and the olefin partner.

β-Hydrogen Elimination

β-Hydrogen elimination is the product-forming step in the mechanism of the Mizoroki-Heck reaction, wherein the carbon–carbon double bond is reformed. Since this process is the microscopic reverse of a migratory insertion reaction of an olefin into a metal–hydride complex, they are frequently followed by reinsertion of the resulting metal–hydride to the coordinated olefin. In many cases, this leads to olefin isomerization which produces isomeric mixtures of olefin products. β-hydrogen elimination does not produce a change in the oxidation state of the metal center, and requires a coordinatively unsaturated metal which is in a *syn*-co-planar orientation with respect to at least one β-hydrogen atom (i.e. complex **1.20**). These eliminations are likely initiated by an agostic interaction between the

[28]Reference [17].

Scheme 1.7 General representation of the β-hydrogen elimination

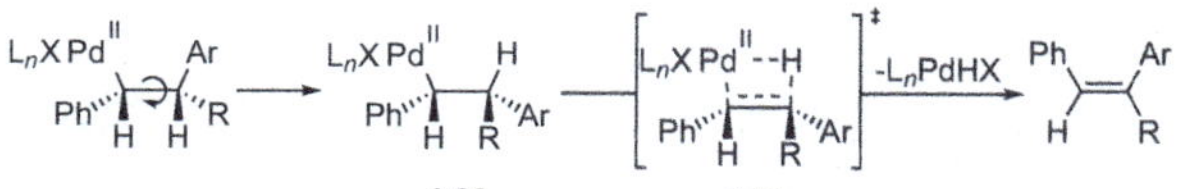

two electrons of the carbon–hydrogen sigma bond and an empty d-orbital on the metal center. This results in a three-center two-electron bond and a net elongation of the carbon–hydrogen bond prior to the elimination which occurs via a 4-membered transition state **1.21** (Scheme 1.7).[29]

1.1.1.2 The Mizoroki-Heck Reaction

In 1969 Heck and Fujiwara independently reported the arylation of olefins using aryl mercury[30] and aryl iodide reagents[31] in the presence of stoichiometric amounts of Pd salts. These advancements led to Mizoroki[32] and Heck[33] independently discovering that these reactions could occur using catalytic quantities of palladium. In the seminal manuscript entitled "*Arylation of Olefin with Aryl Iodide Catalyzed by Palladium*", Mizoroki and co-workers reported the arylation of ethylene, propylene, styrene, and methyl acrylate using iodobenzene and $PdCl_2$ as a pre-catalyst at 120 °C in an autoclave (Scheme 1.8a).[34] Interestingly, this reaction performed with the same level of efficiency when $PdCl_2$ was pre-reduced to Pd black. A full paper was later published in 1973 wherein these authors detailed their complete range of studies.[35] In a manuscript entitled "*Palladium-Catalyzed Vinylic Hydrogen Substitution Reactions with Aryl, Benzyl, and Styryl Halides*" Heck and Nolley showed that $Pd(OAc)_2$ was an efficient pre-catalyst for the vinylation reaction of various organohalides (Scheme 1.8b).[36]

Subsequent studies by Heck outlined the positive effect of phosphine ligands on this process.[37] They found that a 2:1 PPh_3:$Pd(OAc)_2$ ratio was optimal in obtaining satisfactory reaction rate between electron-rich, -neutral and -poor aryl bromides and a series of simple olefins (Scheme 1.9). Aryl chlorides were not reactive under these reactions conditions despite the previous observation that $Pd(PPh_3)_4$ could oxidatively add to the C–Cl bond of chlorobenzene at 135 °C.[38]

---

[29]Reference [17]

[30]References [30–36].

[31]Reference [37].

[32]Reference [38].

[33]Reference [39].

[34]Reference [38]

[35]Reference [40].

[36]Reference [39]

[37]Reference [41].

[38]Reference [42].

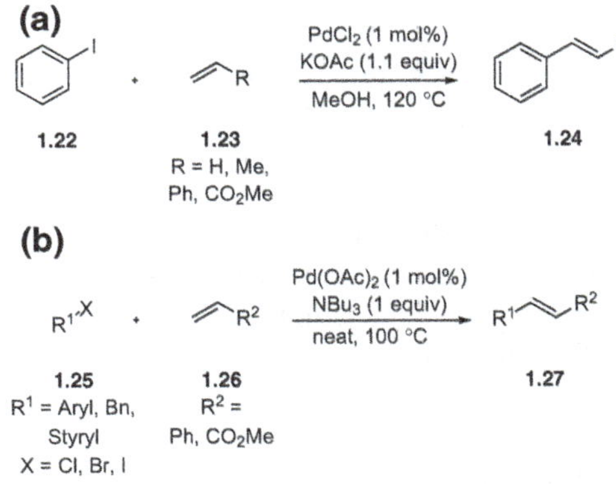

Scheme 1.8 Catalytic arylation of simple olefins catalyzed by Pd salts

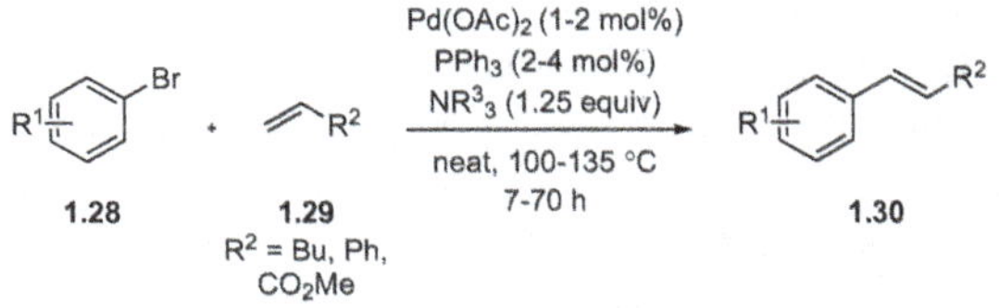

Scheme 1.9 Organophosphinepalladium catalyzed vinylation of aryl bromides

This report contained the proposal of a comprehensive mechanism with PPh_3 as a ligand (Scheme 1.10). The Pd(II) pre-catalyst is reduced to its Pd(0)L_2 species in the presence of the PPh_3 ligand and this can undergo facile oxidative addition to the arly halide to generate ArPd(II)X species **1.31**. Ligand exchange with the olefin component occurs to generate complexes **1.32** which can undergo migratory insertion to the coordinated alkene to generate alkyl Pd(II)X complex **1.33**. This species can undergo β-hydrogen elimination to generate the product olefin bound complex **1.34** which can undergo dissociation of the product via ligand exchange and base-mediated reduction (by removal of the HX byproduct) to regenerate the active Pd(0)L_2 catalyst. Further modifications by these authors led to the inclusion of vinyl bromides into the scope of this process.[39,40]

Shortly thereafter, Ziegler and Heck found that the highly sterically hindered triaryl phosphine ligand P(*o*-Tol)$_3$ was optimal for promoting the vinylation of highly activated, yet previously unreactive aryl bromides such as *p*-bromophenol and *p*-bromoaniline (Scheme 1.11).[41]

---

[39]Reference [43].

[40]Reference [44].

[41]Reference [45].

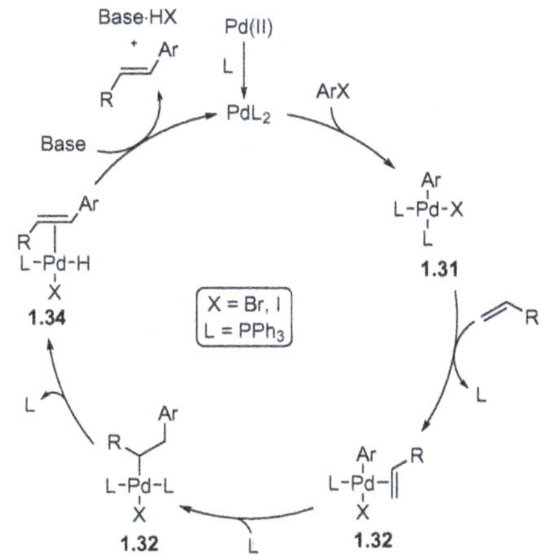

Scheme 1.10 Heck's proposed mechanism for the vinylation of aryl halides with PPh_3 as a ligand

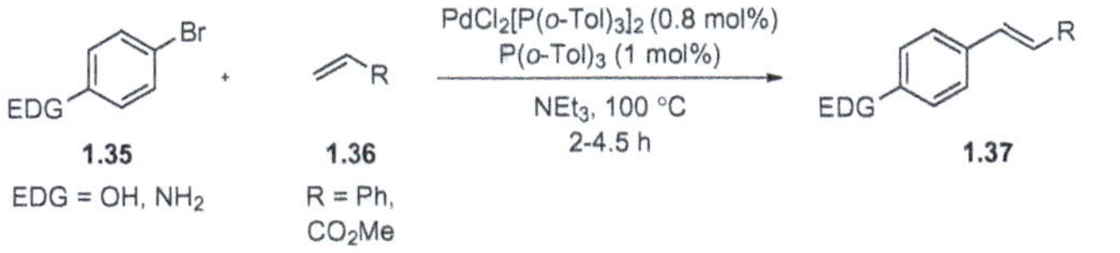

Scheme 1.11 P(*o*-Tol)$_3$ enables the Pd-catalyzed vinylation of electron rich aryl bromides

Due to the low reactivity of aryl chlorides, their use in such reaction have traditionally resulted in low yields.[42,43] Based on the work of Davison,[44] and their own work on the carbonylation[45] and formylation[46] of aryl chlorides using chelating bisphosphines, Milstein and co-workers reported that the dippb can be used as a ligand to promote an efficient catalytic vinylation of chlorobenzene at elevated temperatures (Scheme 1.12).[47,48] If the ArPd(II)Cl complex resulting from oxidative addition to the aryl chloride were to have both phosphines coordinated at the same time, there would be no vacant coordination site for the olefin partner to bind to. Therefore, they proposed that instead of halide dissociation occurring to generate a cationic Pd species, one of the phosphines of dippb may dissociate to

[42]Reference [46].
[43]Reference [47].
[44]Reference [48].
[45]Reference [49].
[46]Reference [50].
[47]Reference [51].
[48]Reference [52].

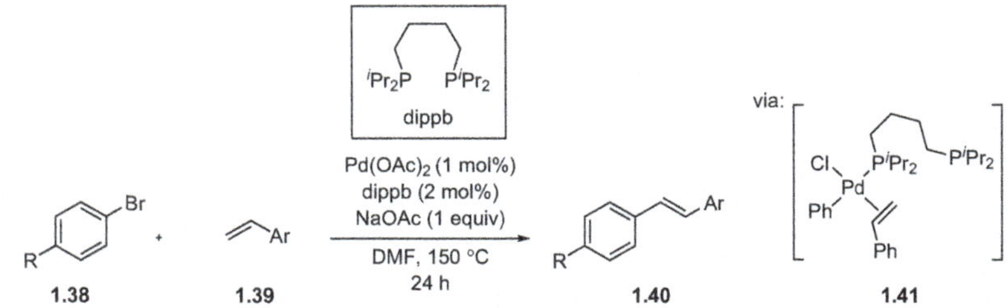

Scheme 1.12 Milstein's vinylation of chlorobenzene using dippb as a labile chelating ligand

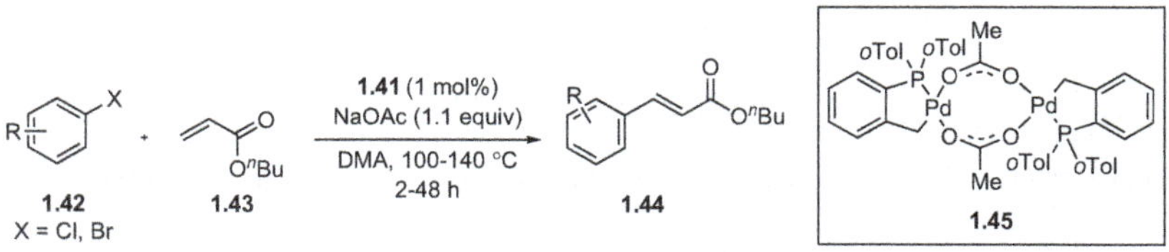

Scheme 1.13 Herrmann's use of palladacycles as catalysts for the vinylation of aryl halides

vacate one coordination site. The authors mentioned that dippb forms a weaker chelate than the 2- and 3-carbon analogs dippe and dippp, and allows for intermediate **1.41** to be more readily formed.

In 1995, Herrmann reported that structurally well-defined palladacycles could be used as catalysts for the vinylation of aryl bromides and chlorides.[49] By carrying out kinetic studies, the authors found that catalyst systems comprised of $Pd(OAc)_2$ and triaryl phosphines undergo rapid deactivation at elevated temperatures by P–carbon bond cleavage at the stage of ArPd(II)X. The authors suggested that this decomposition pathway is what had thwarted the previous use of aryl chlorides and deactivated aryl bromides. By using palladacycle **1.45** the authors were able to efficiently couple both aryl bromides and chlorides with *n*-butyl acrylate at elevated temperatures (Scheme 1.13).

Further advancements have also been made which involve the use of bulky *N*-heterocyclic carbene,[50] phosphite[51] and diadamantyl-*n*-butylphosphane ligands,[52] as well as molten salts as reaction media,[53] and tetraphenylphosphonium salt additives.[54] Nevertheless, the scope of the reaction with respect to both the aryl

[49]Reference [53].

[50]Reference [54].

[51]Reference [55].

[52]Reference [56].

[53]Reference [57].

[54]Reference [58].

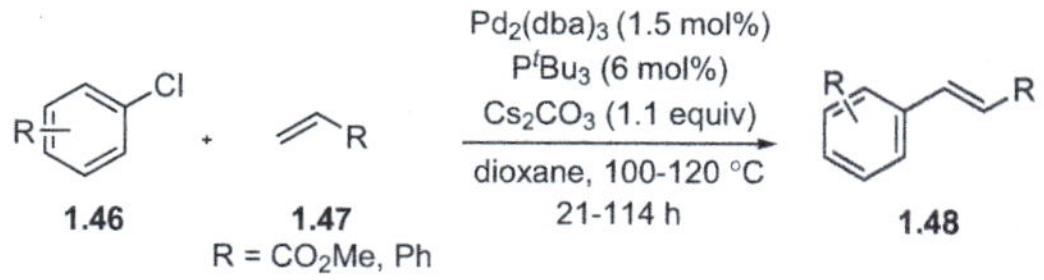

Scheme 1.14 $Pd_2(dba)_3/P^tBu_3$-catalyzed Mizoroki-Heck reaction of aryl chlorides

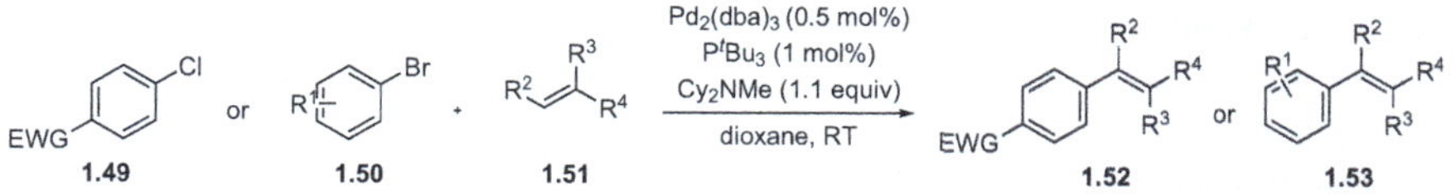

Scheme 1.15 Room temperature Mizoroki-Heck reaction of and aryl bromides and activated aryl chlorides

chloride and the olefin component remained rather limited until Fu and Littke reported that a $Pd_2(dba)_3/P^tBu_3$ combination could catalyze the Mizoroki-Heck coupling of broad range of aryl chlorides at 100–120 °C (Scheme 1.14).[55] Under these new reaction conditions, the authors described the observation of TONs of up to ~400 for the reaction between chlorobenzene and styrene.

Due to the fact that these conditions still involved relatively high temperatures and only tolerated a relatively narrow scope of olefin coupling partners, the same authors reported modified reactions conditions for the vinylation of aryl bromides and chlorides in 2001.[56] These modified conditions involved a 1:1 $Pd{:}P^tBu_3$ ratio in addition to the use of the bulky tertiary amine base Cy_2NMe (Scheme 1.15). However, non-activated aryl chlorides still required higher temperatures (70–120 °C), despite the fact that a broader range of olefin coupling partners could now be employed with these substrates.

This latter study was preceded by a report from Hartwig wherein they described the use of di(*tert*-butyl)phosphinoferrocene as a ligand for the Pd-catalyzed Mizoroki-Heck reaction of aryl chlorides.[57] This discovery was expedited by the use of a fluorescence assay which facilitated rapid screening of over 40 ligands.

This summary of the development of the intermolecular Mizoroki-Heck reaction is by no means comprehensive, yet highlights the key advancements in the field which were made in the hope of generating a more industrially-friendly carbon–carbon coupling reaction. The importance of this key carbon–carbon bond-forming

[55]Reference [59].
[56]Reference [60].
[57]Reference [61].

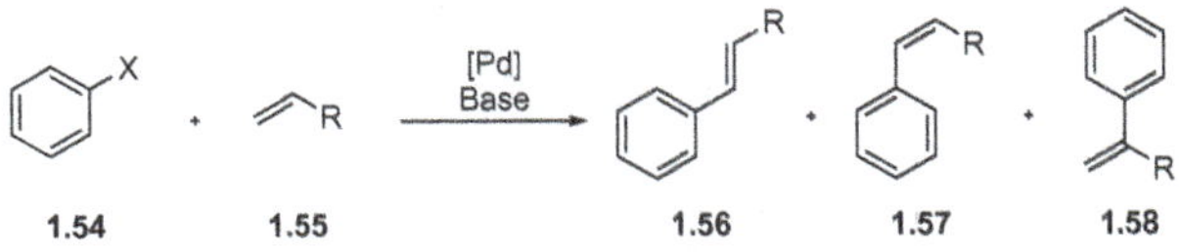

Scheme 1.16 Possible product distribution for the intermolecular Mizoroki-Heck reaction

reaction has been made clear by the significant number of comprehensive reviews and book chapters written on the topic.[58,59,60,61,62,63,64,65]

The Intramolecular Mizoroki-Heck Reaction

Despite the power of the intermolecular Mizoroki-Heck reaction, several selectivity issues exist. Even in cases where a monosubstituted olefin is employed as a coupling partner, multiple products may be observed (Scheme 1.16). The product distribution depends on both the regiochemical outcome of the migratory insertion and the selectivity of the β-hydride elimination. The former factor dictates which regioisomer will be the major product (i.e. **1.56/1.57** vs. **1.58**), whereas the latter dictates which double bond isomer will be predominant (i.e. **1.56** vs. **1.57**). Therefore, biased olefin partners like acrylates and styrenes are typically used to circumvent such issues.

A characteristic which has been the diversifying feature of the Mizoroki-Heck reaction with respect to other classes of cross couplings, is its high aptitude for intramolecular cyclizations (i.e. where the olefin component is tethered to the aryl halide). Carrying out these reactions in an intramolecular manner usually alleviates the aforementioned issues of regioselectivity as certain cyclization modes are highly favored over others (i.e. 5-*exo-trig* over 6-*endo-trig*).[66] In general, these cyclization reactions have been shown in countless examples to produce complex carbo- and heterocyclic products of various ring sizes. These reactions can also be carried out in a diastereoselective fashion, wherein a stereocenter is present on the tether between the aryl halide and the alkene moiety, or in an asymmetric fashion wherein a Pd catalyst that is modified by a chiral ligand is employed (vide infra).

[58]Reference [62].

[59]Reference [63].

[60]Reference [64].

[61]Reference [65].

[62]Reference [66].

[63]Reference [67].

[64]Reference [68].

[65]It it should be noted that other ligands beyond those in the class of PR_3 (i.e. NHC ligands) have been found to be useful in Pd-catalyzed cross couplings. See: Refs. [69, 70].

[66]Reference [71].

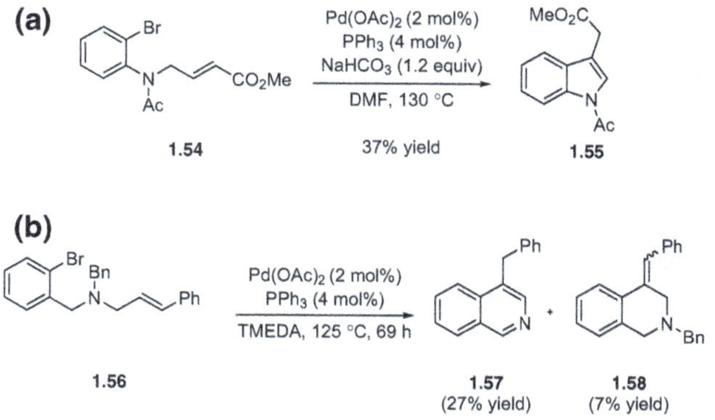

Scheme 1.17 Intramolecular Mizoroki-Heck reaction en route to **a** indoles; and **b** isoquinolines

The development of the intramolecular Mizoroki-Heck reaction began in the late 1970s, and all examples prior to 1983 involved the synthesis of heterocycles.[67] It appears that the first reported heterocycle synthesis via an intramolecular Mizoroki-Heck reaction was in 1977 when Mori reported the synthesis of indole **1.55** (Scheme 1.17a) and quinoline **1.57** (Scheme 1.18b) from either *N*-allyl aniline derivative **1.54** or *N*-allyl benzyl amine derivative **1.56**, respectively.[68] Interestingly, the latter reaction only produced a small amount of the expected product **1.58**, and the major product is **1.57** which arises from a debenzylation/aromatization sequence.

This report was followed up by several others, including Heck,[69] and Hegedus[70] wherein 2-quinolones and indoles were prepared, respectively. These early examples were typically plagued by narrow scope as well as low yields and levels of selectivity, which were all likely factors in the delayed application of this reaction in natural product synthesis. It was not until the mid 1980s were the first examples which constructed carbocyclic cores were reported (Scheme 1.19). In 1983, Heck showed that brominated 1,7-diene **1.59** could undergo intramolecular cyclization and observed the formation of double bond isomers **1.60** and **1.61**, in addition to piperidine trapping products **1.62** and **1.63**.[71] The observation of these two latter amine products would have unforeseen significance on the field of Pd chemistry (Scheme 1.18a). In 1984, Grigg reported that $Pd(OAc)_2/PPh_3$ catalyzed the cyclization of *gem*-diester **1.64**, and the formation of two double bond isomers **1.66** and **1.67** in a 4:1 ratio was observed (Scheme 1.18b).[72] Interestingly, replacing the

[67]Reference [72].
[68]Reference [73].
[69]Reference [74].
[70]Reference [75].
[71]Reference [76].
[72]Reference [77].

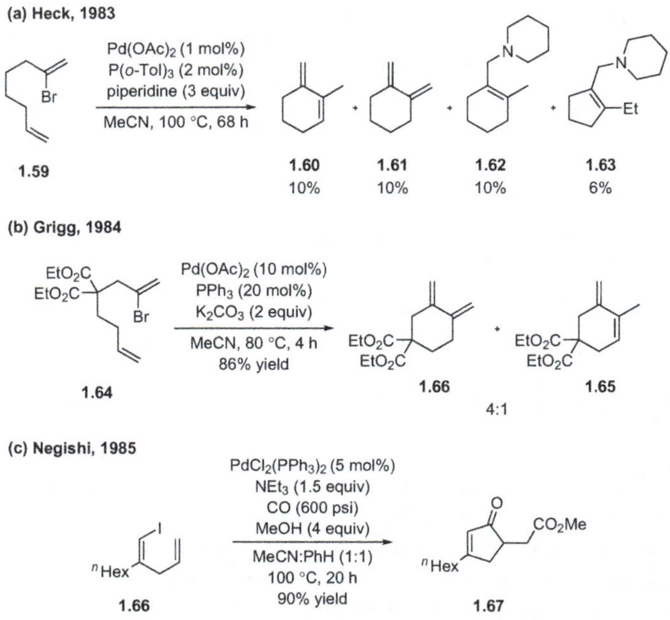

Scheme 1.18 Early examples of intramolecular Mizoroki-Heck reactions leading to carbocycles

piperidine base used by Heck[73] with potassium carbonate circumvented the formation of the amine trapping products. Negishi reported that Mizoroki-Heck reactions could be run in the presence of a high pressure CO atmosphere leading to cyclic ketone products arising from a carbonylative Mizoroki-Heck reaction (Scheme 1.18c).[74] These authors made the important finding that the addition of excess MeOH to the reaction conditions prevented an intractable mixture of oligomeric products from being obtained.

Another intriguing feature of the intramolecular Mizoroki-Heck reaction is its ability to be applied to substrates possessing one or more stereocenters. Through conformational bias, this stereocenter can impart varying levels of selectivity of the olefin carbopalladation step of the reaction mechanism. Although it would be difficult to classify one example as the first of this kind of cyclization, certainly those of Overman,[75] Larock,[76] and Negishi[77] were groundbreaking. Overman reported a series of intramolecular Heck-type cyclizations wherein numerous diastereoselective examples were mentioned. For example, cyclohexene derivative **1.68**

[73]Reference [76]

[74]Reference [78].

[75]Reference [79].

[76]Reference [80].

[77]Reference [81].

Scheme 1.19 Early examples of diastereoselective Mizoroki-Heck reactions leading to carbocycles

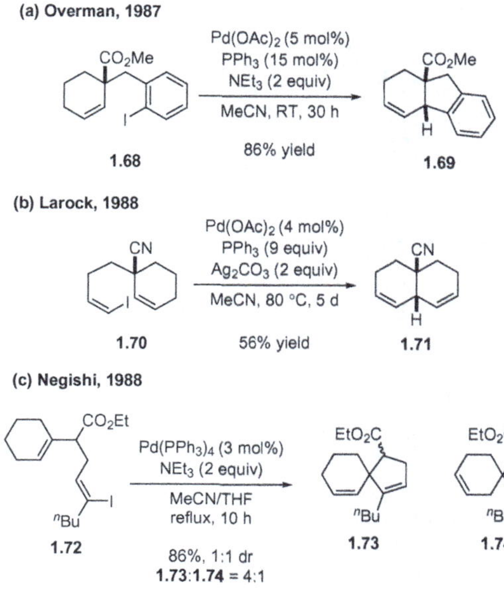

underwent an efficient cyclization to generate tricyclic core **1.69** in >30:1 dr (Scheme 1.19a). Among numerous examples, Larock showed cyclohexene derivative **1.70** could undergo intramolecular cyclization to generate *cis*-decalin-type product **1.71** (Scheme 1.19b). The authors employed Ag_2CO_3 as a stoichiometric additive which likely generates a cationic Pd(II) species after oxidative addition to the vinyl iodide,[78] as well as acts as a base to remove HI which is formed as a byproduct. Negishi showed that 1,5-diene **1.72** could be efficiently cyclized to obtain a 4:1 mixture of spirocycles **1.73** and **1.74** which were both present as a 1:1 mixture of diastereomers with unassigned relative stereochemistries (Scheme 1.19c). These early examples have inspired the application of the intramolecular Mizoroki-Heck reaction in hundreds of unique systems.[79]

The Mizoroki-Heck Reaction in Natural Product Synthesis

The field of complex molecule synthesis has benefited enormously from the use of Pd-catalyzed coupling reactions.[80] Specifically, the Heck reaction has allowed chemists to construct challenging carbon–carbon bonds with high levels of predictability in regards to stereo- and regioselectivity. Among the many elegant

---

[78]Reference [82].

[79]References [66–68]

[80]Reference [83].

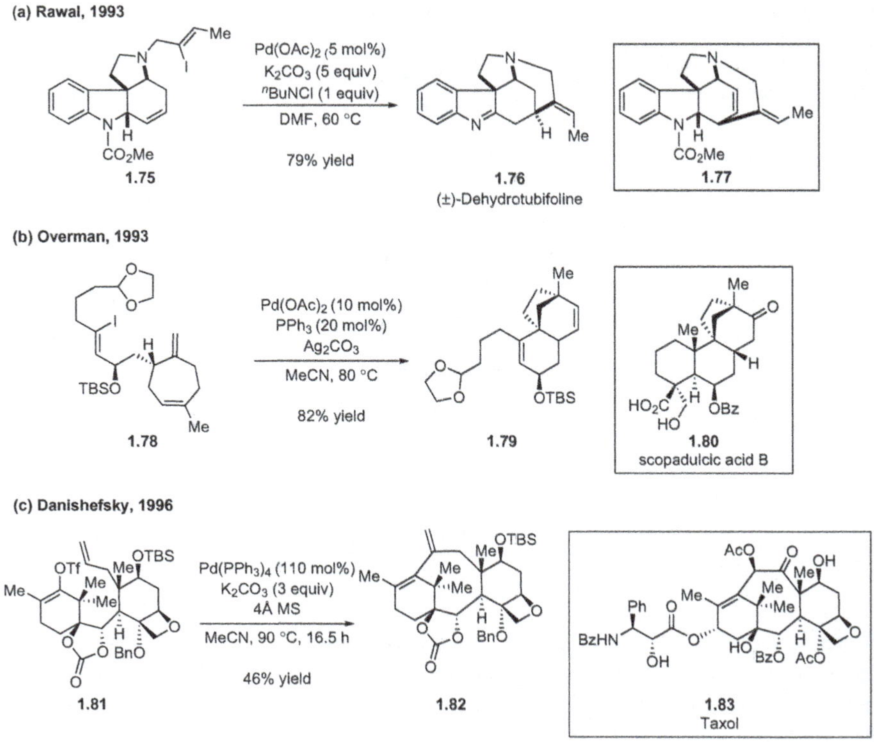

Scheme 1.20 Application of the Mizoroki-Heck reaction in the total synthesis of natural products

applications of the Mizoroki-Heck reaction in the synthesis of natural products, several are notable. In 1993 Rawal and co-workers described the application of a diastereoselective, intramolecular Mizoroki-Heck coupling of vinyl iodide **1.75** as the final synthetic step in the total synthesis of (±)-dehydrotubifoline **1.76** (Scheme 1.20a). The use of Jeffery's conditions which involve the use tetraalkyl ammonium chloride,[81] helped avoid the formation of the undesired product **1.77**. Overman reported the use of an intramolecular Mizoroki-Heck reaction of triene **1.78** which proceeded in domino fashion to generate tricyclic diene **1.79** during their total synthesis of scopadulcic acid B **1.80** (Scheme 1.20b). In 1995, Danishefsky utilized stoichiometric $Pd(PPh_3)_4$ to convert vinyl triflate **1.81** to advanced polycyclic intermediate **1.82** via an intramolecular Mizoroki-Heck cyclization during their total synthesis of Taxol **1.83** (Scheme 1.20c).[82]

[81] Reference [84].

[82] Reference [85].

The Asymmetric Mizoroki-Heck Reaction

Although there were many reports of chiral phosphine ligands and their use in metal catalyzed reactions dating from the early 1970s,[83] there were no attempts to render the Mizoroki-Heck reaction enantioselective until the late 1980s[84,85] when Shibasaki[86] and Overman[87] discovered this process almost simultaneously. Shibasaki studied the 6-exo-trig cyclization of 1,4-cyclohexadiene tethered *cis* vinyl iodide **1.84** which can be obtained via a Birch reduction/α-alkylation strategy. This system yielded triene **1.85** in 74% yield with 46% ee using (*R*)-BINAP as the chiral ligand (Scheme 1.21a). Overman's system involved the domino cyclization of vinyl triflate **1.87** to yield spirocyclic ketone **1.88** in 90% yield and 45% ee using (*R*,*R*)-DIOP as the chiral ligand.

In analogy to previous examples of diastereoselective cyclization, olefin coordination/migratory insertion is the enantiodetermining step of the asymmetric Mizorkoki-Heck reaction mechanism. The selectivity of coordination/insertion to the *Re* versus *Si* face of the olefin is determined by the chirality of the ligand and will determine which enantiomer will be major. Nevertheless, the general mechanistic hypothesis for the asymmetric version is similar to the previously discussed non-enantioselective versions. Both cationic and neutral mechanisms for the olefin coordination/migratory insertion have been proposed for this reaction, and they both have important implications for the success of these reactions (Scheme 1.22). The cationic pathway involves dissociation of X^- from complex **1.90** to generate the cationic 14 electron complex **1.91** with X^- as the counterion. This complex possesses a vacant coordination site where the olefin can bind, thus generating 16 electron η^2 complex **1.92**. This species then undergoes a migratory insertion/β-hydride elimination sequence to yield the final product. By invoking this mechanistic pathway, the ligand can remain bound in a bidentate fashion, thereby more strongly influencing the stereoselectivity of this step by a more efficient transfer of chiral information. This pathway is generally found to provide products with higher levels of ee.[88,89,90]

The X ligand does not dissociate from the Pd(II) center during the neutral pathway. Therefore, the 16 electron complex **1.90** is coordinatively saturated and cannot effectively bind the olefin partner. However, dissociation of a phosphine atom can occur to generate 14 electron complex **1.93** which can then bind the olefin partner and generate complex **1.94**. Presumably due to the less efficient transfer of

---

[83]Reference [86].
[84]Reference [87].
[85]Reference [88].
[86]Reference [89].
[87]Reference [90].
[88]Reference [91].
[89]Reference [92].
[90]Reference [82].

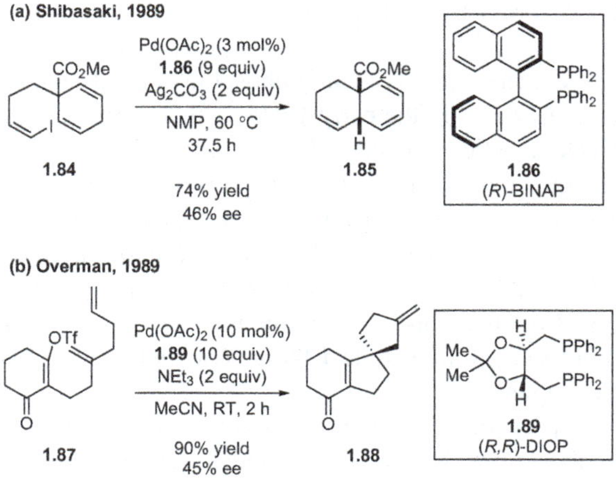

Scheme 1.21 Seminal examples of the asymmetric intramolecular Mizoroki-Heck reaction using chiral bidentate phosphine ligands

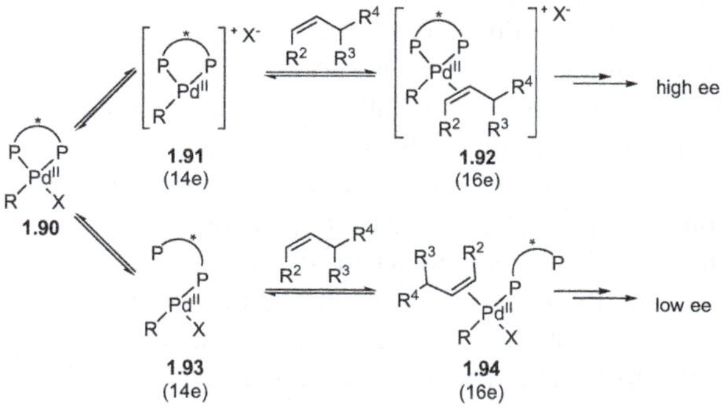

Scheme 1.22 Cationic and neutral pathways for the asymmetric Mizoroki-Heck reaction

chirality, reactions proceeding via the neutral pathway typically provide lower product ee's.

Brown was the first to study the individual steps of the cationic mechanism in detail (Scheme 1.23).[91] They found that aryl iodide **1.95** could undergo oxidative addition with Pd(0) to generate the stable 16 electron intermediate **1.96** which was characterized by X-ray crystallographic analysis. The addition of AgOTf abstracts the iodide ligand from **1.96** rendering cationic complex **1.97** which rapidly converted to η^2 complex **1.98**. Complex **1.98** presumably undergoes insertion when warmed to −40 °C and generates **1.100** as the next spectroscopically detectable

[91]Reference [93].

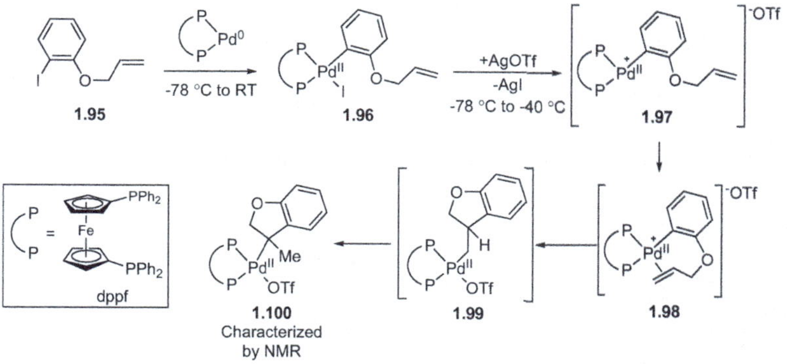

Scheme 1.23 Brown's study on the cationic Mizoroki-Heck reaction mechanism

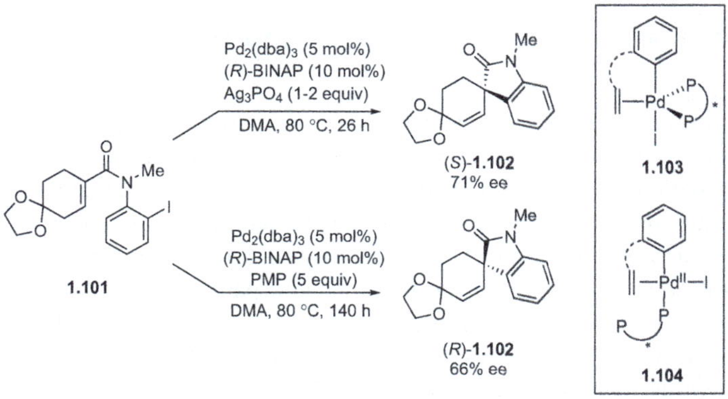

Scheme 1.24 Overman's enantiodivergent protocol to synthesize spirooxindoles

complex. These authors also studied the cationic mechanism for the intermolecular variant of the Mizoroki-Heck reaction.[92,93]

Overman reported an interesting study on the asymmetric intramolecular Mizoroki-Heck reaction which seemed to be an exception to the rule that reaction which proceed via a neutral mechanism are less selective than those which proceed via a cationc mechanism (Scheme 1.24).[94] Therein, by using (*R*)-BINAP, they noted that **1.101** cyclized in the presence of Ag salts to generate (*S*)-**1.102** while (*R*)-**1.102** was obtained as the major enantiomer in the presence of PMP. At the time of this publication, the authors suggested that the insertion step of the base promoted reaction mechanism likely proceeded via either pentacoordinate intermediate **1.103**

---

[92]Reference [94].

[93]Reference [95].

[94]Reference [96].

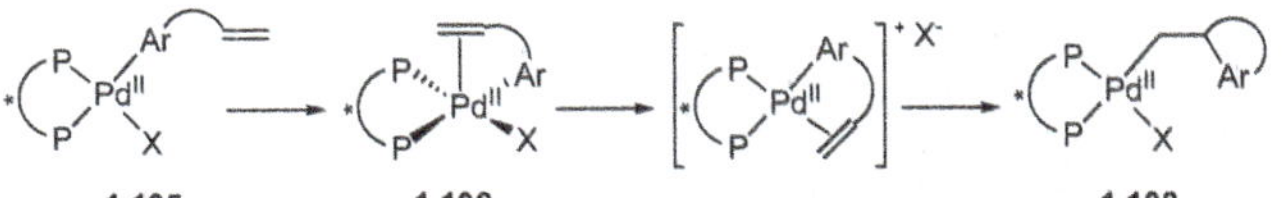

Scheme 1.25 Associative halide ligand replacement mechanism proposed by Overman

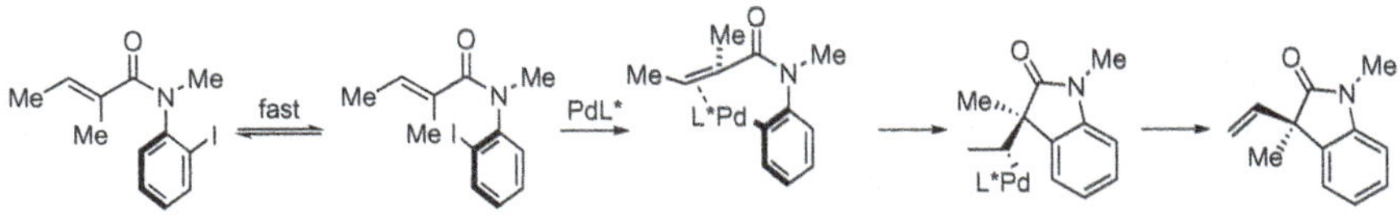

Scheme 1.26 Neutral Mizoroki-Heck reaction of proceeding by dynamic kinetic resolution

or tetracoordinate intermediate **1.104** where the phosphine ligand is bound in a monodentate fashion. Some examples of this reaction the reactions with Ag salts proceeded with lower enantioselectivities than those without halide salts.

The same authors later reported that (a) the addition of halide sources to the intramolecular asymmetric Heck reaction of aryl triflates can increase enantioselectivities by diverting the insertion to the neutral pathway and (b) that the neutral pathway does not proceed via phosphine dissociation.[95] Although they could not rule out the direct migratory insertion from a pentacoordinate intermediate such as **1.103**, previous theoretical studies from Thorn and Hoffman[96] and Samsel and Norton[97] suggest that this would be an energetically unfavorable process. With this in mind, the authors suggested an alternative explanation involving intermediate **1.105** forming pentacoordinate intermediate **1.106** which undergoes an associative halide ligand displacement to generate a cationic tetracoordinate intermediate **1.107**. This intermediate can then undergo a more energetically feasible migratory insertion step to generate alkyl Pd(II)X complex **1.108** which leads to product (Scheme 1.25). The mechanism of the neutral reaction pathway was later studied in greater detail by the same authors.[98]

Curran later published a paper calling the neutral mechanism proposed by Overman into question (Scheme 1.26).[99] The authors suggest that the *o*-iodoanilide substrates (i.e. **1.101**) can undergo rapid rotation around the aryl–nitrogen bond

[95]Reference [97].

[96]Reference [98].

[97]Reference [99].

[98]Reference [100].

[99]Reference [101].

(**1.109–1.110**), and the chiral Pd catalyst can selectively oxidatively add the carbon–iodine bond of one over the other to generate a chiral complex **1.111** which undergoes migratory insertion to generate intermediate **1.112**. By this logic, they proposed that a dynamic kinetic resolution was occurring wherein the enantiodetermining step is oxidative addition instead of asymmetric migratory insertion.

1.1.2 From Mizoroki-Heck Reactivity to the Pd-Catalyzed Alkene Aryliodination

The mechanism for the Mizoroki-Heck reaction has presented several opportunities to explore new modes of reactivity. The alkyl Pd(II)X intermediate **1.115** can interact with an external nucleophile to generate a new carbon–carbon or carbon–heteroatom bond, and is referred to as an anion capture cascade (Scheme 1.27, path a). This mode of reactivity has a long history of development as well as application in complex molecule synthesis. In principle, species **1.115** could directly undergo carbon–halogen bond-forming reductive elimination (Scheme 1.27, path b). However, despite the myriad of examples involving oxidative addition of transition metals to carbon–halogen bonds,[100] reductive elimination to form carbon–halogen bonds is still a rare and challenging transformation.[101,102]

1.1.2.1 Carbon–Halogen Bond-Forming Reductive Elimination

The oxidative addition of transition metals to carbon–halogen bonds has traditionally been considered irreversible due to its high exothermicity.[103,104] In 1968, Ettorre reported the reductive elimination of iodobenzene from the octahedral Pt (IV) complex **1.118** to form *trans*-Pt(II) complex **1.119** (Eq. 1.1).[105] The authors state that the addition of NaI to the reaction slowed the rate of reductive elimination, suggesting that a solvolytic process involving the formation of an iodide ion during the reaction to form complex **1.120.**

---

[100]Reference [102].

[101]Reference [103].

[102]See Ref. [4].

[103]Reference [104].

[104]Reference [105].

[105]Reference [106].

Scheme 1.27 Diverting the Mizoroki-type reaction with novel modes of reactivity

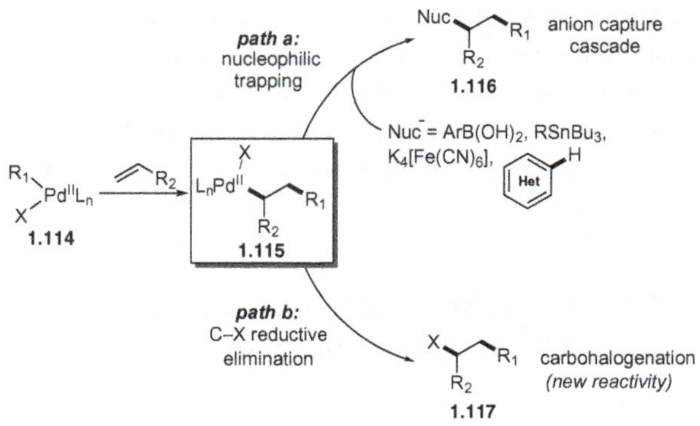

Scheme 1.28 A seminal palladium-catalyzed oxidative C–H halogenation

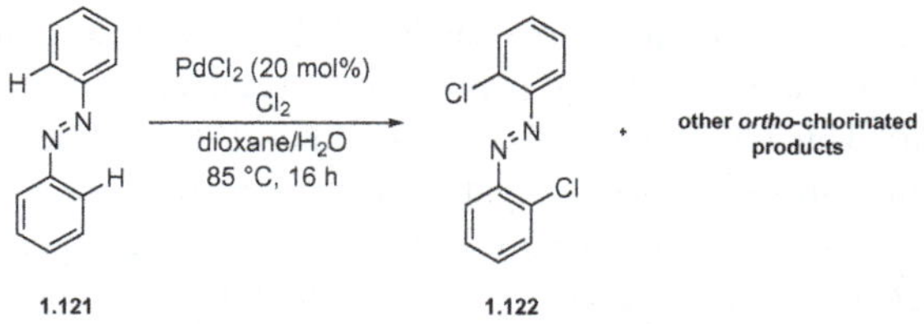

$$\underset{\mathbf{1.118}}{\left[\mathrm{Pt(PEt_3)_2Ph_2I_2}\right]} + \text{solvent} \xrightarrow{-\mathrm{I}^-} \underset{\mathbf{1.120}}{\left[\mathit{trans} - \left[\mathrm{Pt(PEt_3)_2Ph_2I(solvent)}\right]\right]^+} \xrightarrow{-\mathrm{PhI}} \mathit{trans} - \underset{\mathbf{1.119}}{\left[\mathrm{Pt(PEt_3)_2Ph_2I}\right]} \quad (1.1)$$

In 1970, Fahey reported the first Pd-catalyzed directed *ortho* C–H chlorination of azobenzene **1.121** to form **1.122** using Cl_2 (Scheme 1.28).[106,107] Despite the low selectivities observed for the monochlorinated product, this report established a key proof-of-principle.

In 2001, Hartwig conducted kinetic studies regarding carbon–halogen bond reductive elimination from dimeric ArPd(II)X complexes **1.123a–e** using bulky trialkyl phosphine ligands (Table 1.1).[108] Prior to this seminal report, the oxidative addition of Pd(0) complexes to aryl-halide bonds was thought to be irreversible. As a result of this study, the authors rewired the rules for the thermodynamics and kinetics of oxidative addition/reductive elimination of carbon–halogen bonds from Pd(II) centers.

Dimers **1.123a** and **1.123c** which contained $P(o\text{-Tol})_3$ ligands provided higher yields of their respective ArX products **1.124a and 1.124c** as well as 10-fold increases in K_{eq} values than complexes with non-*o*-substituted aryl ligands. Each

[106]Reference [107].

[107]Reference [108].

[108]Reference [109].

Table 1.1 Stoichiometric studies on reductive elimination of aryl halides from ArPd(II)X dimers

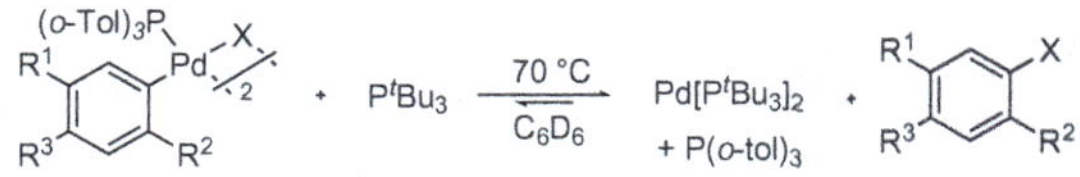

R¹ = ᵗBu, R² = Me, R³ = H: X = Cl: **1.123a**; X = Br: **1.123b**; X = I: **1.123c**

R¹ = R² = H, R³ = ᵗBu: X = Cl: **1.123d**; X = Br: **1.123e**

1.124a-e

X	Yield of **1.124** (%)	K_{eq}
1.123a (X = Cl)	70	$9(3) \times 10^{-2}$
1.123b (X = Br)	70	$2.3(3) \times 10^{-3}$
1.123c (X = I)	39	$3.7(2) \times 10^{-5}$
1.123d (X = Cl)	30	Not measured
1.123e (X = Br)	75	$3.3(6) \times 10^{-4}$

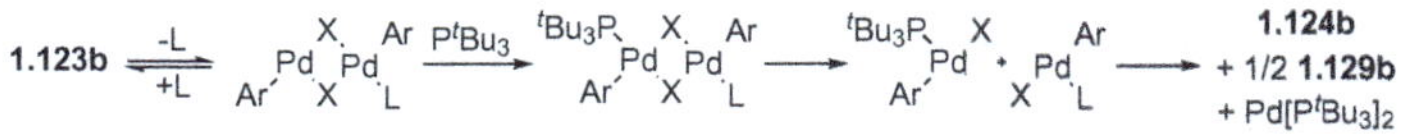

Scheme 1.29 Proposed mechanism of reductive elimination from ArPd(II)X dimers involving ligand dissociation

change in halide ligand (i.e. Cl to Br or Br to I) was associated with a 100-fold change in K_{eq}. All data were consistent with a mechanism involving an irreversible formation of monomers from the dimeric species, which are formed from ligand substitution and subsequent carbon-halogen bond-forming reductive elimination (Scheme 1.29). They noted that the dissociation of P^tBu_3 is slower than formation of monomers and reductive elimination.

The authors reasoned that if carbon–halogen bond strengths were to dominate, reductive elimination from an ArPd(II)Cl species would be faster than the corresponding bromide or iodide analogues. However, if metal–halogen bond strengths were to dominate, reductive elimination from the ArPd(II)Cl would be slower than the corresponding bromide or iodide analogs.[109]

Cross-coupling reactions with bulky phosphine ligands (i.e. P^tBu_3) are commonly proposed to occur via three-coordinate 14-electron complexes such as **1.130**.[110,111] However, until a 2002 report by Hartwig describing their synthesis, characterization, and reactivity, no such species had been isolated and studied (Scheme 1.30).[112] This work led to a subsequent report in 2003 by the same authors

[109] Reference [110].

[110] Reference [111].

[111] Reference [112].

[112] Reference [113].

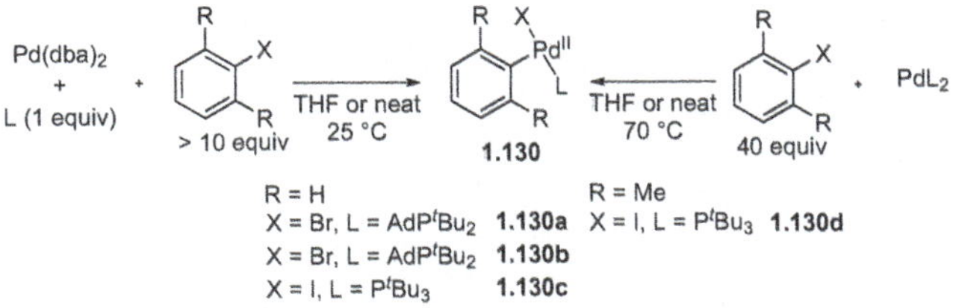

Scheme 1.30 Synthesis of the first isolable three-coordinate 14-electron ArPd(II)X complexes

Table 1.2 Stoichiometric studies on reductive elimination of aryl halides from monomeric ArPd (II)X complexes

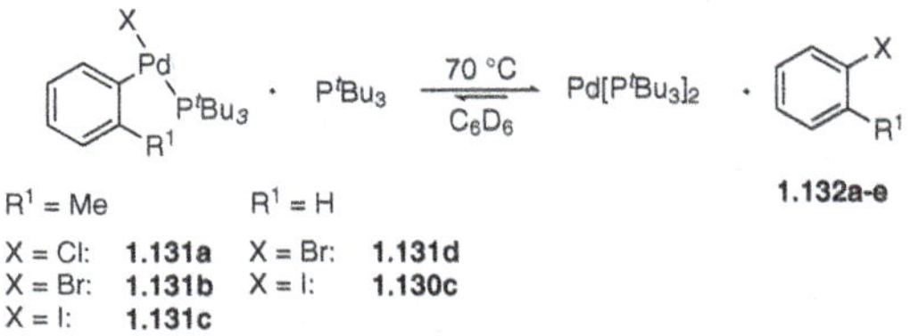

X	Yield of **1.132** (%)	K_{eq}
1.131a (X = Cl)	76	10.9×10^{2}
1.131b (X = Br)	98	32.7×10^{-1}
1.131c (X = I)	79	1.79×10^{-1}
1.131d (X = Br)	68	13.4×10^{-1}
1.130c (X = I)	60	0.51×10^{-1}

which described the first aryl halide reductive elimination from monomeric ArPd(II) X complexes **1.131a–e** (Table 1.2).[113] This observation was a significant advance from their 2001 report,[114] where the species undergoing reductive elimination was not directly observed.

Yields of the respective aryl halide products were higher than those previously reported for the dimeric ArPd(II)X complexes.[115] Reductive elimination from chloride **1.131a** was thermodynamically favored by a factor of 3000 over that of bromide analog **1.131b** and the latter was favored over iodide **1.131c** by approximately a factor of 20. The authors reasoned that this trend was due to the strengths of the resulting carbon–halogen bonds.[116,117] Similarly, the K_{eq} for *ortho*-substituted arenes (**1.131a–c**) were larger than that of their non-substituted analogues (**1.131d–e**). Interestingly, the kinetics did not agree with the thermodynamics, as

[113]Reference [114].

[114]Reference [109]

[115]Reference [110].

[116]Reference [115].

[117]Reference [116].

Scheme 1.31 Mechanism of reductive elimination involving ligand dissociation

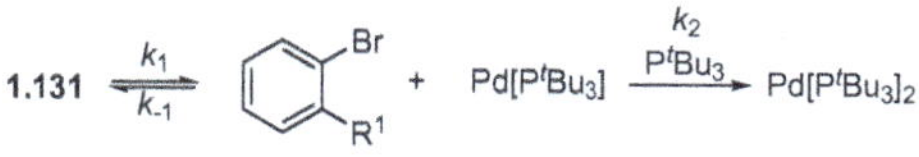

reductive elimination from **1.131a** was slower than from **1.131b**, despite being more energetically favorable. They also stated that metal–halogen bond strength is likely more important than carbon–halogen bond strength. Based on this kinetic data, the authors propose a mechanism involving reversible ArBr reductive elimination followed by trapping of the Pd(0)L intermediate by P^tBu_3 to generate PdL_2 (Scheme 1.31).

1.1.2.2 Alkyl Halide Reductive Elimination from Transition Metal Complexes

The majority of carbon–halogen bond reductive eliminations from late transition metals like Pt(IV),[118] Pd(IV), and Pd(II) involve aryl halides. A few reports of reductive elimination of alkyl halides via thermolysis of Pt(IV) complexes have been reported.[119,120,121] The Rh(III) species $[(CH_3CO)Rh(CO)I_3]^-$ has also been shown to decompose through the loss of MeI during the Monsanto process. In the presence of a CO atmosphere, the newly formed dicarbonyl complex $[(CH_3CO)Rh(CO)_2I_3]^-$ reductively eliminates to from acetyl iodide.[122]

Frech and Milstein reported the first direct observation of reductive elimination,[123,124] of an alkyl halide (MeI) from an isolable Rh(III) pincer complex in 2006 (Scheme 1.32).[125] By reacting η^1 nitrogen complexes **1.133** and **1.134** with an equimolar amount of MeI, complexes **1.135** and **1.136** were obtained, respectively. When complex **1.135** was treated with an excess of CO ($\sim$100 equivalents), slow but nearly quantitative reductive elimination to form Rh(I) carbonyl complex **1.137** occurred. Conversely, when the less bulky ^{i}Pr analog **1.136** was treated under the same conditions, reductive elimination of MeI was not observed. Instead, CO adducts **1.138** and **1.139** were observed. The authors credit this switch in reactivity to the decreased size of the ^{i}Pr-containing pincer ligands, thus echoing the results

---

[118] For a recent example of reductive elimination of an aryl halide from a Pt(IV) complex see: Ref. [117].

[119] Reference [118].

[120] Reference [119].

[121] Reference [120].

[122] Reference [121].

[123] For an example of reductive elimination of aryl chloride from a Rh(III) complex see: Ref. [122].

[124] For an example of reversible oxidative addition of MeI to Rh(I) based on IR spectroscopy see: Ref. [123].

[125] Reference [124].

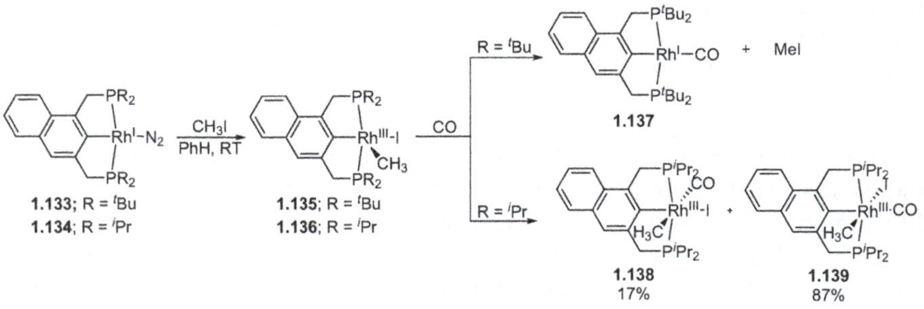

Scheme 1.32 Milstein's synthesis and reactivity of MeRh(III)I pincer complexes. Adapted with permission from Ref. [3]. Copyright (2015) Royal Society of Chemistry

from Hartwig which highlighted the positive impact that steric congestion had on carbon–halogen bond-forming reductive elimination.

1.1.2.3 The Discovery of the Pd(0)-Catalzyed Aryliodination of Alkenes

Based on a report by Newman and Lautens concerning the reversible oxidative addition of carbon–bromine bonds (cf. Chap. 4),[126] these same authors reported the first carbohalogenation reaction invoking carbon–halogen reductive elimination from an alkyl Pd(II) halide intermediate in 2010.[127] To avoid a Heck reaction, their method utilized aryl iodides **1.140** that were bound to a pendent 1,1-disubstituted olefin (Table 1.3). In these systems, the neopentyl Pd(II) halide intermediate **1.142** that is formed after migratory insertion is not in the presence of suitable β-hydrogen atoms resulting in the desired carbon–iodine fond formation. Initial screening of well-defined PdL_2 catalysts at 90 °C revealed that bulky phosphine ligands were essential for reactivity, with QPhos[128] giving the best results (Table 1.3, entries 1–3). Unlike the traditional Mizoroki-Heck reaction, no basic additives were required since the transformation does not generate HX as a byproduct. Less bulky ligands,[129] such a PCy_3 and $P(o\text{-tol})_3$ did not provide any detectable amounts of the desired carbohalogenation product (Table 1.3, entries 4–5).

Under the optimized conditions, a range of oxygen- and nitrogen-containing 5- and 6-membered heterocycles could be prepared in high yield (Scheme 1.33). Aryl chlorides and bromides were unsuccessful in the reaction, suggesting that C–Cl and C–Br reductive elimination may be more difficult. Notably, the intramolecular carboiodination was not hindered by the addition of radical

---

[126]Reference [125]. Dr. Steve Newman was a graduate student in the Lautens group.

[127]Reference [126].

[128]Reference [127].

[129]Reference [128].

Table 1.3 Screening of PdL2 catalysts for the intramolecular carboiodination

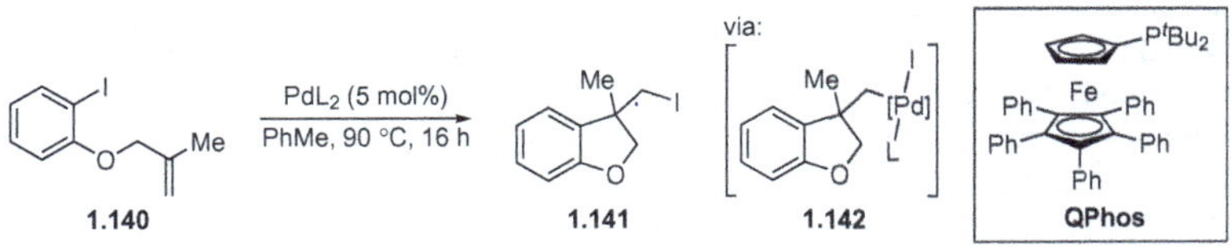

Entry	Ligand	Yield **1.142**
1	QPhos	82 (95)[a]
2	P^tBu_3	61
3	$PhP(^tBu_3)_2$	76
4	PCy_3	0
5	P(*o*-tol)$_3$	0

[a]Optimized conditions: 2.5 mol% Pd(Q-Phos)$_2$, 100 °C

Scheme 1.33 Selected Pd-catalyzed carboiodination products

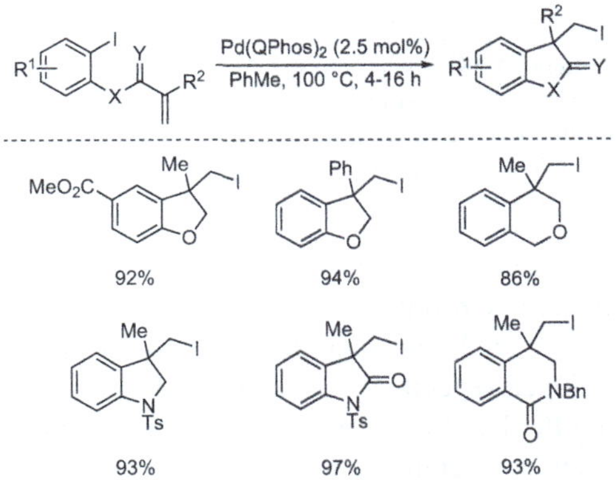

scavengers such as TEMPO or galvinoxyl, therefore discounting the involvement of radical intermediates, while supporting C–I reductive elimination as a key elementary step in the reaction mechanism.

The computationally supported[130] catalytic cycle is shown in Scheme 1.34. Starting from the active PdL_2 complex, substrate binding with concomitant ligand dissociation generates η^2 complex **1.143**. Oxidative addition of the aryl halide then occurs to provide Pd(II) complex **1.144**. Subsequent *cis*-to-*trans* isomerization allows coordination of the pendent olefin to occur forming **1.145**, which is followed by *syn*-carbopalladation to form the new C–C bond in alkyl Pd(II) halide intermediate **1.146**. The C–X reductive elimination proceeds via a three-membered transition state to generate product-bound Pd(0) complex **1.147**. The I → Pd coordination is weak, and **1.147** readily liberates the product **1.141** in order to bind

[130]DFT calculations were used to study the mechanism, ligand effects, and the origins of reactivity and selectivity. See: Ref. [129].

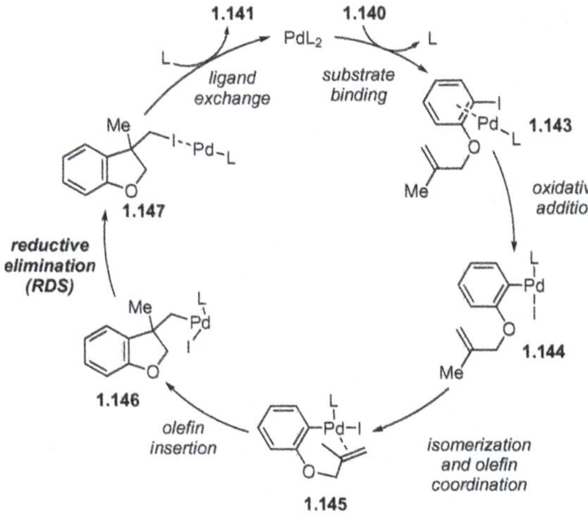

Scheme 1.34 Proposed catalytic cycle for Pd-catalyzed arylhalogenation. Adapted with permission from Ref. [3]. Copyright (2015) Royal Society of Chemistry

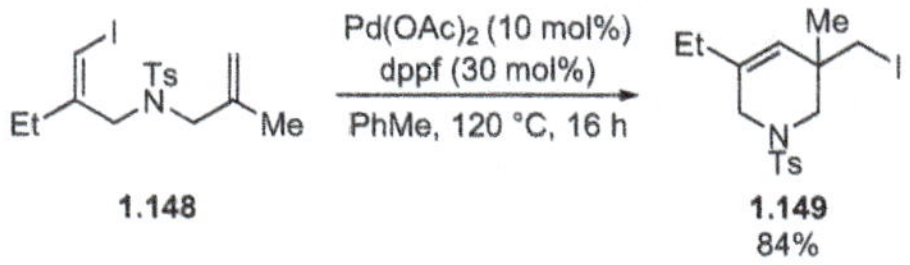

Scheme 1.35 Tong's report of C-I reductive elimination from alkyl Pd(II) halide complexes

another substrate molecule or ligand to regenerate the initial PdL_2 catalyst. Based on the DFT calculations, the rate-determining step of the transformation is carbon–halogen reductive elimination and exhibits a barrier of 24.9 kcal/mol for substrate **1.140**. The high activation energy is related to the endothermicity of the transformation. Overall, the main driving force for the reaction is the formation of a $C(sp^2)$–$C(sp^3)$ σ bond from a C=C π bond.

Soon after the report of Lautens, Tong reported that vinyl iodides such as **1.148** could undergo an intramolecular alkene vinyliodination by using a catalyst system composed of $Pd(OAc)_2$ and dppf (Scheme 1.35).[131] In addition to the variation in catalyst composition, the two methods used different temperatures and substrate classes. Interestingly, Tong's system works best using a bidentate phosphine ligand in three-fold excess relative to the Pd pre-catalyst, whereas Lautens found that a bulky monodentate phosphine ligand was optimal.

Our group reported a Pd-catalyzed aryliodination of aryl bromides **1.150** could occur in the presence of an iodide source (Scheme 1.36).[132] The success of this transformation is based on a halide exchange (bromide to iodide) at Pd, which precedes carbon–iodine reductive elimination. This protocol is reminiscent of the

---

[131]Reference [130].

[132]Reference [131]. Dr. Steve Newman and Jennifer Howell were graduate students in the Lautens group. Dr. Norman Nicolaus was a post-doctroal fellow in the Lautens group.

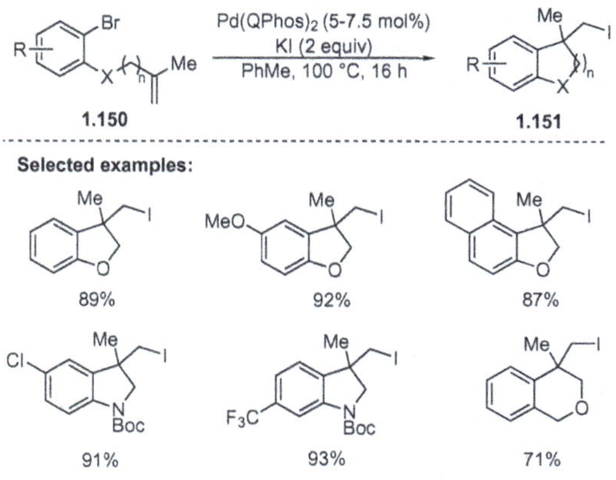

Scheme 1.36 Pd-catalyzed alkene iodoarylation via Br to I exchange. Adapted with permission from Ref. [3]. Copyright (2015) Royal Society of Chemistry

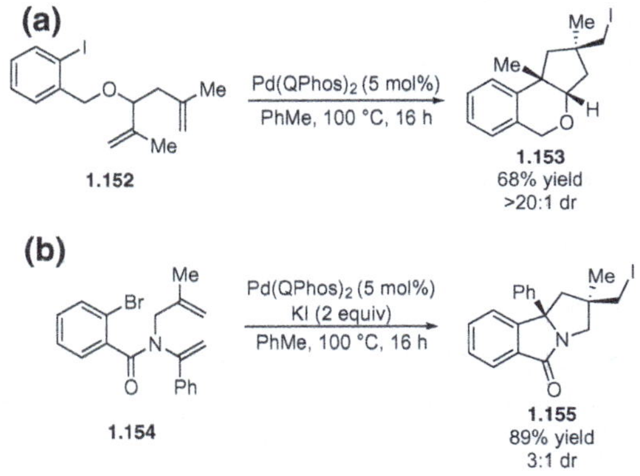

Scheme 1.37 Domino carboiodination with **a** aryl iodides; and **b** aryl bromides

aromatic Finkelstein reaction reported by Buchwald and co-workers,[133] and permits the use of cheaper and more readily-available starting materials.

In the same report, diene-containing aryl iodides were shown to be suitable substrates for aryliodination. Upon oxidative addition of the aryl iodide, two intramolecular migratory insertion steps can proceed before C–I reductive elimination, thus giving rise to complex, polycyclic products containing the alkyl iodide motif. Aryl iodide **1.152** could undergo the domino process to generate isochroman **1.153** with >20:1 dr (Scheme 1.37a). Additionally, aryl bromide **1.154** could also be used in the domino carbohalogenation by utilizing the halide exchange conditions to generate indoline **1.155** in 3:1 dr (Scheme 1.37b).

Lautens and co-workers reported a conjunctive carboiodination of *o*-isopropenyl aryl iodides **1.156** and alkynes which afforded indene products **1.157**

[133]Reference [132].

Scheme 1.38 Synthesis of indenes by a conjunctive Pd-catalyzed aryliodination

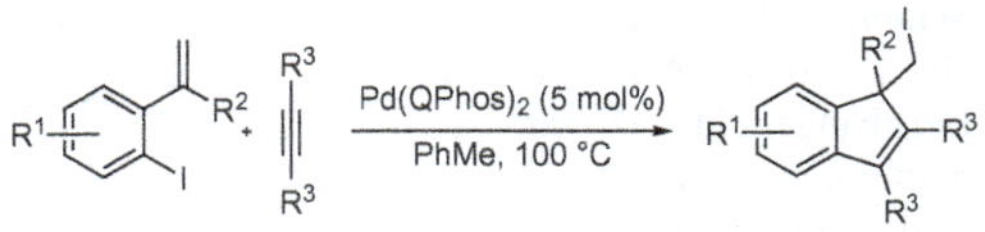

Scheme 1.39 Proposed mechanism for the Pd-catalyzed conjunctive aryliodination. Adapted with permission from Ref. [3]. Copyright (2015) Royal Society of Chemistry

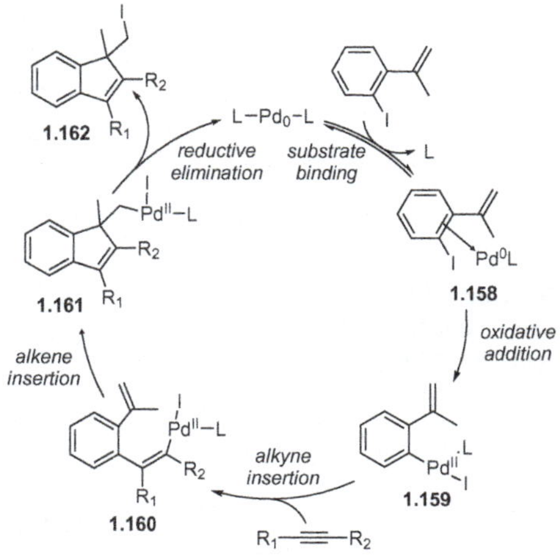

(Scheme 1.38).[134] These studies were undertaken because attempts to achieve direct carboiodination of an alkyne were not successful. The presence of the 2-propenyl group allowed the vinylpalladation species, arising from intermolecular alkyne carbopalladation, to undergo cyclization to from a neopentyl Pd–I intermediate that was known to reductively eliminate.

A mechanism was proposed to rationalize product formation (Scheme 1.39). Substrate binding first occurs to generate Pd(0) complex **1.158**, which undergoes oxidative addition to generate Pd(II) intermediate **1.159**. Intermolecular alkyne insertion occurs to produce **1.160**, which can undergo cyclic alkene carbopalladation to form **1.161**. The indene product **1.162** is formed once intermediate **1.161** undergoes C–I reductive elimination.

1.1.3 Research Goals: Part 1

Recent work from several groups including our own has demonstrated that carbon–halogen bond-forming reductive elimination from Pd(II) is feasible in both stoichiometric and catalytic settings. Dr. Steve Newman and his co-workers Jennifer

[134]Reference [133].

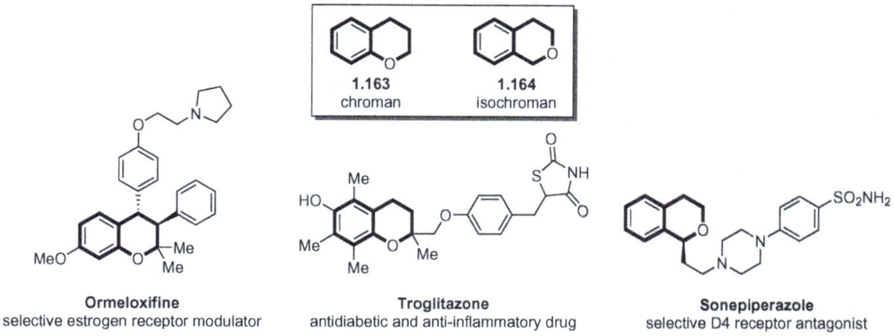

Scheme 1.40 Representations of the chroman and isochroman heterocycles and examples of their occurance

Howell and Dr. Norman Nicolaus had already laid the important groundwork in their initial two reports, including some preliminary results on both the diastereoselective and enantioselective variants. However, the full potential of the aryliodination reaction class had not been reached, nor was it completely clear whether this chemistry could be useful in practical applications.

At the onset of our research, we had two main goals: (1) study the general diastereoselectivity outcomes of intramolecular cyclizations en route to more complex heterocyclic systems, and (2) use this knowledge to correctly apply the diastereoselective arylhalogenation reaction to the synthesis of a natural product target. We decided that the chroman **1.163** and isochroman **1.164** heterocyclic motifs were ideal starting points since many derivatives display interesting biological activity (Scheme 1.40).[135]

1.1.4 Results and Discussion: Chromans and Isochromans via a Diastereoselective Aryliodination

1.1.4.1 Starting Material Preparation

The substrate preparation for both chroman and isochroman could be completed in a straightforward fashion using commercially available materials. The chroman precursors could be prepared via a 2-step protocol involving the 1,2-addition of 2-methylallylmagnesium chloride **1.166** to aldehydes **1.165**, and reaction of the corresponding racemic homoallylic alcohol **1.167** with 2-iodophenol **1.168** under Mitsonobu conditions. The isochroman precursors were prepared via a 2-step protocol reaction involving the 1,2-addition of isopropenyl magnesium bromide

---

[135]Reference [134].

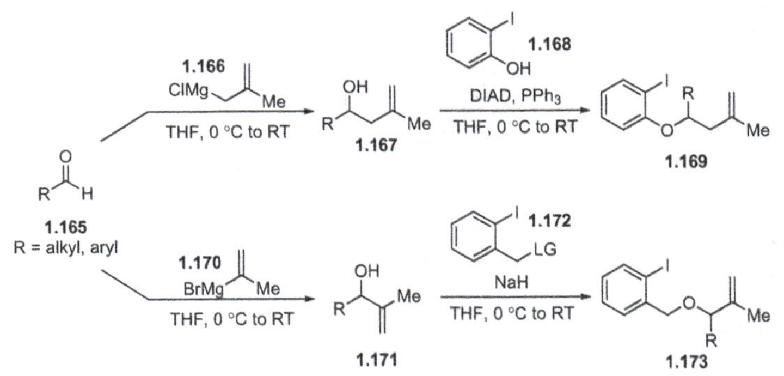

Scheme 1.41 Preparation of starting materials for the chroman and isochroman syntheses

1.170 to aldehydes, and reaction of the corresponding racemic allylic alcohol **1.171** with electrophile **1.172** (where LG = Br, Cl, OMs) to generate precursors **1.173** (Scheme 1.41). By using both substrate classes, we could investigate what effect the distance of the stereocenter to the olefin moiety had on the diastereocontrol of the process. In substrates **1.169** and **1.173** the stereocenter was in a homoallylic and allylic position, respectively.

If the proposed cyclizations were to operate as envisioned, this cyclization would clearly be amenable towards the synthesis of enantioenriched products by using single-enantiomer starting materials. Although the above synthetic routes led to racemic substrates we prepared one enantioenriched substrate **1.174** via a modified route as a means towards this goal (Scheme 1.42a). Naphthaldehyde **1.175** was treated with **1.170** to obtain (±) −**1.178** which was oxidized to enone **1.176** using the Dess-Martin periodinane. The (*R*)-CBS reagent was utilized to stereoselectively reduce **1.176** to the corresponding enantioenriched chiral allylic alcohol (*R*)-**1.178** which was obtained in 99.5:0.5 er. Alcohol **1.178** could be deprotonated with NaH and reacted with mesylate **1.179** to generate (*R*)-**1.180** in 99:1 er.

Since all of the aforementioned substrates possessed vinylic CH_3 groups, substrate **1.181** containing a PMB protected alcohol was prepared (Scheme 1.42b). Alcohol **1.182** is reacted with in situ generated hydrogen iodide to regioselectively generate vinyl iodide **1.183**. Due to the base sensititivity of **1.183**, we found an efficient procedure to execute PMB protection by using trichloroacetimidate **1.184** in the presence of a catalytic amount of PPTS. This could be coupled with benzaldehyde using a Ni-catalyzed Nozaki-Hiyama-Kishi coupling in the presence of stoichiometric anhydrous $CrCl_2$ to obtain allylic alcohol **1.186**. The desired substrate **1.181** was obtained by a benzylation using NaH and 2-iodobenzyl chloride.

1.1.4.2 Optimization

We initially hypothesized that allylic stereocenters would impart a higher level of diastereocontrol than homoallylic stereocenters. Therefore, we began to study the

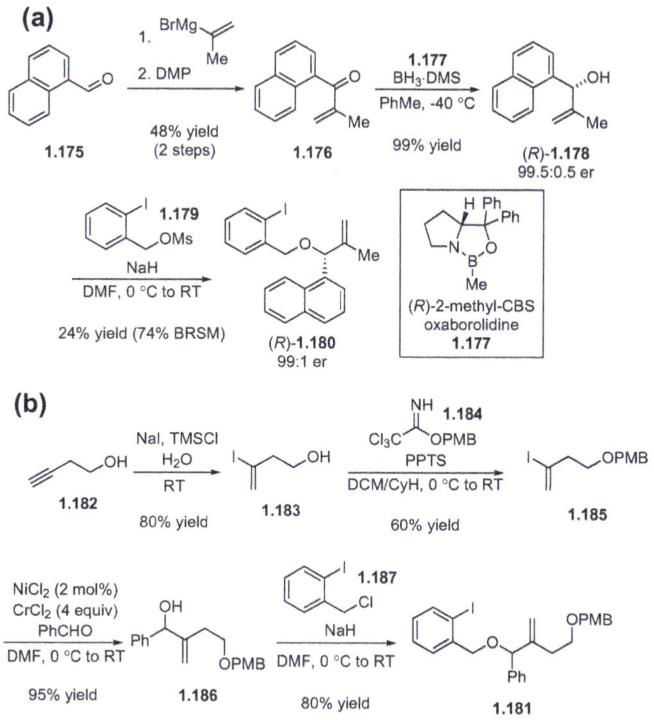

Scheme 1.42 Synthesis of **a** enantioenriched isochroman precursor **1.174**; and **b** PMB protected alcohol **1.181**

cyclization en route to isochromans using precursor **1.173a** which possessed a phenyl group that could easily be tuned by substitution. Newman and Lautens reported that QPhos was the optimal ligand for the aryliodination of alkenes which proceeded at 100 °C in PhMe. Specifically, pre-formed PdL_2 sources were employed which do not require in situ reduction of Pd(II). However, early in our optimization we observed erratic levels of conversion despite the reaction preceding cleanly. We reasoned that the catalyst activity could be decreased due to trace intermolecular Mizoroki-Heck sequences occurring which lead to the production of HPd(II)I. In the absence of base, HI cannot be effectively removed and the catalyst is rendered inactive. Furthermore, in parallel studies concerning larger ring sizes olefin isomerization was observed which suggested that Pd hydrides could in fact be present. A simple solution to this problem was to add one equivalent of NEt_3 to the reaction, which led to much more consistent conversions.

The effectiveness of various bulky phosphine-containing Pd(0) pre-catalysts was tested for the intramolecular cyclization of **1.173a** (Table 1.4). Both $Pd(PCy_3)_2$ (entry 1) and $Pd(P(o\text{-tol})_3)_2$ (entry 2) gave no desired isochroman product. Presumably these ligands were not bulky enough to promote the crucial C–I reductive elimination step. The more bulky ligand P^tBu_2Ph only afforded trace

Table 1.4 Reaction optimization for the diastereoselective synthesis of isochromans

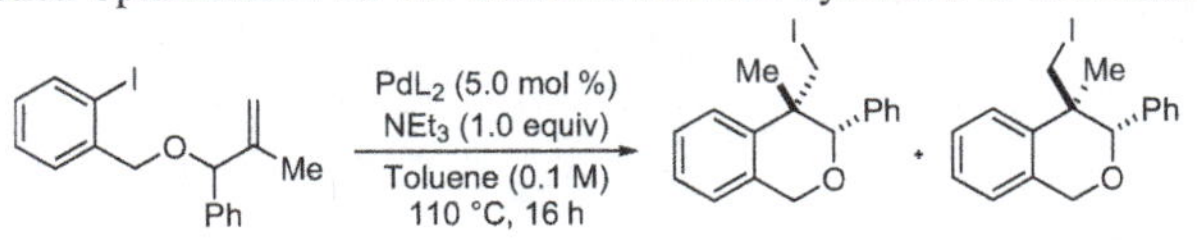

L	mol% PdL_2	Yield of **1.187a** (%)[a]	dr (*syn*:*anti*)[b]
P(*o*-tol)$_3$	5	0	–
PCy$_3$	5	0	–
P^{t}Bu$_2$Ph	5	<2	–
QPhos	5	62	90:10
P^{t}Bu$_3$	**5**	**94**[c]	**94:6**
P^{t}Bu$_3^d$	5	47	94:6
P^{t}Bu$_3$	2	77	91:9

[a]Combined yield of *syn*-**1.187a** and *anti***1.187a** calculated by ^{1}H NMR analysis of the crude reaction mixture using 1,3,5-trimethoxybenzene as an internal standard
[b]Calculated by ^{1}H NMR analysis of the crude reaction mixture
[c]Isolated yield
[d]Reaction run in the absence of NEt_3

amounts of the desired product **2a** (entry 3) despite performing moderately well in previous systems.[136] The use of QPhos afforded 62% of the isochroman product which was present as a 9:1 ratio of diastereomers (entry 4). ^{1}H NMR analysis allowed for effective determination of the diastereomeric ratio, as both sets of methylene protons are adequately resolved in the spectra of the crude reaction mixture. When Pd(P^{t}Bu$_3$)$_2$ was employed, both yield and diastereoselectivity increased to 94% and 94:6, respectively (entry 5). In the absence of NEt_3 we noticed a significant decrease in yield (46%) but no decrease in diastereoselectivity (entry 6). Decreasing the catalyst loading to 2.5 mol% caused a marked decrease in overall yield (77%, entry 7). The optimized reaction conditions were found to be Pd(P^{t}Bu$_3$)$_2$ (5 mol%), NEt_3 (1 equiv) in PhMe (0.1 M) at 110 °C for 16 h.

1.1.4.3 Examination of the Substrate Scope

A diverse set of substituted alkenyl aryl iodides were subjected to the reaction conditions to test the scope and selectivity of this transformation (Table 1.5). When the cyclization of **1a** was conducted on a 2 mmol scale, the corresponding isochroman was afforded in 88% yield with no erosion of diastereoselectivity (entry 1). Dioxolane substituted **1.187b** was afforded in 91% yield with >95:5 dr (entry 2). We were able to grow suitable crystals in order to carry out X-ray crystallographic analysis wherein we were able to unambiguously confirm the *cis* isomer as the

[136]Reference [126].

Table 1.5 Scope of the isochroman synthesis[a]

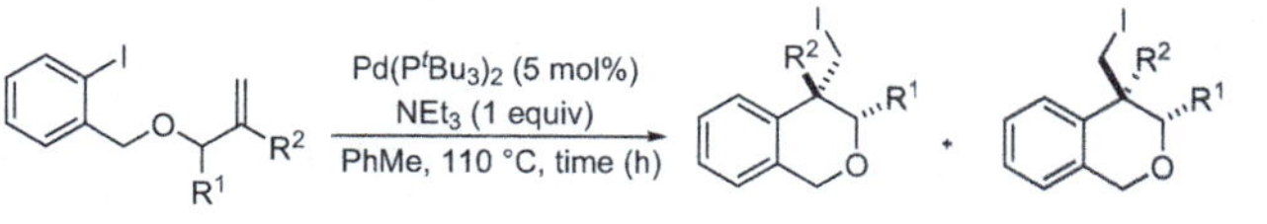

1.173 *syn*-**1.187** *anti*-**1.187**

Entry	Substrate		Product		Yield[b] [time]	dr (*syn*: *anti*)[c]
1	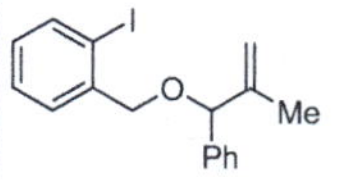	**1.173a**	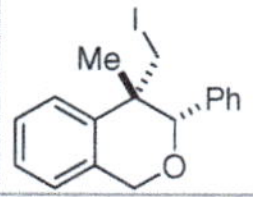	**1.187a**	94% (88%)[d] [10 h]	94:6
2	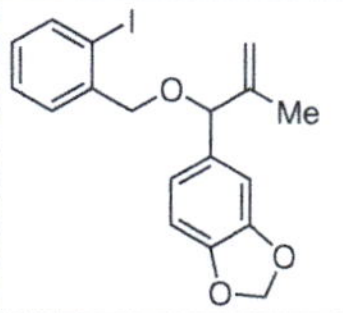	**1.173b**	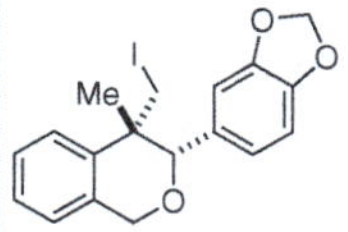	**1.187b**	91% [10 h]	>95:5
3[e]	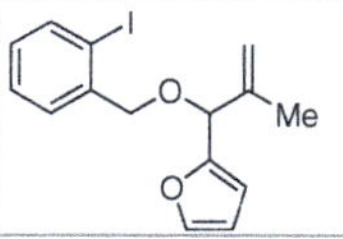	**1.173c**	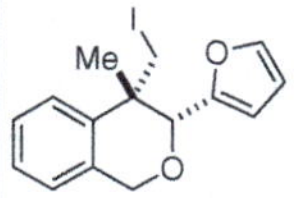	**1.187c**	85% [10 h]	92:8
4[e]	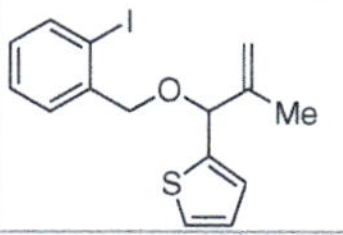	**1.173d**	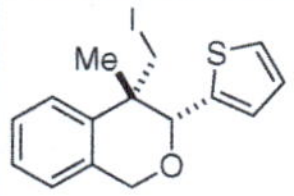	**1.187d**	94% [10 h]	91:9
5	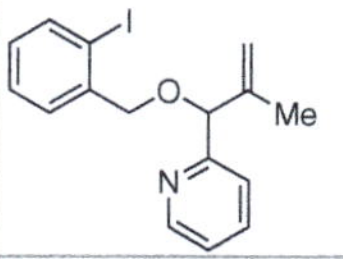	**1.173e**	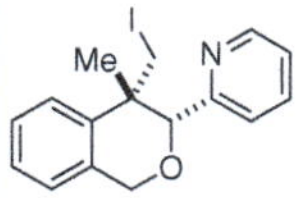	**1.187e**	NR [16 h]	–
6	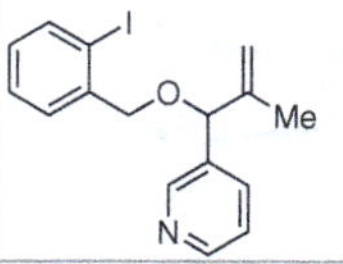	**1.173f**	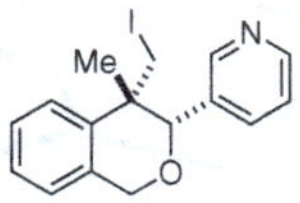	**1.187f**	35% [16 h]	>95:5
7	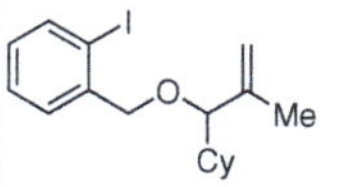	**1.173g**	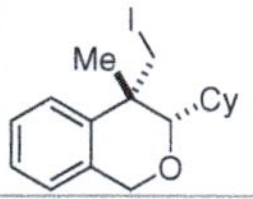	**1.187 g**	81% [4 h]	91:9
8	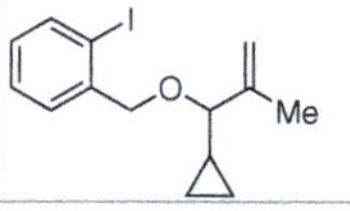	**1.173h**	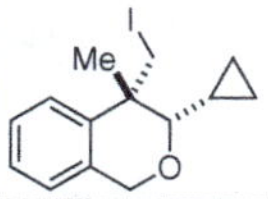	**1.187 h**	93% [10 h]	90:10
9	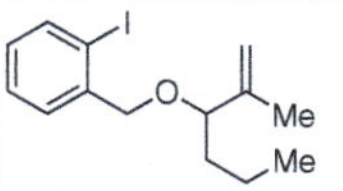	**1.173i**	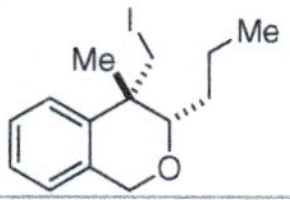	**1.187i**	86% [10 h]	89:11

(continued)

Table 1.5 (continued)

Entry	Substrate		Product		Yield[b] [time]	dr (*syn*: *anti*)[c]
10	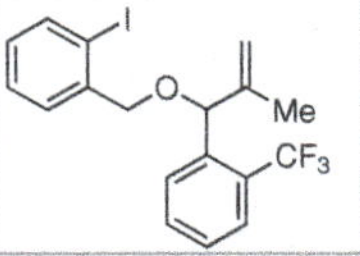	**1.173j**[f]	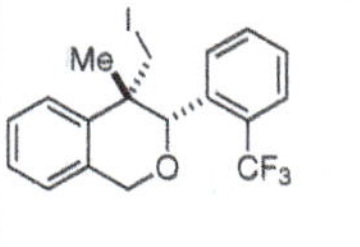	**1.187j**	88% [3 h]	>95:5
11	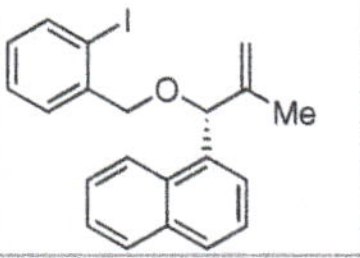	(*R*)-**1.180**	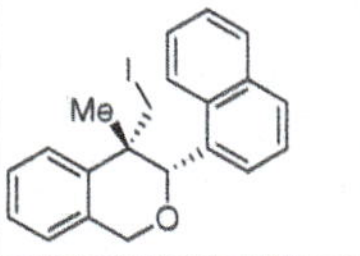	(*S*,*S*)-**1.187k**	70% 98% ee [14 h]	>95:5 dr
12	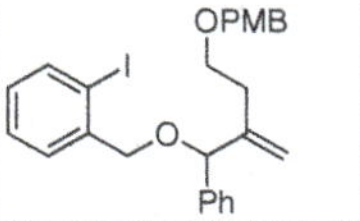	**1.181**	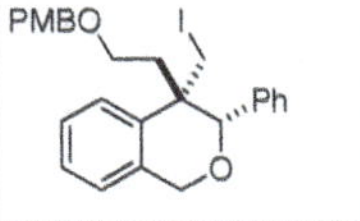	**1.187l**	89% [24 h]	>95:5 dr

[a]Reactions were run on a 0.2 mmol scale
[b]Combined isolated yield of both the *syn* and *anti* diastereomers
[c]Determined by ^{1}H NMR analysis of the crude reaction mixture
[d]Isolated yield of the reaction run on a 2 mmol scale
[e]1.0 equivalents of $^{i}Pr_2NEt$ was used instead of NEt_3
[f]This compound was prepared by Antonin Clemenceau

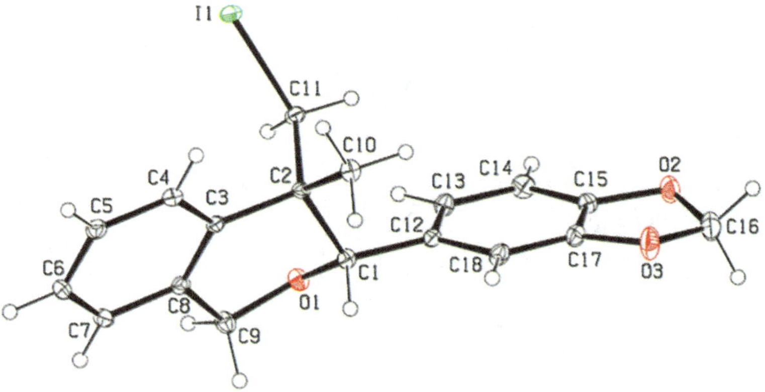

Fig. 1.1 Crystal structure of the *syn*-aryliodination product **1.187b**

major isomer (Fig. 1.1). Furan and thiophene containing substrates **1.173c** and **1.173d** underwent the transformation and afforded the desired isochromans **1.187c** and **1.187d** in 85 and 94% yields with 92:8 and 91:9 dr, respectively (entries 3 and 4). The 2-pyridyl analog **1.187e** did not undergo the desired transformation, yet, 3-pyridyl substituted **1.187f** afforded in 35% yield with >95:5 dr (entries 5 and 6). It is possible that the pyridyl moiety in **1.187e** can coordinate to the Pd(II) species after carbopalladation, leading to the formation of a stable palladacycle which

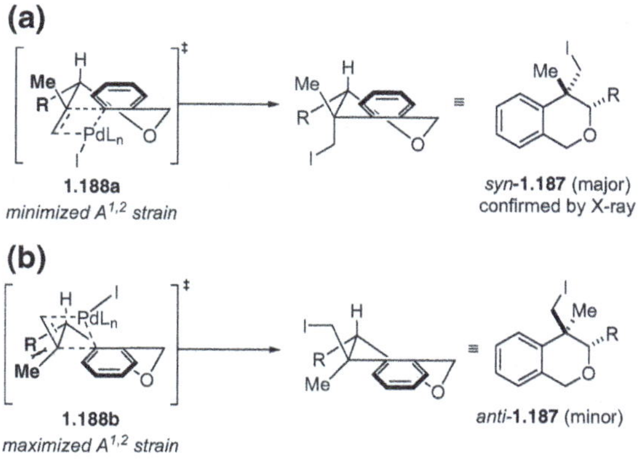

Scheme 1.43 Stereochemical models describing the migratory insertion step for isochroman synthesis leading to the **a** major isomer *syn*-**1.187**; and the **b** minor isomer *anti*-**1.187**. Adapted with permission from Ref. [1]. Copyright (2012) American Chemical Society

cannot undergo reductive elimination. However, **1.187f** still cyclized in relatively low yield which can suggest that pyridine heterocycles may act as a catalyst poison in this reaction. Cyclohexyl containing **1.173g** was isolated in 81% yield with 91:9 dr, while cyclopropyl (**1.173h**) and *n*-propyl (**1.173i**) analogs were obtained in yields of 93 and 86%, and with 90:10 and 89:11 dr (entries 7–9), respectively. At this point it became evident that stereoselectivity was related to the size of the ring tether substituent. In support of this, *o*-CF_3 functionalized **1.173j** was synthesized and underwent the cyclization to afford **1.187j** in 88% yield and >95:5 dr (entry 10). Enantiomerically enriched precursor (*R*)-**1.180** underwent cyclization to afford the corresponding isochroman (*S*,*S*)-**1.187k** in 80% yield with >95:5 dr and no apparent erosion of enantiomeric excess (entry 11). PMP protected **1.181** underwent clean cyclization to afford product **1.187l** in 89% yield with >95:5 dr (entry 12).

This stereoselectivity of isochroman formation is thought to arise from the decreased $A^{1,2}$ strain of the tether substituent and the vinylic CH_3 prior to carbopalladation. We hypothesize that in the mode of alkene coordination leading to the *syn* isomer, the $A^{1,2}$ strain between the vinylic CH_3 and the R group is reduced (**1.188a**, Scheme 1.43a). Conversely, in the mode of alkene coordination leading to the *anti* isomer, there is steric clash between the vinylic CH_3 and the R group (**1.188a**, Scheme 1.43b). This model can rationalize why larger R groups provide increased levels of diastereoselectivity.

We next sought to examine the generality of these reaction conditions on the preparation of chroman cores **1.189** using alkene substrates **1.169** (Table 1.6). Phenyl chroman **1.189a** was obtained in 81% yield with 93:7 dr (entry 1). These conditions afforded naphthalene analog **1.189b** in 96% yield as a single observable diastereomer (entry 2). The increased size of the naphthalene is likely resulting in the increased level of dr with respect to **1.189a**. This reaction is exceptionally tolerant towards heteroatoms, halogens, and electron withdrawing groups as the

Table 1.6 Scope of the chroman synthesis[a]

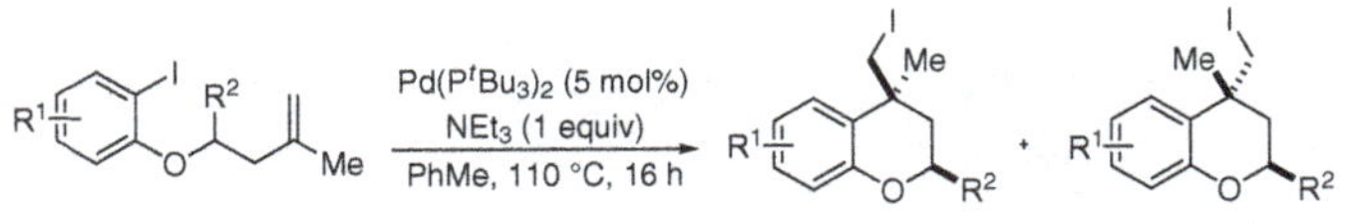

1.169 *syn*-**1.189** *anti*-**1.189**

Entry	Substrate		Product		Yield (%)[b]	dr (*anti*: *syn*)[c]
1	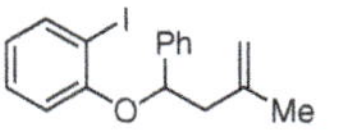	**1.169a**	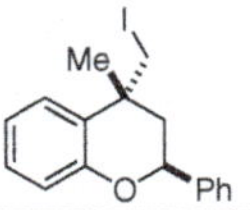	**1.189a**	81	93:7
2	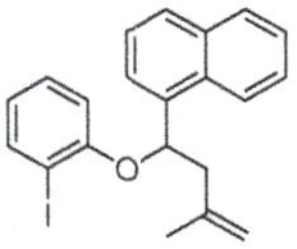	**1.169b**	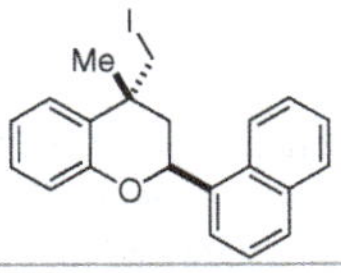	**1.189b**	96	>95:5
3	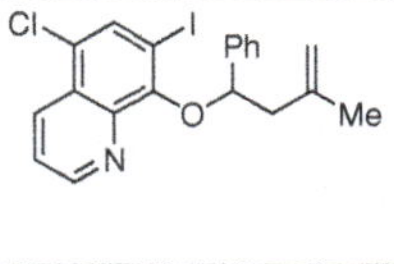	**1.169c**[d]	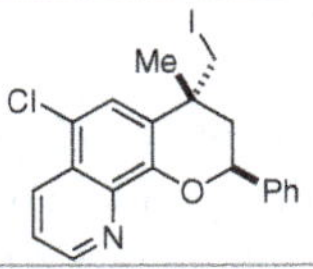	**1.189c**	92	90:10
4	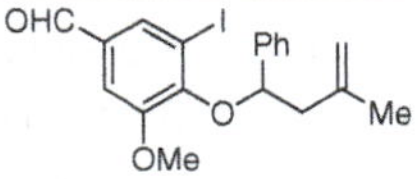	**1.169d**[d]	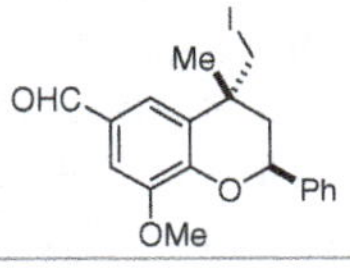	**1.189d**	86	92:8
5	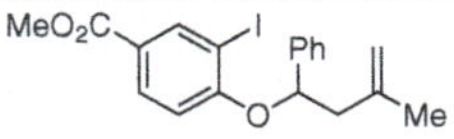	**1.169e**	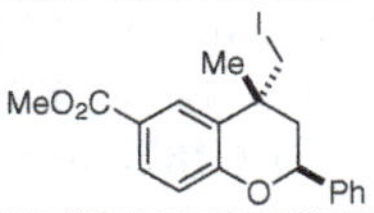	**1.189e**	98	91:9
6	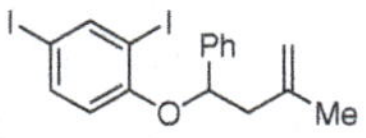	**1.169f**	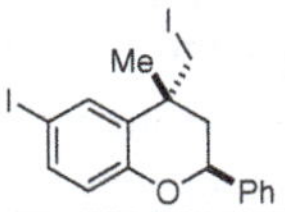	**1.189f**	70	91:9

[a]Reactions were run on a 0.2 mmol scale
[b]Combined isolated yield of both the *syn* and *anti* diastereomers
[c]Determined by ^{1}H NMR analysis of the crude reaction mixtureu
[d]This compound was prepared by Antonin Clemenceau

desired chromans were obtained with excellent yields and diastereoselectivities (**1.189c–e**). Remarkably, di-iodinated substrate **1.169f** was converted to the corresponding chroman **1.189f** in 70% yield and 91:9 diastereoselectivity without deleterious byproduct formation (entry 6). This result is consistent with an earlier report that highlights the possibility of reversible oxidative addition to C–I bonds by a Pd catalyst containing P^{t}Bu$_3$ as a ligand (cf. Chap. 4), since the iodine atom at the 4-position could be anticipated to react before the more hindered iodide atom at the 2-position.

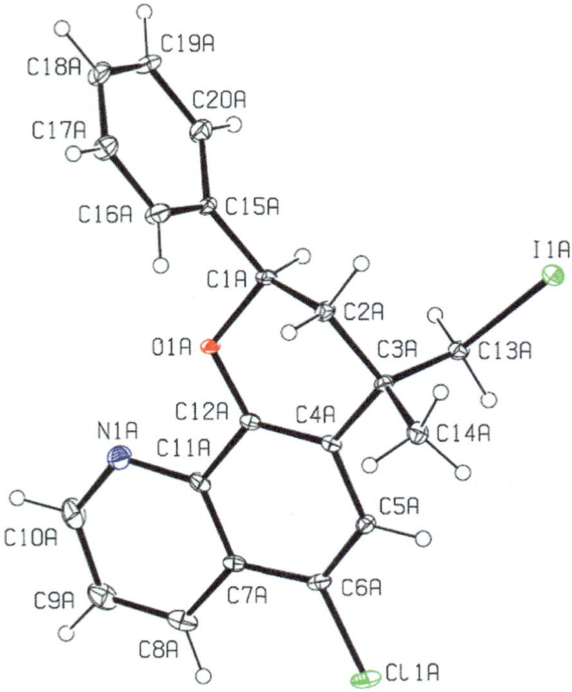

Fig. 1.2 Crystal structure of the *anti*-aryliodination product **1.189c**

In the case of the chroman synthesis, a switch in stereochemistry occurs with respect to the isochromas series, and the *anti* stereoisomer is obtained as the major product. This stereochemical outcome was unambiguously determined through x-ray analysis (Fig. 1.2). This stereoselectivity in chroman formation is thought to arise from the minimization of axial-axial interactions in the migratory insertion step. We hypothesize that in the mode of alkene coordination leading to the *trans* isomer, the axial-axial interaction between the vinylic CH_3 and the R group is reduced (Scheme 1.44a). Conversely, in the mode of alkene coordination leading to the *cis* isomer, this interaction is much greater (Scheme 1.44b). Again, this model rationalizes the observation that larger R groups provide increased levels of diastereocontrol.

Isochroman **1.187l** was successfully employed in a series of synthetic modifications (Scheme 1.45). The PMB protecting group could be removed in high yield under standard conditions using DDQ to generate primary alcohol **1.191**. Alcohol **1.191** could be carried forward to spirocycle **1.192** in 87% yield via an intramolecular S_N2 displacement of the neopentyl iodide under surprisingly mild conditions. Furthermore, **1.191** could be oxidized using DMP under mild conditions to generate aldehyde **1.193** in 92% yield.

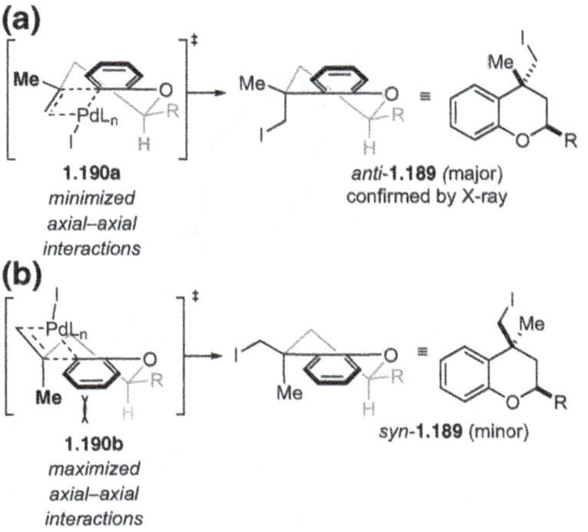

Scheme 1.44 Stereochemical models describing the migratory insertion step for isochroman synthesis leading to the **a** major isomer *anti*-**1.189**; and the **b** minor isomer *syn*-**1.189**. Adapted with permission from Ref. [1]. Copyright (2012) American Chemical Society

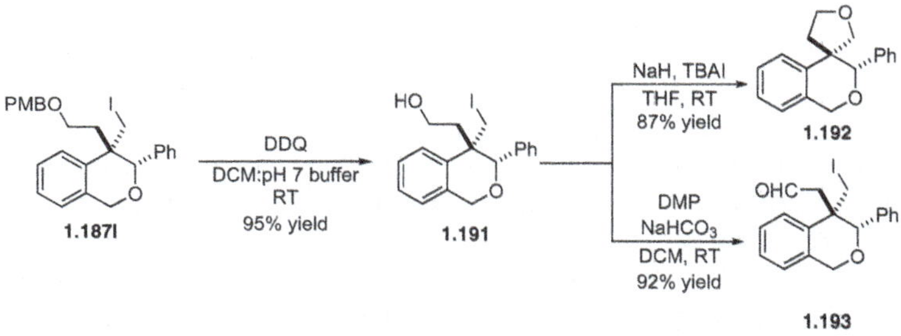

Scheme 1.45 Derivatizations of isochroman **1.187l**

1.1.4.4 Limitations

We attempted the cyclization of substrate **1.194** which would provide access to dihydrobenzofuran **1.195** (Scheme 1.46). This substrate cyclized in high yields under the optimized conditions, however, the product was obtained as only a 2:1 mixture of diastereomers. Despite many attempts, these products were not separable, but the ^{1}H shifts of the two diastereomers are well resolved in all key alkyl regions. Therefore, a 2D NOE NMR experiment was useful in confirming the stereochemistry of the major and minor diastereomers. By this method, *anti* **1.195** was assigned as the major product due to the presence of an observed NOE between the H_a region (5.63 ppm) and both the H_b and H_b' regions (3.5 and 3.42 ppm). This effect was absent in the minor *syn* stereoisomer.

Scheme 1.46 Attempt at the synthesis of 5-membered ring products

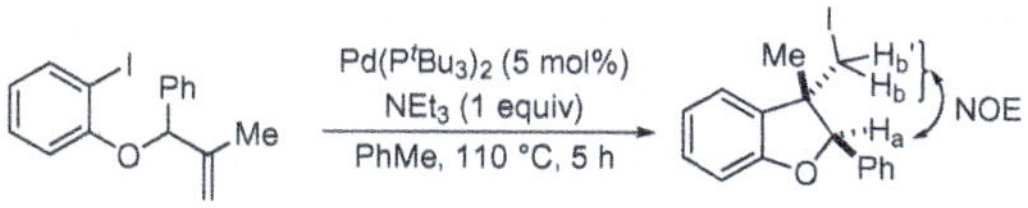

1.194

anti-1.195
95% yield, 2:1 *dr*

Scheme 1.47 Substrates which failed to provide aryliodination product

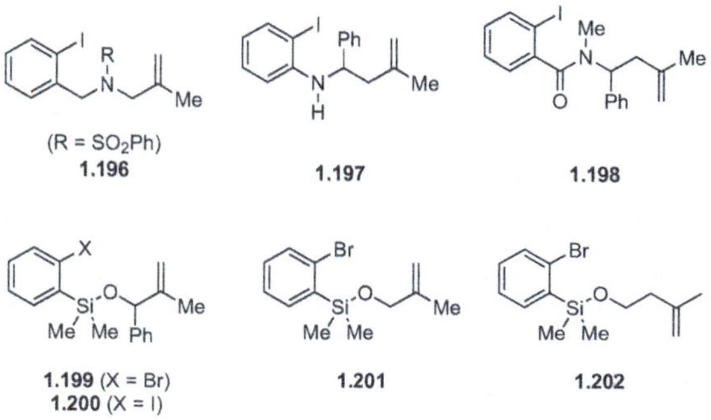

In addition to substrates aimed at providing isochromans and chromans, numerous other classes of substrates were prepared and failed to give any of the desired products under the optimized reaction conditions (Scheme 1.47). Nitrogen-containing substrates were consistently problematic. Both 6-(**1.196** and **1.197**) and 7-membered ring precursors (**1.198**) only provided varying amounts arising from isomerization of the alkene in addition to unreacted starting material. We were hopeful that silicon-tethered substrates would undergo this transformation, as they would provide a useful way to access the formal intermolecular aryliodination products by various forms of Si deprotection. These substrates either were completely unreactive (**1.199** and **1.200**) or led exclusively to products arising from alkene isomerization (**1.201** and **1.202**).

1.1.5 Research Goals: Part 2

We sought to apply the diastereoselective aryliodination to the total synthesis of a natural product as it would allow us to highlight the capabilities of the methodology. A former post-doctoral fellow Dr. Harasld Weinstabl became aware of the stereoselectivity observed in the synthesis of chromans and isochromans and began to seek out natural product candidates which had a similar arrangement of stereocenters on the benzofused heterocyclic ring. A literature search revealed (+)-corynoline (+)-**1.203**, which appeared to be accessible by our diastereoselective aryliodination. This hexacyclic natural product has exhibited diverse pharmacological effects including inhibition of cell adhesion[137] and acetyl-cholinesterase,[138] as well as fungi- and cytotoxicity,[139] yet only a limited number of total synthesis of this molecule have been reported.[140,141,142,143,144] (+)-Corynoline belongs to the family of hydroxy benzo[*c*]phenanthridine alkaloids and was first isolated from the aerial parts of *Corydalis incis* in Japan by the group of Takao.[145] They found that 3.7 kg of dry weight plant material gave 3.5 g of (±)-corynoline (0.095% dry weight). It possesses a *cis*-fused aza-decalin core containing three stereogenic centers, one of which is quaternery.

Naito reported the first total synthesis of (±)-corynoline in 1976 (Scheme 1.48).[146] A base-mediated amide bond-forming reaction between substituted benzoyl chloride **1.204** and imine **1.205** afforded amide **1.206** in 66% yield. By using a high pressure Hg lamp, **1.206** could be photocyclized under high-dilution conditions to afford lactam **1.207** in 20% yield which accompanied by a phenol-containing byproduct (10%). The aminal functionality in **1.207** could be reduced using Pd/C and H_2 to afford both *anti*–**1.208** in 60% yield and *syn*–**1.208** in 21% yield, the latter of which was desired. Dehygrogenative oxidation of *syn* –**1.208** using DDQ afforded alkene **1.209** in 28% yield. Hydride reduction of the lactam carbonyl was done using $LiAlH_4$ which afforded amine **1.210** in 56% yield. A two-step protocol involving epoxidation of the alkene in **1.210** and subsequent epoxide opening using hydroxide afforded *trans* diol **1.211** in 91% yield. (±)-Corynoline could be obtained by a selective hydrogenolysis of the benzylic alcohol moiety using Pd/C in 35% yield.

Cushman and co-workers reported a greatly improved route to (±)-**1.203** (Scheme 1.49) [139]. Their synthesis began with the condensation of anhydride **1.212** (prepared in 8 steps) and imine **1.213** which generated a mixture of *anti*-**1.214** (66% yield) and *syn*-**1.214** (29% yield). The desired, yet minor, *cis* isomer was treated with thionyl chloride to generate the corresponding acid chloride, which was treated with an ethereal solution of diazomethane to generate advanced diazo intermediate **1.215**. When treated with trifluoroacetic acid at low temperature, **1.215** could undergo an intramolecular denitrogenative cyclization to generate ketone **1.216**. Finally, treatment with $LiAlH_4$ in refluxing THF afforded (±)-corynoline via reduction of both and the amide and ketone functionalities.

---

137 Reference [135].
138 Reference [136].
139 Reference [137].
140 Reference [138].
141 Reference [139].
142 Reference [140].
143 Reference [141].
144 Reference [142].
145 Reference [143].
146 Reference [138].

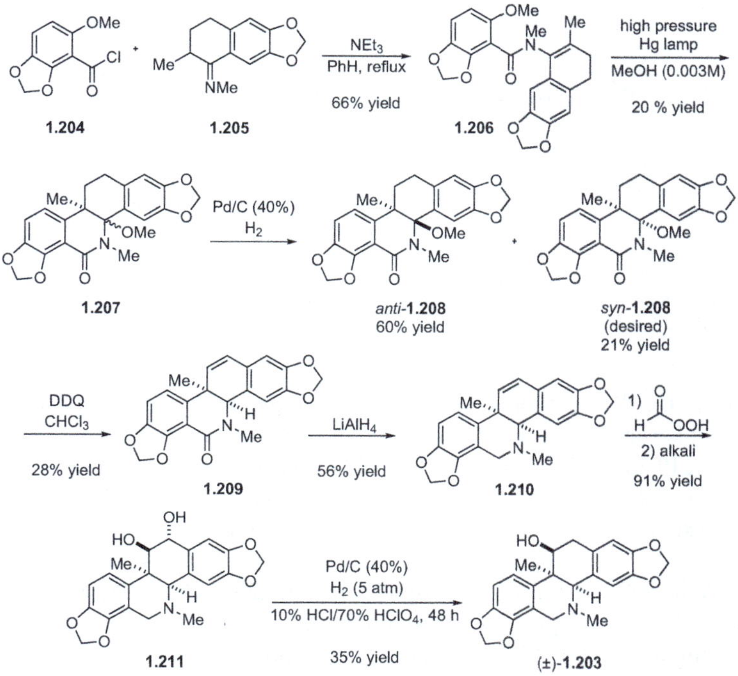

Scheme 1.48 Total synthesis of (±)-corynoline by Naito

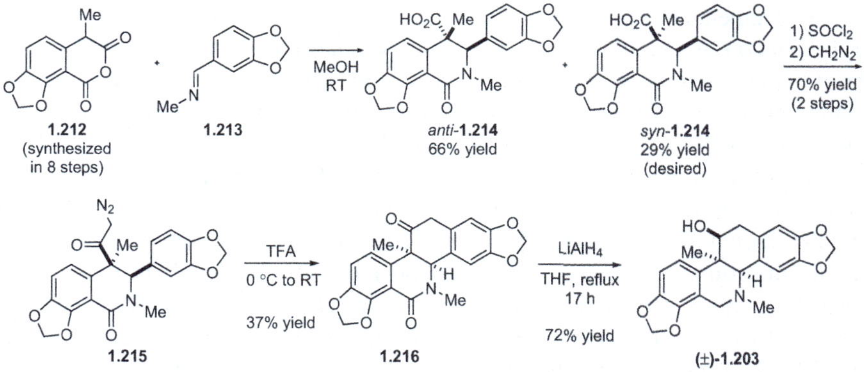

Scheme 1.49 Total synthesis of (±)-corynoline by Cushman

Although this synthesis clearly presented many improvements over that of Naito, it still led to racemic material and involved a key step where the desired diastereoisomer turned out to be the minor product. The same authors reported an improvement to this route which permitted them access to (+)-corynoline

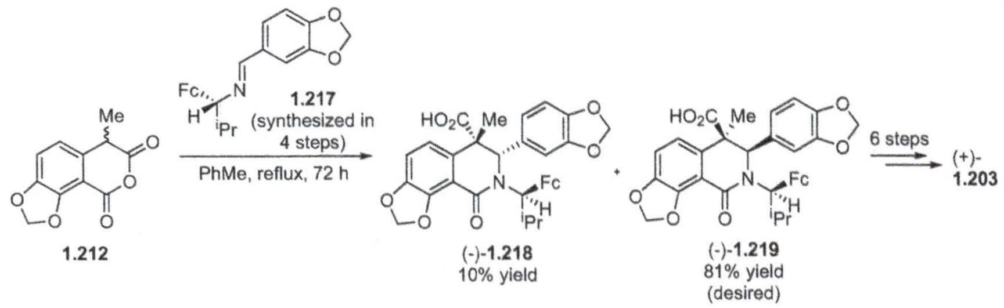

Scheme 1.50 Key step in the synthesis of (+)-corynoline employing a chiral auxiliary by Cushman (Fc = ferrocenyl)

(Scheme 1.50). The authors prepared chiral imine **1.217** and employed it in a similar condensation with **1.212** to afford a mixture of two diastereomers (−)-**1.218** and (−)-**1.219** in 10% yield and 81% yield, respectively. Fortunately, the major product was the one required to access (+)-corynoline, and this was accomplished using a 6-step sequence reminiscent of their previous end game (Scheme 1.50) [139].

Dr. Weinstabl developed a retrosynthesis (+)-**1.203** which began with the formation of the natural product by a simultaneous amide reduction/regioselective epoxide ring opening of **1.220** (Scheme 1.51). Epoxide **1.220** can be accessed by a stereoselective α-bromination followed by reduction of the ketone in **1.221**. We envisioned that **1.221** could be accessed by an intramolecular Houben-Hoesch reaction of nitrile **1.222**. The required nitrile functionality could be installed by a direct S_N2 displacement of neopentyl iodide **1.223**. Obtaining **1.223** would rely on the development and application of a diastereoselective intramolecular Pd-catalyzed aryliodination of enantioenriched a substrate similar to carboxamide **1.224**. As per the previous section, we had little success with nitrogen-containing heterocycles containing amines. However, preliminary results obtained suggested that linear amides could in fact undergo a cyclization. This linear starting material could be obtained by an amide bond coupling between acid chloride **1.225** (available in three steps from commercially available material) and chiral secondary amine **1.226**. We initially thought that amine **1.226** could be accessed by an asymmetric Ir-catalyzed allylic amination, yet, due to the early stage that this reaction would be required, we opted for a more scalable and inexpensive route. Therefore, amine **1.226** could be obtained by a diastereoselective addition of isopropenyl magnesium bromide **1.227** to **1.228**. Sulfoninamide **1.228** could be accessed by a straightforward condensation between commercially available (yet in some cases restricted) piperonal **1.229** and the *S*-enantiomer of Ellman's auxiliary **1.230**.

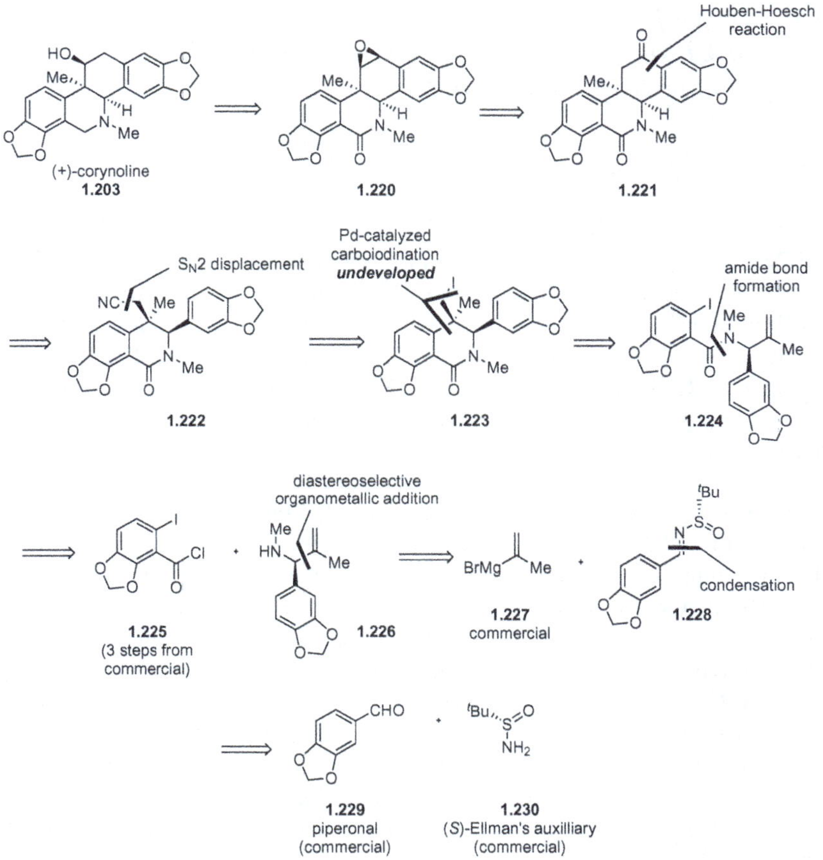

Scheme 1.51 Retrosynthetic analysis of the alkaloid natural product (+)-corynoline

This route has several qualitative advantages over those previously discussed. Namely, this strategy has its roots in common and inexpensive commercial materials including the Ellman's auxiliary which is available in large quantities for both enantiomers. The amide bond disconnection promotes convergence in this route and allows for derivatives to be rapidy prepared from common intermediates. The use of the Houben-Hoesch reaction allows the final cyclization to proceed directly from nitrile **1.222** which avoids the need to hydrolyze to the acid and further conversion to the α-diazo compound (i.e. **1.215**). Finally, this route is protecting group free which should help decrease the number of steps required to access to the target.

1.1.6 Development of the Diastereoselective Aryliodination of Chiral Amides: Application to the Chemical Synthesis of (+)-Corynoline

1.1.6.1 Starting Material Preparation

Aldehydes **1.231** underwent condensation with (*S*)-2-methylpropane-2-sulfinamide (*S*)-**1.230** when refluxed in DCM in the presence of stoichiometric Cs_2CO_3 to generate the enentiopure sulfinamides **1.232**.[147] In a process initially developed by Hon Wei Lam,[148] **1.232** was reacted with isopropenyl magnesium bromide in the presence of stoichiometric $ZnMe_2$ in THF at −78 °C to generate 2° allylic amines sulfinamides **1.233** as virtually single diastereomers. These compounds were alkylated using LiHMDS in combination with either iodomethane or benzylbromide in DMF at 0 °C to generate the corresponding 3° allylic sulfinamides **1.234**. The SO^{t}Bu moiety was easily removed using dry HCl in dioxane at 0 °C followed by treatment with ammonium hydroxide to cleanly obtain the corresponding free 2° allylic amines **1.235**. The desired linear amides **1.237** were obtained by a Schotten-Baumann-type amide coupling using the corresponding acid chloride **1.236**. The required benzoyl chlorides were commercially available in only a few cases, yet they were easily prepared from the carboxylic acids using oxalyl chloride with a catalytic amount of DMF in DCM at 0 °C (Scheme 1.52).

Acid **1.225**, was prepared in a three step route from commercially available aniline derivative **1.238**. Iodination of **1.238** under standard Sandmeyer conditions afforded iodobenzene derivative **1.239** in 66% yield. By using a regioselective directed-*ortho* lithiation/formylation sequence,[149] aryl iodide **1.239** could be converted to aldehyde **1.240** which is carried on to acid **1.241** by a Pinnick oxidation (Scheme 1.53).

The route described in Scheme 1.52 was applied to the synthesis of chiral carboxamide (*S*)-**1.224** (Scheme 1.54). Piperonal underwent condensation with (*S*)-**1.230** in 92% yield to generate chiral sulfonamide (*S*)-**1.228** which reacted with isopropenyl magnesium bromide and $ZnMe_2$ to access (*S,S*)-**1.242** as a single diastereomer. Chrial allylic amine (*S*)-**1.235** was obtained after subsequent alkylation and deprotection of (*S,S*)-**1.242**, and then by reacting with benzoyl chloride derivative **1.225** (generated from **1.241**) to afford amide (*S*)-**1.124**.

---

[147]Reference [144].

[148]Reference [145].

[149]Reference [146].

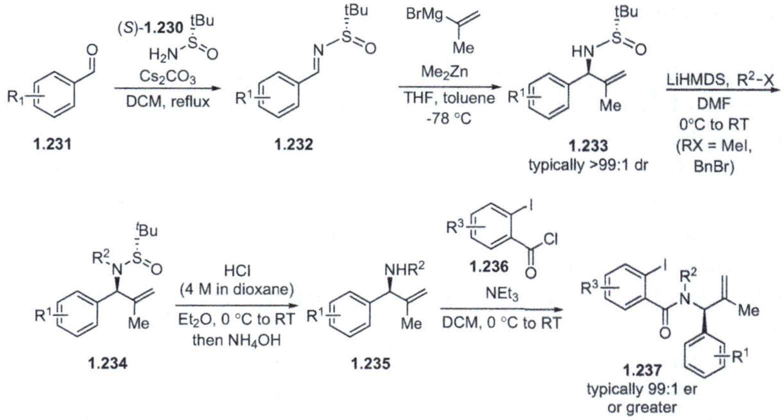

Scheme 1.52 General route to enantioenriched *N*-allyl carboxamides

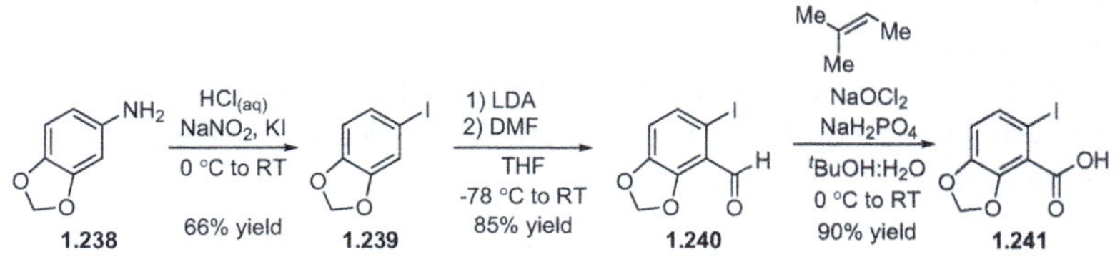

Scheme 1.53 Synthetic route to acid **1.225**

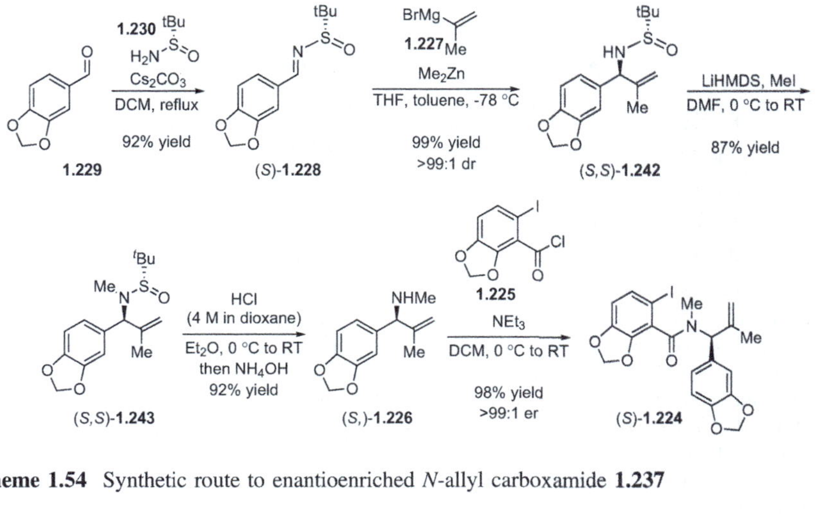

Scheme 1.54 Synthetic route to enantioenriched *N*-allyl carboxamide **1.237**

1.1.6.2 Optimization

Chiral amide (*S*)-**1.237a** was chosen for the optimization of the reaction conditions instead of (*S*)-**1.224** since it was a more electron neutral substrate. Optimizing the reaction conditions for such a substrate would allow for a better chance for electron deficient derivatives to be effectively incorporated into the substrate scope. Conversely, optimizing for a electron rich substrate such as (*S*)-**1.224** could potentially have led to a skewed representation of effective reaction conditions. Since the nature of the ligand had the greatest effect on reaction efficiency, optimization began by examining a series of Pd catalysts possessing different ligands (Table 1.7).

We initiated our efforts testing the aryliodination of (*S*)-**1.237a** using $Pd(P^tBu_3)_2$ (10 mol%) and NEt_3 (2 equiv) in PhMe (0.05 M) at 100 °C for 20 h. This led to 85% conversion of starting material with the desired product being obtained in 28% yield with 94:6 dr (entry 1). Switching the catalyst to $Pd(QPhos)_2$ led to improved results and the product was obtained in 42% yield with >95:5 dr (entry 2). The Pd (II) pre-catalyst Pd(crotyl)(QPhos)Cl in the presence of a catalytic amount of KO^tBu provided worse results (entry 3). This catalyst system was tried due to the fact that it has a 1:1 M:L ratio instead of $Pd(QPhos)_2$ which has a 1:2 M:L ratio. The combination of $Pd(OAc)_2$ and $P^tBu_3{\cdot}HBF_4$ failed to give the desired product, however, combining $Pd(dba)_2$ and $P^tBu_3{\cdot}HBF_4$ afforded the product in 46% yield, albeit only with 87:13 dr (entries 4 and 5).

Table 1.7 Optimization of the intramolecular aryliodination of (*S*)-**1.237**: screening Pd-sources[a]

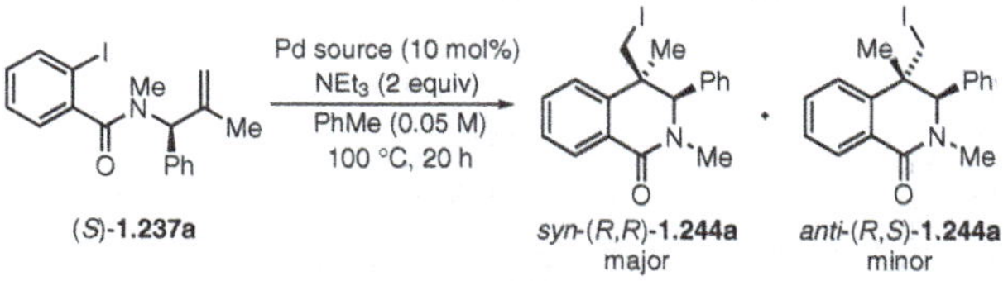

Entry	Pd source	Conv. (%)[b]	Yield[b] (*syn* + *anti*)	dr (*syn:anti*)[c]
1[g]	$Pd(P^tBu_3)_2$	85	28	94:6
2[g]	$Pd(QPhos)_2$	>95	42	>95:5
3[d]	PdCrotyl(QPhos)Cl	>95	29	85:15
4[e]	$Pd(OAc)_2/P^tBu_3{\cdot}HBF_4$	77	N/A	N/A
5[f]	$Pd(dba)_2/P^tBu_3{\cdot}HBF_4$	>95	46	83:17

[a]Reactions run on 0.05 mmol scale
[b]Determined by ^1H NMR analysis of the crude reaction mixture using 1,3,5-trimethoxybenzene as an internal standard (both diastereomers combined)
[c]Determined by ^1H NMR analysis of the crude reaction mixture
[d]KO^tBu added to generate the active Pd(0) source
[e]20 mol% of $P^tBu_3{\cdot}HBF_4$ salt used
[f]10 mol% of $P^tBu_3{\cdot}HBF_4$ salt used
[g]Reaction carried out by Hyung Yoon

Table 1.8 Optimization of the intramolecular aryliodination of (*S*)-1.237: screening of bases[a]

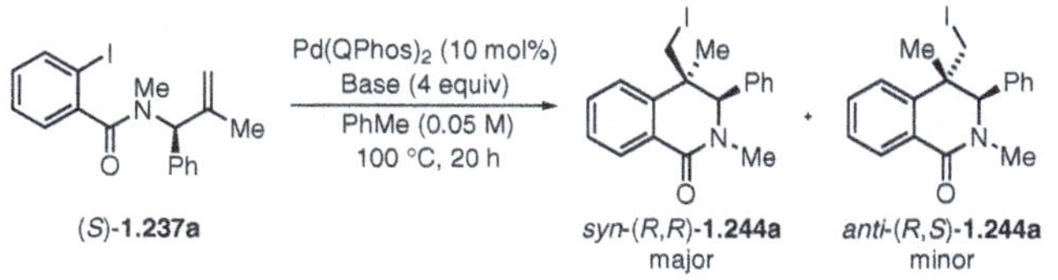

Entry	Base	Conv.[b]	Yield[b] (*syn* + *anti*)	dr[c] (*syn*:*anti*)
1	NEt_3	>95	46	>95:5
2[c]	$^iPr_2NEt$	>95	67	>95:5
3	Cy_2NMe	>95	40	95:5
4	$BuNMe_2$	>95	43	>95:5
5[d]	PMP	>95	71	>95:5
6[c]	Cs_2CO_3	N/A	10	N/A
7	none	>95	79	81:19

[a]Reactions run on 0.05 mmol scale by Hyung Yoon
[b]Determined by ^{1}H NMR analysis of the crude reaction mixture using 1,3,5-trimethoxybenzene as an internal standard (both diastereomers combined)
[c]Determined by ^{1}H NMR analysis of the crude reaction mixture
[d] Average values over 3 experiments

We screened as series of bases because these additives had beneficial effects on the reaction in previous studies (Table 1.8). Increasing the amount of base to 4 equivalents had a slight positive effect on the yield of the reaction (entry 1). Employing Hünig's base led to a marked improvement to the reaction conditions, and the desired product was obtained in 67% yield in >95:5 dr. Both Cy_2NMe and $BuNMe_2$ led to decreased yields (entries 3 and 4) whereas the addition of PMP gave the product in 71% yield in >95:5 dr (entry 5). The use of Cs_2CO_3 led to an intractable crude reaction mixture as judged by ^{1}H NMR (entry 6). In the absence of base, the yield of the reaction was increased to 79%, but most interestingly, the product was obtained with 81:19 dr (entry 7). This result intrigued us, as previously the absence of base only negatively affected the product yield and not the diastereomeric ratio. At the time, the origin of this result was not clear so further studies were done to attempt to understand the effect (vide infra).

We next sought to lower the catalyst loading, increase the reaction concentration and decrease the amount of PMP since it is one of the more costly commercially available tertiary amine bases. Previous optimizations were carried out on a 0.05 mmol scale (19.5 mg of (*S*)-**1.237a**) in order to preserve the chiral material. However, as we began approaching the final stages of optimization, we opted to increase the reaction scale to 0.2 mmol (78 mg of (*S*)-**1.237a**). This change let to slight changes in observed yields and diastereoselectivities (Table 1.9). We found that this change led to an increase in product yield to 90% after isolation, yet the dr dropped slightly from >95:5 (Table 1.8, entry 1) to 92:8 (Table 1.9, entry 1). Decreasing the loading of $Pd(QPhos)_2$ to 5 mol% led to a marked decrease in yield with no change in dr (entry 2). However, decreasing the amount of PMP from 4

Table 1.9 Optimization of the intramolecular aryliodination of (*S*)-1.237: catalyst/base loading[a]

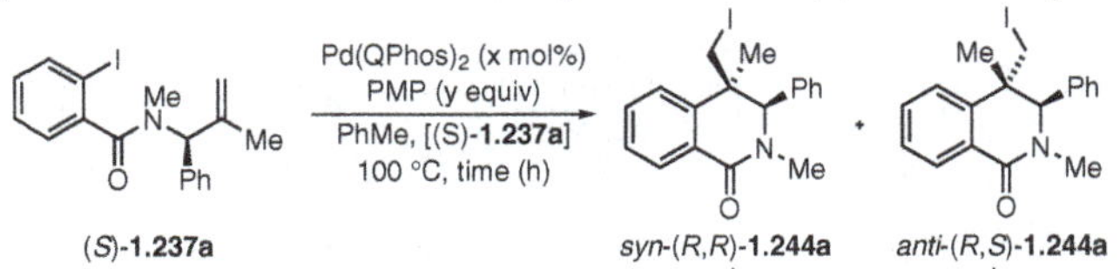

(*S*)-**1.237a** *syn*-(*R*,*R*)-**1.244a** major *anti*-(*R*,*S*)-**1.244a** minor

Entry	Catalyst loading (x mol%)	equiv PMP	([(*S*)-**1.237a**])	Yield[b,c] (*syn* + *anti*)	dr[d] (*syn*:*anti*)
1[e]	10 (10 h)	4	0.05 M	quant (90)	92:8
2	5.0 (30 h)	4	0.05 M	69	92:8
3[e]	5.0 (22 h)	2	0.1 M	quant (88)	>95:5
4	5.0 (22 h)	1	0.2 M	90	>95:5
5	5.0 (22 h)	0	0.1 M	95 (97)	83:17

[a]Reactions run on 0.2 mmol scale
[b]Determined by ^{1}H NMR analysis of the crude reaction mixture using 1,3,5-trimethoxybenzene as an internal standard (both diastereomers combined)
[c]Values in parentheses represent isolated yields
[d]Determined by ^{1}H NMR analysis of the crude reaction mixture
[e]Reaction conducted by Hyung Yoon

equivalents to 2 equivalents, while simultaneously increasing the reaction concentration from 0.05 to 0.1 M, afforded a quantitative yield by ^{1}H NMR and an 88% isolated yield with an increase to >95:5 dr (entry 3). Decreasing the loading of PMP from 2 equivalents to 1 equivalent led to a slight decrease in yield (entry 4), whereas complete removal of the PMP base actually increased the yield to 95% albeit with a marked decrease to 83:17 dr (entry 5). The final optimized conditions were found to be Pd(QPhos)$_2$ (5 mol%) with PMP (2 equiv) in PhMe (0.1 M) at 100 °C for 22 h.

1.1.6.3 The Effect of Base on Diastereoselectivity

At the onset, we were perplexed by the significant positive effect that added base had on the diastereoselectivity of the reaction. Since the reaction run in the absence of PMP led to a higher yield than the reaction run with PMP, one possible explanation for the change in diastereoselectivity was that the minor diastereomer was decomposing under the reaction conditions. If this were to be true, an artificial increase in the ratio of *cis*:*trans* in favor of the *cis* isomer would result, and the dr would appear enhanced. Second, since the stereodetermining step of the proposed mechanism is assumed to be the intramolecular carbopalladation, the bulky tertiary amine base could perhaps alter the geometry of the active catalyst by weak coordination at this stage and influence the facial selectivity of the olefin insertion. Third, we previously saw no effect on selectivity by added base in the synthesis of chromans and isochromans. One of the main distinguishing spectroscopic features between the

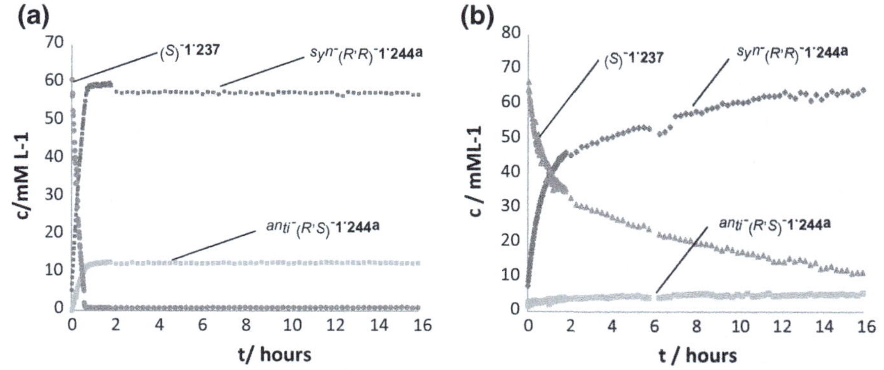

Fig. 1.3 In situ reaction monitoring for the intramolecular carboiodination of (*S*)-**1.237**. **a** Reaction conditions: $Pd(QPhos)_2$ (0.05 equiv), $[(S)\text{-}1.237]_0 = 0.1$ M, PhMe-d_8, 368 K, 600 MHz. **b** Reaction conditions: $Pd(QPhos)_2$ (0.05 equiv), PMP (2 equiv), $[(S)\text{-}\mathbf{1.237}]_0 = 0.1$ M, [PMP] = 0.2 M, PhMe-d_8, 368 K, 600 MHz. Adapted with permission from Ref. [2]. Copyright (2013) John Wiley & Sons

oxygen-containing substrates (**1.169** and **1.170**) and these amides is the latter are present in solution as a complex mixture of rotamers (up to 4). Was the bulky amine base coordinating to the Pd catalyst prior to carbon–iodine oxidative addition and influencing which rotamers the catalyst undergoing oxidative addition, with similar to the Curran's argument concerning a DKR operating in the Pd-catalyzed asymmetric Mizoroki-Heck reaction? This could possibly lead to oxidative addition complexes more primed to undergo carbopalladation leading to the *cis* isomer.

We believed that a deeper understanding of these results could be gained by using in situ NMR reaction monitoring.[150] By running the reaction under standard conditions in the NMR spectrometer, we were able to track the progress of the reaction with and without added PMP base. During these experiments we observed that full conversion of (*S*)-**1.237** had been reached after 35 min in the absence of PMP base (Fig. 1.3a), whereas only about 60% conversion of (*S*)-**1.237** had been reached after 35 min in the presence of 2 equivalents of PMP base (Fig. 1.3b).

The aforementioned possibility of selective decomposition of the *trans* diastereomer was ruled out since the concentration of *trans*-(*R*,*S*)-**1.244a** consistently increased with no loss even at high substrate conversions. However, these experiments add support to the hypothesis that the amine base can coordinate to the catalyst, as its presence decreases the overall rate of the reaction, while imparting nearly a five-fold increase in diastereoselectivity with respect to the base-free reaction.

The facial selectivity of alkene coordination and the subsequent insertion step determine the diastereoselectivity of this reaction. However, in order for all three substrate-derived ligands (aryl, iodide, and alkene) in addition to both a single

[150] Reference [147].

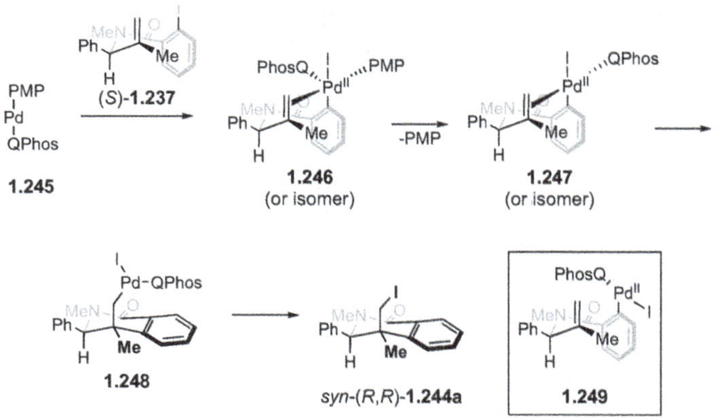

Scheme 1.55 Possible modes in which PMP can influence diastereoselectivity

phosphine from QPhos and the nitrogen of PMP to be bound during this step, Pd(II) would need to exist in a pentacoordinate mode.[151] However, olefin insertion reactions of pentacoordinate Pd(II) species are normally considered to be energetically unfavorable [98, 99]. As previously discussed, Overman reported that a pentacoordinate Pd(II) species can undergo associative ionization to form a cationic tetracoordinate Pd(II)-BINAP species in polar solvents like DMA during the asymmetric Mizoroki-Heck reaction [96]. However, the use of non-polar PhMe in the arlyiodination casts a level of doubt on a similar cationic process occurring. Mikami reported that four- or five-coordinate Pd(II) species can undergo carbopalladation during cyclization of 1,6-enyes when a polar (DMSO) or non-polar solvent (PhH) is used, respectively.[152]

Initially it was proposed that proposed that a Pd(0) species (i.e. **1.245**) could undergo oxidative addition to a substrate molecule to generate pentacoordinate Pd(II) species **1.246**. Associative displacement of the PMP ligand of **1.246** with the alkene would generate tetracoordinate species **1.247** which goes on to form product via complex **1.248.** However, the extreme steric crowding of a tetracoordinate Pd (II) species containing substrate-derived aryl and ioidide ligands, in addition to both QPhos and PMP becomes evident after simply constructing a physical or in silico 3D model. Therefore, the involvement of a pentacoordinate complex such as **1.246** would be even less likely. Although the exact origin of the effect of PMP of diastereoselectivity is not known, it is possible that a Pd(0) complex containing PMP as a ligand could be involved in the selective oxidative addition to one rotamer to generate tetracoordinate Pd(II) species **1.249** (Scheme 1.55). From our base screening studies it appears that all soluble amine bases resulted in nearly identical increases in diastereoselectivity. Since the structural identity of the amine

[151]References [148, 149].

[152]Reference [150].

additive does not appear to have a direct influence on the diastereoselecitivty within this data set, an alternative possibly could be that the amine base is not acting to create a more selectivite aryliodination catalyst, yet instead deactivates a $L_nPd(0)$ species which catalyzes this transformation with a lower level of selectivity.

1.1.6.4 Examination of the Substrate Scope

The generality of the optimized conditions was tested on a series of chiral *N*-allyl carboxamide substrates (Table 1.10). In all cases the enantioenriched substrates could be cyclized to the desired products with no erosion of enantiomeric excess. As previously mentioned, either enantiomer could be accessed and (*R*)-**1.237** could be prepared from the *R*-enantiomer of the Ellman's auxiliary and cyclized to the corresponding product in 90% yield with >95:5 dr (entry 6). Electron-rich aryl iodides (entries 2, 3 and 7) could be efficiently cyclized under the standard

Table 1.10 Scope of the dihydroisoquinolinone synthesis[a]

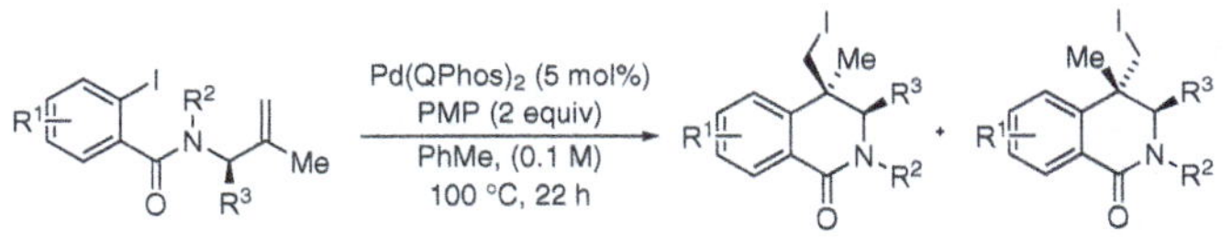

(*S*)-**1.237** syn-(*R*,*R*)-**1.244** major anti-(*R*,*S*)-**1.244** minor

Entry	Substrate		Product		Yield (%)[b]	dr (*syn*:*anti*)[c] (er)[d]
1	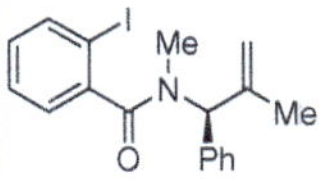	(*S*)-**1.237a**	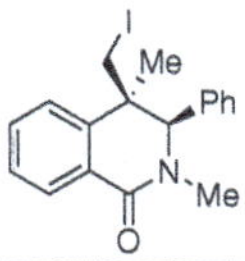	(*R*,*R*)-**1.244a**	88	>95:5 (>99:1)
2	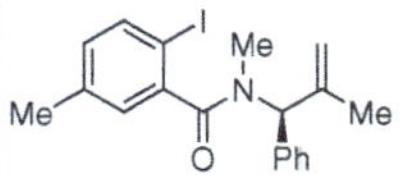	(*S*)-**1.237b**	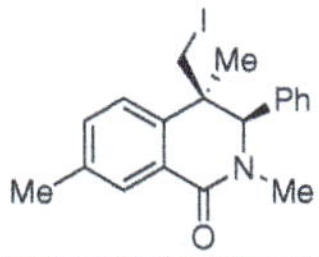	(*R*,*R*)-**1.244b**[e]	88	94:6 (>99:1)
3	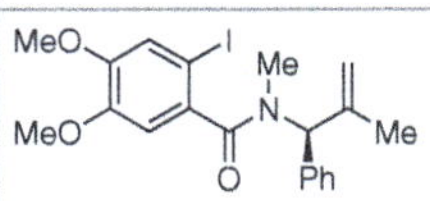	(*S*)-**1.237c**	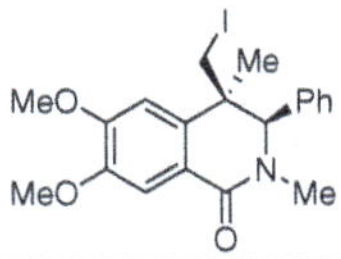	(*R*,*R*)-**1.244c**	82	93:7 (>99:1)
4[f]	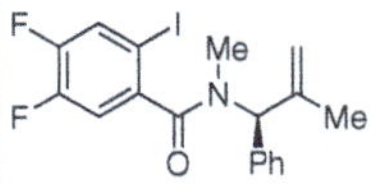	(*S*)-**1.237d**[e]	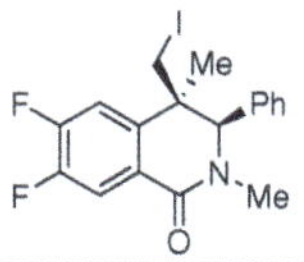	(*R*,*R*)-**1.244d**[e]	54	>95:5 (>99:1)

(continued)

Table 1.10 (continued)

Entry	Substrate		Product		Yield (%)[b]	dr (*syn*:*anti*)[c] (er)[d]
5	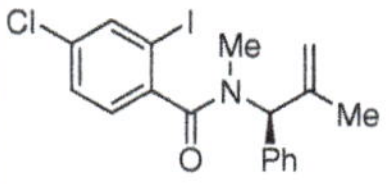	(*S*)-**1.237e**	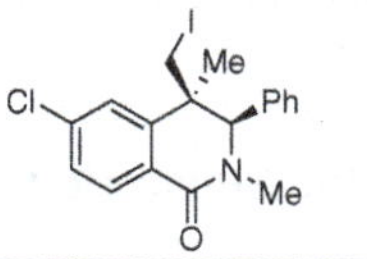	(*R*,*R*)-**1.244e**[e]	76	>95:5 (>99:1)
6	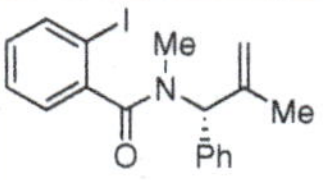	(*R*)-**1.237f**	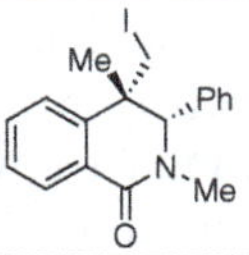	(*S*,S)-**1.244f**	90	>95:5 (>99:1)
7	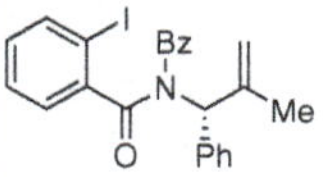	(*R*)-**1.237g**	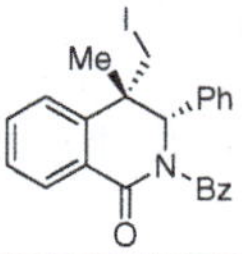	(*S*,*S*)-**1.244g**[e]	61	65:35 (>99:1)
8	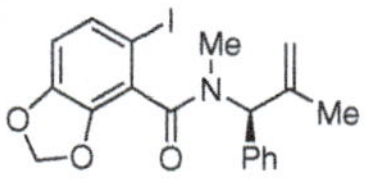	(*S*)-**1.237h**	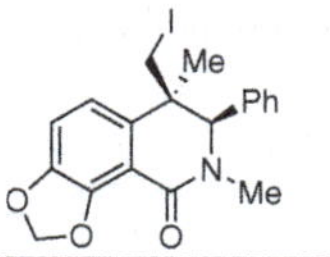	(*R*,*R*)-**1.244h**	82	94:6 (>99:1)
9	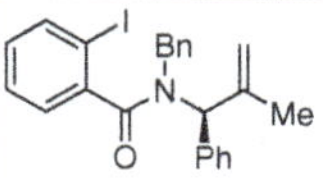	(*S*)-**1.237i**[e]	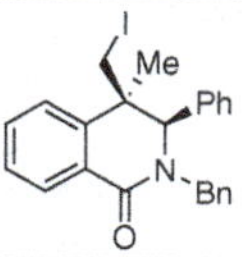	(*R*,*R*)-**1.244i**[e]	95	>95:5 (>99:1)
10	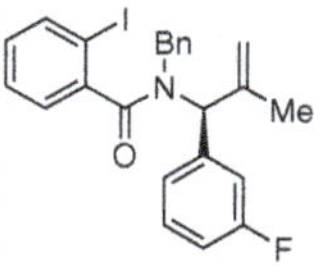	(*S*)-**1.237j**	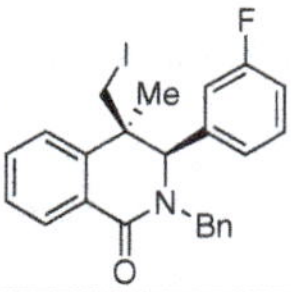	(*R*,*R*)-**1.244j**	87	>95:5 (>99:1)
11	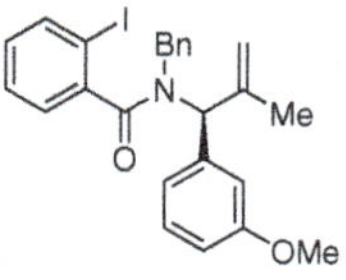	(*S*)-**1.237k**	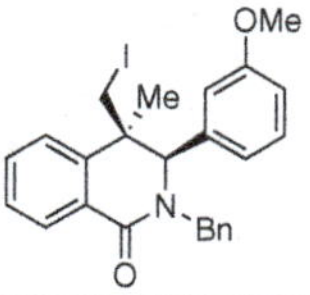	(*R*,*R*)-**1.244k**	90	>95:5 (>99:1)
12	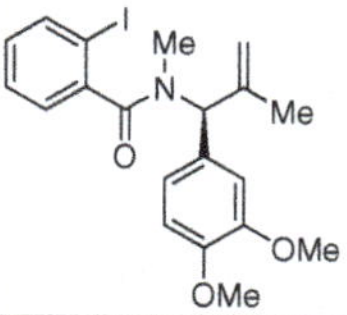	(*S*)-**1.237l**	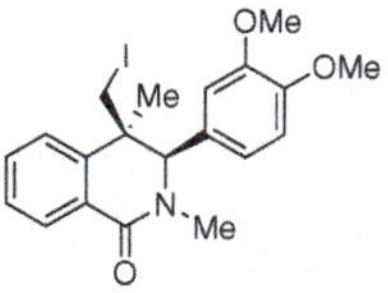	(*R*,*R*)-**1.244l**	76	93:7 (>99:1)
13	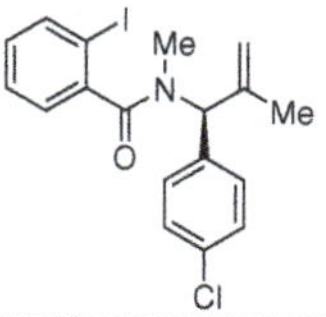	(*S*)-**1.237m**[e]	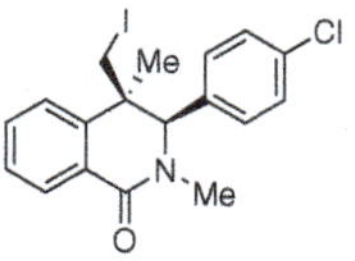	(*R*,*R*)-**1.244m**[e]	82	94:6 (>99:1)

(continued)

Table 1.10 (continued)

Entry	Substrate		Product		Yield (%)[b]	dr (*syn*: *anti*)[c] (er)[d]
14	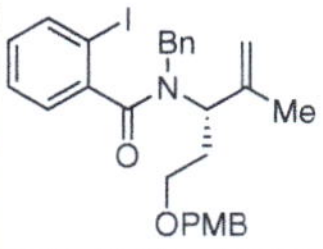	(*S*)-**1.237n**	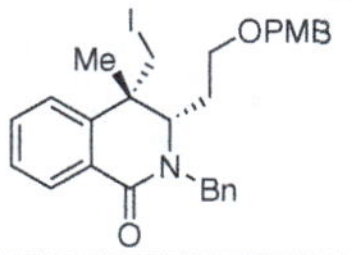	(*S*,*S*)-**1.244n**	91	80:20 (>99:1)
15	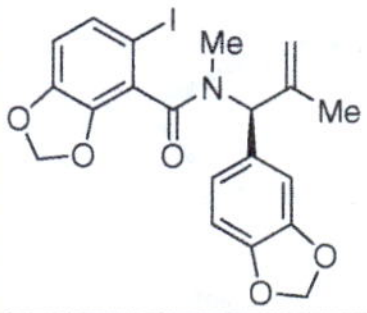	(*S*)-**1.224**	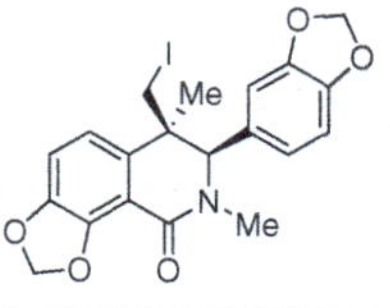	(*R*,*R*)-**1.223**	84 (48 h)	92:8 (>99:1)
16	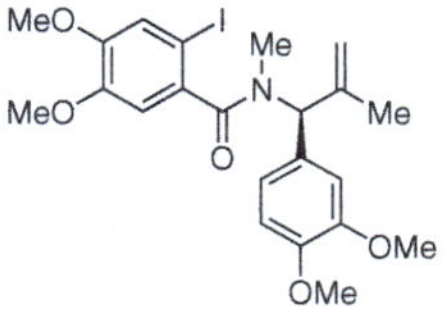	(*S*)-**1.237o**	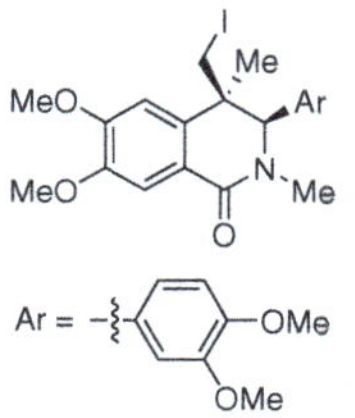	(*R*,*R*)-**1.244o**	69	90:10 (>99:1)
17	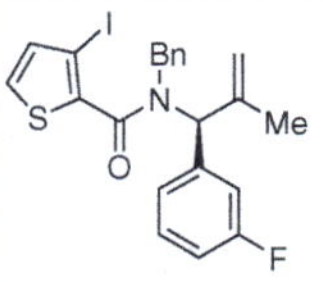	(*S*)-**1.237p**[e]	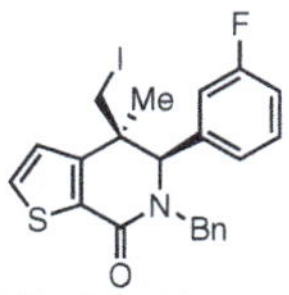	(*R*,*R*)-**1.244p**[e]	70	90:10 (>99:1)

[a]Reactions were run on a 0.2 mmol scale
[b]Combined isolated yield of both the *cis* and *trans* diastereomers
[c]Determined by ^{1}H NMR analysis of the crude reaction mixture
[d]Determined by HPLC using a chiral stationary phase
[e]This compound was prepared by Hyung Yoon
[f]Reaction was run with 7.5 mol% $Pd(QPhos)_2$

conditions to yield the desired products in good yields (82–88%). Electron-deficient substrates (*S*)-**1.237d** and (*S*)-**1.237e** could be cyclized with >95:5 dr in 54 and 76% yield, respectively. Converting the *N*-protecting group from Me to Bn had no significant impact on diastereoselectivity, and led to products being obtained with comparable yields (87–95%, entries 9–11). However, converting the *N*-protecting group from Me to Bz led to a marked decrease in both yield and dr (entry 7). Furthermore, substitution on the non-iodinated aromatic ring had no serious deleterious effects on yield or selectivity (entries 12 and 13). The yield was not affected when the non-iodinated aromatic ring was converted to an alkyl chain, but the selectivity dropped significantly to 80:20 dr. Electron rich substrates (*S*)-**1.224** and (*S*)-**1.237o** could be converted to the desired dihydroisoquinolinones products in 84 and 69% with 92:8 and 90:10 dr, respectively (entries 15 and 16). We were pleased to see that the former substrate was participating in this chemistry as it was the one chosen to undergo the key Pd-catalyzed step in the proposed synthesis of

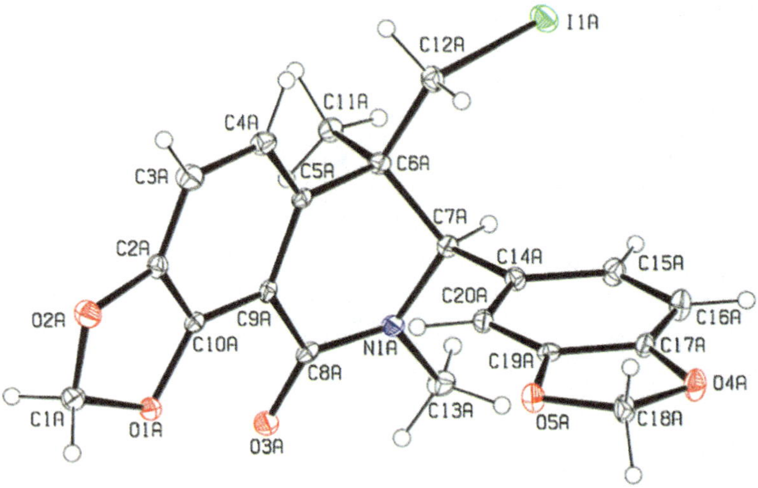

Fig. 1.4 Crystal structure of the *syn*-aryliodination product (*R*,*R*)-**1.223** (Crystals for the X-ray structure were obtained by Dr. Harald Weinstabl.)

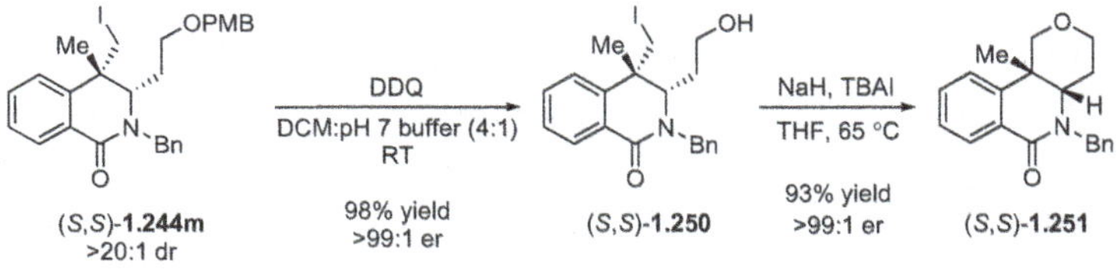

Scheme 1.56 Structural elaboration of PMB protected alcohol (*S*,*S*)-**1.244m**

(+)-corynoline. We were able to unambiguously confirm the stereochemistry of the product from these cyclizations by X-ray crystallographic analysis of suitable single crystals of (*R*,*R*)-**1.223** (Fig. 1.4). Heteroaromatic substrate (*S*)-**1.237p** could also be efficiently cyclized to the corresponding lactam **1.244p** in 70% yield with 90:10 dr. This example represents the first reported heteroaryliodination reaction.

(*S*,*S*)-**1.244m** presented us with an interesting opportunity to test the reactivity of the neopentyl iodide functionality which we proposed could be involved in an S_N2 displacement by cyanide in our retrosynthetic analysis. The PMP group could be effectively removed using DDQ at room temperature to generate alcohol (*S*,*S*)-**1.250** which could then undergo an intramolecular 6-*exo*-tet cyclization using NaH to generate tetrahydropyran-containing (*S*,*S*)-**1.251** (Scheme 1.56).

1.1.6.5 Limitations

There were a few substrates which failed to undergo the desired aryliodination reaction and provided incomplete conversions and complex mixtures of products (Scheme 1.57). It seems that substitution on the alkene other than CH_3 in not tolerated as (*R*)-**1.237p** led to an intractable mixture of products arising from isomerization of the alkene among other processes. Vinyl iodide (*S*)-**1.237q** also led to a complex mixture of products including those which appeared to originate from dehydrohalogenation. Trifluoromethyl substitution at the *ortho* position of the noniodinated aromatic ring greatly inhibited the reaction. However, the quantity of product actually formed appeared to be present with a high dr, but possibly as a mixture of multiple atropisomers (around bond *a* and *b*), although the structures of these products could not be fully elucidated. Indole substrate (*R*)-**1.237s** led to a complex mixture of products including an isolable, yet unstable product arising from nucleophilic attack of the resulting neopentyl Pd(II) species by the C3 position of the indole moiety.

As per previous reports from our group [126, 131], substrate (*S*)-**1.237t** possessing a monosubstituted alkene does not produce the desired aryliodination product (Scheme 1.58). Instead, reacting it under the standard conditions leads to a clean mixture of three products in a 0.47:1:0.57 ratio. Alkene (*S*)-**1.252** is formed by the expected β-hydride elimination pathway, and undergoes isomerization under the reaction conditions to generate achiral olefin isomer **1.253**. Five-membered

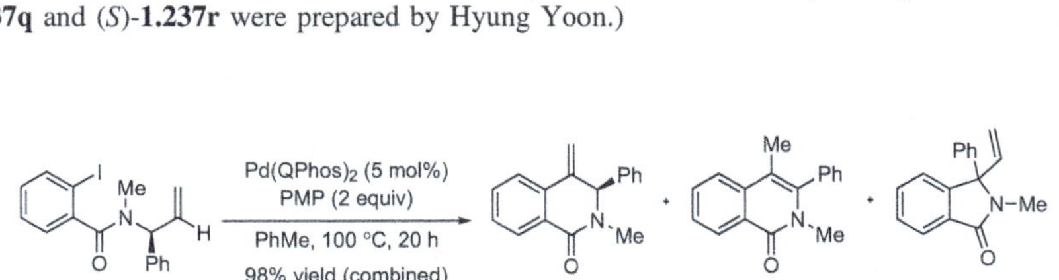

(*R*)-**1.237p** (*S*)-**1.237q**[a] (*S*)-**1.237r**[a] (*S*)-**1.237s**

Scheme 1.57 Substrates which failed to undergo the desired Pd-catalyzed aryliodination. ((*S*)-**1.237q** and (*S*)-**1.237r** were prepared by Hyung Yoon.)

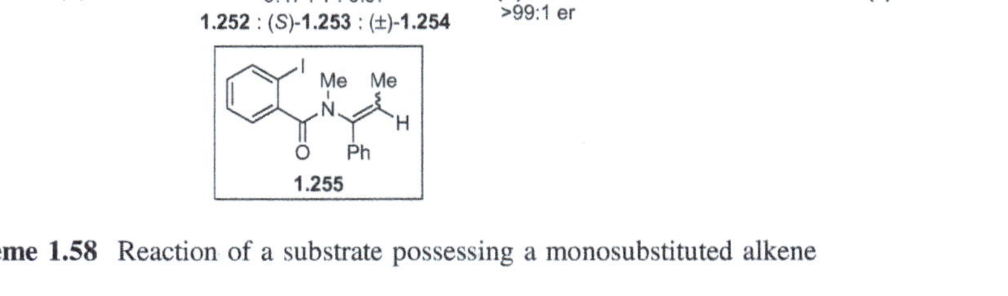

Scheme 1.58 Reaction of a substrate possessing a monosubstituted alkene

lactam **1.254a** could also be isolated albeit as a racemic mixture. We hypothesized that this product was obtained by initial isomerization of the olefin of (*S*)-**1.237t** to render olefin **1.255** which can then undergo a 5-exo-trig type carbopalladation/ β-hydride elimination sequence.

1.1.6.6 Formal Synthesis of (+)-Corynoline

We had proven that (*S*)-**1.244** could undergo an efficient diastereoselective (92:8 dr) aryliodination reaction on small scale (0.2 mmol) at this point. Due to the scalability of the route used to obtain these enantioenriched substrates, we were able to obtain the quantities of (*S*)-**1.244** needed to work our way through our proposed synthesis. First we attempted a significant scale up of the aryliodination reaction under slightly modified conditions (Scheme 1.59). The changes made included higher catalyst (10 mol%) and base loading (3.5 equiv), the addition of 4 Å MS (0.3 g/mmol (*S*)-**1.244**) and decreased concentration (0.05 M) and time (6 h). It should be noted that the diastereomeric mixture obtained from the aryliodination reaction could be easily separated using standard chromatographic techniques.

We proposed that a nucleophilic cyanation of the neopentyl iodide motif found in (*R*,*R*)-**1.223** could install the final carbon atom found in (+)-corynoline. In practice, this reaction required the use of excess KCN and 18-crown-6 in DMF at 100 °C for 2 days in order to achieve full conversion. However, product can be obtained cleanly with concomitant formation of separable baseline impurities. This reaction was carried out on a diastereomerically and enantiomerically pure sample of (*R*,*R*)-**1.223**, and afforded the desired nitrile (*R*,*R*)-**1.222** in 67% yield on gram scale (Scheme 1.60). The structure of this compound was also confirmed by X-ray crystallographic analysis (Fig. 1.5).

We originally planned to access ketone **1.221** via an intramolecular Houben-Hoesch reaction between the nitrile moiety and the electron-rich aromatic ring. However, under numerous standard activating conditions including $ZnCl_2$/ HCl, Au(I)-catalysis, MeOTf, no desired product was obtained, and starting material was recovered (Scheme 1.61).

In light of these setbacks, we took inspiration from Cushman's synthesis [139] which utilized an intramolecular denitrogenative cyclization of α-diazo ketone **1.215** under acidic conditions. In this regard, we hypothesized that if the nitrile

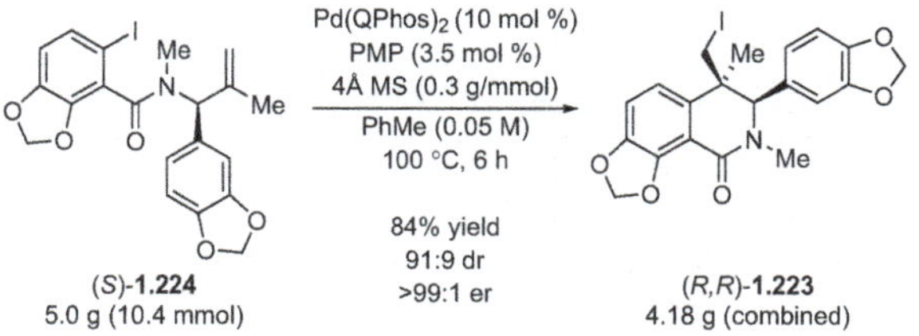

Scheme 1.59 Mutli-gram-scale aryliodination reaction of (+)-corynoline precursor (*S*)-**1.224**

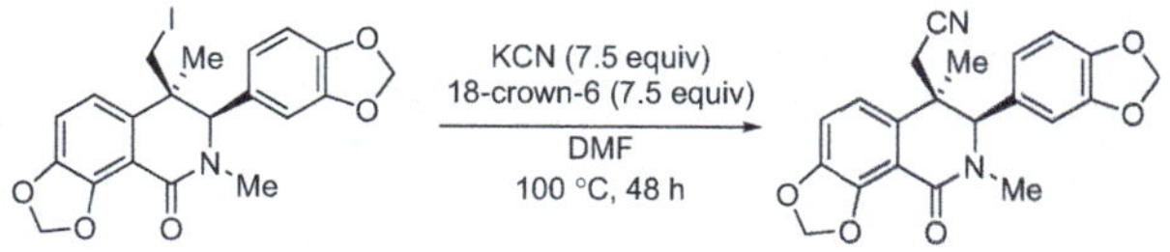

Scheme 1.60 Nucleophilic cyanation of (*R*,*R*)-**1.223** using KCN

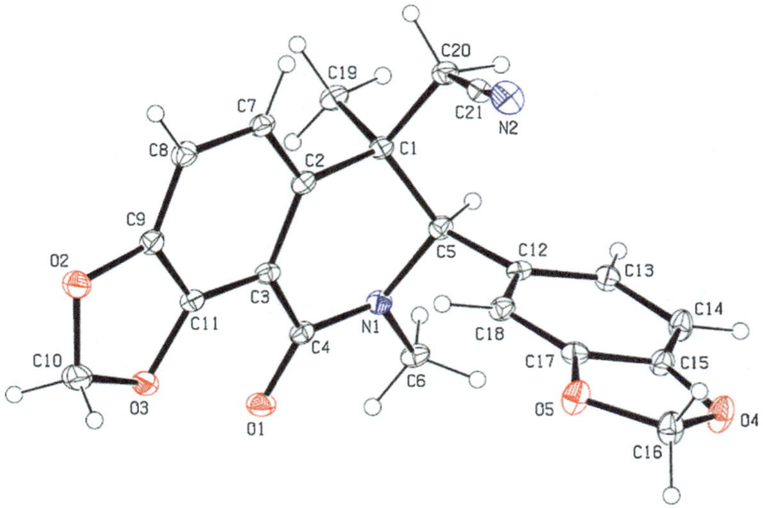

Fig. 1.5 Crystal structure of nitrile (*R*,*R*)-**1.222**

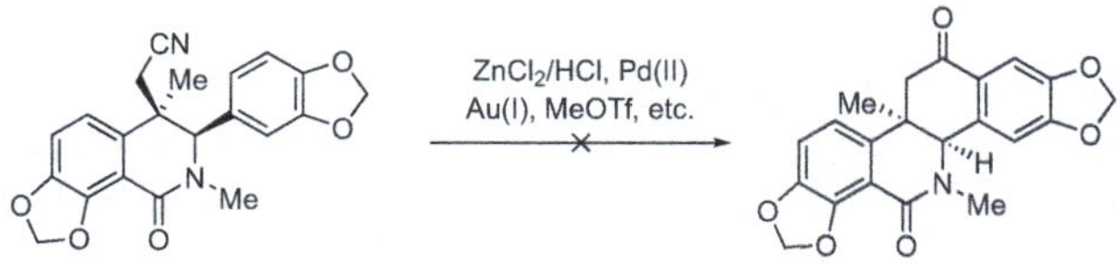

Scheme 1.61 Failed attempts at the Houben-Hoesch reaction of nitrile (*R*,*R*)-**1.222**

could be hydrolyzed to the carboxylic acid, an efficient intramolecular dehydrative Friedel-Crafts-type cyclization could be accomplished. However, even using forcing conditions (i.e. H_2O_2 and NaOH or KOH with $HOCH_2CH_2OH$ using μW), we were not able to completely hydrolyze nitrile (*R*,*R*)-**1.222** to acid (*R*,*R*)-**1.255**.

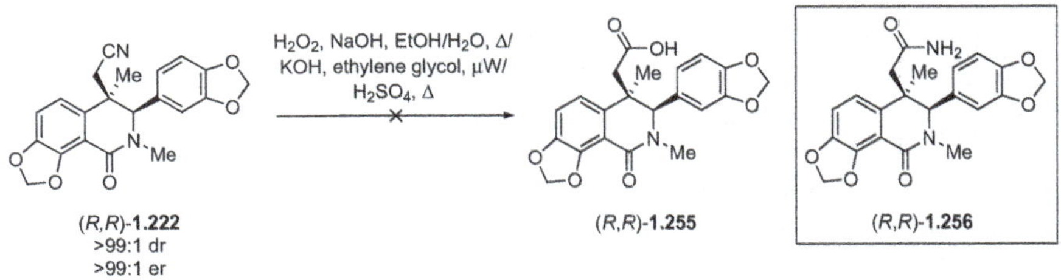

Scheme 1.62 Failed attempts at the complete hydrolysis of nitrile (*R*,*R*)-**1.222**

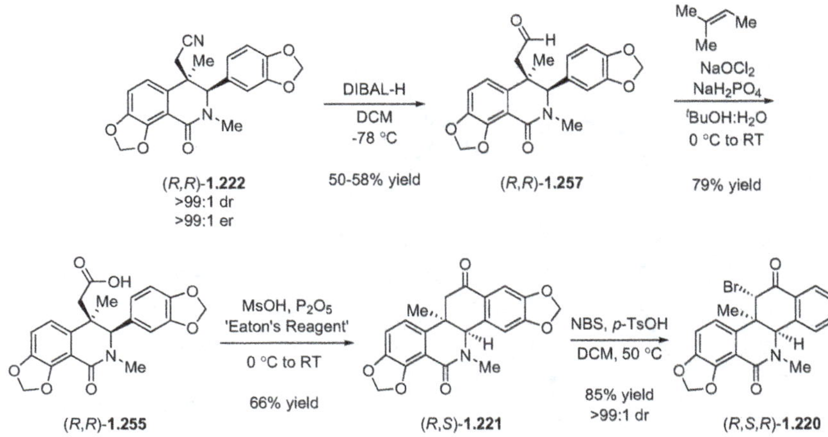

Scheme 1.63 Synthetic route to α-bromo ketone (*R*,*S*,*R*)-**1.220**

Instead, the intermediate primary amide (*R*,*R*)-**1.256** was isolated cleanly in most cases (Scheme 1.62). We were under the impression that the hydrolysis was halting at the amide stage due to the significant steric congestion around this functional group that was evident in the solid state.

Instead, we opted for a two-step procedure including DIBAL reduction of (*R*,*R*)-**1.222** to aldehyde (*R*,*R*)-**1.257** in 50–58% yield which could undergo a Pinnick oxidation to generate acid (*R*,*R*)-**1.255** in 79% yield. This acid then successfully underwent the proposed dehydrative cyclization reaction in neat Eaton's reagent (MsOH, P_2O_5) to generate ketone (*R*,*R*)-**1.221** (Scheme 1.63) in 66% yield. Although one could image multiple nucleophilic sites on the aromatic ring, we were sure that the reaction proceeded with the correct regiochemistry from X-ray crystallographic analysis (Fig. 1.6a). A highly stereoselective α-bromination could be achieved using NBS with catalytic *p*-TsOH in DCM at 50 °C to generate α-halo ketone (*R*,*S*,*R*)-**1.220** in 85% yield as a single diasteroemer. X-ray crystallographic analysis of a single crystal of (*R*,*S*,*R*)-**1.220** confirmed that the correct stereoisomer had been obtained (Fig. 1.6b).

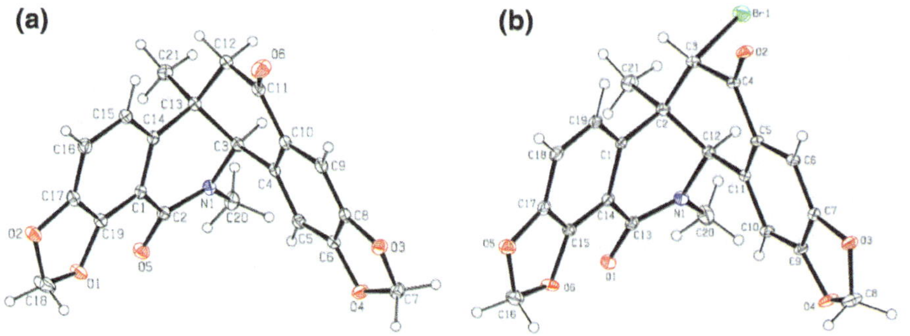

Fig. 1.6 Crystal structure of **a** ketone (*R*,*S*)-**1.221**; and **b** α-bromo ketone (*R*,*S*,*R*)-**1.220**

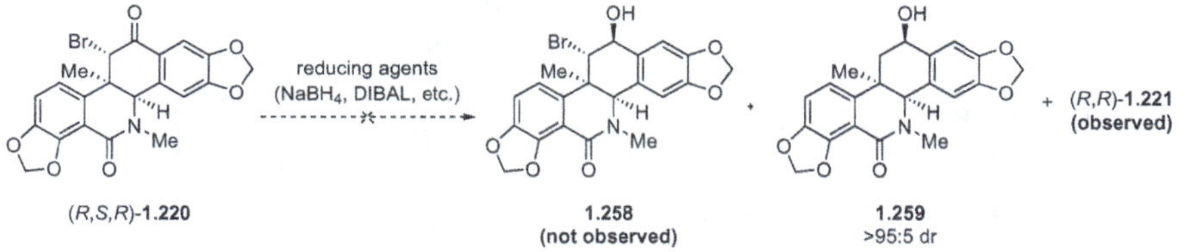

Scheme 1.64 Unsuccessful attempts at the synthesis of *trans* halohydrin **1.258**

The proposed endgame of the synthesis involved a stereoselective reduction of the ketone functionality of (*R*,*S*,*R*)-**1.220**. We thought that there was a good chance that reaction could proceed in a highly stereoselective fashion from the convex face to generate *trans* halohydrin **1.258** despite the presence of the large bromine atom. However, using many different reducing agents, halohydrin **1.258** was not formed. Instead, **1.259** resulting from both bromide and ketone reduction as well as ketone (*R*,*S*)-**1.221** were observed. Furthermore, the longer the reaction time, the more **1.259** was observed. Furthermore, this stereoselective reduction failed in a similar fashion using the α-chloride analog. Perhaps it is not surprising that the increased reactivity of the carbon–bromine bond imparted by the adjacent carbonyl functionality is observed (Scheme 1.64).[153]

At this point, we sought to complete a formal synthesis of (+)-corynoline by employing ketone (*R*,*S*)-**1.221** to access Naito's racemic alkene intermediate **1.201**. This sequence would represent a good compromise as our present route could afford direct access to both enantiomers of this compound. Since we already had established that the ketone moiety can be reduced using standard conditions, we attempted this reaction directly on (*R*,*S*)-**1.221** using $NaBH_4$ under standard conditions (Scheme 1.65). This reaction afforded highly stereoselective access to benzylic alcohol **1.259** whose structure was confirmed by X-ray crystallographic

[153]Reference [151].

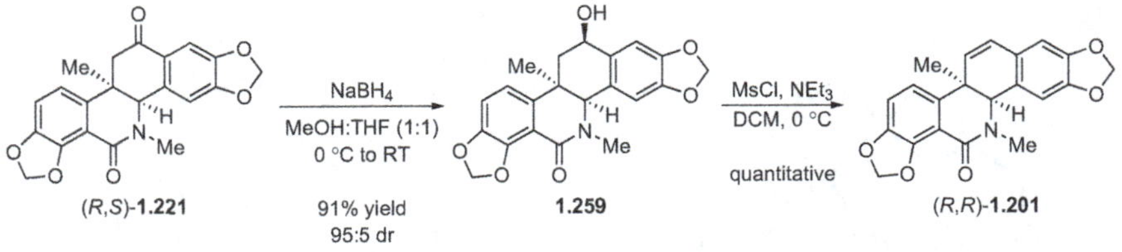

Scheme 1.65 Completion of the formal synthesis of (+)-corynoline

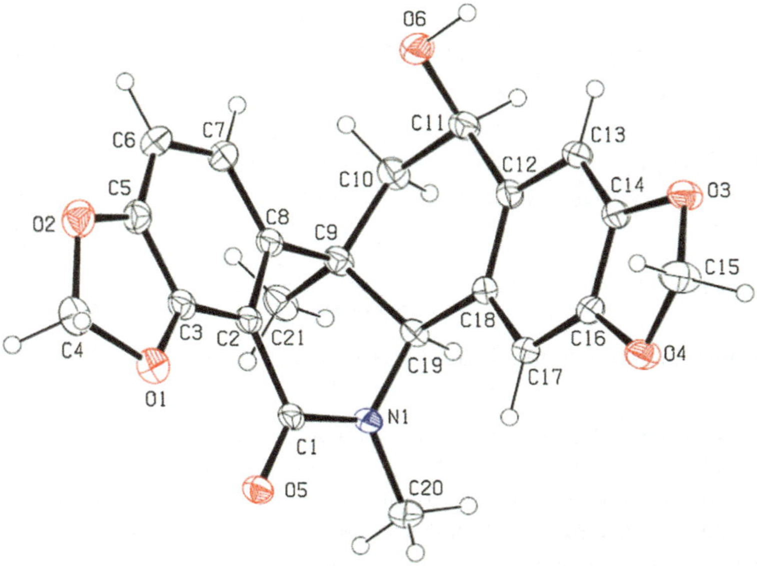

Fig. 1.7 Crystal structure of benzylic alcohol **1.259**

analysis (Fig. 1.7). This alcohol could undergo an efficient dehydration reaction using a mesylation/elimination sequence employing MsCl and NEt_3 which generates highly reactive sulfene in situ. Alkene **1.201** was shown by Naito to access corynoline in 3 steps [138].

1.1.7 Summary

The aryliodination has been shown to be applicable to the highly diastereoselective synthesis of variously substituted chroman and isochroman scaffolds with high levels of efficiency under a single set of conditions. We found that amine base was important to maintain catalyst lifespan, but found that it had no bearing on the stereoselectivity of the transformation. We have designed and tested models which

Scheme 1.66 Alkaloid natural products potentially accessible by the aryliodination of amides

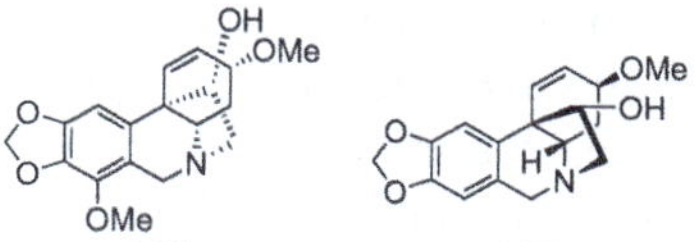

help to rationalize the diastereoselectivity in both the isochroman and chroman series. We believe that the high selectivity for the major *cis* isochroman isomer originates from the minimization of allylic-1,3 strain during the olefin insertion step, whereas the major *trans* chroman isomer originates from the minimization of axial-axial interaction during the same step in the mechanism.

Furthermore, dihydroisoquinolinones cores can also be accessed as virtually single enantiomers with high levels of efficiency using enantioenriched substrates generated via a scalable route. The diastereoselectivity greatly benefits from the addition of a stoichiometric amount of a tertiary amine base with respect to the base-free reaction. We hypothesize that the amine can interact with the active catalyst via coordination, and influence the stereochemical outcome as a result. Although the exact nature of this interaction is not clear at the moment, it is plausible that the catalyst/amine combination can selectively interact with one reactive rotamer of the amide substrate, leading to a diastereoenriched olefin coordination complex. The slowing of the reaction rate by the presence of base, as judged by in situ NMR experiments, adds some support to this notion. It would be wise to undertake a computational campaign to study the origin of this base effect. Not only would it help us learn more about the origin of this interesting affect, it could have further reaching implication on other reactions including the asymmetric Mizoroki-Heck reaction of rotameric amide substrates.

Although our proposed endgame of the synthesis of (+)-corynoline was met with adversity, we were able to show that this new class of aryliodination reaction could be applied to the formal synthesis of this compound. This new route clearly provides numerous marked improvements to previously outlined syntheses of this alkaloid natural product, and could inspire other synthesis of similar complex alkaloid natural products such as ambelline **1.260** and haomanthamine **1.261** in the future (Scheme 1.66).

1.1.8 Experimental

General Considerations. Unless otherwise stated, all catalytic reactions were carried out under an inert atmosphere of dry argon or dry nitrogen utilizing glassware that was oven (120 °C) or flame dried and cooled under argon, whereas the work-up and isolation of the products from the catalytic reactions were conducted on the bench-top using standard techniques. Reactions were monitored using thin-layer chromatography (TLC) on EMD Silica Gel 60 F254 plates. Visualization of the developed plates was performed under UV light (254 nm) or using $KMnO_4$, *p*-anisaldehyde, or ceric ammonium molybdate (CAM) stain. Organic solutions were concentrated under reduced pressure on a Büchi rotary evaporator. Tetrahydrofuran was distilled over sodium, toluene was distilled over sodium and degassed via 5 freeze-pump-thaw cycles and stored over activated 4 Å molecular sieves, triethyl amine was distilled over potassium hydroxide, DCM was distilled over calcium hydride, and anhydrous *N,N*-dimethylformamide and diethyl ether were purchased from Aldrich and used as recieved. Silica gel flash column chromatography was performed on Silicycle 230–400 mesh silica gel. All standard reagents including benzaldehyde derivatives, (*S*)-2-methylpropane-2-sulfinamide and (*R*)-2-methylpropane-2-sulfinamide, isopropenyl magnesium bromide, and dimethylzinc, were purchased from Sigma Aldrich, Alfa Aesar, Combi-Blocks, or Oakwood and were used without further purification. The 'racemic' assays for HPLC determination of enantiomeric excess for the final dihydroquinolinone products were synthesized using **general procedures 5–10**, except the 'racemic' 2-methylpropane-2-sulfinamide was employed. NOTE: it was later found that 2-methylpropane-2-sulfinamide from Combi-Blocks product # **QB-2211** and batch number **L77452** was present with a slight enantiomeric excess of the (*S*) enantiomer (confirmed by the company, and our independent analysis). This explains the slight enantiomeric excess in our 'racemic' HPLC assays of some linear starting materials and the corresponding dihydroisoquinolinone products. We have simultaneously confirmed that this is not due to an overlapped impurity causing an apparent ee. All 2-iodobenzoic acid derivatives were purchased from commercial sources except 2-iodo-4-chlorobenzoic acid,[154] and 2-iodo-4,5-dimethoxybenzoic acid[155] whose synthetic procedures can be found at the indicated references, and 5-iodobenzo[*d*][1, 3]dioxole-4-carboxylic acid whose synthesis has been outlined herein. All PdL_2 pre-catalysts were obtained from Johnson Matthey, and are available commercially from Strem Chemicals or Sigma-Aldrich.

Instrumentation. NMR characterization data was collected at 296 K on a Varian Mercury 300, Varian Mercury 400, Varian 600, or a Bruker Avance III spectrometer operating at 300, 400, 500, or 600 MHz for ^{1}H NMR, and 75, 100, 125, for 150 MHz or ^{13}C NMR. ^{1}H NMR spectra were internally referenced to the residual solvent signal ($CDCl_3$ = 7.26 ppm, Toluene-d_8 = 2.3 ppm) or TMS. ^{13}C NMR spectra were internally referenced to the residual solvent signal ($CDCl_3$ = 77.0 ppm) and are reported as observed. Data for ^{1}H NMR are reported as follows: chemical shift (δ ppm), multiplicity (s = singlet, d = doublet, t = triplet, q = quartet, m = multiplet, br = broad), coupling constant (Hz), integration. Coupling constants have been rounded to the nearest 0.5 Hz. Melting point ranges were determined on a Fisher-Johns Melting Point Apparatus and are reported uncorrected. All reported diasteromeric ratios in data section are those obtained from ^{1}H NMR analysis of the crude reaction mixtures using 5 s delay. NMR yields

[154] Reference [152].

[155] Reference [153].

for the optimization section were obtained by ^{1}H NMR analysis of the crude reaction mixture using 5 s delay and 1,3,5-trimethoxybenzene as an internal standard. Enantiomertic ratios were determined using HPLC on a chiral stationary phase using a complete Agilent HPLC system. All retention times have been rounded to the nearest 0.1 min. High resolution mass spectra (HRMS) were obtained on a micromass 70S-250 spectrometer (EI) or an ABI/Sciex QStar Mass Spectrometer (ESI) or a JEOL AccuTOF medel JMS-T1000LC mass spectrometer equipped with an IONICS® Direct Analysis in real Time (DART) ion source at Advanced Instrumentation for Molecular Structure (AIMS) in the Department of Chemistry at the University of Toronto.

General Procedures—Diastereoselective synthesis of chromans and isochromans

General procedure 1) Synthesis of allylic and homoallylic alcohols

Neat aldehyde (5 mmol, 1 equiv) was added dropwise via syringe to a flask containing a THF solution (12 mL) of isopropenylmagnesium bromide (6 mmol, 1.2 equiv) or 2-methallyl magnesium chloride (6 mmol, 1.2 equiv) at 0 °C. The resulting solution was allowed to stir at this temperature for 5 min before being warmed to room temperature. The reaction was quenched with saturated ammonium chloride (10 mL) after 15 min, and extracted with diethyl ether (3 × 25 mL). The combined organic layers were sequentially washed with saturated aqueous sodium bicarbonate (10 mL) and brine (15 mL), dried over anhydrous Na_2SO_4, filtered, and concentrated in vacuo. The crude alcohols were either used without further purification, or purified on silica gel flash column chromatography using the indicated mobile phase.

General procedure 2) Synthesis of choman precursors via a Mitsunobu reaction

A 2 M THF solution of DIAD (1 equiv) was added dropwise to a stirred THF solution (0.66 M) of homoallylic alcohol (1 equiv), 2-iodophenol derivative (1 equiv) and triphenyl phosphine (1 equiv) at 0 °C. The reaction was immediately warmed to room temperature and allowed to stir for 12 h, at which time the THF was removed via rotary evaporation. The crude residue was adsorbed onto silica gel using DCM and purified using silica gel flash column chromatography using the indicated mobile phase.

General procedure 3) Synthesis of isochroman precursors via an S_N2 reaction

A DMF solution (5 mL) of an allylic alcohol (1 equiv) was slowly added to a flask containing NaH (60% in mineral oil, 1 equiv) in THF (5 mL) at room temperature. When the evolution of H_2 gas ceased, a DMF solution (4 mL) of 2-iodobenzyl chloride, 2-iodobenzyl methanesulfonate, or 2-bromobenyl bromide (1.0 or 1.2 equiv) was added dropwise over 30 min. After 18 h, the reaction was quenched by addition of saturated NH_4Cl (5 mL) and water (20 mL). The aqueous layer was extracted with ether (3 × 20 mL) and the combined organic layers were washed sequentially with saturated $NaHCO_3$ (15 mL), water (3 × 15 mL), and brine

(15 mL), dried over anhydrous Na_2SO_4, filtered, and concentrated *in vacuo* to afford the crude alkenyl ethers. Crude products were purified via silica gel flash column chromatography using the indicated mobile phase.

General procedure 4) Diastereoselective carboiodination of alkenyl ethers

An oven dried 2 dram equipped with a magnetic stir bar was charged with Pd $(P^tBu_3)_2$ (5.11 mg, 0.01 mmol, 5 mol%). The vial was purged with argon for 5 min after which time a PhMe solution (0.1 M) of the aryl iodide (1 equiv) and triethylamine (1 equiv) was via syringe. The vial was capped with a Teflon® line screw cap and placed in an oil bath pre-heated to 110 °C. After stirring at this temperature for the indicated amount of time, the vial was cooled, and the contents filtered over a short silica plug and eluted with 100% ethyl acetate. The concentrated crude material was purified by silica gel flash column chromatography using the indicated solvent.

General Procedures—Diastereoselective synthesis of enantioenriched dihydroisoquinoloinones

General procedure 5) Sulfinamide synthesis

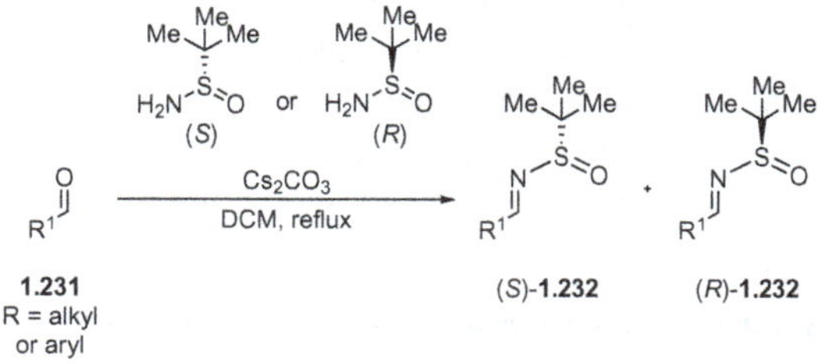

5a—Liquid Aldehydes: (*S*)-2-methylpropane-2-sulfinamide or (*R*)-2-methyl propane-2-sulfinamide (1.2 equiv) and anhydrous Cs_2CO_3 (2 equiv) were added to a flame dried flask, and the solid mixture was purged with argon for 10 min. Freshly distilled DCM, followed by the liquid aldehyde (1 equiv, individual concentration indicated) were added in that order via syringe through a rubber septum. The reaction vessel was fitted with a dry reflux condenser, and the reaction was heated at reflux for the indicated amount of time. Once TLC analysis indicated full conversion of aldehyde starting material the mixture was cooled to room temperature, and the contents were filtered over Celite® which was subsequently washed through with DCM. The collected organic fraction was concentrated in vacuo using a rotary evaporator. The purity of the crude sulfinamide was upgraded by silica gel flash column chromatography using the indicated mobile phase.

5b—Solid aldehydes: Aldehyde (1 equiv), (*S*)-2-methylpropane-2-sulfinamide or (*R*)-2-methylpropane-2-sulfinamide (1.2 equiv), and anhydrous Cs_2CO_3 (2 equiv) were added to a flame dried flask, and the solid mixture was purged with argon for 10 min. Freshly distilled DCM (concentration indicated) was added via syringe through a rubber septum. The reaction vessel was fitted with a dry reflux condenser, and the reaction was heated at reflux for the indicated amount of time. Once TLC analysis indicated full conversion of aldehyde starting material the mixture was cooled to room temperature, and the contents were filtered over Celite® which was subsequently washed through with DCM. The collected organic fraction was concentrated in vacuo using a rotary evaporator. The purity of the crude sulfinamide was upgraded by silica gel flash column chromatography using the indicated mobile phase.

General procedure 6) synthesis of allylic sulfinamides via a diastereoselective addition

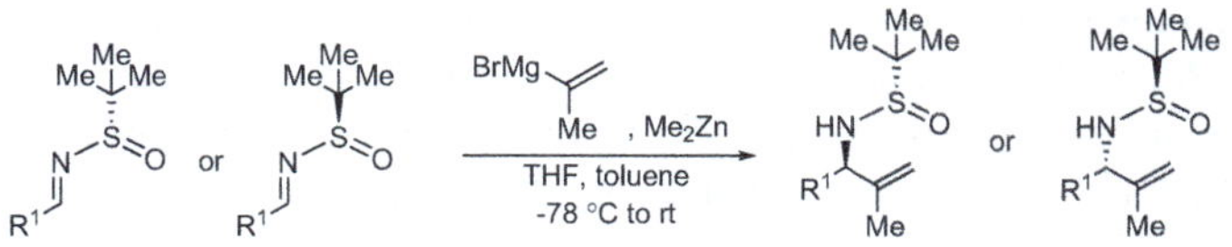

(*S*)-**1.232** (*R*)-**1.232** **1.233**

Synthesized according to a modified procedure of H. W. Lam and co-workers.[156] Two flame dried flasks were cooled under a stream of argon, and one flask was charged with isopropenyl magnesium bromide (0.5 M in THF, 1.8 equiv). This solution then had a solution of dimethylzinc (2 M in toluene, 1.8 equiv) added drop-wise over 10 min at room temperature. The resulting solution was stirred at this temperature for 30 min before cooling to −78 °C. The second flask was charged with the sulfinamide (**1.232**, 1 equiv), and purged with argon for 10 min. The sulfinamide was then dissolved in dry THF (individual concentration indicated) and the resulting solution of substrate was cooled to −78 °C. The generated zinc reagent contained in the first flask was added slowly drop wise via cannula to the second flask containing the substrate, using a positive pressure of argon. This resulting solution was stirred at this temperature for 3 h before slowly warming to room temperature. At this time TLC analysis indicated full and clean conversion of starting material, and the reaction was cooled to 0 °C and slowly quenched by the addition of a saturated aqueous solution of ammonium chloride. The resulting biphasic mixture was diluted with water, and when persistent, the insoluble inorganic salts were filtered. The resulting clear aqueous layer was extracted with EtOAc (3x), and the combined organic layers were washed with brine, dried over $MgSO_4$, filtered, and concentrated *in vacuo*. Diastereomeric ratios were determined by ^1H NMR analysis of the crude reaction mixture. The pure allylic sulfinamides **1.233** were purified via silica gel flash column chromatography using the indicated mobile phase.

[156]Reference [145].

General procedure 7) Alkylation of allylic sulfinamides

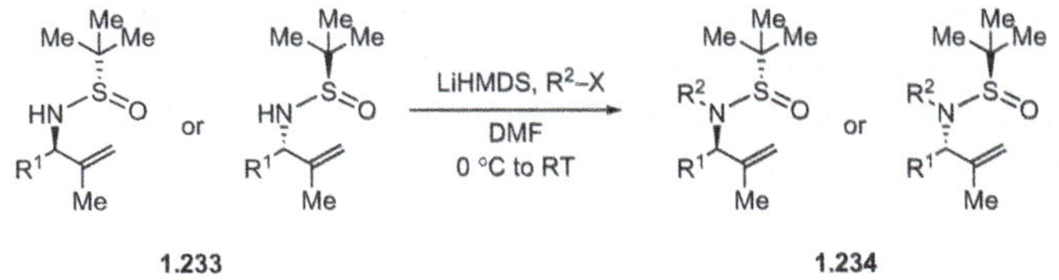

A flame dried flask was charged with the allylic sulfinamide (1 equiv, **1.233**) and was purged with argon for 10 min. The amine was then dissolved in anhydrous DMF (0.3 M) and the resultant solution was cooled to 0 °C. At this time, LiHMDS (1 M in THF, 1.2 equiv) was added drop wise over 10 min. The solution was allowed to stir for 30 min at this temperature before warming to room temperature. The solution was re-cooled to 0 °C and the neat alkyl halide (2 equiv) was added drop wise over 10 min. At this time, the resulting solution was warmed to room temperature and was stirred until TLC analysis indicated full conversion of starting material. The reaction was slowly quenched at 0 °C by the addition of a saturated aqueous solution of ammonium chloride, the layers of the resulting biphasic mixture were separated and aqueous layer was extracted with EtOAc (3x). The combined organic layers were washed with water (4x) and brine (2x), dried over Na_2SO_4, filtered, and concentrated *in vacuo*. The pure alkylated allylic sulfinamides **1.234** were obtained via silica gel flash column chromatography using the indicated mobile phase.

General procedure 8) Acid-mediated cleavage of the *tert*-butyl sulfinyl auxiliary

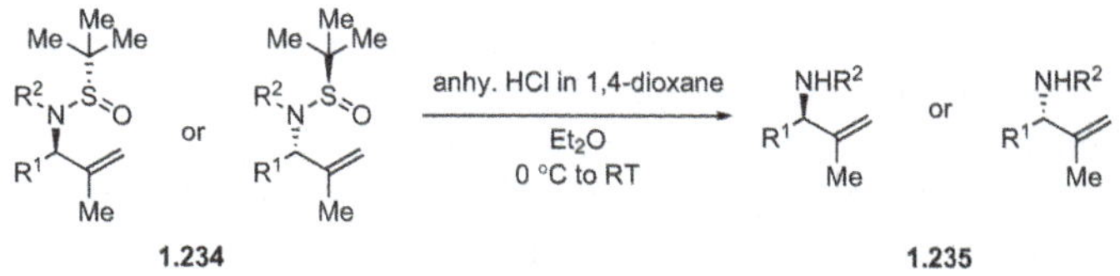

A flame dried flask was charged with the alkylated allylic sulfinamide (1 equiv, **1.234**) and was purged with argon for 10 min. The contents of the flask were then taken up in anhydrous Et_2O (individual concentration indicated) and the resulting solution was cooled to 0 °C. At this time, anhydrous HCl in 1,4-dioxane (4 M, 2 equiv) was added drop wise via syringe, and the resulting mixture was stirred at this temperature for 30 min. At this time the reaction mixture was warmed to room temperature and was handled in one of the two following ways:

8a—Work up for precipitated amine. HCl salts: The white precipitate was filtered through a fritted funnel using a water aspirator and the filter cake was washed with a small amount of ice cooled Et_2O. The filter cake was dissolved through the frit into

a new flask using distilled water. The aqueous filtrate had its volume doubled with Et_2O and with stirring, was slowly basified to pH 10 using 1 M aqueous NH_4OH. The layers were separated, and the aqueous layer was extracted with Et_2O (2x). The combined organic layers were washed with brine, dried over Na_2SO_4, filtered and concentrated *in vacuo*. The chiral allylic amines **1.235** were characterized and used without further purification.
8b—Work up for soluble amine·HCl salts: The volume of the reaction mixture was doubled with distilled water and with stirring, basified to pH 10 with a 1 M aqueous solution of NH_4OH. The layers were separated, and the aqueous layer was extracted with Et_2O (2x). The combined organic layers were washed with brine, dried over Na_2SO_4, filtered and concentrated *in vacuo*. These crude chiral allylic amines were found to be significantly less pure than their precipitated analogs, and were purified using the indicated measures.

General procedure 9) Synthesis of linear enantioenriched *N*-ally carboxamides

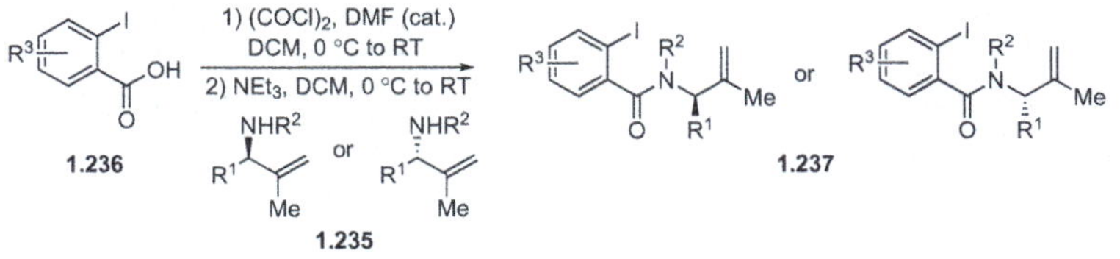

A flame dried flask was charged with the *o*-iodobenzoic acid derivative (1 equiv) and was purged with argon for 10 min. The contents of the flask were taken up in dry DCM (10 mL), and anhydrous DMF (2 drops) was added. The resulting suspension was cooled to 0 °C and oxalyl chloride (2 equiv) was added drop wise over 5 min. Once gas evolution had ceased, the reaction mixture was warmed to room temperature, and stirred vigorously. Once the slow gas evolution ceased at this temperature, the reaction mixture was concentrated to dryness on a rotary evaporator in a fume hood, followed by high-vacuum to remove any excess oxalyl chloride. The flask containing the crude acid chloride was fitted with a septum and purged with argon for 10 min. The contents of the flask were taken up in dry DCM (10 mL) and cooled to 0 °C. To this solution was added dry NEt_3 (2 equiv) which immediately resulted in the appearance of a persistent orange-to-red colour. This then had a DCM solution (2.5 mL) of chiral allylic amine (1.1 equiv, **1.235**) and dry NEt_3 (2 equiv) added drop wise over 10 min. The reaction was stirred at this temperature for 10 min and then warmed to room temperature where it was stirred for an additional 10 min. At this time, the reaction was quenched by doubling the reaction volume with a concentrated aqueous solution of $NaHCO_3$. The layers were separated and the aqueous layer was extracted with DCM (3x). The combined organic layers were washed with brine, dried over Na_2SO_4, filtered, and concentrated *in vacuo*. The pure chiral *N*-allyl carboxamides **1.237** were purified using silica gel flash column chromatography using the indicated mobile phase.

General procedure 10) Pd-Catalyzed carboiodination of linear enantioenriched *N*-allyl cabroxamides

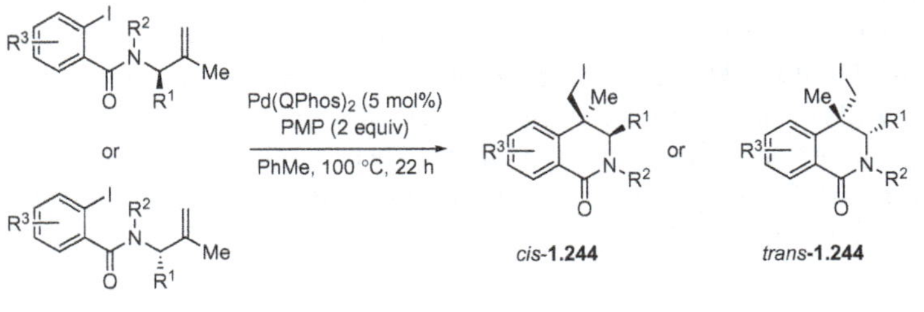

1.237 *cis*-**1.244** *trans*-**1.244**

10a—Using liquid amides: One oven-dried 2 dram vial was cooled under argon, charged with the liquid chiral *N*-allyl carboxamides (0.2 mmol, 1 equiv, **1.237**), and purged with argon for 10 min. A second oven-dried vial was cooled under argon, charged with $Pd(QPhos)_2$ (0.01 mmol, 5 mol% [Pd]), and purged with argon for 10 min. The amide was dissolved in dry and degassed PhMe (2 mL, 0.1 M) and 1,2,2,6,6-pentamethylpiperidine (72.4 µL, 0.4 mmol, 2 equiv) was added via a dry microsyringe. This well-mixed solution of amide and base was added into the vial containing the catalyst. The vial was fitted with a Teflon® lined screw cap under a stream of argon passed through a large inverted glass funnel, sealed using Teflon tape, and placed in a pre-heated oil bath at 100 °C for 22 h. At this time either TLC analysis or ^1H NMR analysis of a small aliquot indicated full conversion of starting material, and the reaction vial was cooled, and its contents were filtered through a 2 cm plug of silica gel in a pipette eluting with 100% EtOAc. The pure dihydroisoquinolinones **1.244** were obtained via silica gel flash column chromatography using the indicated mobile phase.

10b—Using solid amides: An oven-dried 2 dram vial was cooled under argon, charged with the solid chiral linear amide (0.2 mmol, 1 equiv, **1.237**) and Pd $(QPhos)_2$ (0.01 mmol, 5 mol% [Pd]), and purged with argon for 10 min. The contents of the vial were taken up in dry and degassed PhMe (2 mL, 0.1 M) and 1,2,2,6,6-pentamethylpiperidine (72.4 µL, 0.4 mmol, 2 equiv) was added via a dry microsyringe. The vial was fitted with a Teflon® lined screw cap under a stream of argon passed through a large inverted glass funnel, sealed using Teflon tape, and placed in a pre-heated oil bath at 100 °C for 22 h. At this time either TLC analysis or ^1H NMR analysis of a small aliquot indicated full conversion of starting material, and the reaction vial was cooled, and its contents were filtered through a 2 cm plug of silica gel in a pipette eluting with 100% EtOAc. The pure dihydroisoquinolinones **1.244** were obtained via silica gel flash column chromatography using the indicated mobile phase.

Starting material and product experimental: Diastereoselective synthesis of chromans and isochromans

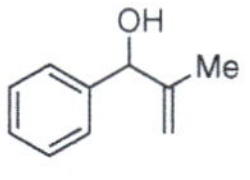

(±)-2-methyl-1-phenylprop-2-en-1-ol—Synthesized using **general procedure 1**. After purification on silica gel using pentanes:EtOAc (5:1 v:v) the desired alcohol was afforded as a clear and colourless oil (0.732 g, 4.92 mmol, 98%). Characterization data was in accordance with that of the reported literature.[157] **^{1}H NMR** (400 MHz, $CDCl_3$) δ 7.38–7.24 (m, 5H), 5.19 (d, *J* = 0.7 Hz, 1H), 5.11 (s, 1H), 4.95 (s, 1H), 2.04 (d, *J* = 11.7 Hz, 1H), 1.60 (s, 3H). **^{13}C NMR** (100 MHz, $CDCl_3$) δ 146.8, 141.9, 128.3, 127.6, 126.4, 111.1, 77.8, 18.2.

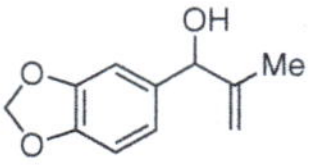

(±)-1-(benzo[*d*][1, 3]dioxol-5-yl)-2-methylprop-2-en-1-ol—Synthesized using **general procedure 1**. The crude allylic alcohol was sufficiently pure after work up, and was afforded as a light yellow oil (0.9447 g, 4.95 mmol, 98%) upon concentration. **^{1}H NMR** (400 MHz, $CDCl_3$) δ 6.87–6.81 (m, 1H), 6.78–6.75 (m, 1H), 5.94 (s, 1H), 5.19 (s, 1H), 5.02 (s, 1H), 4.94 (d, *J* = 0.7 Hz, 1H), 2.06 (s, 1H), 1.60 (s, 2H). **^{13}C NMR** (100 MHz, $CDCl_3$) δ 147.7, 147.0, 146.8, 136.0, 120.0, 110.8, 108.0, 106.9, 101.0, 77.5, 18.5. **IR** (cm^{-1}, thin film) 3385, 2891, 1653, 1440, 1244, 1039, 808. **HRMS** (ESI+): Calc'd for $C_{11}H_{16}NO^+$, 210.11302; found, 201.11314.

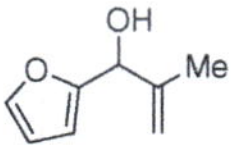

(±)-1-(furan-2-yl)-2-methylprop-2-en-1-ol—Synthesized using **general procedure 1**. The crude allylic alcohol was sufficiently pure after work up, and was afforded as a light yellow oil (0.692 g, 4.99 mmol, 99%) upon concentration. Characterization data was in accordance with that of the reported literature.[158] **^{1}H NMR** (400 MHz, $CDCl_3$) δ 7.38 (dd, J = 2.0, 1.0 Hz, 1H), 6.40–6.31 (m, 1H), 6.26 (d, J = 3.0 Hz, 1H), 5.19 (d, J = 0.5 Hz, 1H), 5.15 (d, J = 4.5 Hz, 1H), 5.02 (dd, J = 2.5, 1.5 Hz, 1H), 2.12 (d, J = 5.0 Hz, 1H), 1.74 (s, 3H). **^{13}C NMR** (100 MHz, $CDCl_3$) δ 154.8, 144.3, 142.0, 111.9, 110.1, 106.8, 71.3, 18.5.

---

[157] Reference [154].

[158] Reference [155].

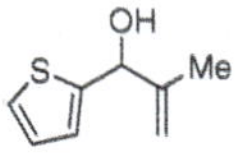

(±)-2-methyl-1-(thiophen-2-yl)prop-2-en-1-ol—Synthesized using **general procedure 1**. The crude allylic alcohol was sufficiently pure after work up, and was afforded as a yellow oil (0.769 g, 4.99 mmol, 99%) upon concentration. **^{1}H NMR** (400 MHz, $CDCl_3$) δ 7.21–7.17 (m, 1H), 6.93–6.87 (m, 2H), 5.31 (s, 1H), 5.16 (s, 1H), 4.96–4.89 (m, 1H), 2.03 (s, J = 1.0 Hz, 1H), 1.65 (s, 3H). **^{13}C NMR** (100 MHz, $CDCl_3$) δ 146.3, 146.2, 126.7, 125.1, 124.7, 111.5, 73.8, 18.2. **IR** (cm^{-1}, thin film) 3373, 2971, 2862, 1437, 1373, 1074, 1028, 904, 700. **HRMS** (EI +): Calc'd for $C_8H_{10}OS$, 154.0452; found, 154.0451.

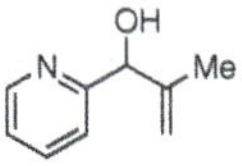

(±)-2-methyl-1-(pyridin-2-yl)prop-2-en-1-ol—Synthesized using **general procedure 1**. After purification on silica gel using 60% EtOAc in pentanes the desired alcohol was isolated as a light yellow oil (0.626 g, 4.19 mmol, 84%). **^{1}H NMR** (300 MHz, $CDCl_3$) δ 8.55 (d, J = 5.0 Hz, 1H), 7.67 (td, J = 7.5, 1.5 Hz, 1H), 7.29 (d, J = 8.0 Hz, 1H), 7.24–7.18 (m, 1H), 5.19 (d, J = 0.5 Hz, 1H), 5.17 (s, 1H), 5.02–4.97 (m, 1H), 1.54 (s, 3H). **^{13}C NMR** (75 MHz, $CDCl_3$) δ 159.5, 147.7, 146.4, 136.7, 122.4, 120.7, 113.8, 76.9, 16.5. **IR** (cm^{-1}, thin film) 3377, 1593, 1435, 1060, 1049, 904, 759. **HRMS** (ESI+): Calc'd for $C_9H_{12}NO$, 150.09189; found, 150.09179.

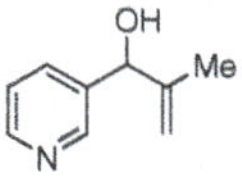

(±)-2-methyl-1-(pyridin-3-yl)prop-2-en-1-ol—Synthesized using **general procedure 1**. After purification on silica gel using EtOAc:hexanes (6:1 v:v) the desired alcohol was afforded as a light yellow oil (0.731 g, 4.81 mmol, 98%). **^{1}H NMR** (400 MHz, $CDCl_3$) δ 8.46 (s, 1H), 8.36 (d, J = 4.0 Hz, 1H), 7.71 (dt, J = 8.0, 1.5 Hz, 1H), 7.22 (dd, J = 78.0, 45.0 Hz, 1H), 5.16 (d, J = 7.0 Hz, 1H), 5.14 (s, 1H), 4.93 (s, 1H), 1.57 (s, 3H). **^{13}C NMR** (100 MHz, $CDCl_3$) δ 148.0, 147.8, 146.4, 138.2, 134.4, 123.3, 111.9, 75.2, 17.7. **IR** (cm^{-1}, thin film) 3206, 2856, 1593, 1581, 1425, 1060, 1041, 904, 713. **HRMS** (ESI+): Calc'd for $C_9H_{12}NO$, 150.09189; found, 150.09140.

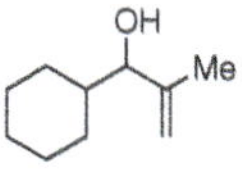

(±)-1-cyclohexyl-2-methylprop-2-en-1-ol—Synthesized using **general procedure 1**. The crude allylic alcohol was sufficiently pure after work up, and was afforded as a clear and colourless oil (0.500 g, 3.24 mmol, 65%) upon concentration. Characterization data was in accordance with that of the reported literature.[159] **^{1}H NMR** (300 MHz, $CDCl_3$) δ 4.91–4.84 (m, 2H), 3.74 (dd, J = 7.5, 3.5 Hz, 1H), 1.94 (dd, J = 8.5, 7.0 Hz, 1H), 1.82–1.59 (m, 6H), 1.55–1.35 (m, 3H), 1.35–0.80 (m, 6H). **^{13}C NMR** (100 MHz, $CDCl_3$) δ 146.4, 112.1, 80.8, 40.4, 29.5, 28.4, 26.4, 26.1, 25.9, 17.2.

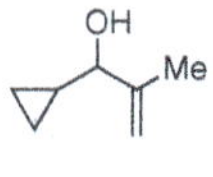

(±)-1-cyclopropyl-2-methylprop-2-en-1-ol—Synthesized using general procedure **1**. The crude allylic alcohol was sufficiently pure after work up, and was concentrated to afford a light yellow oil (0.536 g, 4.75 mmol, 95%). **^{1}H NMR** (400 MHz, $CDCl_3$) δ 4.98 (dd, J = 1.5, 1.0 Hz, 1H), 4.83 (s, 1H), 3.30 (d, J = 8.5 Hz, 1H), 1.84 (s, 3H), 1.62 (s, 1H), 1.11–0.95 (m, 1H), 0.60–0.55 (m, 2H), 0.42–0.23 (m, 2H). **^{13}C NMR** (100 MHz, $CDCl_3$) δ 147.3, 110.5, 80.3, 18.4, 16.6, 3.2, 3.0. **IR** (cm^{-1}, thin film) 3372, 3084, 1653, 997, 898. **HRMS** (EI+): Calc'd for $C_7H_{12}O$, 112.0888; found, 112.0893.

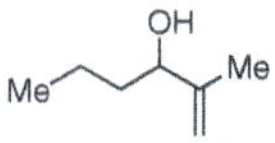

(±)-2-methylhex-1-en-3-ol—Synthesized using **general procedure 1**. The crude allylic alcohol was purified using 30% Et_2O in pentanes and was afforded as a light yellow oil (0.5698 g, 4.00 mmol, 80%). Characterization data was in accordance with that of the reported literature.[160] **^{1}H NMR** (400 MHz, $CDCl_3$) δ 4.95–4.88 (m, 1H), 4.85–4.77 (m, 1H), 4.05 (t, J = 6.5 Hz, 1H), 1.71 (s, 3H), 1.67 (s, 1H), 1.55–1.47 (m, 2H), 1.45–1.26 (m, 2H), 0.92 (t, J = 7.3 Hz, 3H). **^{13}C NMR** (100 MHz, $CDCl_3$) δ 147.7, 110.8, 75.7, 37.1, 18.8, 17.4, 13.9.

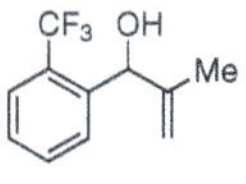

(±)-2-methyl-1-(2-(trifluoromethyl)phenyl)prop-2-en-1-ol—Synthesized using **general procedure 1**. The crude allylic alcohol was sufficiently pure after work up, and was afforded as a white semi-solid (MP 32–34 °C) (0.980 g, 4.55 mmol, 91%) upon concentration. **^{1}H NMR** (400 MHz, $CDCl_3$) δ 7.69–7.64 (m, 2H), 7.56 (ddd, J = 8.0, 1.0, 0.5 Hz, 1H), 7.42–7.37 (m, 1H), 5.51 (s, 1H), 5.23 (dq, J = 2.5, 1.0 Hz, 1H), 5.06–5.03 (m, 1H), 2.22 (s, 1H), 1.60 (s, 2H). **^{13}C NMR** (100 MHz, $CDCl_3$) δ 145.64 (s), 140.6 (s), 132.2 (d, J = 1.1 Hz), 128.6 (s), 127.9 (s), 125.7

159 Reference [156].

160 Reference [157].

(s), 125.5 (q, J = 5.8 Hz), 124.4 (q, J = 275.2 Hz), 123.0 (s), 111.3 (s), 71.8 (q, J = 2.4 Hz), 19.5 (s). **^{19}F NMR** (376 MHz, $CDCl_3$) δ-57.7 (s). **IR** (cm^{-1}, thin film) 3335, 3080, 2970, 2920, 1456, 1311, 1240, 1159, 1124, 1112, 1033, 908, 769. **HRMS** (ESI+): Calc'd for $C_{11}H_{15}FNO$, 234.11057; found, 234.11104.

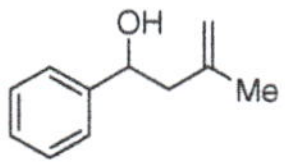

(±)-3-methyl-1-phenylbut-3-en-1-ol—Synthesized using **general procedure 1**. The crude homo allylic alcohol was purified using 10% EtOAc in pentanes and was afforded as a clear and colourless oil (0.874 g, 5.29 mmol, 92%). Characterization data was in accordance with that of the reported literature.[161] **^{1}H NMR** (300 MHz, $CDCl_3$) δ 7.43–7.22 (m, 1H), 4.94–4.90 (m, 1H), 4.85 (dd, J = 2.0, 1.0 Hz, 1H), 4.81 (t, J = 7.0 Hz, 1H), 2.45–2.40 (m, 1H), 2.15 (s, 1H), 1.80 (s, 1H). **^{13}C NMR** (100 MHz, $CDCl_3$) δ 144.0, 142.4, 128.4, 127.5, 125.7, 114.1, 71.4, 48.4, 22.3.

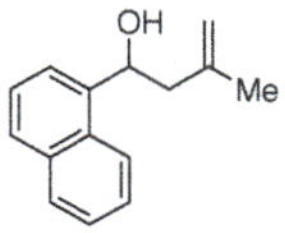

(±)-3-methyl-1-(naphthalen-1-yl)but-3-en-1-ol—Synthesized using general procedure 1. The crude homo allylic alcohol was purified using 10% EtOAc in pentanes and was afforded as a clear and colourless oil (0.957 g, 4.5 mmol, 90%). Characterization data was in accordance with that of the reported literature.[162] **^{1}H NMR** (300 MHz, $CDCl_3$) δ 8.13–8.06 (m, 1H), 7.90 (dd, J = 7.5, 2.5 Hz, 1H), 7.80 (d, J = 8.0 Hz, 1H), 7.74 (dd, J = 7.0, 1.0 Hz, 1H), 7.58–7.46 (m, 3H), 5.64 (dd, J = 2.0, 3.0 Hz, 1H), 5.04–5.00 (m, 1H), 4.98 (d, J = 1.0 Hz, 1H), 2.70 (dd, J = 14.5, 3.0 Hz, 1H), 2.52 (ddd, J = 14.5, 10.0, 1.0 Hz, 1H), 1.93 (dd, J = 1.5, 0.5 Hz, 3H). **^{13}C NMR** (300 MHz, $CDCl_3$) δ 144.8, 139.6, 133.8, 130.2, 129.0, 127.9, 126.0, 125.5, 125.4, 122.8, 122.7, 114.0, 68.1, 47.4, 22.4.

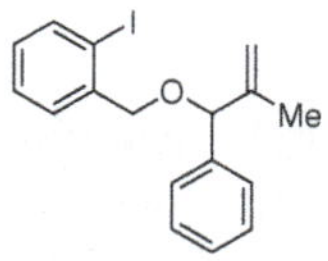

(±)-1-iodo-2-(((2-methyl-1-phenylallyl)oxy)methyl)benzene (1.173a)—The title compound was synthesized from 2-methyl-1-phenylprop-2-en-1-ol (0.200 g, 1.34 mmol) and 2-iodobenzyl methanesulfonate (0.503 g, 1.61 mmol) using **general procedure 3**, and was afforded as a clear and colourless oil (0.262 g,

[161] Reference [158].

[162] Reference [159].

0.723 mmol, 54%) after purification on silica gel using 3% EtOAc in hexanes as the mobile phase. **^{1}H NMR** (400 MHz, $CDCl_3$) δ 7.79 (dd, *J* = 8.0, 1.0 Hz, 1H), 7.53 (dd, *J* = 7.0, 0.5 Hz, 1H), 7.44–7.40 (m, 2H), 7.36–7.31 (m, 3H), 7.27 (ddd, *J* = 7.0, 3.5, 1.0 Hz, 1H), 7.12–6.81 (m, 1H), 5.33–5.14 (m, 1H), 5.14–4.98 (m, 1H), 4.86 (s, 1H), 4.52 (d, *J* = 13.0 Hz, 1H) 4.43 (d, *J* = 13.0 Hz, 1H), 1.61 (s, 3H). **^{13}C NMR** (100 MHz, $CDCl_3$) δ 144.7, 140.8, 140.2, 139.0, 128.9, 128.6, 128.2, 128.1, 127.4, 126.6, 113.6, 97.4, 84.9, 74.1, 17.7. **IR** (cm^{-1}, thin film) 3063, 3026, 2972, 2916, 2860, 2856, 1456, 1448, 1089, 1004, 904, 748, 1698. **HRMS** (ESI+): Calc'd for $C_{17}H_{21}INO^+$, 382.06678; found, 382.06716.

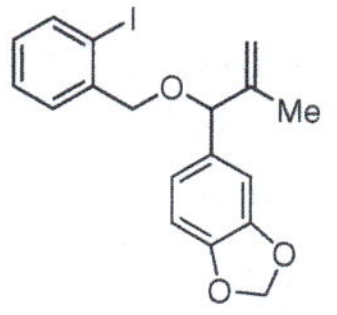

(±)-5-(1-((2-iodobenzyl)oxy)-2-methylallyl)benzo[d][1, 3]dioxole (1.173b)—The title compound was synthesized from 1-(benzo[*d*][1, 3]dioxol-5-yl)-2-methylprop-2-en-1-ol (0.400 g, 2.1 mmol) and 2-iodobenzyl methanesulfonate (0.637 g, 2.5 mmol) using **general procedure 3**, and was isolated as a clear and colourless oil (0.409 g, 1.01 mmol, 48%) after purification on silica gel using hexanes:EtOAc (20:1 v:v) as the mobile phase. **^{1}H NMR** (400 MHz, $CDCl_3$) δ 7.81 (dd, *J* = 8.0, 1.0 Hz, 1H), 7.52 (ddd, *J* = 13.0, 7.0, 6.0 Hz, 1H), 7.36 (td, *J* = 7.5, 1.2 Hz, 1H), 6.98 (qd, *J* = 8.0, 4.0 Hz, 1H), 6.94 (d, *J* = 1.5 Hz, 1H), 6.87 (tt, *J* = 7.5, 3.0 Hz, 1H), 6.79 (d, *J* = 8.0 Hz, 1H), 5.96 (s, 2H), 5.20–5.18 (m, 1H), 5.12–4.91 (m, 1H), 4.77 (s, 1H), 4.50 (d, *J* = 13.0 Hz, 1H), 4.43 (d, *J* = 13.0 Hz, 1H), 1.63 (s, *J* = 1.0 Hz, 3H). **^{13}C NMR** (100 MHz, $CDCl_3$) δ 147.7, 146.9, 144.8, 140.8, 139.0, 134.3, 129.0, 128.6, 128.2, 120.3, 113.2, 107.9, 107.2, 100.9, 97.4, 84.6, 74.1, 17.9. **IR** (cm^{-1}, thin film) 3068, 2879, 1564, 1500, 1440, 1244, 1085, 1041, 933. **HRMS** (ESI+): Calc'd for $C_{18}H_{21}INO^+$, 426.05661; found, 426.05713.

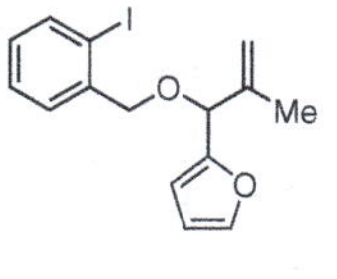

(±)-2-(1-((2-iodobenzyl)oxy)-2-methylallyl)furan (1.173c)—The title compound was synthesized from 1-(furan-2-yl)-2-methylprop-2-en-1-ol (0.101 g, 0.721 mmol) and 2-iodobenzyl methanesulfonate (0.290 g, 0.94 mmol) using **general procedure 3**, and was isolated as a clear and colourless oil (0.200 g, 0.560 mmol, 79%) after purification on silica gel using hexanes:EtOAc (10:1 v: v) + 1% NEt_3. **^{1}H NMR** (400 MHz, $CDCl_3$) δ 7.81 (dd, *J* = 8.0, 1.0 Hz, 1H), 7.54–7.50 (m, 1H), 7.43–7.40 (m, 1H), 7.35 (td, *J* = 7.5, 1.0 Hz, 1H), 6.98 (td, *J* = 8.0, 1.5 Hz, 1H), 6.39–6.34 (m, 2H), 5.25–5.19 (m, 1H), 5.10 (s, 1H), 4.89

(s, 1H), 4.54 (d, J = 13.0 Hz, 1H), 4.48 (d, J = 13.0 Hz, 1H), 1.76 (s, 3H). **^{13}C NMR** (100 MHz, $CDCl_3$) δ 153.1, 142.4, 142.3, 140.5, 139.1, 129.0, 128.7, 128.2, 113.9, 110.2, 108.1, 97.5, 78.6, 74.3, 18.5. **IR** (cm^{-1}, thin film) 2972, 2918, 2852, 1564, 1437, 1141, 1114, 1012, 908, 746. **HRMS** (ESI+): Calc'd for $C_{15}H_{19}INO^+$, 372.04605; found, 372.04658.

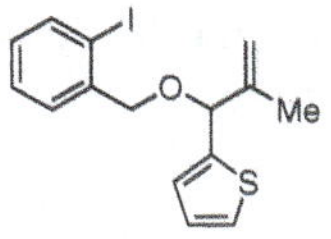

(±)-2-(1-((2-iodobenzyl)oxy)-2-methylallyl)thiophene (1.173d)—The title compound was synthesized from 2-methyl-1-(thiophen-2-yl)prop-2-en-1-ol (0.200 g, 1.30 mmol) and 2-iodobenzyl methanesulfonate (0.506 g, 1.62 mmol) using **general procedure 3**, and was isolated as a light yellow oil (0.381 g, 1.02 mmol, 79%) after purification on silica gel using hexanes:EtOAc (20:3 v:v) + 1% NEt_3 as the mobile phase. **^{1}H NMR** (400 MHz, $CDCl_3$) δ 7.74 (dd, J = 8.0, 1.0 Hz, 1H), 7.46 (dd, J = 7.5, 1.5 Hz, 1H), 7.28 (ddd, J = 8.5, 5.5, 1.0 Hz, 1H), 7.22–7.17 (m, 1H), 6.94–6.88 (m, 3H), 5.16 (d, J = 0.5 Hz, 1H), 5.04–5.00 (m, J = 3.0 Hz, 2H), 4.50–4.44 (m, 1H), 4.41 (d, J = 13.0 Hz, 1H), 1.66 (s, 3H). **^{13}C NMR** (100 MHz, $CDCl_3$) δ 144.3, 144.2, 140.5, 139.1, 129.1, 128.7, 128.2, 126.5, 125.0, 124.8, 114.0, 97.5, 81.2, 74.2, 17.7. **IR** (cm^{-1}, thin film) 3068, 2918, 2852, 1437, 1085, 1012, 906, 748, 700, 648. **HRMS** (ESI+): Calc'd for $C_{15}H_{19}INOS^+$, 388.02320; found, 388.02364.

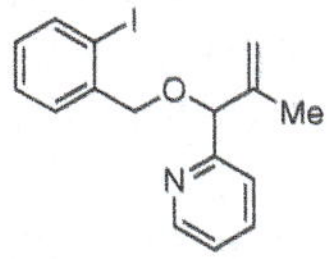

(±)-2-(1-((2-iodobenzyl)oxy)-2-methylallyl)pyridine (1.173e)—The title compound was synthesized from 2-methyl-1-(pyridin-2-yl)prop-2-en-1-ol (0.1 g, 0.67 mmol) and 2-iodobenzyl chloride (0.209 mg, 0.83 mmol) using **general procedure 3**, and was afforded as a light yellow oil (0.108 g, 0.295 mmol, 44%) after purification silica gel using hexanes:EtOAc (2:1 v:v) as the mobile phase. **^{1}H NMR** (400 MHz, $CDCl_3$) δ 8.61–8.54 (m, 1H), 7.80 (dd, J = 8.0, 1.0 Hz, 1H), 7.70 (tt, J = 7.5, 4.0 Hz, 1H), 7.61 (d, J = 8.0 Hz, 1H), 7.55–7.50 (m, 1H), 7.34 (td, J = 7.5, 1.0 Hz, 1H), 7.18 (ddd, J = 7.5, 5.0, 1.0 Hz, 1H), 6.97 (td, J = 7.5, 1.5 Hz, 1H), 5.32–5.24 (m, 1H), 5.14–5.09 (m, 1H), 5.00 (s, 1H), 4.59 (d, J = 13.0 Hz, 1H), 4.47 (d, J = 13.0 Hz, 1H), 1.64 (s, 3H). **^{13}C NMR** (100 MHz, $CDCl_3$) δ 160.0, 148.9, 143.4, 140.5, 139.0, 136.5, 129.0, 128.7, 128.1, 122.3, 120.9, 114.8, 97.5, 86.1, 74.3, 17.7. **IR** (cm^{-1}, thin film) 3063, 3053, 3009, 2972, 2943, 2916, 2868, 1577, 1566, 1471, 1433, 1099, 1014, 904, 748. **HRMS** (ESI+): Calc'd for $C_{16}H_{17}INO^+$, 366.03548; found, 366.03533.

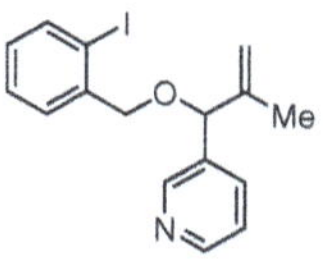

(±)-3-(1-((2-iodobenzyl)oxy)-2-methylallyl)pyridine (1.173f)—The title compound was synthesized from 2-methyl-1-(pyridin-3-yl)prop-2-en-1-ol (0.300 g, 2.01 mmol) and 2-iodobenzyl chloride (0.608 g, 2.41 mmol) using **general procedure 3**, and was afforded as a light yellow oil (0.300 g, 0.820 mmol, 41%) after purification silica gel using hexanes:EtOAc (2:1 v:v) as the mobile phase. **^{1}H NMR** (400 MHz, $CDCl_3$) δ 8.66 (d, J = 1.5 Hz, 1H), 8.54 (dd, J = 4.5, 1.5 Hz, 1H), 7.82 (dd, J = 8.0, 1.0 Hz, 1H), 7.76 (dt, J = 8.0, 1.5 Hz, 1H), 7.54–7.47 (m, 1H), 7.35 (td, J = 7.5, 1.5 Hz, 1H), 7.29 (dd, J = 6.5, 1.5 Hz, 1H), 6.99 (td, J = 8.0, 1.5 Hz, 1H), 5.26–5.22 (m, 1H), 5.15–5.09 (m, 1H), 4.91 (s, 1H), 4.56 (d, J = 13.0 Hz, 1H), 4.44 (d, J = 13.0 Hz, 1H), 1.62 (d, J = 1.0 Hz, 3H). **^{13}C NMR** (100 MHz, $CDCl_3$) δ 148.9, 148.5, 143.7, 140.3, 139.2, 135.7, 134.1, 129.2, 128.7, 128.2, 123.2, 114.8, 97.6, 82.9, 74.2, 17.3. **IR** (cm^{-1}, thin film) 3063, 2868, 1587, 1433, 1116, 1014, 904, 748. **HRMS** (ESI+): Calc'd for $C_{16}H_{17}INO^+$, 366.03548; found, 366.03524.

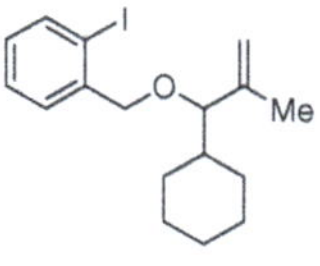

(±)-1-(((1-cyclohexyl-2-methylallyl)oxy)methyl)-2-iodobenzene (1.173 g)—The title compound was synthesized from 1-cyclohexyl-2-methylprop-2-en-1-ol (0.415 g, 2.7 mmol) and 2-iodobenzyl chloride (0.682 g, 2.7 mmol) using **general procedure 3**, and was afforded a clear and colourless oil (0.211 g, 0.567 mmol, 21%) after purification on silica gel using hexanes:DCM (5:1 v:v) as the mobile phase. **^{1}H NMR** (400 MHz, $CDCl_3$) δ 7.81 (dd, J = 8.0, 1.0 Hz, 1H), 7.48 (dd, J = 7.5, 1.5 Hz, 1H), 7.34 (td, J = 7.5, 1.0 Hz, 1H), 6.97 (td, J = 7.5, 1.5 Hz, 1H), 5.06–5.02 (m, 1H), 4.93 (s, 1H), 4.45 (d, J = 12.5 Hz, 1H), 4.23 (d, J = 12.5 Hz, 1H), 3.42 (d, J = 9.0 Hz, 1H), 2.26–2.14 (m, 1H), 1.81–1.61 (m, 6H), 1.61–1.43 (m, 2H), 1.36–0.78 (m, 6H). **^{13}C NMR** (100 MHz, $CDCl_3$) δ 143.3, 141.2, 139.0, 129.1, 128.9, 128.1, 115.2, 98.0, 89.0, 73.9, 39.5, 30.1, 29.4, 26.6, 26.1, 26.0, 16.7. **IR** (cm^{-1}, thin film) 2922, 2850, 1448, 1437, 1114, 1085, 1070, 1043, 1012, 902, 746. **HRMS** (ESI+): Calc'd for $C_{17}H_{27}INO^+$, 388.11373; found, 388.11299.

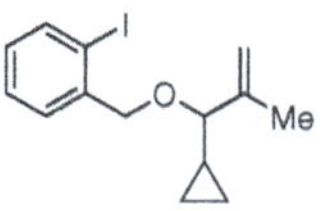

(±)-1-(((1-cyclopropyl-2-methylallyl)oxy)methyl)-2-iodobenzene (1.173h)—The title compound was synthesized from 1-cyclopropyl-2-methylprop-2-en-1-ol

(0.200 g, 1.78 mmol) and 2-iodobenzyl methanesulfonate (0.54 g, 2.14 mmol) using **general procedure 3**, and was isolated as a clear and colourless oil (0.249 g, 0.76 mmol, 43%) after purification on silica gel using hexanes:EtOAc (20:1 v:v) as the mobile phase. **^{1}H NMR** (400 MHz, $CDCl_3$) δ 7.81 (dd, *J* = 8.0, 1.0 Hz, 1H), 7.51–7.46 (m, 1H), 7.34 (td, *J* = 7.5, 1.0 Hz, 1H), 6.99–6.93 (m, 1H), 4.98–4.94 (m, 1H), 4.94–4.92 (m, 1H), 4.45 (d, *J* = 12.9 Hz, 1H), 4.33 (d, *J* = 13.0 Hz, 1H), 3.11 (d, *J* = 8.0 Hz, 1H), 1.86–1.83 (m, 3H), 1.19–1.04 (m, 1H), 0.72–0.60 (m, 1H), 0.56–0.39 (m, 2H), 0.29–0.14 (m, 1H). **^{13}C NMR** (100 MHz, $CDCl_3$) δ 144.9, 141.0, 139.0, 128.9, 128.9, 128.1, 113.2, 97.9, 87.9, 73.8, 17.5, 14.5, 4.5, 2.1. **IR** (cm^{-1}, thin film) 3072, 2852, 1450, 1437, 1085, 902. **HRMS** (ESI+): Calc'd for $C_{14}H_{18}IO^+$, 329.04023; found, 329.04077.

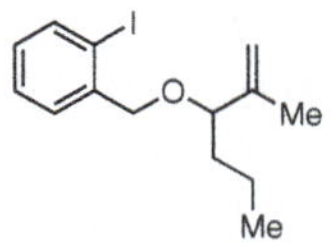

(±)-1-iodo-2-(((2-methylhex-1-en-3-yl)oxy)methyl)benzene (1.173i)—The title compound was synthesized from 2-methylhex-1-en-3-ol (0.75 g, 6.57 mmol) and 2-iodobenzyl methanesulfonate (2.25 g, 7.23 mmol) using **general procedure 3**, and was isolated as a clear and colourless oil (0.76 g, 2.29 mmol, 35%) after purification on silica gel using hexanes:DCM (5:1 v:v) as the mobile phase. **^{1}H NMR** (400 MHz, $CDCl_3$) δ 7.81 (dd, *J* = 8.0, 1.0 Hz, 1H), 7.50–7.44 (m, 1H), 7.34 (td, *J* = 7.5, 1.1 Hz, 1H), 6.97 (td, *J* = 7.5, 1.5 Hz, 1H), 4.99 (dt, *J* = 3.0, 1.5 Hz, 1H), 4.97 (d, *J* = 0.5 Hz, 1H), 4.45 (d, *J* = 13.0 Hz, 1H), 4.27 (d, *J* = 13.0 Hz, 1H), 3.79 (dd, *J* = 7.5, 6.0 Hz, 1H), 1.77–1.65 (m, 4H), 1.59–1.25 (m, 3H), 0.92 (t, *J* = 7.5 Hz, 3H). **^{13}C NMR** (100 MHz, $CDCl_3$) δ 144.6, 141.1, 139.1, 128.9, 128.9, 128.1, 113.7, 97.9, 83.8, 73.9, 35.8, 19.1, 16.7, 14.0. **IR** (cm^{-1}, thin film) 2956, 2933, 2870, 1647, 1464, 1112, 1087, 1043, 1012, 900, 746, 3068. **HRMS** (ESI+): Calc'd for $C_{14}H_{23}INO^+$, 348.08243; found, 348.08193.

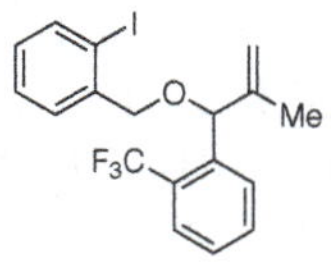

(±)-1-iodo-2-(((2-methyl-1-(2-(trifluoromethyl)phenyl)allyl)oxy)methyl)benzene (1.173j)—The title compound was synthesized from 2-methyl-1-(2-(trifluoromethyl)phenyl)prop-2-en-1-ol (0.276 g, 1.00 mmol) and 2-iodobenzyl methanesulfonate (0.312 g, 1.00 mmol) using **general procedure 3**, and was isolated as a clear and colourless oil (0.213 g, 0.493 mmol, 49%) after purification on silica gel using hexanes:DCM (10:1 → 5:1 v:v) as the mobile phase. **^{1}H NMR** (300 MHz, $CDCl_3$) δ 7.81 (dd, *J* = 8.0, 1.0 Hz, 2H), 7.69 (dd, *J* = 8.0, 0.5 Hz, 1H), 7.59 (t, *J* = 7.5 Hz, 1H), 7.52–7.47 (m, 1H), 7.42 (t, *J* = 7.5 Hz, 1H), 7.35 (td, *J* = 7.5, 1.0 Hz, 1H), 6.98 (td, *J* = 8.0, 2.0 Hz, 1H), 5.31 (s, 1H), 5.09 (s, 1H),

5.05 (dt, J = 2.5, 1.5 Hz, 1H), 4.52 (d, J = 12.5 Hz, 1H), 4.38 (d, J = 12.5 Hz, 1H), 1.74 (s, 3H). **^{13}C NMR** (100 MHz, $CDCl_3$) δ 144.01 (s), 140.54 (s), 139.13 (d, J = 1.5 Hz), 139.0 (s), 132.0 (d, J = 1.1 Hz), 129.2 (s), 129.0 (s), 128.6 (d, J = 21.4 Hz), 128.1 (s), 127.8 (s), 125.7 (s), 125.4 (q, J = 5.9 Hz), 124.4 (q, J = 274.86 Hz), 123.0 (s), 113.3 (s), 97.5 (s), 79.6 (d, J = 2.0 Hz), 74.8 (s), 19.7 (s). **^{19}F NMR** (376 MHz, $CDCl_3$) δ-57.4 (s). **IR** (cm^{-1}, thin film) 3066, 2974, 2947, 2918, 2870, 1606, 1585, 1566, 1452, 1311, 1159, 1123, 1087, 1066, 908, 769, 748. **HRMS** (ESI+): Calc'd for $C_{18}H_{20}F_3INO$, 450.05417; found, 450.05501.

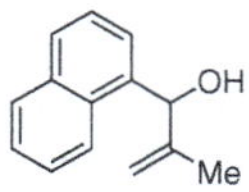

(±)-2-methyl-1-(naphthalen-1-yl)prop-2-en-1-ol—Synthesized **using general procedure 1**. The crude allylic alcohol was sufficiently pure after work up, and was afforded as a clear and colourless oil (0.900 g, 4.54 mmol, 91%) upon concentration. Characterization data was in accordance with that of the reported literature.[163] **^{1}H NMR** (300 MHz, $CDCl_3$) δ 8.23–8.16 (m, 1H), 7.91–7.85 (m, 1H), 7.81 (d, J = 8.0 Hz, 1H), 7.67–7.61 (m, 1H), 7.57–7.44 (m, 3H), 5.85 (s, 1H), 5.31 (d, J = 0.5 Hz, 1H), 5.14–5.07 (m, 1H), 2.13 (br s), 1.72–1.68 (m, 3H). **^{13}C NMR** (100 MHz, $CDCl_3$) δ 146.4, 137.2, 133.9, 131.1, 128.7, 128.5, 126.0, 125.6, 125.3, 124.4, 123.7, 112.4, 74.8, 19.2.

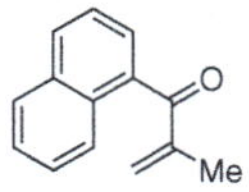

2-methyl-1-(naphthalen-1-yl)prop-2-en-1-one (1.176)—To a stirred CH_2Cl_2 solution (50 mL) of (±)-2-methyl-1-(naphthalen-1-yl)prop-2-en-1-ol (0.505 g, 2.54 mmol, 0.051 M) at room temperature was added Dess-Martin periodinane (2.03 g, 4.78 mmol, 1.88 equiv) and solid sodium bicarbonate (0.601 g, 7.17 mmol, 2.82 equiv). This was allowed to stir for 2 h at this temperature. At this time the solid precipitate was removed by filtrate through Celite, and concentrated to afford the crude ketone as a clear oil (0.265 g, 1.34 mmol, 53%) upon concentration, which was purified by silica gel chromatography using pentanes:Et_2O (2:1 v:v) as the mobile phase. Characterization data was in accordance with that of the reported literature.[164] **^{1}H NMR** (400 MHz, $CDCl_3$) δ 8.03–7.97 (m, 1H), 7.97–7.91 (m, 1H), 7.91–7.86 (m, 1H), 7.56–7.43 (m, 4H), 6.00–5.97 (m, 1H), 5.67–5.63 (m, 1H), 2.20–2.13 (m, 3H). **^{13}C NMR** (100 MHz, $CDCl_3$) δ 199.9, 145.5, 136.6, 133.5, 130.7, 130.5, 130.2, 128.24, 126.9, 126.6, 126.2, 125.4, 124.1, 17.5.

---

[163] Reference [154].

[164] Reference [160].

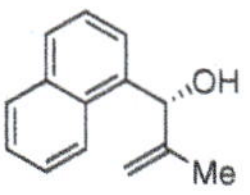

(*R*)-2-methyl-1-(naphthalen-1-yl)prop-2-en-1-ol ((*R*)-**1.178**)—To a stirred THF solution (22 mL) of 2-methyl-1-(naphthalen-1-yl)prop-2-en-1-one (0.265 g, 2.33 mmol, 0.060 M) at −40 °C was added (*R*)-2-methyl-CBS-oxazaborolidine in toluene (2.66 mL, 2.66 mmol, 1 M, 2 equiv) dropwise over 2 min. The resulting solution was allowed to stir at this temperature for 15 min, at which time $BH_3{\cdot}SMe_2$ (0.65 mL, 6.65 mmol, 5 equiv, 10 M) was added dropwise over 10 min. The resultant solution was stirred at this temperature for 3 h, at which time 2 mL of EtOH was added. The solution was allowed to stir until at room temperature until gas evolution ceased. The mixture had water (15 mL) added and then was extracted with Et_2O (3 × 15 mL) and the combined organic layers were washed with water and brine, dried over anhydrous $NaSO_4$, filtered and concentrated to afford the crude allylic alcohol as a clear oil. This was purified by Si Gel chromatography using pentanes:EtOAc (5:1 v:v) as the mobile phase to afford the allylic alcohol as a clear and colourless oil (0.226 g, 1.33 mmol, 99%) upon concentration. $[a]_D^{25.2} = +24.07$ (c = 0.54, $CHCl_3$). The er value was determined to be 99.5:0.5 by chiral HPCL with a Daicel chiral OD-H column. Eluent: hexane:^{i}PrOH = 90:10, flow rate: 1.0 mL/min, retention time: minor peak 8.47 min, major peak 13.92 min. NMR spectroscopic data was in accordance with that of the reported literature.[165] **^{1}H NMR** (400 MHz, $CDCl_3$) δ 8.23–8.16 (m, 1H), 7.92–7.86 (m, 1H), 7.82 (d, J = 8.2 Hz, 1H), 7.67–7.61 (m, 1H), 7.56–7.45 (m, 3H), 5.82 (s, 1H), 5.31 (dd, J = 2.1, 1.7 Hz, 1H), 5.11 (dq, J = 3.0, 1.5 Hz, 1H), 2.32 (s, 1H), 1.72–1.66 (m, 3H). **^{13}C NMR** (100 MHz, $CDCl_3$) δ 146.3, 137.15, 133.8, 131.1, 128.7, 128.4, 126.0, 125.5, 125.3, 124.4, 123.7, 112.3, 74.6, 19.2.

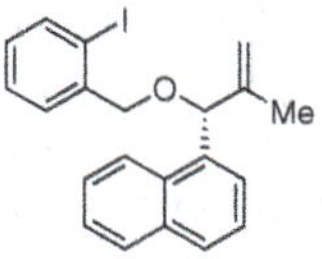

(*R*)-1-(1-((2-iodobenzyl)oxy)-2-methylallyl)naphthalene ((*R*)-**1.180**)[166]—The title compound was synthesized from (*R*)-2-methyl-1-(naphthalen-1-yl)prop-2-en-1-ol (0.250 g, 1.26 mmol) and 2-iodobenzyl methanesulfonate (0.472 g, 1.51 mmol) using **general procedure 3**, and was afforded as a clear and colourless oil (0.129 g, 0.312 mmol, 24, 74% BRSM) after purification on silica gel using 3% EtOAc in hexanes as the mobile phase. $[a]_D^{23.9} = +5.37$ (c = 1.19, $CHCl_3$). The er value was determined to be 99:1 by chiral HPCL with a Daicel chiral OD-H column. Eluent: hexane:^{i}PrOH = 98:2, flow rate: 1.0 mL/min, retention time: minor peak 4.252 min,

[165] Reference [154].

[166] Racemic assay prepared by Antonin Clemenceau.

major peak 5.463 min. **^{1}H NMR** (400 MHz, $CDCl_3$) δ 8.20 (dq, *J* = 6.8, 3.2 Hz, 1H), 7.94–7.87 (m, 1H), 7.86–7.80 (m, 2H), 7.76 (t, *J* = 7.2 Hz, 1H), 7.62–7.45 (m, 4H), 7.37 (td, *J* = 7.6, 1.0 Hz, 1H), 7.00 (td, *J* = 7.7, 1.6 Hz, 1H), 5.57 (s, 1H), 5.30 (d, *J* = 2.6 Hz, 1H), 5.14 (s, 1H), 4.59 (q, *J* = 13.0 Hz, 2H), 1.71 (s, 3H). **^{13}C NMR** (100 MHz, $CDCl_3$) δ 144.2, 140.8, 139.0, 135.3, 133.9, 131.2, 129.0, 128.7, 128.7, 128.3, 128.2, 125.8, 125.4, 125.3, 125.3, 124.0, 113.9, 97.6, 82.7, 74.6, 19.1. **IR** (cm^{-1}, thin film) 3063, 3051, 2976, 2947, 2914, 2866, 1564, 1510, 1464, 1446, 1435, 1371, 1112, 1087, 1043, 1012, 906, 798, 779, 748, 734. **HRMS** (ESI+): Calc'd for $C_{21}H_{23}INO^+$, 432.08243; found, 432.08064.

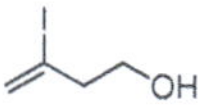

3-iodobut-3-en-1-ol (**1.183**). The title compound was synthesized in accordance with a known literature procedure and spectra data were identical to that reported therein.[167] **^{1}H NMR** (400 MHz, $CDCl_3$) δ 6.18 (q, *J* = 1.5 Hz, 1H), 5.86 (d, *J* = 1.5 Hz, 1H), 3.76 (q, *J* = 6.0 Hz, 2H), 2.64 (td, *J* = 6.0, 1.0 Hz, 2H). **^{13}C NMR** (100 MHz, $CDCl_3$) δ 128.3, 107.39, 60.8, 48.0.

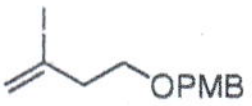

1-(((3-iodobut-3-en-1-yl)oxy)methyl)-4-methoxybenzene (**1.185**). The title compound was synthesized in accordance with a known literature procedure[168] and spectra data were identical to that of the reported literature.[169] **^{1}H NMR** (400 MHz, $CDCl_3$) δ 7.28–7.25 (m, 2H), 6.92–6.83 (m, 2H), 6.14–6.10 (m, 1H), 5.82–5.75 (m, 1H), 4.47 (s, 2H), 3.80 (s, 3H), 3.57 (t, *J* = 6.5 Hz, 2H), 2.69–2.64 (m, 2H). **^{13}C NMR** (100 MHz, $CDCl_3$) δ 159.2, 130.2, 129.3, 127.3, 113.8, 107.4, 72.7, 68.4, 55.3, 45.3.

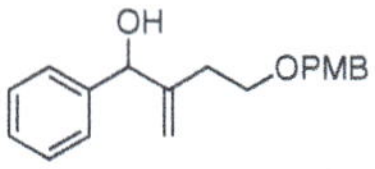

(±)-**4-((4-methoxybenzyl)oxy)-2-methylene-1-phenylbutan-1-ol** (**1.186**). In an argon filled glovebox a 25 mL round bottom flask was charged with $NiCl_2$ (13.5 mg, 0.104 mmol, 2 mol%) and anhydrous $CrCl_2$ (2.55 g, 20.8 mmol, 4 equiv). The flask was sealed with a rubber septum and removed from the glovebox. Its contents were placed under positive argon pressure and dissolved in anhydrous DMF (15 mL). The resultant emerald green solution was cooled to 0 °C at which time a DMF solution (5 mL) of benzaldehyde (0.546 g, 5.2 mmol, 1 equiv) and (±)-4-

[167]Reference [161].

[168]Reference [162].

[169]Reference [163].

((4-methoxybenzyl)oxy)-2-methylene-1-phenylbutan-1-ol (1.5 g, 5.2 mmol, 1 equiv) was added drop wise over 20 min. The reaction was allowed to slowly reach room temperature. After 18 h at this temperature the reaction was quenched with saturated $NaHCO_3$ (30 mL) and extracted with Et_2O (3 × 30 mL). The combined organic layers were washed with water (2 × 20 mL) and brine (30 mL), dried over Na_2SO_4, filtered, and concentrated *in vacuo* to afford the crude allylic alcohol. Purification via silica gel chromatography using hexanes:EtOAc (10:3 v:v) as the mobile phase afforded the title compound as a colourless oil (1.47 g, 4.94 mmol, 95%). **^{1}H NMR** (400 MHz, $CDCl_3$) δ 7.39–7.28 (m, 4H), 7.27–7.20 (m, 3H), 6.91–6.83 (m, 2H), 5.23 (s, 1H), 5.19 (d, *J* = 4.5 Hz, 1H), 5.02 (d, *J* = 1.0 Hz, 1H), 4.42 (s, 2H), 3.80 (s, 3H), 3.61–3.37 (m, 3H), 2.33–2.10 (m, 2H). **^{13}C NMR** (100 MHz, $CDCl_3$) δ 159.3, 149.0, 142.6, 129.8, 129.4, 128.2, 127.2, 126.3, 113.8, 77.2, 72.8, 69.9, 55.3, 31.9. **IR** (cm^{-1}, neat) 3401, 2933, 2918, 2858, 2873, 1514. **HRMS** (ESI +, $M+NH_4^+$): Calc'd for $C_{19}H_{26}NO_3$, 316.19127; found, 316.19284.

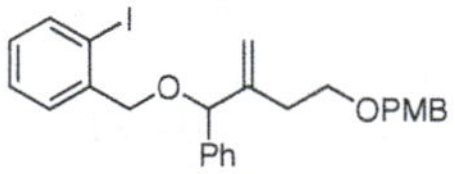

(±)-1-iodo-2-((4-((4-methoxybenzyl)oxy)-2-methylene-1-phenylbutoxy)methyl) benzene (1.181). The title compound was synthesized from (±)-4-((4-methoxybenzyl)oxy)-2-methylene-1-phenylbutan-1-ol (0.414 g, 1.39 mmol) and 2-iodobenzyl chloride (0.350 g, 1.39 mmol) using **general procedure 3** and was isolated as a colourless oil (0.587 g, 1.11 mmol, 80%) after purification on silica gel using hexanes:EtOAc (2:1 v:v) as the mobile phase. **^{1}H NMR** (400 MHz, $CDCl_3$) δ 7.80 (dd, *J* = 8.0, 1.0 Hz, 1H), 7.50 (dd, *J* = 7.5, 1.5 Hz, 1H), 7.42–7.37 (m, 2H), 7.35–7.29 (m, 3H), 7.29–7.25 (m, 1H), 7.22–7.16 (m, 2H), 6.97 (td, *J* = 7.5, 1.5 Hz, 1H), 6.91–6.78 (m, 2H), 5.28 (s, 1H), 5.15–5.05 (m, 1H), 4.89 (s, 1H), 4.49 (d, *J* = 13.0 Hz, 1H), 4.42 (d, *J* = 13.0 Hz, 1H), 4.35 (s, 2H), 3.79 (s, 3H), 3.56–3.32 (m, 2H), 2.38–2.28 (m, 1H), 2.27–2.16 (m, 1H). **^{13}C NMR** (100 MHz, $CDCl_3$) δ 159.1, 145.8, 140.8, 140.0, 139.0, 130.5, 129.2, 128.9, 128.6, 128.2, 128.1, 127.5, 126.9, 113.7, 113.7, 97.4, 84.7, 74.3, 72.5, 68.7, 55.3, 31.5. **IR** (neat, cm^{-1}) 2931, 2906, 2856, 2835, 1612, 1512, 1448, 1438, 1301, 1247, 1172, 1112, 1093. **HRMS** (ESI+): Calc'd for $C_{26}H_{31}INO_3$, 532.13486; found, 532.13442.

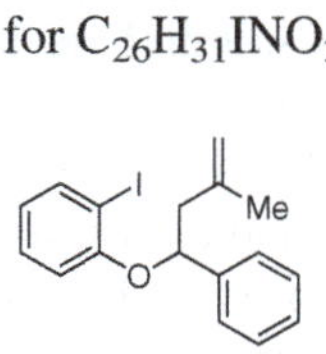

(±)-1-iodo-2-((3-methyl-1-phenylbut-3-en-1-yl)oxy)benzene (1.169a)—The title compound was synthesized from 3-methyl-1-phenylbut-3-en-1-ol (1.57 mmol) using **general procedure 2** and was isolated as a clear and colourless oil (0.345 g, 0.950 mmol, 63%) after purification on silica gel using hexanes:DCM (5:1 v:v) as the mobile phase. **^{1}H NMR** (400 MHz, $CDCl_3$) δ 7.64 (dd, *J* = 7.5, 1.5 Hz, 1H),

7.30–7.19 (m, 4H), 7.18–7.11 (m, 1H), 7.01–6.95 (m, 1H), 6.53–6.47 (m, 2H), 5.22 (dd, J = 8.0, 5.0 Hz, 1H), 4.77–4.74 (m, 1H), 4.72 (dd, J = 2.0, 1.0 Hz, 1H), 2.75 (ddd, J = 14.0, 8.0, 1.0 Hz, 1H), 2.48–2.42 (m, 1H), 1.71–1.70 (m, 3H). **^{13}C NMR** (100 MHz, $CDCl_3$) δ 156.3, 141.41, 140.9, 139.4, 129.0, 128.5, 127.7, 126.0, 122.3, 113.9, 113.6, 87.1, 80.2, 46.7, 23.3. **IR** (cm^{-1}, thin film) 3064, 3030, 2969, 2941, 2912, 1581, 1471, 1438, 1373, 1354, 1273, 1242, 1120, 1049, 993, 895, 746, 700. **HRMS** (ESI+): Calc'd for $C_{17}H_{21}INO$, 382.06678; found, 382.06871.

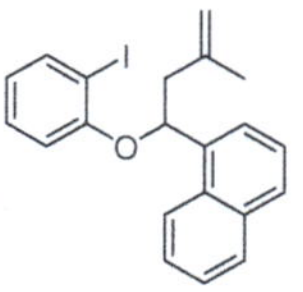

(±)-1(1-(2-iodophenoxy)-3-methylbut-3-en-1-yl)naphthalene (1.169b)—The title compound was synthesized from 3-methyl-1-(naphthalen-1-yl)but-3-en-1-ol (1.64 mmol) using **general procedure 2** and was isolated as a clear and colourless oil (0.305 g, 0.738 mmol, 45%) after purification on silica gel using hexanes: DCM (5:1 v:v) as the mobile phase. **^{1}H NMR** (400 MHz, $CDCl_3$) δ 8.18 (d, J = 8.5 Hz, 1H), 7.93–7.86 (m, 1H), 7.80–7.72 (m, 2H), 7.55 (dddd, J = 15.0, 9.5, 7.0, 1.5 Hz, 3H), 7.40 (td, J = 8.0, 4.5 Hz, 1H), 6.94 (ddd, J = 8.5, 7.5, 1.5 Hz, 1H), 6.56 (qd, J = 7.0, 1.5 Hz, 1H), 6.43 (dd, J = 8.5, 1.0 Hz, 1H), 6.03 (dd, J = 9.0, 3.0 Hz, 1H), 4.93 (dd, J = 2.0, 1.0 Hz, 1H), 4.90 (dd, J = 3.0, 1.5 Hz, 1H), 2.92 (ddd, J = 14.5, 9.0, 0.5 Hz, 1H), 2.72 (dd, J = 14.5, 3.0 Hz, 1H), 1.93–1.88 (m, 3H). **^{13}C NMR** (100 MHz, $CDCl_3$) δ 156.3, 142.1, 139.5, 136.3, 133.9, 129.9, 129.2, 129.11, 128.2, 126.4, 125.7, 125.7, 123.7, 122.4, 122.2, 113.7, 113.1, 86.6, 45.8, 23.4. **IR** (cm^{-1}, thin film) 3065, 2966, 2943, 2916, 1579, 1568, 1467, 1438, 1240, 1020, 798, 665, 767. **HRMS** (ESI+): Calc'd for $C_{21}H_{23}INO$, 432.08243; found, 432.08246.

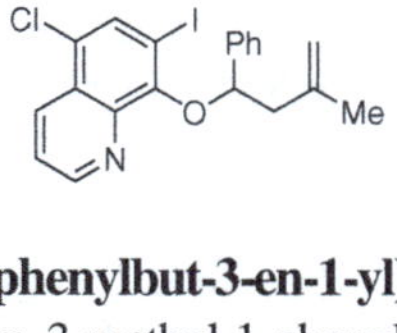

(±)-5-chloro-7-iodo-8-((3-methyl-1-phenylbut-3-en-1-yl)oxy)quinoline (1.169c)[170]—The title compound was synthesized from 3-methyl-1-phenylbut-3-en-1-ol (1.57 mmol) using **general procedure 2** and was isolated as a yellow oil (0.243 g, 0.541 mmol, 34%) after purification on silica gel using 5% EtOAc in hexanes as the mobile phase. **^{1}H NMR** (400 MHz, $CDCl_3$) δ 8.95 (dd, J = 4.0, 1.5 Hz, 1H), 8.42 (dd, J = 8.5, 1.5 Hz, 1H), 7.88 (s, 1H), 7.52–7.45 (m, 3H), 7.25–7.17 (m, 3H), 6.81 (dd, J = 7.5, 7.0 Hz, 1H), 4.79–4.75 (m, 2H), 3.11 (dd, J = 14.0, 6.4 Hz, 1H), 2.79 (ddd, J = 14.0, 8.0, 0.5 Hz, 1H), 1.75–1.74 (m, 3H). **^{13}C NMR** (100 MHz, $CDCl_3$) δ 153.4, 149.0, 141.9, 141.6, 140.1, 135.2, 133.3,

[170]Substrate prepared by Antonin Chemenceau.

128.1, 127.9, 127.8, 127.1, 125.0, 121.9, 113.4, 91.2, 84.6, 45.2, 22.8. **IR** (cm^{-1}, thin film) 3068, 3032, 2966, 2933, 2918, 1653, 1570, 1483, 1440, 1384, 1367, 1350, 1239, 1207, 1087, 727. **HRMS** (ESI+): Calc'd for $C_{20}H_{18}ClINO$, 450.01216; found, 450.01242.

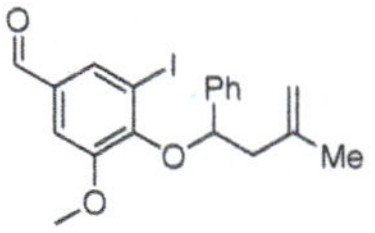

(±)-3-iodo-5-methoxy-4-((3-methyl-1-phenylbut-3-en-1-yl)oxy)benzaldehyde (1.169d)[171]—The title compound was synthesized from 3-methyl-1-phenylbut-3-en-1-ol (1.85 mmol) using **general procedure 2** and was isolated as a clear and colourless oil (0.330 g, 0.781 mmol, 42%) after purification on silica gel using hexanes:Et_2O (2:1, v:v) the mobile phase. **^{1}H NMR** (400 MHz, $CDCl_3$) δ 9.72 (s, 1H), 7.77 (d, J = 2.0 Hz, 1H), 7.42–7.39 (m, 2H), 7.29–7.20 (m, 4H), 5.91 (t, J = 7.0 Hz, 1H), 4.79–4.77 (m, 1H), 4.75 (dd, J = 2.0, 1.0 Hz, 1H), 3.84 (s, 3H), 2.96 (dd, J = 14.0, 7.0 Hz, 1H), 2.66 (ddd, J = 14.0, 7.0, 0.5 Hz, 1H), 1.75–1.71 (m, 3H). **^{13}C NMR** (100 MHz, $CDCl_3$) δ 189.53, 151.85, 151.71, 141.30, 139.63, 135.23, 132.87, 128.24, 127.98, 127.67, 113.72, 110.86, 93.01, 83.60, 55.73, 45.44, 22.79. **IR** (cm^{-1}, thin film) 3072, 2966, 2937, 2916, 2845, 1693, 1583, 1452, 1413, 1383, 1273, 1141, 1041, 968, 893, 858, 759, 736, 700. **HRMS** (ESI+, M+Na^+): Calc'd for $C_{19}H_{19}INaO_3$, 445.0271; found, 4445.0261.

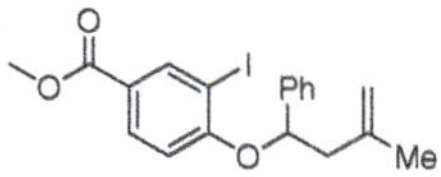

(±)-methyl 3-iodo-4-((3-methyl-1-phenylbut-3-en-1-yl)oxy)benzoate (1.169d)—The title compound was synthesized from 3-methyl-1-phenylbut-3-en-1-ol (1.85 mmol) using **general procedure 2** and was isolated as a clear and colourless oil (0.590 g, 1.40 mmol, 76%) after purification on silica gel using 10% EtOAc in hexanes as the mobile phase. **^{1}H NMR** (400 MHz, $CDCl_3$) δ 8.43 (d, J = 2.0 Hz, 1H), 7.79 (dd, J = 8.5, 2.0 Hz, 1H), 7.37–7.22 (m, 5H), 6.63–6.59 (m, 1H), 5.38 (dd, J = 8.5, 5.0 Hz, 1H), 4.87–4.85 (m, 1H), 4.82 (d, J = 1.0 Hz, 1H), 3.83 (s, 3H), 2.85 (dd, J = 14.0, 8.0 Hz, 1H), 2.55 (dd, J = 14.0, 4.5 Hz, 1H), 1.80 (s, 3H). **^{13}C NMR** (100 MHz, $CDCl_3$) δ 165.4, 160.0 141.0, 141.0, 140.2, 131.1, 128.7, 128.0, 125.9, 124.1, 114.2, 112.6, 86.4, 80.7, 52.0, 46.6, 23.2. **IR** (cm^{-1}, thin film) 2966, 2949, 2914, 1712, 1651, 1543, 1502, 1452, 1435, 1301, 1288, 1259, 1116, 991, 896, 763, 700. **HRMS** (ESI+): Calc'd for $C_{19}H_{20}IO$, 423.04571; found, 423.04666.

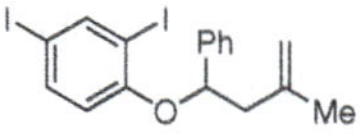

---

[171]Substrate prepared by Antonin Chemenceau.

(±)-2,4-diiodo-1-((3-methyl-1-phenylbut-3-en-1-yl)oxy)benzene (1.169e)—The title compound was synthesized from 3-methyl-1-phenylbut-3-en-1-ol (1.85 mmol) using **general procedure 2** and was isolated as a clear and colourless oil (0.320 g, 0.652 mmol, 42%) after purification on silica gel using hexanes:DCM (5:1, v:v) the mobile phase. **^{1}H NMR** (400 MHz, $CDCl_3$) δ 8.00 (d, *J* = 2.0 Hz, 1H), 7.35–7.21 (m, 6H), 6.34 (d, *J* = 8.5 Hz, 1H), 5.25 (dd, *J* = 8.5, 5.0 Hz, 1H), 4.86–4.81 (m, 1H), 4.79 (d, *J* = 1.0 Hz, 1H), 2.82 (dd, *J* = 14.0, 8.0 Hz, 1H), 2.52 (dd, *J* = 14.0, 4.5 Hz, 1H), 1.78 (s, 3H). **^{13}C NMR** (100 MHz, $CDCl_3$) δ 156.4, 146.6, 141.2, 140.3, 137.7, 128.7, 127.9, 125.9, 115.4, 114.1, 88.6, 83.2, 80.5, 46.6, 23.2. **IR** (cm^{-1}, thin film) 3063, 3028, 2966,2939, 2914, 1564,1464, 1452,1373,1354, 1276, 1242, 1215, 1144,1045, 1026, 991,895,873, 800, 756, 698, 671, 622. **HRMS** (ESI+, $M+NH_4^+$): Calc'd for $C_{17}H_{20}I_2NO$, 507.96342; found, 507.96386.

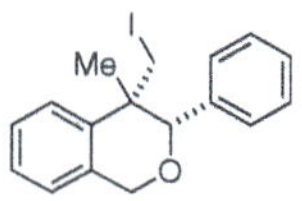

(±)-4-(iodomethyl)-4-methyl-3-phenylisochroman (1.187a)—The title compound was synthesized from **1.173a** (72.8 mg, 0.2 mmol) using **general procedure 4** (0.05 M, 10 h), and was isolated as a clear and colourless oil (68.4 mg, 0.18 mmol, 94%, 94:6 dr (*cis:trans*)) after purification on silica gel using hexanes:DCM (5:1 v: v) as the mobile phase. **^{1}H NMR** (400 MHz, $CDCl_3$) δ 7.48–7.40 (m, 1H), 7.37–7.09 (m, 7H), 6.99–6.92 (m, 1H), 4.99 (q, *J* = 15.5 Hz, 2H), 4.68 (s, 1H), 3.52 (d, *J* = 10.0 Hz, 1H), 3.25 (d, *J* = 10.0 Hz, 1H), 1.25 (s, 3H). **^{13}C NMR** (100 MHz, $CDCl_3$) δ 139.3, 137.5, 133.4, 128.2, 128.0, 127.9, 127.8, 126.9, 125.9, 124.0, 83.2, 69.1, 39.6, 23.9, 18.5. **IR** (cm^{-1}, thin film) 3060, 3030, 2972, 2933, 2839, 1492, 1444, 1369, 1211, 1112, 1093, 1080, 1035, 908, 754, 725, 700. **HRMS** (ESI+): Calc'd for $C_{17}H_{18}IO^+$, 365.04023; found, 365.03888.

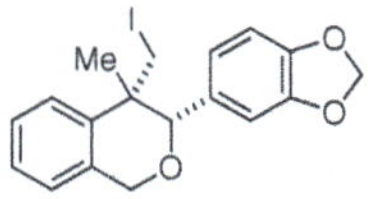

(±)-3-(benzo[d][1, 3]dioxol-5-yl)-4-(iodomethyl)-4-methylisochroman (1.187b)—The title compound was synthesized from **1.173b** (81.6 mg, 0.2 mmol) using **general procedure 4** (0.05 M, 10 h), and was isolated a white foam (74.2 mg, 0.182 mmol, 91%, >95:5 dr (*cis:trans*)) after purification on silica gel using hexanes:EtOAc (20:1 v:v) as the mobile phase. Suitable X-ray quality crystals were grown by slow evaporation of a minimal volume of toluene over 24 h, and were isolated as colourless blocks. **^{1}H NMR** (400 MHz, CDC_3) δ 7.53–7.46 (m, 1H), 7.30–7.22 (m, 2H), 7.06–6.97 (m, 1H), 6.94 (d, *J* = 1.5 Hz, 1H), 6.86–6.77 (m, 2H), 5.97 (s, 2H), 5.12–5.05 (d, *J* = 15.5 Hz, 1H), 5.02 (d, *J* = 15.5 Hz, 1H), 4.68 (s, 1H), 3.56 (d, *J* = 10.0 Hz, 1H), 3.35 (d, *J* = 10.0 Hz, 1H), 1.32 (s, 3H). **^{13}C NMR** (100 MHz, $CDCl_3$) δ 147.4, 147.2, 139.3, 133.3, 131.3, 127.8, 126.9,

125.9, 123.9, 121.7, 108.6, 107.6, 101.1, 83.0 69.0, 39.7, 24.0, 18.4. **IR** (cm^{-1}, thin film) 2887, 1502, 1487, 1251, 1240, 1087, 1039, 935, 725. **HRMS** (ESI+): Calc'd for $C_{18}H_{18}IO$, 409.03006; found, 409.03083.

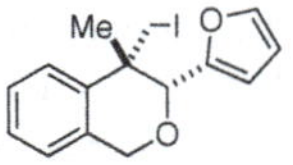

(±)-3-(furan-2-yl)-4-(iodomethyl)-4-methylisochroman (1.187c)—The title compound was synthesized from **1.173c** (70.8 mg, 0.2 mmol) using **general procedure 4** (0.05 M, 10 h), and was isolated a light yellow oil (60.2 mg, 0.17 mmol, 85%, 92:8 dr (*cis:trans*)) after purification on silica gel using hexanes:EtOAc (10:1 v:v) + 1% NEt_3 as the mobile phase. **^{1}H NMR** (400 MHz, $CDCl_3$) δ 7.49–7.42 (m, 2H), 7.30–7.22 (m, 2H), 7.06–6.97 (m, 1H), 6.39 (dd, *J* = 3.0, 2.0 Hz, 1H), 6.33 (d, *J* = 3.0 Hz, 1H), 5.00 (s, 2H), 4.84 (s, 1H), 3.67 (d, *J* = 10.0 Hz, 1H), 3.52 (d, *J* = 10.0 Hz, 1H), 1.41 (s, 3H). **^{13}C NMR** (100 MHz, $CDCl_3$) δ 151.1, 142.3, 138.7, 133.0, 127.1, 127.0, 126.2, 124.0, 110.3, 109.4, 77.6, 68.4, 39.4, 24.5, 17.7. **IR** (cm^{-1}, thin film) 3064, 3018, 2968, 2933, 2839, 1491, 1444, 1379, 1369, 1209, 1149, 1112, 1085, 1010, 1808, 758. **HRMS** (ESI+): Calc'd for $C_{15}H_{16}IO$, 355.01950; found, 355.0204804.

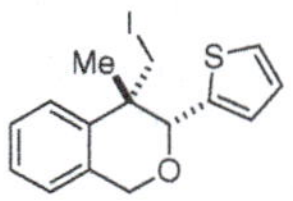

(±)-4-(iodomethyl)-4-methyl-3-(thiophen-2-yl)isochroman (1.187d)—The title compound was synthesized from **1.173d** (73.8 mg, 0.2 mmol) using **general procedure 4** (10 h), and was isolated a light yellow oil (69.4 mg, 0.188 mmol, 94%, 91:9 *dr* (*cis:trans*)) after purification on silica gel using hexanes:EtOAc (10:1 v:v) + 1% NEt_3 as the mobile phase. **Major: ^{1}H NMR** (400 MHz, $CDCl_3$) δ 7.54–7.44 (m, 1H), 7.34–7.30 (m, 1H), 7.30–7.26 (m, 2H), 7.06–6.98 (m, 3H), 5.11–5.00 (m, 3H), 3.62 (d, *J* = 10.0 Hz, 1H), 3.46 (d, *J* = 10.0 Hz, 1H), 1.44 (s, 2H). **^{13}C NMR** (100 MHz, $CDCl_3$) δ 139.8, 138.7, 133.1, 127.5, 127.0, 126.6, 126.3, 126.2, 125.4, 124.0, 79.9, 68.6, 39.7, 24.5, 17.8. **IR** (cm^{-1}, thin film) 2972, 2929, 2937, 1491, 1444, 1367, 1209, 758, 702. **HRMS** (ESI+): Calc'd for $C_{15}H_{16}IOS$, 370.99665; found, 370.99553.

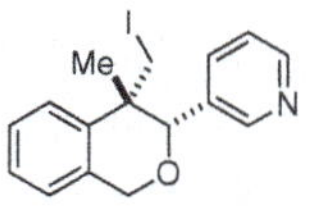

(±)-4-(iodomethyl)-4-methylisochroman-3-yl)pyridine (1.187f)—The title compound was synthesized from **1.173f** (36.5 mg, 0.1 mmol) using **general procedure 4** (16 h, 0.1 M), and was isolated a dark yellow oil (12.6 mg, 0.035 mmol, 35%, >95:5 dr (*cis:trans*)) after purification via preparative TLF using 60% EtOAc in hexanes as the mobile phase. **^{1}H NMR** (500 MHz, $CDCl_3$) δ 7.85 (d, *J* = 8.0 Hz, 1H), 7.56–7.39 (m, 2H), 7.29 (dt, *J* = 7.5, 4.0 Hz, 2H), 7.12–7.00 (m, 2H), 5.08 (q, *J* = 15.5 Hz, 2H), 4.80 (s, 1H), 3.52 (d, *J* = 10.0 Hz, 1H), 3.26 (d, *J* = 10.0 Hz, 1H), 1.36 (s, 3H). **^{13}C NMR** (125 MHz, $CDCl_3$) δ 149.3, 138.7, 135.6, 133.1, 132.4, 127.6, 127.1, 126.2, 124.0, 81.9, 69.0, 39.5, 24.0, 16.8. **IR** (cm^{-1}, thin film) 2982, 2929, 2850, 1616, 1261, 1236, 1207, 1176, 1130, 1091, 1045, 912, 748. **HRMS** (ESI+): Calc'd for $C_{16}H_{17}INO$, 366.03548; found, 366.03606.

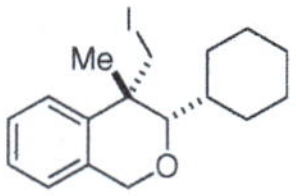

(±)-3-cyclohexyl-4-(iodomethyl)-4-methylisochroman (1.187 g) The title compound was synthesized from **1.173 g** (74.0 mg, 0.2 mmol) using **general procedure 4** (0.05 M, 4 h), and was isolated as a clear and colourless oil (59.9 mg, 0.162 mmol, 81%, 91:9 dr (*cis:trans*) after purification of silica gel using hexanes: DCM (5:1 v:v) as the mobile phase. **Major:** ^{1}H NMR (400 MHz, $CDCl_3$) δ 7.52–7.39 (m, 1H), 7.25–7.17 (m, 2H), 6.99–6.92 (m, 1H), 4.94 (d, *J* = 15.0 Hz, 1H), 4.80 (d, *J* = 15.0 Hz, 1H), 3.64 (d, *J* = 10.0 Hz, 1H), 3.55 (d, *J* = 10.0 Hz, 1H), 3.38 (d, *J* = 2.5 Hz, 1H), 1.92–1.73 (m, 4H), 1.71–1.59 (m, 2H), 1.50–1.41 (m, 1H), 1.40 (s, 3H), 1.38–1.13 (m, 4H). **^{13}C NMR** (100 MHz, $CDCl_3$) δ 140.4, 133.9, 127.4, 126.6, 125.8, 123.8, 85.7, 69.5, 40.2, 38.0, 34.6, 28.6, 26.8, 26.5, 26.3, 23.8, 18.7. **IR** (cm^{-1}, thin film) 2926, 2848, 1616, 1377, 1338, 1207, 1111, 758. **HRMS** (ESI+): Calc'd for $C_{17}H_{24}IO$, 371.08718; found, 371.08616.

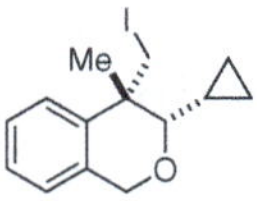

(±)-3-cyclopropyl-4-(iodomethyl)-4-methylisochroman (1.187 h)—The title compound was synthesized from **1.173 h** (65.6 mg, 0.2 mmol) using **general procedure 4** (0.05 M, 10 h), and was isolated a light yellow oil (64.5 mg, 0.186 mmol, 93%, 89:11 dr (*cis:trans*)) after purification on silica gel using hexanes:EtOAc (20:1 v:v) as the mobile phase. **^{1}H NMR** (300 MHz, $CDCl_3$) δ 7.47–7.38 (m, 1H), 7.26–7.18 (m, 2H), 7.00–6.93 (m, 1H), 4.96 (d, *J* = 15.5 Hz, 1H), 4.80 (dd, *J* = 15.5, 0.5 Hz, 1H), 3.74 (d, *J* = 10.0 Hz, 1H), 3.66 (d, *J* = 10.0 Hz, 1H), 3.02 (d, *J* = 8.0 Hz, 1H), 1.48 (s, 3H), 1.09 (qt, *J* = 8.0, 5.0 Hz, 1H), 0.73–0.57 (m, 2H), 0.57–0.40 (m, 2H). **^{13}C NMR** (75 MHz, $CDCl_3$) δ 139.4, 134.3, 133.7, 126.9, 126.8, 126.6, 126.5, 126.0, 125.8, 124.0, 123.9, 84.6, 83.6, 68.3, 66.1, 40.7, 40.0, 24.9, 23.3, 20.0, 18.0, 10.6, 10.2, 4.7, 2.1, 1.7. **IR** (cm^{-1}, thin film) 3065, 2837, 1491, 1377, 1209, 1093, 758. **HRMS** (ESI+): Calc'd for $C_{14}H_{18}IO$, 329.04023; found, 329.04023.

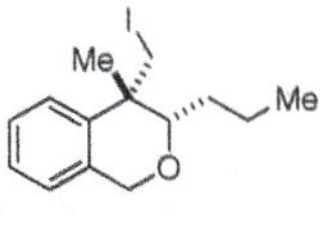

(±)-4-(iodomethyl)-4-methyl-3-propylisochroman (1.187i)—The title compound was synthesized from **1.173i** (66.0 mg, 0.2 mmol) using **general procedure 4** (0.1 M, 10 h), and was isolated a clear and colourless oil (57.0 mg, 0.173 mmol, 87%, 90:10 *dr* (*cis:trans*)) after purification on silica gel using hexanes:DCM (10:1 → 5:1 v:v) as the mobile phase. **^{1}H NMR** (400 MHz, $CDCl_3$) δ 7.38–7.34 (m, 1H), 7.18–7.10 (m, 2H), 6.92–6.87 (m, 1H), 4.78 (q, *J* = 15.5 Hz, 2H), 1.71–1.44 (m, 3H), 1.41–1.32 (m, 1H), 1.30 (s, 3H), 0.90 (t, *J* = 7.0 Hz, 3H). **^{13}C NMR** (100 MHz, $CDCl_3$) δ 139.5, 134.0, 127.0, 126.6, 126.1, 123.9, 81.6, 68.1, 38.5, 31.2, 24.5, 20.1, 17.2, 14.0. **IR** (cm^{-1}, thin film) 2956, 2931, 2870, 2837, 1491, 1444, 1379, 1209, 1099, 1087, 1041, 758, 723, 651. **HRMS** (ESI+): Calc'd for $C_{14}H_{20}IO$, 331.05588; found, 331.05537.

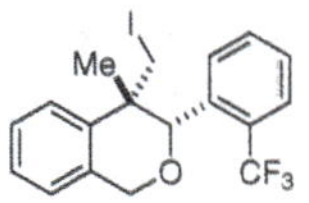

(±)-4-(iodomethyl)-4-methyl-3-(2-(trifluoromethyl)phenyl)isochroman (1.187j)—The title compound was synthesized from **1.173j** (61.6 mg, 0.140 mmol) using **general procedure 4** (0.14 M, 3 h), and was isolated a light yellow oil (55.1 mg, 0.125 mmol, 89%, >95:5 dr (*cis:trans*)) after purification on silica gel using hexanes:DCM (5:1 v:v) as the mobile phase. **^{1}H NMR** (Major) (400 MHz, $CDCl_3$) δ 7.77 (d, *J* = 8.0 Hz, 1H), 7.64 (d, *J* = 8.0 Hz, 1H), 7.53 (t, *J* = 7.5 Hz, 1H), 7.40 (t, *J* = 7.5 Hz, 1H), 7.37–7.32 (m, 1H), 7.24–7.18 (m, 2H), 7.00–6.92 (m, 1H), 5.22 (s, 1H), 5.00 (s, 2H), 3.75 (d, *J* = 10.0 Hz, 1H), 3.46 (d, *J* = 10.0 Hz, 1H), 1.17 (s, 3H). **^{13}C NMR** (100 MHz, $CDCl_3$) δ 140.0 (s), 136.3 (d, *J* = 1.4 Hz), 133.0 (s), 131.4 (d, *J* = 1.0 Hz), 130.2 (s), 129.19 (s), 128.9 (s), 128.6 (s), 127.4 (s), 127.1 (s), 126.4 (q, *J* = 5.9 Hz), 126.0 (s), 124.6 (q, *J* = 274.5 Hz), 123.9 (s, *J* = 5.9 Hz), 122.8 (s), 77.5 (q, *J* = 2.4 Hz), 69.1 (s), 40.3 (s), 23.4 (q, *J* = 2.6 Hz), 18.7 (s). **IR** (cm^{-1}, thin film) 3066, 2974, 2939, 2837, 1492, 1448, 1381, 1307, 1163, 1122, 1089, 754, 725. **HRMS** (ESI+): Calc'd for $C_{18}H_{17}F_3IO$, 433.02762; found, 433.02741.

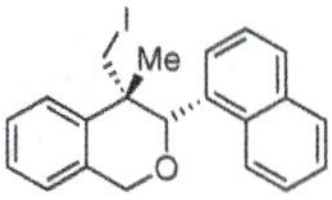

(3*S*,4*S*)-4-(iodomethyl)-4-methyl-3-(naphthalen-1-yl)isochroman (1.187k)—The title compound was synthesized from (*R*)-**1.80k** (61.0 mg, 0.147 mmol) **using general procedure 4** (0.1 M, 14 h), and an inseperarable mixture of diastereomers was isolated a white foam (42.6 mg, 0.103 mmol, 70%, >95:5 dr (*cis:trans*))

after purification on silica gel using hexanes:DCM (5:1 v:v) as the mobile phase. **^{1}H NMR** (400 MHz, $CDCl_3$) δ 7.99 (s br, 1H), 7.83–7.76 (m, 2H), 7.70 (d, *J* = 7.0 Hz, 1H), 7.49–7.35 (m, 4H), 7.26–7.19 (m, 2H), 7.04–6.97 (m, 1H), 5.65 (s, 1H), 5.07 (d, *J* = 2.0 Hz, 2H), 3.78 (d, *J* = 10.0 Hz, 1H), 3.48 (d, *J* = 10.0 Hz, 1H), 1.21 (s, 3H). **^{13}C NMR** (100 MHz, $CDCl_3$) δ 140.0, 133.8, 133.6, 133.4, 131.9, 129.0, 128.7, 127.7, 127.0, 126.6, 126.0, 125.9, 125.4, 125.0, 124.0, 123.9, 69.3, 40.8, 24.6, 19.3. $[a]_D^{23.9} = +144.47$ (*c* = 1.00, $CHCl_3$). The *ee* value was determined to be 98% by chiral HPCL with a Daicel chiral OD-H column. Eluent: hexane:^{i}PrOH = 98:2, flow rate: 1.0 mL/min, t_r 17.71 min (major), t_r 23.72 min (minor). **IR** (cm^{-1}, thin film) 3063, 3018, 2972, 2835, 1597 1491, 1444, 1369, 1338,1234, 1209, 1111, 1093, 1037, 1018, 970, 906, 798, 758, 731. **HRMS** (ESI +): Calc'd for $C_{21}H_{20}IO$, 415.05588; found, 415.05522.

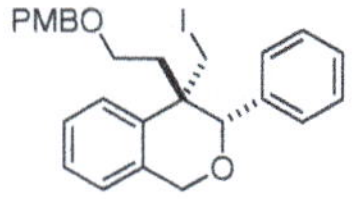

(±)-4-(iodomethyl)-4-(2-((4-methoxybenzyl)oxy)ethyl)-3-phenylisochroman (1.187l)—The title compound was synthesized from **1.173l** (205.6 mg, 0.4 mmol) using general procedure 4 (0.1 M, 24 h), and was isolated a colourless oil (181.1 mg, 0.352 mmol, 89%, >95:5 dr (*cis:trans*)) after purification on silica gel using hexanes:EtOAc (20:3 v:v) as the mobile phase. **^{1}H NMR** (400 MHz, $CDCl_3$) δ 7.46 (dd, *J* = 7.5, 2.0 Hz, 2H), 7.42–7.38 (m, 1H), 7.37–7.30 (m, 3H), 7.28–7.22 (m, 2H), 7.20–7.14 (m, 2H), 7.08–6.99 (m, 1H), 6.89–6.78 (m, 2H), 4.99–4.86 (m, 3H), 4.40 (d, *J* = 11.5 Hz, 1H), 4.29 (d, *J* = 11.5 Hz, 1H), 3.79 (s, 3H), 3.63 (d, *J* = 10.0 Hz, 1H), 3.61–3.53 (m, 1H), 3.39 (ddd, *J* = 9.5, 8.0, 5.0 Hz, 1H), 3.26 (d, *J* = 10.0 Hz, 1H), 2.33 (ddd, *J* = 15.0, 8.0, 5.0 Hz, 1H), 1.86 (ddd, *J* = 15.0, 8.0, 7.0 Hz, 1H). **^{13}C NMR** (100 MHz, CDC_3) δ 159.1, 137.4, 136.2, 134.6, 130.2, 129.1, 128.3, 128.2, 127.9, 127.9, 126.8, 125.9, 124.2, 113.7, 80.3, 72.6, 68.8, 66.3, 55.3, 41.7, 35.3, 18.9. IR (cm^{-1}, neat) 2933, 2860, 2835, 1612, 1512, 1494, 1452, 1367, 1301, 1247, 1211, 1172. **HRMS** (ESI +): Calc'd for $C_{26}H_{28}IO_3$, 515.10831; found, 515.10855.

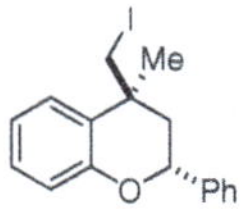

(±)-4-(iodomethyl)-4-methyl-2-phenylchroman (1.189a)—The title compound was synthesized from **1.169a** (72.0 mg, 0.2 mmol) using **general procedure 4** (0.1 M, 15 h), and was isolated a clear and light yellow oil (60.1 mg, 0.161 mmol, 81%, 93:7 dr (*trans:cis*)) after purification silica gel using hexanes:DCM (5:1 v:v) as the mobile phase. **^{1}H NMR** (400 MHz, $CDCl_3$) δ 7.51–7.46 (m, 2H), 7.46–7.40 (m, 2H), 7.39–7.30 (m, 2H), 7.19 (ddd, *J* = 8.0, 7.5, 1.5 Hz, 1H), 6.99–6.92 (m, 2H), 4.99 (dd, *J* = 12.5, 2.0 Hz, 1H), 3.67 (d, *J* = 10.5 Hz, 1H), 3.58 (dd, *J* = 10.5,

1.5 Hz, 1H), 2.40 (dd, *J* = 14.5, 2.0 Hz, 1H), 1.88 (ddd, *J* = 14.0, 9.5, 1.5 Hz, 1H), 1.45 (d, *J* = 0.5 Hz, 3H). **^{13}C NMR** (100 MHz, $CDCl_3$) δ 154.2, 141.0, 128.6, 128.0, 126.7, 125.9, 125.2, 120.8, 117.6, 73.8, 42.6, 34.6, 27.5, 22.1. **IR** (cm^{-1}, thin film) 3063, 3030, 2964, 2916, 1604, 1577, 1481, 1448, 1226, 1176, 1060, 1008, 745, 696. **HRMS** (ESI+): Calc'd for $C_{17}H_{16}IO$, 365.04023; found, 365.04013.

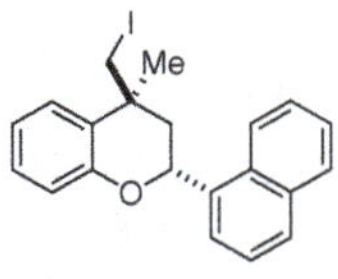

(±)-4-(iodomethyl)-4-methyl-2-(naphthalen-1-yl)chroman (1.189b)—The title compound was synthesized from **1.169b** (82.0 mg, 0.2 mmol) using **general procedure 3** (0.1 M, 13 h), and was isolated a clear and light yellow oil (79.4 mg, 0.192 mmol, 96%, >95:5 dr (*trans:cis*)) after purification silica gel using hexanes: DCM (5:1 v:v) as the mobile phase. **^{1}H NMR** (400 MHz, $CDCl_3$) δ 8.27–8.21 (m, 1H), 7.93 (dd, *J* = 8.5, 1.0 Hz, 1H), 7.86 (d, *J* = 8.0 Hz, 1H), 7.81 (d, *J* = 7.0 Hz, 1H), 7.65–7.48 (m, 3H), 7.37 (dd, *J* = 8.0, 1.5 Hz, 1H), 7.25 (ddd, *J* = 8.0, 7.0, 156 Hz, 1H), 7.09–6.98 (m, 2H), 5.84 (dd, *J* = 12.0, 2.0 Hz, 1H), 3.88 (d, *J* = 10.5 Hz, 1H), 3.67 (dd, *J* = 10.5, 2.0 Hz, 1H), 2.67 (dd, *J* = 14.5, 2.0 Hz, 1H), 1.98 (ddd, *J* = 14.5, 12.0, 2.0 Hz, 1H), 1.48 (s, 3H). **^{13}C NMR** (100 MHz, $CDCl_3$) δ 154.3, 136.5, 133.7, 130.1, 128.9, 128.6, 128.3, 126.8, 126.3, 125.7, 125.5, 125.5, 123.2, 123.2, 120.9, 117.7, 70.9, 41.7, 35.1, 27.8, 21.6. **IR** (cm^{-1}, thin film) 3176, 3151, 3051, 2964, 2918. 1599, 1577, 1483, 1448, 1257, 1230, 1176,1120, 1084, 906, 777, 756, 732. **HRMS** (ESI+): Calc'd for $C_{21}H_{20}IO$, 415.05588; found, 415.05481.

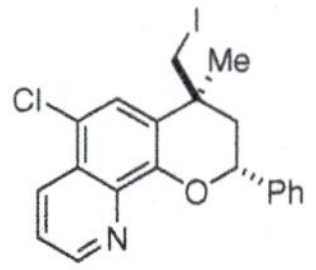

(±)-6-chloro-4-(iodomethyl)-4-methyl-2-phenyl-3,4-dihydro-2H-pyrano[3,2-h] quinoline (1.189c)—The title compound was synthesized from **1.169c** (89.4 mg, 0.2 mmol) using general procedure 4 (0.1 M, 15 h), and was isolated a light yellow solid (MP 181–183 °C) (82.2 mg, 0.184 mmol, 92%, 90:10 dr (*trans:cis*)) after purification silica gel using 25% EtOAc in hexanes + 1% NEt_3 as the mobile phase. The major diastereomer was sufficiently separated to be characterized. Crystals suitable for X-ray analysis were obtained by slow evaporation of a minimal volume of toluene over 24 h, and were isolated as colourless plates. **^{1}H NMR** (400 MHz, $CDCl_3$) δ 8.98 (dd, *J* = 4.0, 1.5 Hz, 1H), 8.50 (dd, *J* = 8.5, 1.5 Hz, 1H), 7.59–7.50 (m, 5H), 7.45–7.40 (m, 2H), 7.38–7.32 (m, 1H), 5.20 (dd, *J* = 12.0, 2.0 Hz, 1H), 3.67 (s, 2H), 2.52 (dd, *J* = 14.5, 2.5 Hz, 1H), 2.05–1.95 (m, 1H), 1.50 (s, 3H). **^{13}C**

NMR (100 MHz, $CDCl_3$) δ 150.3, 149.1, 140.6, 140.3, 132.7, 128.7, 128.2, 126.2, 126.2, 124.5, 123.1, 122.3, 122.1, 75.0, 42.6, 35.1, 27.6, 20.7. **IR** (Major) (cm^{-1}, neat) 3064, 3032, 2966, 2922, 2895, 1593, 1496, 1448, 1369, 1334, 1219, 1163, 1128, 1099, 1003, 912, 750, 732. **HRMS** (Major) (ESI+): Calc'd for $C_{20}H_{18}ClINO$, 450. 01216; found, 450.01078.

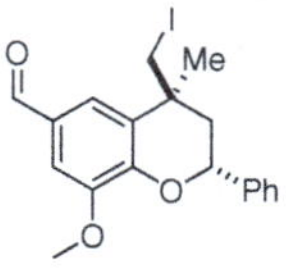

(±)-4-(iodomethyl)-8-methoxy-4-methyl-2-phenylchroman-6-carbaldehyde (1.189d). The title compound was synthesized from **1.169d** (84.0 mg, 0.2 mmol) using **general procedure 4** (0.1 M, 12 h), and was isolated a white foam (72.0 mg, 0.172 mmol, 86%, 92:8 dr (*trans:cis*) after purification silica gel using hexanes:Et_2O (2:1 v:v) as the mobile phase. The major diastereomer was sufficiently separated to be characterized. **^{1}H NMR** (Major) (400 MHz, $CDCl_3$) δ 9.85 (s, 1H), 7.52–7.46 (m, 3H), 7.45–7.39 (m, 2H), 7.39–7.34 (m, 1H), 7.33 (d, J = 1.7 Hz, 1H), 5.09 (dd, J = 12.3, 2.2 Hz, 1H), 3.92 (s, J = 7.0 Hz, 3H), 3.65–3.55 (m, 2H), 2.46 (dd, J = 14.5, 2.4 Hz, 1H), 1.94–1.85 (m, 1H), 1.49 (s, 3H). **^{13}C NMR** (Major) (100 MHz, $CDCl_3$) δ 190.7, 149.6, 149.6, 140.0, 129.3, 128.7, 128.3, 126.0, 125.8, 123.6, 108.6, 74.9, 56.2, 42.2, 34.7, 27.4, 20.7. **IR** (Major) (cm^{-1}, neat) 2966, 2935, 1661, 1577, 1477, 1464, 1452, 1427, 1394, 1284, 1253, 1234, 1145, 1074, 698. **HRMS** (Major) (ESI+): Calc'd for $C_{19}H_{20}IO_3$, 423.0462; found, 423.0463.

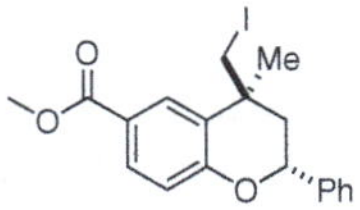

(±)-methyl 4-(iodomethyl)-4-methyl-2-phenylchroman-6-carboxylate (1.189e) —The title compound was synthesized from **3e** (84.54 mg, 0.2 mmol) using **general procedure 4** (0.1 M, 13 h), and was isolated a light yellow oil (84.0 mg, 0.196 mmol, 98%, 91:9 *dr* (*trans:cis*)) after purification silica gel using hexanes: EtOAc (15:2 v:v) as the mobile phase. The major diastereomer was sufficiently separated to be characterized. **^{1}H NMR** (400 MHz, $CDCl_3$) δ 8.05 (d, J = 2.0 Hz, 1H), 7.89–7.84 (m, 1H), 7.50–7.40 (m, 4H), 7.40–7.33 (m, 1H), 6.98 (dd, J = 8.5, 4.0 Hz, 1H), 5.03 (dd, J = 12.5, 2.0 Hz, 1H), 3.90 (s, 2H), 3.62 (d, J = 10.5 Hz, 1H), 3.57 (dd, J = 10.5, 1.5 Hz, 1H), 2.43 (dd, J = 14.5, 2.0 Hz, 1H), 1.87 (ddd, J = 14.0, 12.5, 1.5 Hz, 1H), 1.48 (s, 3H). **^{13}C NMR** (100 MHz, $CDCl_3$) δ 166.6, 158.2, 140.3, 130.2, 129.1, 128.6, 128.2, 125.8, 125.2, 122.7, 117.6, 74.4, 51.9, 42.3, 34.6, 27.3, 21.1. **IR** (Major) (cm^{-1}, neat) 2966, 2944, 2914, 1712, 1610, 1575, 1456, 1435, 1290, 1255, 1240, 1226, 1132, 1112, 1064, 1008, 906, 769, 757, 732. **HRMS** (Major) (ESI+): Calc'd for $C_{19}H_{20}IO$, 423.04571; found, 423.04594.

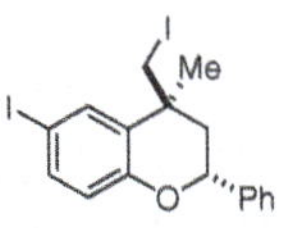

(±)-6-iodo-4-(iodomethyl)-4-methyl-2-phenylchroman (1.189f)—The title compound was synthesized from **1.169f** (97.9 mg, 0.2 mmol) using **general procedure 4** (0.1 M, 12 h), and was isolated a white semi-solid (67.9 mg, 0.140 mmol, 70%, 91:9 dr (*trans:cis*)) after purification silica gel using hexanes:DCM (5:1 v:v) as the mobile phase. The major diastereomer was sufficiently separated to be characterized (white solid). **^{1}H NMR** (400 MHz, $CDCl_3$) δ 7.57 (d, J = 2.0 Hz, 1H), 7.49–7.38 (m, 5H), 7.38–7.32 (m, 1H), 6.72 (d, J = 8.5 Hz, 1H), 4.96 (dd, J = 12.5, 2.0 Hz, 1H), 3.61 (d, J = 10.5 Hz, 1H), 3.52 (dd, J = 10.5, 1.5 Hz, 1H), 2.37 (dd, J = 14.5 2.0 Hz, 1H), 1.83 (ddd, J = 14.0, 12.5, 1.5 Hz, 1H), 1.42 (s, 3H). **^{13}C NMR** (100 MHz, $CDCl_3$) δ 154.1, 140.5, 137.4, 135.5, 128.6, 128.2, 128.0, 125.8, 119.9, 82.7, 74.0, 42.2, 34.6, 27.4, 21.2. **MP** = (152–154°C). **IR** (Major) (cm^{-1}, neat) 3030, 2964, 2918, 2850, 1334, 1273, 1219, 1172. **HRMS** (Major) (ESI+): Calc'd for $C_{17}H_{17}I_2O$, 490.93688; found, 490.93666.

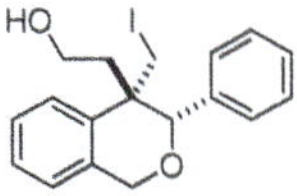

(±)-2-(4-(iodomethyl)-3-phenylisochroman-4-yl)ethanol (1.191)—DDQ (95 mg, 0.415 mmol, 1.18 equiv) was added at once to a 4.5 mL solution (9:1 DCM:pH 7 buffer) of **1.187l** (181 mg, 0.352 mmol, 1 equiv). This was allowed to stir for 90 min at which time the reaction was poured into 20 mL of 10:1 H_2O:saturated aqueous $NaHCO_3$. The aqueous layer was extracted with EtOAc (3 × 15 mL), and the combined organic layers were washed with brine, dried over $MgSO_4$, filtered and concentrated *in vacuo*. The crude carbinol was purified via silica gel chromatography using hexanes:EtOAc (3:1 v:v) as the mobile phase to afford the title compound was a white foam (131 mg, 0.334 mmol, 95%). **^{1}H NMR** (400 MHz, $CDCl_3$) δ 7.50–7.42 (m, 3H), 7.36 (dd, J = 9.0, 3.6 Hz, 3H), 7.32–7.26 (m, 2H), 7.16–6.99 (m, 1H), 4.98 (s, 1H), 4.94 (s, 2H), 3.84–3.70 (m, 1H), 3.68 (d, J = 10.0 Hz, 1H), 3.64–3.52 (m, 1H), 3.28 (d, J = 10.0 Hz, 1H), 2.25 (ddd, J = 14.5, 8.0, 5.0 Hz, 1H), 1.95 (ddd, J = 15.0, 8.0, 7.0 Hz, 1H), 1.78 (s, 1H). **^{13}C NMR** (100 MHz, $CDCl_3$) δ 136.9, 136. 1, 134.5, 128.4, 128.1, 128.1, 127.8, 127.0, 126.3, 124.3, 80.9, 68.0, 59.3, 41.33 39.4, 18.5. **IR** (cm^{-1}, neat) 3375, 2941, 2891, 2870, 1492, 1450, 1213, 1105, 1087, 1031. **HRMS** (ESI+): Calc'd for $C_{18}H_{20}IO_3$, 395.05080; found, 395.04914.

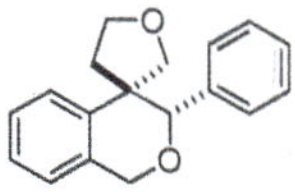

(±)-3′-phenyl-4,5-dihydro-2*H*-spiro[furan-3,4′-isochroman] (1.192)—To a stirred THF solution (6 mL) of **1.191** (33.6 mg, 0.0852 mmol 1 equiv) was added NaH (7.4 mg, 0.222 mmol, 2.6 equiv) and TBAI (32 mg, 0.08527, 1 equiv) at once. The resulting suspension was allowed to stir overnight at room temperature at which time the reaction was quenched with saturated NH_4Cl (5 mL). The aqueous layer was extracted with Et_2O (3 × 5 mL) and the combined organic layers were washed with brine, dried over Na_2SO_4, filtered, and concentrated. The crude spirocycle was purified via silica gel chromatography using hexanes:EtOAc (3:1 v:v) as the mobile phase to afford the title compound as a white foam (19.7 mg, 0.0740 mmol, 87%). **^{1}H NMR** (400 MHz, $CDCl_3$) δ 7.41 (dd, *J* = 7.9, 1.2 Hz, 1H), 7.26 (s, 6H), 7.17 (td, *J* = 7.5, 1.5 Hz, 1H), 7.01–6.87 (m, 1H), 4.87 (d, *J* = 15.5 Hz, 1H), 4.76–4.67 (m, 2H), 4.04 (d, *J* = 9.5 Hz, 1H), 3.91 (td, *J* = 8.5, 6.5 Hz, 1H), 3.76 (d, *J* = 9.5 Hz, 1H), 3.51 (td, *J* = 8.5, 6.0 Hz, 1H), 2.43–2.14 (m, 2H). **^{13}C NMR** (100 MHz, $CDCl_3$) δ 140.9, 138.0, 133.8, 128.3, 128.1, 128.0, 127.31 126.2, 126.1, 123.8, 82.8, 76.1, 68.2, 66.9, 49.2, 39.7. IR (cm^{-1}, neat) 3028, 2972, 2955, 2931, 2848, 1556, 1492, 1452, 1076. **HRMS** (ESI +): Calc'd for $C_{18}H_{19}IO_2$, 267.13850; found, 267.13862.

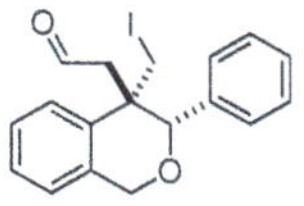

(±)-2-(4-(iodomethyl)-3-phenylisochroman-4-yl)acetaldehyde (1.192)—To a stirred DCM solution of (±)-2-(4-(iodomethyl)-3-phenylisochroman-4-yl)ethanol (33.6 mg, 0.0852 mmol, 1 equiv) was added DMP (72.3 mg, 0.17 mmol, 2 equiv) and solid $NaHCO_3$ (72 mg, 0.852 mmol, 10 equiv). The resulting suspension was allowed to stir for 1 h, at which time the reaction was filtered over a pipette of Celite to remove all solids. The filtrate was washed with water (3 mL) and brine (3 mL), dried over Na_2SO_4, filtered, and concentrated *in vacuo*. The crude aldehdye was purified via SI gel chromatography using hexanes:EtOAc (5:1 v:v) as the mobile phase to afford the title compound as a white foam (30.7 mg, 0.0783 mmol, 92%). **^{1}H NMR** (300 MHz, $CDCl_3$) δ 9.66 (dd, *J* = 2.5, 1.5 Hz, 1H), 7.51–7.32 (m, 6H), 7.32–7.26 (m, 2H), 7.12–7.03 (m, 1H), 5.15 (s, 1H), 5.07 (s, 2H), 3.74 (d, *J* = 10.0 Hz, 1H), 3.44–3.26 (m, 2H), 2.47 (dd, *J* = 17.5, 2.5 Hz, 1H). **^{13}C NMR** (75 MHz, $CDCl_3$) δ 201.1, 136.5, 135.6, 134.0, 128.3, 128.3, 128.1, 127.8, 127.4, 126.1, 124.6, 80.3, 69.0, 49.2, 41.4, 17.7. **IR** (cm^{-1}, neat) 3063, 3030, 2958, 2931, 2841, 1716, 1494, 1450, 1082. **HRMS** (ESI+): Calc'd for $C_{18}H_{18}IO_2$, 393.03515; found, 393.03436.

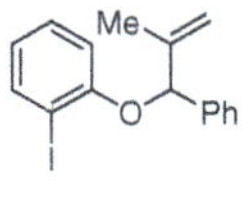

(±)-1-iodo-2-((2-methyl-1-phenylallyl)oxy)benzene (1.194)—The title compound was synthesized from 2-methyl-1-phenylprop-2-en-1-ol (6.70 mmol) using general procedure 2 and was isolated as a clear and colourless oil (0.908 g, 2.61 mmol, 38%) after purification on silica gel using hexanes:DCM (5:1 v:v) as the mobile phase. **^{1}H NMR** (400 MHz, $CDCl_3$) δ 7.66 (dd, *J* = 8.0, 1.5 Hz, 1H), 7.40 (d, *J* = 7.5 Hz, 2H), 7.27–7.21 (m, 2H), 7.19–7.14 (m, 1H), 7.05 (ddd, *J* = 9.0, 7.5, 1.5 Hz, 1H), 6.65 (dd, *J* = 8.5, 1.0 Hz, 1H), 6.56–6.51 (m, 1H), 5.56 (s, 1H), 5.19 (s, 1H), 4.92–4.90 (m, 1H), 1.61 (s, 3H). **^{13}C NMR** (100 MHz, $CDCl_3$) δ 156.0, 144.1, 139.4, 139.0, 129.0, 128.3, 127.7, 126.5, 122.5, 113.7, 113.7, 87.1, 83.9, 17.7. **IR** (cm^{-1}, thin film) 3063, 3028, 2974, 2947, 2914, 1651, 1579, 1491, 1467, 1453, 1271, 1242, 1120, 1049, 1019, 906, 746, 698. **HRMS** (ESI+): Calc'd for $C_{16}H_{19}INO$, 368.05113; found, 368.05187.

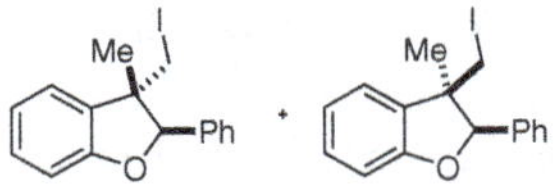

trans-(*trans*-**1.195**) and ***cis*-(±)-1-iodo-2-((2-methyl-1-phenylallyl)oxy)benzene** (*cis*-**1.195**). The title compound was synthesized from **1.194** using general procedure 5 (0.1 M, 10 h) and was isolated as a clear and colourless oil (66.6 mg, 0.19 mmol, 95%, 2:1 dr (*trans*:*cis*)) after purification on silica gel using hexanes: DCM (10:3 v:v) as the mobile phase.

Partial ^{1}H NMR (400 MHz, $CDCl_3$):
Major: δ 5.63 (s, 1H), 3.50 (d, *J* = 10.0 Hz, 1H), 3.42 (d, *J* = 10.0 Hz, 1H), 0.98 (s, 3H).
Minor: δ 5.46 (s, 1H), 3.05 (dd, *J* = 10.0, 0.7 Hz, 1H), 2.66 (d, *J* = 10.0 Hz, 1H), 1.59 (d, *J* = 0.5 Hz, 3H).
^{13}C NMR (100 MHz, $CDCl_3$) δ 159.0, 158.6, 137.9, 135.4, 133.0, 131.6, 129.4, 129.0, 128.5, 128.5, 128.2, 128.1, 126.3, 126.1, 126.1, 125.1, 123.3, 121.0, 120.6, 110.3, 109.9, 109.9, 49.4, 24.8, 23.1, 20.6, 17.4.
HRMS (ESI+): Calc'd for $C_{16}H_{16}IO$, 351.02458; found, 351.02458.

Starting material and product experimental: Diastereoselective synthesis of enantioenriched dihydroisoquinoloinones

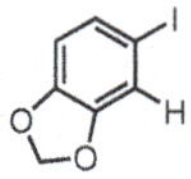

5-iodobenzo[*d*][1, 3]dioxole (1.239)—In a 1000 mL Erlenmeyer flask benzo[*d*][1,3]dioxol-5-amine **1.238** (5.0 g, 36.4 mmol, 1 equiv) was dissolved in a mixture of ice (100 g) and concentrated HCl (32 mL) at 0 °C. To this solution of protonated amine was added $NaNO_2$ (5.30 g, 76.8 mmol, 2.1 equiv) portion wise. The resulting foamy solution was stirred for 10 min at this temperature before slowly adding solid urea (1.66 g, 27.6 mmol, 0.75 equiv) (*Note*: Intense effervescence occurs during this stage and care must be taken to not over flow the reaction mixture). The resulting dark brown and foamy mixture was stirred for 10 min at 0 ° C at which time an aqueous solution (20 mL) of KI (17.9 g, 107.8 mmol, 2.96 equiv) was added drop wise over 10 min. The mixture was allowed to slowly warm to room temperature and was rapidly stirred for 12 h. The reaction was quenched with 10 mL of a saturated aqueous solution of Na_2SO_3 was added extracted with DCM (3 × 50 mL). The combined organic layers were washed with brine (30 mL), dried over anhydrous $MgSO_4$, filtered, and concentrated *in vacuo* to afford the title compound as a deep brown oil (5.98 g, 24.1 mmol, 66%) which was taken forward without further purification. Spectral data was in good accordance to that of the reported literature.[172] **^{1}H NMR** (400 MHz, $CDCl_3$) δ 7.18–7.02 (m, 2H), 6.59 (d, *J* = 8.0 Hz, 1H), 5.95 (s, 2H). **^{13}C NMR** (100 MHz, $CDCl_3$) δ 148.7, 147.8, 130.6, 117.7, 110.5, 101.4, 82.2.

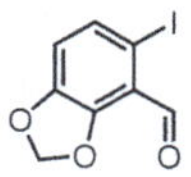

5-iodobenzo[*d*][1, 3]dioxole-4-carbaldehyde (1.240)—Prepared in accordance to a published procedure.[173] (*Note*: The THF used to prepare the LDA solution, was distillation over excess sodium metal and benzophenone, and collected immediately prior to use. It was always transferred via dry, oxygen-free syringes, and LDA was prepared by standard methods with care taken to exclude oxygen and moisture). To a solution of stirred solution of LDA (0.357 M, 80 mL, 1.25 equiv) at −78 °C was added neat **1.239** (5.7 g, 22.9 mmol, 1 equiv) drop wise such that the internal reaction temperature did not exceed −70 °C. This was stirred at −78 °C for 2 h, at which time anhydrous DMF (4.23 mL, 54.9 mmol, 2.4 equiv) was added drop wise such that the internal temperature did not exceed −70 °C. The resulting solution was allowed to warm to room temperature over night at which time saturated aqueous NH_4Cl (10 mL) and water (30 mL) were added sequentially at 0 °C. The aqueous layer was extracted with EtOAc (3 × 30 mL), and the combined organic layers were washed with brine (30 mL), dried over anhydrous $MgSO_4$, filtered, and concentrated to afford a deep yellow powder which was recrystallized from hexanes:EtOAc to afford the aryl aldehydes (single regioisomer) as fluffy, deep yellow crystals (5.34 g, 19.4 mmol, 85%, MP = 161–163 °C). Spectral data was in good accordance to that of the reported literature.[174] **^{1}H NMR** (400 MHz, $CDCl_3$) δ 10.05 (s, 1H), 7.41 (d, *J* = 8.1 Hz, 1H), 6.73 (d, *J* = 8.1 Hz, 1H), 6.16 (s, 2H). **^{13}C NMR** (100 MHz, $CDCl_3$) δ 193.9, 149.9, 149.2, 133.4, 118.2, 114.3, 103.1, 86.2.

[172]Reference [164].
[173]Reference [165].
[174]Reference [166].

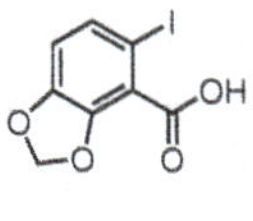

5-iodobenzo[*d*][1, 3]dioxole-4-carboxylic acid (1.241)–1.240 was suspended in a solution of ^{t}BuOH (233 mL) and 2-methyl-2-butene (40.1 mL, 380 mmol, 15 equiv) and the resulting mixture was cooled to 0 °C. To this was added an aqueous solution (153 mL) of $NaClO_2$ (22.9 g, 253 mmol, 10 equiv) and NaH_2PO_4 (22.9 g, 190 mmol, 8.4 equiv) (*Note*: to dissolve $NaOCl_2$ and NaH_2PO_4 in water, sonicate in a well-operative fume hood, and handle with care). The mixture was allowed to warm to room temperature, and after stirring at this temperature for 2 h, TLC analysis showed full and clean conversion of starting material. The reaction was cooled to 0 °C and was quenched with 150 mL of saturated Na_2SO_3 and was diluted with water (100 mL). The aqueous layer was extracted with EtOAc (3 × 500 mL), washed with brine (200 mL), dried over anhydrous $MgSO_4$, filtered, and concentrated *in vacuo* to afford a yellow solid with was dissolved in 50 mL of 2 M NaOH. The dissolved carboxylic acid was washed with EtOAc (30 mL), and re-acidified with 5% HCl. The filtered precipitate was washed with water, and dried to a constant weight *in vacuo*. The title carboxylic acid was afforded as a light yellow solid (5.3 g, 18.2 mmol, 72%, MP = 196–198 °C). Spectral data was in good accordance to that of the reported literature.[175] **^{1}H NMR** (400 MHz, DMSO-d_6) δ 13.63 (s, 1H), 7.38 (d, $J = 8.1$ Hz, 1H), 6.85 (d, $J = 8.1$ Hz, 1H), 6.14 (s, 2H). **^{13}C NMR** (100 MHz, DMSO-d_6) δ 166.8, 149.0, 146.7, 133.2, 122.1, 112.4, 103.2, 82.5.

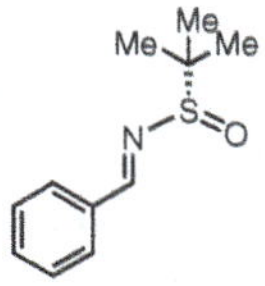

(*S*,*E*)-*N*-benzylidene-2-methylpropane-2-sulfinamide (1.232a)—Was synthesized according to **general procedure 5a** using benzaldehyde (3.5 g, 33 mmol, 0.47 M), and the (*S*) sulfinamide auxiliary, and was refluxed for 4 h. The crude sulfinamide was obtained in analytically pure form by silica gel flash column chromatography using hexanes:EtOAc (10:1 v:v) as the mobile phase. The title compound was obtained as a clear and light yellow oil (5.68 g, 27.1 mmol, 82%). All spectra data were in accordance to the reported literature.[176] **^{1}H NMR** (400 MHz, $CDCl_3$) δ 8.58 (s, 1H), 7.86–7.82 (m, 2H), 7.53–738 (m, 3H), 1.26 (s, 9H). **^{13}C NMR** (100 MHz, $CDCl_3$) δ 162.66, 134.03, 132.37, 129.31, 128.89, 57.67, 22.57. $[\alpha]_D^{20}$ + 122.38 (c = 1.0, $CHCl_3$).

[175]Reference [167].

[176]Reference [168].

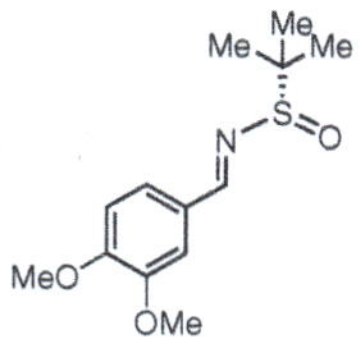

(*S*,*E*)-*N*-(3-fluorobenzylidene)-2-methylpropane-2-sulfinamide (1.232b)—Was synthesized according to **general procedure 5b** using 3-fluorobenzaldehyde (1.63 g, 12 mmol, 0.2 M) and the (*S*) sulfinamide auxiliary, and was refluxed for 9 h. The crude sulfinamide was obtained in analytically pure form by silica gel flash column chromatography using hexanes:EtOAc (10:2 v:v) as the mobile phase. The title compound was obtained as a clear and yellow oil (2.66 g, 11.1 mmol, 93%). **^{1}H NMR** (400 MHz, $CDCl_3$) δ 8.55 (d, *J* = 1.5 Hz, 1H), 7.62–7.54 (m, 2H), 7.44 (td, *J* = 8.0, 5.5 Hz, 1H), 7.20 (tdd, *J* = 8.0, 2.5, 1.5 Hz, 1H), 1.26 (s, 9H). **^{13}C NMR** (100 MHz, $CDCl_3$) δ 163.02 (d, *J* = 247.5 Hz), 161.60 (d, *J* = 3.0 Hz), 136.17 (d, *J* = 7.5 Hz), 130.61 (d, *J* = 8.0 Hz), 125.75 (d, *J* = 3.0 Hz), 119.31, 115.03 (d, *J* = 22.5 Hz), 57.99, 22.63. **^{19}F{^{1}H} NMR** (377 MHz, $CDCl_3$) δ-111.85. **IR** (neat, cm^{-1}) 2979, 2961, 2927, 2900, 2867, 1605, 1579, 1488, 1475, 1450, 1365, 1284, 1268, 1260, 1243, 1178, 1164, 1139, 1086. **HRMS** (ESI+) Calc'd for $C_{11}H_{15}FNOS$ 228.0856, found 228.0853. $[\alpha]_D^{20}$ + 111.2 (c = 1.0, $CHCl_3$).

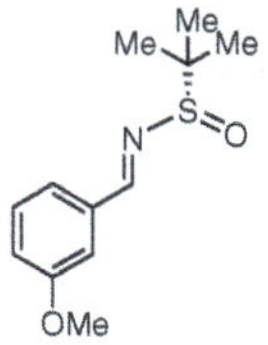

(*S*,*E*)-*N*-(3-methoxybenzylidene)-2-methylpropane-2-sulfinamide (1.232c)—Was synthesized according to **general procedure 5a** using 3-methoxybenzaldehyde (1.48 g, 12 mmol, 0.2 M) and the (*S*) sulfinamide auxiliary, and was refluxed for 5 h. The crude sulfinamide was obtained in analytically pure form by silica gel flash column chromatography using hexanes:EtOAc (10:2 v:v) as the mobile phase. The title compound was obtained as a translucent and yellow oil (2.46 g, 10.8 mmol, 90%). **^{1}H NMR** (400 MHz, $CDCl_3$) δ 8.56 (s, 1H), 7.46–7.35 (m, 3H), 7.12–7.04 (m, 1H), 3.87 (s, 3H), 1.28 (s, 9H). **^{13}C NMR** (100 MHz, $CDCl_3$) δ 162.48, 159.83, 135.23, 129.82, 122.34, 118.60, 113.02, 77.32, 77.00, 76.68, 57.62, 55.22, 22.46. **IR** (neat, cm^{-1}) 2977, 2959, 2927, 1607, 1599, 1576, 1455, 1430, 1363, 1323,1287, 1268, 1252, 1193, 1170, 1155, 1084, 1044. **HRMS** (ESI+) Calc'd for $C_{12}H_{18}NO_2S$ 240.1056, found 240.1053. $[\alpha]_D^{20}$ +76.7 (c = 1.0, $CHCl_3$).

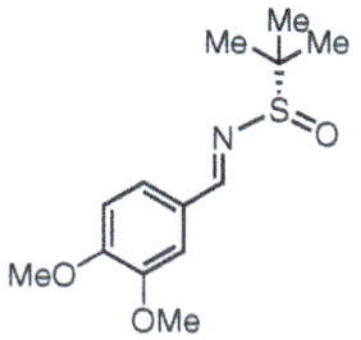

(*S*,*E*)-*N*-(3,4-dimethoxybenzylidene)-2-methylpropane-2-sulfinamide (1.232d)—Was synthesized according to **general procedure 5b** using 3,4-dimethoxybenzaldehyde (2.19 g, 12 mmol, 0.2 M) and the (*S*) sulfinamide auxiliary, and was refluxed for 9 h. The crude sulfinamide was obtained in analytically pure form by silica gel flash column chromatography using hexanes:EtOAc (10:4 v:v) as the mobile phase. The title compound was obtained as a white solid (2.38 g, 8.4 mmol, 70%, MP = 75–77 °C (Lit = 77–78 °C[177]). All spectra data were in accordance to the reported literature.[178] **^{1}H NMR** (400 MHz, $CDCl_3$) δ 8.49 (s, 1H), 7.44 (d, *J* = 2.0 Hz, 1H), 7.37 (dd, *J* = 8.5, 2.0 Hz, 1H), 6.93 (d, *J* = 8.5 Hz, 1H), 3.94 (d, *J* = 2.0 Hz, 6H), 1.26 (s, 9H). **^{13}C NMR** (100 MHz, $CDCl_3$) δ 161.89, 152.81, 149.38, 127.38, 124.97, 110.64, 109.73, 57.52, 56.00, 55.87, 22.53. $[\alpha]_D^{20}$ +41.7 (c = 1.0, $CHCl_3$).

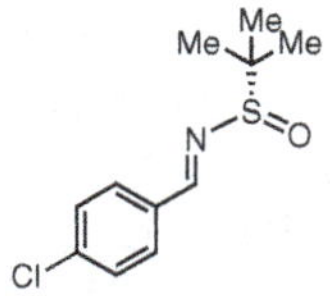

(*S*,*E*)-*N*-(4-chlorobenzylidene)-2-methylpropane-2-sulfinamide (1.232e)[179]—Was synthesized according to **general procedure 5b** using 4-chlorobenzaldehyde (1.686 g, 12 mmol, 0.2 M) and the (*S*) sulfinamide auxiliary, and was refluxed for 7 h. The crude sulfinamide was obtained in analytically pure form by silica gel flash column chromatography using hexanes:EtOAc (10:4 v:v) as the mobile phase. The title compound was obtained as a white solid (2.8 g, 11.5 mmol, 95%, MP = 37–38 °C (Lit = 41–42 °C[180]). All spectra data were in accordance to the reported literature.[181] **^{1}H NMR** (400 MHz, $CDCl_3$) δ 8.55 (s, 1H), 7.81–7.77 (m, 2H), 7.48–7.42 (m, 2H), 1.26 (s, 9H). **^{13}C NMR** (100 MHz, $CDCl_3$) δ 161.3, 138.5, 132.4, 130.4, 129.2, 57.8, 22.5. $[\alpha]_D^{20}$ +77.1 (c = 1.0, $CHCl_3$).

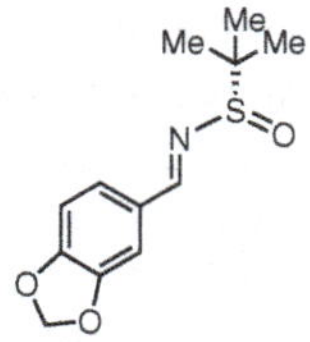

(*S*,*E*)-*N*-(benzo[*d*][1, 3]dioxol-5-ylmethylene)-2-methylpropane-2-sulfinamide (1.228)—Was synthesized according to **general procedure 5b** using piperonal

[177]Reference [169].
[178]Reference [169].
[179]Substrate prepared by Hyung Yoon.
[180]Reference [170].
[181]Reference [170].

(7.34 g, 60.6 mmol, 0.5 M) and the (*S*) sulfinamide auxiliary, and was refluxed for 3 h. At this time the reaction mixture was filtered through a short plug of silica and was eluted with 50 mL of DCM. The clear and colourless solution was concentrated *in vacuo* to afford the title compound as a white solid (14.1 g, 55.6 mmol, 92%, MP = 58–60 °C) which was used without further purification. **^{1}H NMR** (400 MHz, $CDCl_3$) δ 8.45 (s, 1H), 7.40 (d, *J* = 1.6 Hz, 1H), 7.28 (dd, *J* = 8.0, 1.5 Hz, 1H), 6.88 (d, *J* = 8.0 Hz, 1H), 6.04 (q, *J* = 1.5 Hz, 2H), 1.24 (s, 9H). **^{13}C NMR** (100 MHz, $CDCl_3$) δ 161.6, 151.4, 148.5, 129.1, 126.9, 108.4, 107.2, 101.8, 57.6, 22.6. **IR** (neat, cm^{-1}) 2978, 2960, 2924, 2901, 2868, 1581, 1558, 1504, 1491, 1448, 1313, 1257, 1082, 1035. **HRMS** (DART) Calc'd for $C_{12}H_{16}NO_3S$ 254.08509, found 254.08520. $[\alpha]_D^{20}$ +4.4 (c = 1.0, $CHCl_3$).

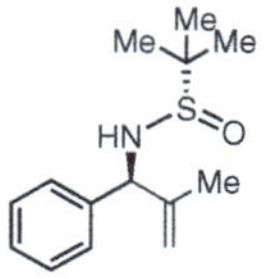

(*S*)-2-methyl-*N*-((*S*)-2-methyl-1-phenylallyl)propane-2-sulfinamide (1.233a)—Was synthesized according to **general procedure 6** using **1.232a** (4.0 g, 19.11 mmol, 0.13 M). ^{1}H NMR analysis of the crude product determined the diastereoselectivity of the addition to be >99:1. The crude allylic sulfinamide was obtained in analytically pure form by silica gel flash column chromatography using hexanes:EtOAc (10:7 v:v) as the mobile phase. The title compound was obtained as a clear and light yellow oil (3.91 g, 15.6 mmol, 82%). **^{1}H NMR** (400 MHz, $CDCl_3$) δ 7.38–7.26 (m, 5H), 5.29 (dt, *J* = 2.0, 1.0 Hz, 1H), 5.06–4.99 (m, 1H), 4.98 (d, *J* = 2.0 Hz, 1H), 3.43 (d, *J* = 2.0 Hz, 1H), 1.61 (dd, *J* = 2.0, 1.0 Hz, 3H), 1.27 (s, 9H). **^{13}C NMR** (100 MHz, $CDCl_3$) δ 143.7, 140.6, 128.5, 127.8, 127.0, 114.4, 63.6, 55.5, 22.6, 18.2. **IR** (neat, cm^{-1}) 2977, 2951, 2923, 2905, 2868, 1491, 1474, 1452, 1371, 1363, 1062, 900. **HRMS** (DART) Calc'd for $C_{14}H_{22}NOS$ 252.14221 found 252.14208. $[\alpha]_D^{20}$ +108.1 (c = 1.0, $CHCl_3$).

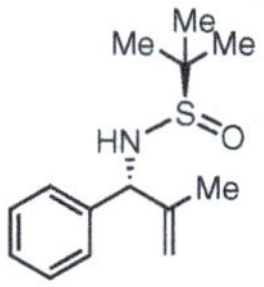

(*R*)-2-methyl-*N*-((*R*)-2-methyl-1-phenylallyl)propane-2-sulfinamide (*ent*-1.233a)—Was synthesized according to **general procedure 6** using *ent*-**1.232a** (2.0 g, 9.5 mmol, 0.13 M). ^{1}H NMR analysis of the crude product determined the diastereoselectivity of the addition to be >99:1. The crude allylic sulfinamide was obtained in analytically pure form by silica gel flash column chromatography using hexanes: EtOAc (1:1 v:v) as the mobile phase. The title compound was obtained as a clear and yellow oil (1.98 g, 7.88 mmol, 83%). All spectroscopic data was consistent with that of the opposite enantiomer **1.233a**.

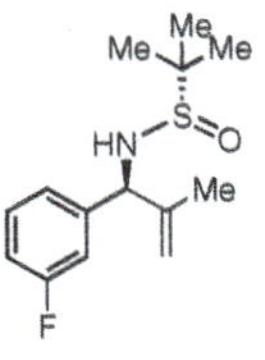

(*S*)-*N*-((*S*)-1-(3-fluorophenyl)-2-methylallyl)-2-methylpropane-2-sulfinamide (1.233b)—Was synthesized according to **general procedure 6** using **1.232b** (1.79 g, 6.06 mmol, 0.2 M). ^{1}H NMR analysis of the crude product determined the diastereoselectivity of the addition to be >99:1. The crude allylic amine was obtained in analytically pure form by silica gel flash column chromatography using hexanes:EtOAc (5:1 v:v) as the mobile phase. The title compound was obtained as a colourless solid (1.79 g, 5.07 mmol, 84%, MP = 50–51 °C). **^{1}H NMR** (400 MHz, $CDCl_3$) δ 7.34 (td, *J* = 8.0, 6.0 Hz, 1H), 7.20–7.14 (m, 1H), 7.09 (dt, *J* = 10.0, 2.0 Hz, 1H), 7.01 (tdd, *J* = 8.5, 2.5, 1.0 Hz, 1H), 5.32 (dt, *J* = 2.0, 1.0 Hz, 1H), 5.08 (p, *J* = 1.5 Hz, 1H), 5.00 (d, *J* = 3.0 Hz, 1H), 3.44 (d, *J* = 3.0 Hz, 1H), 1.64–1.63 (m, 3H), 1.31 (s, 9H). **^{13}C NMR** (100 MHz, $CDCl_3$) δ 162.85 (d, *J* = 246.5 Hz), 143.29, 143.23, 130.15 (d, *J* = 8.5 Hz), 122.86 (d, *J* = 3.0 Hz), 115.1, 114.8 (d, *J* = 21.0 Hz), 113.9 (d, *J* = 22.5 Hz), 63.4 (d, *J* = 2.0 Hz), 55.7, 22.7, 18.2. **^{19}F{^{1}H} NMR** (377 MHz, $CDCl_3$) δ-112.3. **IR** (neat, cm^{-1}) 3198, 2978, 2957, 2922, 1614, 1590, 1486, 1476, 1444, 1371, 1245, 1055. **HRMS** (DART) Calc'd for $C_{14}H_{21}FNOS$ 270.13279, found 270.13229. $[\alpha]_D^{20}$ + 106.3 (c = 1.0, $CHCl_3$).

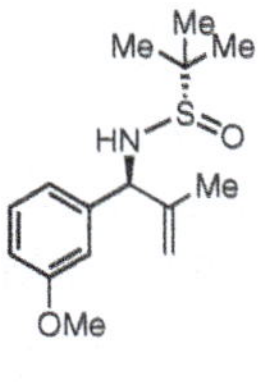

(*S*)-*N*-((*S*)-1-(3-methoxyphenyl)-2-methylallyl)-2-methylpropane-2-sulfinamide (1.233c)—Was synthesized according to **general procedure 6** using **1.232c** (1.67 g, 6.98 mmol, 0.175 M). ^{1}H NMR analysis of the crude product determined the diastereoselectivity of the addition to be >99:1. The crude allylic amine was obtained in analytically pure form by silica gel flash column chromatography using hexanes:EtOAc (2:1 v:v) as the mobile phase. The title compound was obtained as a clear and light yellow oil (1.53 g, 5.44 mmol, 78%). **^{1}H NMR** (400 MHz, $CDCl_3$) δ 7.30–7.21 (m, 1H), 6.98–6.91 (m, 1H), 6.90 (t, *J* = 2.0 Hz, 1H), 6.82 (ddd, *J* = 8.0, 2.5, 1.0 Hz, 1H), 5.29–5.27 (m, 1H), 5.03 (p, *J* = 1.5 Hz, 1H), 4.95 (d, *J* = 2.5 Hz, 1H), 3.80 (s, 3H), 3.43–3.41 (m, 1H), 1.62–1.61 (m, 3H), 1.27 (s, 9H). **^{13}C NMR** (100 MHz, $CDCl_3$) δ 159.7, 143.7, 142.3, 129.6, 119.4, 114.5, 113.0, 112.9, 63.6, 55.6, 55.2, 22.7, 18.3. **IR** (neat, cm^{-1}) 3201, 2977, 2921, 2868, 1600, 1585, 1488, 1466, 1450, 1436, 1370, 1364, 1320, 1260, 1148, 1016. **HRMS**

(ESI+) Calc'd for $C_{15}H_{24}NO_2S$ 282.1522, found 282.1523. $[\alpha]_D^{20}$ +84.6 (c = 1.0, $CHCl_3$).

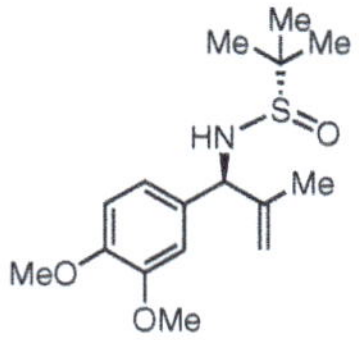

(*S*)-*N*-((*S*)-1-(3,4-dimethoxyphenyl)-2-methylallyl)-2-methylpropane-2-sulfinamide (1.233d)—Was synthesized according to **general procedure 6** using **1.232d** (2.16 g, 8.03 mmol, 0.16 M). ^{1}H NMR analysis of the crude product determined the diastereoselectivity of the addition to be >99:1. The crude allylic amine was obtained in analytically pure form by silica gel flash column chromatography using hexanes: EtOAc (1:3 v:v) as the mobile phase. The title compound was obtained as clear and light yellow oil which solidifies upon storage at −20 °C (2.07 g, 6.64 mmol, 83%). **^{1}H NMR** (400 MHz, $CDCl_3$) δ 6.95–6.84 (m, 2H), 6.83 (d, *J* = 8.0 Hz, 1H), 5.29–5.24 (m, 1H), 5.03 (p, *J* = 1.5 Hz, 1H), 4.90 (d, *J* = 3.0 Hz, 1H), 3.87 (s, 3H), 3.87 (s, 3H), 3.37 (d, *J* = 3.0 Hz, 1H), 1.65–1.60 (m, 3H), 1.27 (s, 9H). **^{13}C NMR** (100 MHz, $CDCl_3$) δ 149.0, 148.6, 144.1, 133.2, 119.3, 113.9, 111.0, 110.6, 63.4, 55.8 (2), 55.6, 22.7, 18.7. **IR** (neat, cm^{-1}) 3251, 2975, 2956, 2868, 2835, 1593, 1515, 1464, 1456, 1418, 1363, 1262, 1247, 1239, 1153, 1030, 1029. **HRMS** (DART) Calc'd for $C_{16}H_{26}NO_3S$ 312.16334, found 312.16413. $[\alpha]_D^{20}$ +65.3 (c = 1.0, $CHCl_3$).

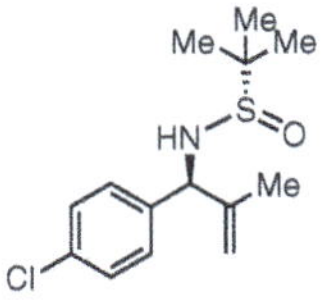

(*S*)-*N*-((*S*)-1-(4-chlorophenyl)-2-methylallyl)-2-methylpropane-2-sulfinamide (1.233e)[182]—Was synthesized according to **general procedure 6** using **1.232e** (1.94 g, 7.98 mmol, 0.16 M). ^{1}H NMR analysis of the crude product determined the diastereoselectivity of the addition to be >99:1. The crude allylic amine was obtained in analytically pure form by silica gel flash column chromatography using hexanes:EtOAc (1:1 v:v) as the mobile phase. The title compound was obtained as a clear and light yellow oil (2.12 g, 7.44 mmol, 93%). **^{1}H NMR** (400 MHz, $CDCl_3$) δ 7.34–7.27 (m, 4H), 5.30–5.24 (m, 1H), 5.04 (p, *J* = 1.5 Hz, 1H), 4.94 (d, *J* = 3.0 Hz, 1H), 3.38–3.36 (m, 1H), 1.62–1.57 (m, 3H), 1.27 (s, 9H). **^{13}C NMR** (100 MHz, $CDCl_3$) δ 143.4, 139.2, 133.7, 128.8, 128.5, 114.8, 63.3, 55.7, 22.6, 18.3. **IR** (neat, cm^{-1}) 3193. 2978, 2958, 1911, 1906, 2867,

[182]Substrate prepared by Hyung Yoon.

1492, 1455, 1405, 1364, 1065, 1014, 904, 832. **HRMS** (DART) Calc'd for $C_{14}H_{21}ClNOS$ 286.10324, found 286.10319. $[\alpha]_D^{20}$ +83.0 (c = 1.0, $CHCl_3$).

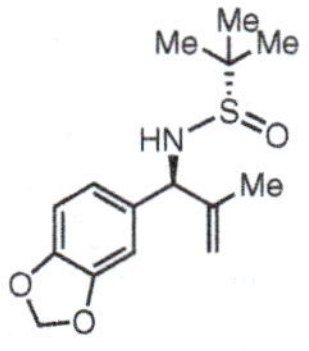

(*S*)-*N*-((*S*)-1-(3,4-dimethoxyphenyl)-2-methylallyl)-2-methylpropane-2-sulfinamide (1.242)—Was synthesized according to **general procedure 6** using **1.228** (5.7 g, 22.5 mmol, 0.15 M). ^{1}H NMR analysis of the crude product determined the diastereoselectivity of the addition to be >99:1. The crude allylic amine was obtained in analytically pure form by silica gel flash column chromatography using hexanes:EtOAc (10:9 v:v) as the mobile phase, to afford the diastereomerically pure title compound as a clear and light yellow oil (6.57 g, 22.3 mmol, 99%). **^{1}H NMR** (400 MHz, $CDCl_3$) δ 6.87–6.79 (m, 2H), 6.75 (d, *J* = 8.4 Hz, 1H), 5.94 (s, 2H), 5.34–5.18 (m, 1H), 5.07–4.97 (m, 1H), 4.86 (d, *J* = 2.7 Hz, 1H), 3.36 (d, *J* = 2.8 Hz, 1H), 1.60 (s, 1H), 1.26 (s, 5H). **^{13}C NMR** (100 MHz, $CDCl_3$) δ 147.8, 147.2, 144.0, 134.6, 120.7, 114.1, 108.2, 107.5, 101.1, 63.4, 55.7, 22.7, 18.6. **IR** (neat, cm^{-1}) 3277, 2976, 2960, 2922, 2902, 2872, 1504, 1489, 1440, 1363, 1247, 1236, 1097, 1062. **HRMS** (DART) Calc'd for $C_{15}H_{22}NO_3S$ 296.13204, found 296.13195. $[\alpha]_D^{20}$ +54.5 (c = 0.7, $CHCl_3$).

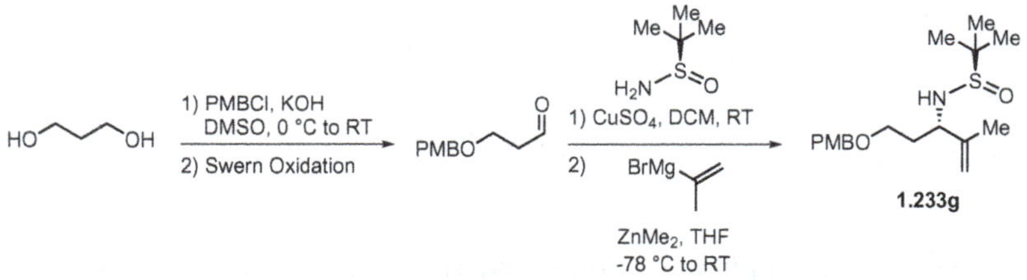

1.233g

(*R*)-*N*-((*S*)-5-((4-methoxybenzyl)oxy)-2-methylpent-1-en-3-yl)-2-methylpropane-2-sulfinamide (1.233 g)—3-((4-methoxybenzyl)oxy)propan-1-ol was synthesized according to the procedure by Tashiro and co-workers,[183] which was oxidized to 3-((4-methoxybenzyl)oxy)propanal according to the procedure by Skrydstrup and co-workers.[184] A flame dried flask was charged with (*R*)-2-methylpropane-2-sulfinamide (0.96 g, 7.9 mmol, 1.1 equiv) and anhydrous $CuSO_4$ (3.43 g, 21.6 mmol, 3 equiv), and the solid mixture was purged with argon for 10 min. The contents of the flask were taken up in dry DCM (25 mL), and with

[183]Reference [171].

[184]Reference [172].

stirring, a DCM solution (2.5 mL) of 3-((4-methoxybenzyl)oxy)propanal (1.5 g, 7.2 mmol, 1 equiv) was added drop wise over 5 min. After 4 h, TLC indicated incomplete conversion of the aldehyde and (*R*)-2-methylpropane-2-sulfinamide (87 mg, 0.72 mmol, 0.1 equiv) and $CuSO_4$ (3.43 g, 21.6 mmol, 3 equiv) were added as solids at once, in that order. After 4 additional hours, the reaction mixture was filtered through a plug of Celite® flushing with DCM. The collected organic fraction was washed with water (1x) and brine (1x), dried over Na_2SO_4, filtered, and concentrated *in vacuo*. ^{1}H NMR analysis of the crude reaction mixture indicated a single sulfinamide isomer as a 92:8 ratio with the aldehyde starting material. This mixture was carried forward without any additional purification.

The crude aldehyde/sulfinamide mixture (2.44 g, 6.62 mmol (sulfinamide only, assuming 100% mass conversion, and considering a 92:8 ratio), 1 equiv was added to a flame dried flask which was purged with argon for 10 min. The contents of the flask were taken up on dry THF (50 mL). This solution was cooled to −78 °C, and a pre-stirred −78 °C mixture of isopropenylmagnesium bromide (27.2 mL, 13.0 mmol, 1.8 equiv) and $ZnMe_2$ (6.48 mL, 13.0 mmol, 1.8 equiv) were added drop wise to the substrate via cannula using a positive argon pressure. After stirring at this temperate for 3 h, the reaction was warmed to room temperature, and TLC analysis indicated full conversion of the starting material mixture. The reaction mixture was cooled to 0 °C and was carefully quenched with a saturated aqueous solution of NH_4Cl. The insoluble salts were filtered, and the filtrate was transferred to a separatory funnel. The aqueous layer was extracted with EtOAc (3x) and the combined organic layers were washed with brine (1x), dried over Na_2SO_4, filtered and concentrated *in vacuo*. ^{1}H NMR analysis of the crude reaction mixture indicated a 96:4 mixture of allylic amine diasteromers. The crude material was purified using silica gel flash column chromatography using hexanes:EtOAc (1:2 v:v) as the mobile phase. The purified chiral allylic amine was isolated as a 96:4 mixture of diastereomers as a clear and light yellow oil. **^{1}H NMR (Major)** (400 MHz, $CDCl_3$) δ 7.34–7.26 (m, 2H), 6.92–6.87 (m, 2H), 5.06–5.04 (m, 1H), 4.94–4.92 (m, 1H), 4.50–4.39 (m, 3H), 4.04 (ddd, J = 8.5, 4.5, 2.0 Hz, 1H), 3.83 (s, 3H), 3.69–3.54 (m, 2H), 2.10–1.89 (m, 1H), 1.88–1.75 (m, 1H), 1.73–1.67 (m, 3H), 1.16 (s, 9H). **^{13}C NMR (Major)** (100 MHz, $CDCl_3$) δ 159.2, 144.5, 129.9, 129.6, 113.9, 113.7, 73.0, 68.8, 60.1, 55.2, 54.93 34.3, 22.6, 17.5. **IR** (neat, cm^{-1}) 3261, 2977, 2922, 2863, 1612, 1513, 1363, 1248, 1173, 1092, 1067, 1037. **HRMS** (ESI+) Calc'd for $C_{18}H_{30}NO_3S$ 340.1941, found 340.1942. $[\alpha]_D^{20}$ +63.8 (c = 1.0, $CHCl_3$).

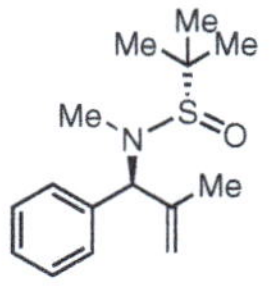

(*S*)-*N*,2-dimethyl-*N*-((*S*)-2-methyl-1-phenylallyl)propane-2-sulfinamide (1.234a)— Was synthesized according to **general procedure 7** using **1.233a** (2.12 g, 8.46 mmol)

and iodomethane (1.05 mL, 16.92 mmol). The crude alkylated amine was obtained in analytically pure form by silica gel flash column chromatography using hexanes: EtOAc (10:2 v:v) as the mobile phase. The title compound was obtained as clear and light yellow oil (2.11 g, 7.99 mmol, 94%). **^{1}H NMR** (400 MHz, $CDCl_3$) δ 7.35–7.21 (m, 5H), 5.11–5.08 (m, 1H), 5.07 (s, 1H), 4.92 (s, 1H), 2.52 (s, 3H), 1.69 (s, 3H), 1.18 (s, 9H). **^{13}C NMR** (100 MHz, $CDCl_3$) δ 143.1, 138.5, 128.6, 128.3, 127.5, 115.7, 72.9, 58.5, 29.5, 23.8, 20.9. **IR** (neat, cm^{-1}) 2974, 2956, 2923, 2867, 1492, 1475, 1451, 1361, 1075, 922. **HRMS** (DART) Calc'd for $C_{15}H_{24}NOS$ 266.15744, found 266.15786. $[\alpha]_D^{20}$ −27.9 (c = 1.0, $CHCl_3$).

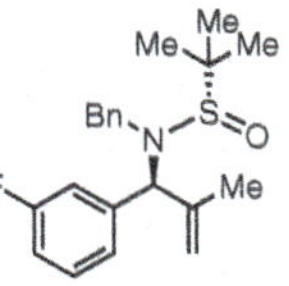

(*S*)-*N*-benzyl-*N*-((*S*)-1-(3-fluorophenyl)-2-methylallyl)-2-methylpropane-2-sulfinamide (1.234b)—Was synthesized according to **general procedure 7** using **1.233b** (1.6 g, 6.06 mmol) and benzyl bromide (1.44 mL, 12.12 mmol). The crude alkylated amine was obtained in analytically pure form by silica gel flash column chromatography using hexanes:EtOAc (5:1 v:v) as the mobile phase. The title compound was obtained as clear and light yellow oil (1.79 g, 5.07 mmol, 84%). **^{1}H NMR** (400 MHz, $CDCl_3$) δ 7.45–7.24 (m, 6H), 7.11–6.94 (m, 3H), 5.70 (s, 1H), 5.30 (s, 1H), 4.68 ('d', *J* = 16.0 Hz, 2H [overlapped s, δ 4.66, 1H]), 3.99 (d, *J* = 16.0 Hz, 1H), 1.55 (s, 3H), 0.95 (s, 9H). **^{13}C NMR** (100 MHz, $CDCl_3$) δ 162.5 (d, *J* = 246.0 Hz), 141.3 (d, *J* = 6.5 Hz), 139.6, 137.9, 129.5 (d, *J* = 8.0 Hz), 128.7, 127.9, 127.3, 126.1 (d, *J* = 3.0 Hz), 117.3 (d, *J* = 21.5 Hz), 115.2, 114.8 (d, *J* = 21.0 Hz), 68.9, 58.4, 47.3, 23.1, 22.0. **^{19}F {^{1}H} NMR** (376 MHz, $CDCl_3$) δ-113.3. **IR** (neat, cm^{-1}) 3320, 3064, 3028, 2971, 2917, 2830, 1645, 1615, 1589, 1484, 1452, 1446, 13.71, 1262, 1245, 1137, 1112, 1029. **HRMS** (DART) Calc'd for $C_{21}H_{27}FNOS$ 360.17803, found 360.17974. $[\alpha]_D^{20}$ −232.4 (c = 1.0, $CHCl_3$).

(*S*)-*N*-benzyl-*N*-((*S*)-1-(3-methoxyphenyl)-2-methylallyl)-2-methylpropane-2-sulfinamide (1.234c)—Was synthesized according to **general procedure 7** using **1.233c** (1.08 g, 3.86 mmol) and benzyl bromide (0.916 mL, 7.7 mmol). The crude alkylated amine was obtained in analytically pure form by silica gel flash column chromatography using hexanes:EtOAc (2:1 v:v) as the mobile phase. The title compound was obtained as clear and light yellow oil (1.18 g, 3.19 mmol, 83%).

^{1}H NMR (400 MHz, $CDCl_3$) δ 7.44–7.19 (m, 6H), 6.92–6.79 (m, 3H), 5.69 (s, 1H), 5.27 (s, 1H), 4.73–4.63 (m, 2H), 4.73–4.63 (m, 1H), 1.55 (s, 3H), 0.94 (s, 9H). **^{13}C NMR** (100 MHz, $CDCl_3$) δ 159.3, 140.2, 140.1, 138.3, 129.0, 128.6, 127.9, 127.2, 122.9, 116.1, 114.6, 113.3, 69.5, 58.3, 55.2, 47.2, 23.1, 22.1. **IR** (neat, cm^{-1}) 2983, 2956, 2917, 1599, 1584, 1493, 1476, 1454, 1263, 1062. **HRMS** (ESI+) Calc'd for $C_{22}H_{30}NO_2S$ 372.1993, found 372.1992. $[\alpha]_D^{20}$ −235.9 (c = 1.0, $CHCl_3$).

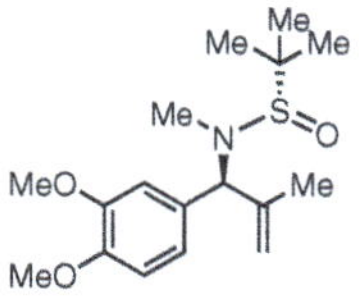

(*S*)-*N*-((*S*)-1-(3,4-dimethoxyphenyl)-2-methylallyl)-N,2-dimethylpropane-2-sulfinamide (1.234d)—Was synthesized according **to general procedure 7** using **1.233d** (2.0 g, 6.4 mmol) and iodomethane (0.798 mL, 12.8 mmol). The crude alkylated amine was obtained in analytically pure form by silica gel flash column chromatography using hexanes:EtOAc (3:1 v:v) as the mobile phase. The title compound was obtained as clear and light yellow oil (1.83 g, 5.6 mmol, 88%). **^{1}H NMR** (400 MHz, $CDCl_3$) δ 6.83 (m, 3H), 5.10 (ddt, *J* = 2.5, 2.0, 1.0 Hz, 1H), 5.09 (s, 1H), 4.89 (s, 1H), 3.86 (d, *J* = 0.5 Hz, 6H), 2.51 (s, 3H), 1.72 (s, 3H), 1.19 (s, 9H). **^{13}C NMR** (100 MHz, $CDCl_3$) δ 148.7, 148.3, 143.1, 130.8, 120.9, 115.3, 111.9, 110.7, 72.9, 58.5, 55.8, 55.8, 28.9, 23.8, 21.1. **IR** (neat, cm^{-1}) 2956, 2867, 2835, 1591, 1517, 1404, 1456, 1416, 1361, 1263, 1255, 1241, 1140, 1071, 1029. **HRMS** (DART) Calc'd for $C_{17}H_{28}NO_3S$ 326.17937, found 326.17899. $[\alpha]_D^{20}$ −37.8 (c = 1.0, $CHCl_3$).

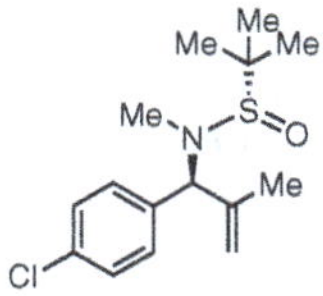

(*S*)-*N*-((*S*)-1-(4-chlorophenyl)-2-methylallyl)-*N*,2-dimethylpropane-2-sulfinamide (1.234e)[185]—Was synthesized according to **general procedure 7** using **1.233e** (2.0 g, 7.02 mmol) and iodomethane (0.890 mL, 14.04 mmol). The crude alkylated amine was obtained in analytically pure form by silica gel flash column chromatography using hexanes:EtOAc (10:4 v:v) as the mobile phase. The title compound was obtained as clear and light yellow oil (2.01 g, 6.7 mmol, 96%). **^{1}H NMR** (400 MHz, $CDCl_3$) δ 7.36–7.23 (m, 4H), 5.15 (td, *J* = 1.5, 1.0 Hz, 1H), 5.10–5.07 (m, 1H), 4.94 (s, 1H), 2.55 (s, 3H), 1.75–1.71 (m, 3H), 1.22 (s, 9H). **^{13}C NMR** (100 MHz, $CDCl_3$) δ 142.7, 137.2, 133.3, 130.0, 128.5, 116.2, 72.1, 58.7, 29.4, 23.8,

[185]Substrate prepared by Hyung Yoon.

21.0. **IR** (neat, cm^{-1}) 2974, 2956, 2922, 2866, 1455, 1406, 1362, 1091, 1074, 1015, 981, 924, 879, 847, 819. **HRMS** (DART) Calc'd for $C_{15}H_{23}ClNOS$ 300.11875, found 300.11889. $[\alpha]_D^{20}$ −36.8 (c = 1.0, $CHCl_3$).

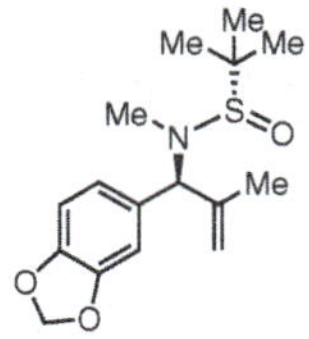

(*S*)-*N*-((*S*)-1-(benzo[*d*][1, 3]dioxol-5-yl)-2-methylallyl)-*N*,2-dimethylpropane-2-sulfinamide (1.243)—Was synthesized according to **general procedure 7** using **1.242** (6.5 g, 22 mmol) and methyl iodide (2.7 mL, 44 mmol). The crude alkylated amine was obtained in analytically pure form by silica gel flash column chromatography using hexanes:EtOAc (5:2 v:v) as the mobile phase. The title compound was obtained as clear yellow oil (5.89 g, 19.1 mmol, 87%) ^{1}**H NMR** (400 MHz, $CDCl_3$) δ 6.76 (s, br, 3H), 5.93 (s, 2H), 5.11–5.02 (m, 2H), 4.81 (s, 1H), 2.50 (s, 3H), 1.68 (s, 3H), 1.18 (s, 9H). ^{13}**C NMR** (100 MHz, $CDCl_3$) δ 147.6, 146.9, 143.1, 132.4, 122.1, 115.4, 109.0, 108.0, 101.0, 72.9, 58.5, 29.2, 23.9, 20.9. **IR** (neat, cm^{-1}) 2955, 2922, 2887, 2789, 1502, 1485, 1438, 1244, 1132, 1105, 1039. **HRMS** (DART) Calc'd for $C_{16}H_{24}NO_3S$ 310.14769, found 310.14772. $[\alpha]_D^{20}$ −26.2 (c = 1.0, $CHCl_3$).

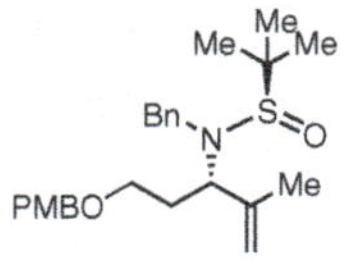

(*R*)-*N*-benzyl-N-((*S*)-5-((4-methoxybenzyl)oxy)-2-methylpent-1-en-3-yl)-2-methyl propane-2-sulfinamide (1.234 g)—Was synthesized according to **general procedure 7** using **1.234 g** (1.03 g, 3.03 mmol, 96:4 dr) and benzyl bromide (720 μL, 6.06 mmol). The crude alkylated amine was obtained in analytically pure form as a single diastereomer by silica gel flash column chromatography using hexanes:EtOAc (2:1 v:v) as the mobile phase. The title compound was obtained as a clear yellow oil (920 mg, 2.14 mmol, 71%) ^{1}**H NMR** (400 MHz, $CDCl_3$) δ 7.38–7.22 (m, 5H), 7.17–7.09 (m, 2H), 6.86–6.78 (m, 2H), 4.97 (p, *J* = 1.5 Hz, 1H), 4.61–4.54 (m, 2H (overlapped)), 4.37 (d, *J* = 11.5 Hz, 1H), 4.29 (d, *J* = 11.5 Hz, 1H), 3.79 (s, 3H), 3.74 (d, *J* = 17.0 Hz, 1H), 3.67 (dd, *J* = 11.0, 4.0 Hz, 1H), 3.41 (ddd, *J* = 9.5, 7.0, 5.0 Hz, 1H), 3.27 (ddd, *J* = 9.5, 8.5, 6.0 Hz, 1H), 2.27–2.11 (m, 1H), 1.99 (dddd, *J* = 13.5, 8.5, 7.0, 4.0 Hz, 1H), 1.91–1.86 (m, 3H), 1.23 (s, 9H). ^{13}**C NMR** (100 MHz, $CDCl_3$) δ 159.1, 141.3, 137.9, 130.3, 129.1, 128.6, 127.4, 126.8, 117.7, 113.7, 72.6, 67.1, 63.9, 58.3, 55.2, 45.2, 32.3, 23.7, 17.4. **IR** (neat, cm^{-1}) 2956, 2863, 1613, 1513, 1495, 1454, 1359, 1301, 1247, 1172, 1096, 1071, 1035. **HRMS**

(DART) Calc'd for $C_{25}H_{36}NO_3S$ 430.24159, found 430.24245. $[\alpha]_D^{20}$ +100.5 (c = 1.0, $CHCl_3$).

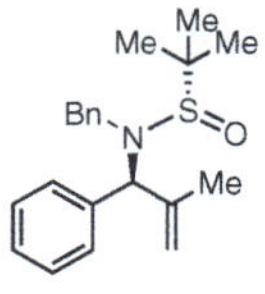

(*S*)-*N*-benzyl-2-methyl-*N*-((*S*)-2-methyl-1-phenylallyl)propane-2-sulfinamide (1.234 h)[186]—Was synthesized according to **general procedure 7** using **1.233a** (1.35 g, 5.4 mmol) and benzyl bromide (1.28 mL, 10.8 mmol). The crude alkylated amine was obtained in analytically pure form by silica gel flash column chromatography using hexanes:EtOAc (10:2 v:v) as the mobile phase. The title compound was obtained as clear and light yellow oil (1.33 g, 3.9 mmol, 72%). **^{1}H NMR** (400 MHz, $CDCl_3$) δ 7.46–7.35 (m, 4H), 7.35–7.23 (m, 6H), 5.72 (s, 1H), 5.29 (s, 1H), 4.76–4.64 (m, 2H [overlapping s, 1H]), 4.00 (d, *J* = 16.5 Hz, 1H), 1.54 (s, 3H), 0.90 (s, 9H). **^{13}C NMR** (100 MHz, $CDCl_3$) δ 134.0, 138.5, 138.3, 130.5, 128.6, 128.0, 127.8, 127.1, 114.5, 69.7, 58.2, 47.1, 29.7, 23.0, 22.1. **IR** 3086, 3063, 3028, 2982, 2957, 2917, 2863, 2849, 1495, 1476, 1456, 1436, 1389, 1374, 1363, 1194, 1179, 1106, 1027, 936 (neat, cm^{-1}). **HRMS** (ESI+) Calc'd for $C_{21}H_{28}NOS$ 342.1882, found 342.1886. $[\alpha]_D^{20}$ −225.4 (c = 1.0, $CHCl_3$).

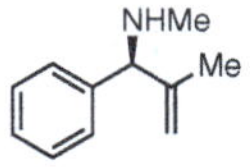

(*S*)-*N*,2-dimethyl-1-phenylprop-2-en-1-amine (1.235a)—Was synthesized according to **general procedure 8b** using **1.234a** (1.83 g, 6.9 mmol, 0.38 M) and anhydrous HCl in 1,4-dioxane (3.45 mL, 13.8 mmol). The title compound was obtained as clear and light yellow oil (0.89 g, 5.36 mmol, 78%). **^{1}H NMR** (400 MHz, $CDCl_3$) δ 7.39–7.31 (m, 4H), 7.31–7.21 (m, 1H), 5.21–5.15 (m, 1H), 4.92 (p, *J* = 1.5 Hz, 1H), 4.05 (s, 1H), 2.39 (s, 3H), 1.63–1.57 (m, 3H), 1.45 (s (br)), 1H). **^{13}C NMR** (100 MHz, $CDCl_3$) δ 146.5, 142.1, 128.1, 127.2, 126.9, 110.9, 70.8, 34.6, 18.7. **IR** (neat, cm^{-1}) 3333, 3082, 3062, 3026, 2971, 2949, 2848, 2788, 1645, 1491, 1486, 1451, 1371, 1130, 1105, 1074. **HRMS** (DART) Calc'd for $C_{11}H_{16}N$ 162.12795, found 162.12827. $[\alpha]_D^{20}$ −27.9 (c = 1.0, $CHCl_3$).

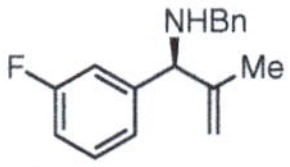

(*S*)-*N*-benzyl-1-(3-fluorophenyl)-2-methylprop-2-en-1-amine (1.235b)—Was synthesized according to **general procedure 8b** using **1.234b** (1.7 g, 4.8 mmol,

[186]Substrate prepared by Hyung Yoon.

0.3 M) and anhydrous HCl in 1,4-dioxane (2.4 mL, 9.6 mmol). The title compound was obtained as clear and light yellow oil (0.90 g, 3.5 mmol, 74%). **^{1}H NMR** (400 MHz, $CDCl_3$) δ 7.39–7.11 (m, 8H), 6.94 (ddq, *J* = 10, 7.0, 1.0 Hz, 1H), 5.18 (dq, *J* = 2.0, 1.0 Hz, 1H), 4.93 (h, *J* = 1.5 Hz, 1H), 4.20 (s, 1H), 3.76–3.64 (m, 2H), 1.60 (m, 4H, overlapped with NH). **^{13}C NMR** (100 MHz, $CDCl_3$) δ 163.0 (d, *J* = 245.0 Hz), 146.1, 145.1 (d, *J* = 6.5 Hz), 140.4, 129.6 (d, *J* = 8.0 Hz), 128.2 (d, *J* = 24.5 Hz), 126.9, 123.0 (d, *J* = 3.0 Hz), 114.1 (d, *J* = 21.5 Hz), 113.9 (d, *J* = 21.5 Hz), 112.0, 6748, 67.4, 51.4, 18.5. **^{19}F{^{1}H} NMR** (376 MHz, $CDCl_3$) δ-113.6. **IR** (neat, cm^{-1}) 3320, 3064, 3028, 2971, 2917, 2830, 1645, 1613, 1589, 1484, 1452, 1446, 1371, 1262, 1245, 1137, 1112, 1029. **HRMS** (DART) Calc'd for $C_{17}H_{19}FN$ 256.15006, found 256.15015. $[\alpha]_D^{20}$ −14.4 (c = 1.0, $CHCl_3$).

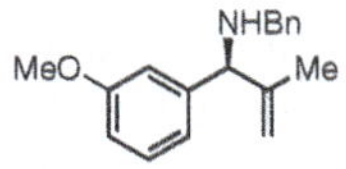

(*S*)-*N*-benzyl-1-(3-methoxyphenyl)-2-methylprop-2-en-1-amine (1.235c)—Was synthesized according to **general procedure 8b** using **1.234c** (1.11 g, 3.0 mmol, 0.23 M) and anhydrous HCl in 1,4-dioxane (1.5 mL, 6.0 mmol). The title compound was obtained as clear and light yellow oil (0.73 g, 2.74 mmol, 91%). **^{1}H NMR** (400 MHz, $CDCl_3$) δ 7.40–7.32 (m, 4H), 7.32–7.21 (m, 2H), 7.04–6.96 (m, 2H), 6.83 (ddd, *J* = 8.0, 2.5, 1.0 Hz, 1H), 5.23–5.19 (m, 1H), 4.97–4.94 (m, 1H), 4.21 (s, 1H), 3.84 (s, 3H), 3.77 (d, *J* = 13.5 Hz, 1H), 3.72 (d, *J* = 13.5 Hz, 1H), 1.66–1.64 (m, 3H), 1.64 (s (br) 1H). **^{13}C NMR** (100 MHz, $CDCl_3$) δ 159.6, 146.5, 144.0, 140.6, 129.1, 128.3, 128.1, 126.8, 119.8, 112.9, 112.3, 111.4, 67.7, 55.1, 51.4, 18.8. **IR** (neat, cm^{-1}) 3320, 3026, 2937, 2919, 2833, 1599, 1485, 1404, 1452, 1437, 1317, 1256, 1150, 1113, 1049. **HRMS** (DART) Calc'd for $C_{18}H_{22}NO$ 268.16911, found 268,17014. $[\alpha]_D^{20}$ −20.4 (c = 1.0, $CHCl_3$).

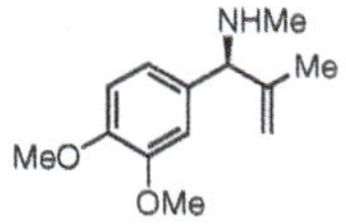

(*S*)-1-(3,4-dimethoxyphenyl)-*N*,2-dimethylprop-2-en-1-amine (1.235d)—Was synthesized according to **general procedure 8a** using **1.234d** (1.71 g, 5.26 mmol, 0.31 M) and anhydrous HCl in 1,4-dioxane (4 mL, 15.78 mmol, 3 equiv). The title compound was obtained as clear and light yellow oil (1.09 g, 4.91 mmol, 93%). ^{1}H NMR (400 MHz, $CDCl_3$) δ 6.90–6.85 (m, 2H), 6.83–6.78 (m, 1H), 5.14–5.11 (m, 1H), 4.89–4.86 (m, 1H), 3.96 (s, 1H), 3.88 (s, 3H), 3.87 (s, 3H), 2.35 (s, 3H), 1.61–1.53 (m, 3H), 1.42 (s (br), 1H). **^{13}C NMR** (100 MHz, $CDCl_3$) δ 148.8, 148.0, 146.7, 134.8, 119.5, 110.71, 110.7, 110.1, 70.5, 55.8, 55.8, 34.6, 18.9. **IR** (neat, cm^{-1}) 3332, 2948, 2838, 2788, 1645, 1604, 1591, 1513, 1464, 1441, 1412, 1342, 1258, 1231, 1138, 1030. **HRMS** (DART) Calc'd for $C_{13}H_{20}O_2N$ 222.14939, found 222.14940. $[\alpha]_D^{20}$ −35.9 (c = 1.0, $CHCl_3$).

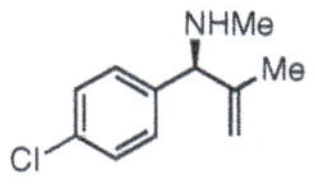

(*S*)-1-(4-chlorophenyl)-*N*,2-dimethylprop-2-en-1-amine (1.235e)[187]—Was synthesized according to **general procedure 8a** using **1.234e** (1.9 g, 6.36 mmol, 0.32 M) and anhydrous HCl in 1,4-dioxane (3.18 mL, 12.7 mmol). The title compound was obtained as clear and light yellow oil (1.21 g, 6.14 mmol, 97%). **^{1}H NMR** (400 MHz, $CDCl_3$) δ 7.29 (s, 4H), 5.14 (dt, *J* = 1.5, 1.0 Hz, 1H), 4.92–4.89 (m, 1H), 4.01 (s, 1H), 2.35 (s, 3H), 1.99 (s (br), 1H), 1.57–1.53 (m, 3H). **^{13}C NMR** (100 MHz, $CDCl_3$) δ 146.2, 140.7, 132.6, 128.6, 128.3, 111.4, 70.2, 34.5, 18.6. **IR** (neat, cm^{-1}) 3073, 3025, 2973, 2949, 2937, 2916, 2871, 2847, 2790, 1484, 1371, 1330, 1288, 1238, 1129, 1091, 1015, 901. **HRMS** (DART) Calc'd for $C_{11}H_{16}ClN$ 196.08973, found 196.08930. $[\alpha]_D^{20}$ −28.6 (c = 1.0, $CHCl_3$).

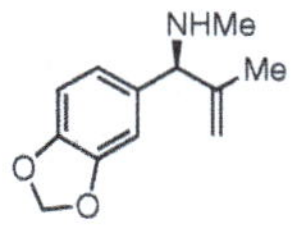

(*S*)-1-(benzo[*d*][1, 3]dioxol-5-yl)-*N*,2-dimethylprop-2-en-1-amine (1.226)—Was synthesized according to **general procedure 8b** using **1.243** (5.89 g, 19.1 mmol,0.32 M) and anhydrous HCl in Et_2O (58 mL, 57.4 mmol, 3 equiv, 1 M). The title compound was obtained as clear and light yellow oil (3.61 g, 17.6 mmol, 92%). **^{1}H NMR** (400 MHz, $CDCl_3$) δ 6.84 (d, *J* = 1.7 Hz, 1H), 6.81–6.68 (m, 2H), 5.91 (s, 2H), 5.13–5.08 (m, 1H), 4.89–4.84 (m, 1H), 3.91 (s, 1H), 2.33 (s, 3H), 1.62–1.30 (m, 3H), 1.47 (s (br), 1 H). **^{13}C NMR** (100 MHz, $CDCl_3$) δ 147.6, 146.67 146.5, 136.2, 120.5, 110.7, 107.8, 107.5, 100.8, 70.4, 34.5, 18.9. **IR** (neat, cm^{-1}) 2972, 2949, 2887, 2789, 1502, 1485, 1438, 1244, 1132, 1105, 1039. **HRMS** (DART+) Calc'd for $C_{12}H_{16}NO_2$ 206.11810, found 206.11826. $[\alpha]_D^{20}$ −55.2 (c = 1.0, $CHCl_3$).

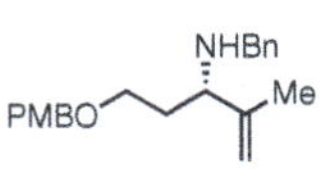

(*S*)-*N*-benzyl-5-((4-methoxybenzyl)oxy)-2-methylpent-1-en-3-amine (1.235g)—Was synthesized according to **general procedure 8a** using **1.234g** (0.91 g, 2.12 mmol, 0.35 M) and anhydrous HCl in 1,4-dioxane (1.06 mL, 8.48 mmol, 4 equiv). The crude alkylated amine was obtained in analytically pure form by silica gel flash column chromatography using hexanes:EtOAc (1:1 v:v) as the mobile phase. The title compound was obtained as clear and light yellow oil (0.47 g, 1.44 mmol, 68%). **^{1}H NMR** (400 MHz, $CDCl_3$) δ 7.32–7.27 (m, 4H), 7.25–7.19

[187]Substrate prepared by Hyung Yoon.

(m, 3H), 6.89–6.83 (m, 2H), 4.94–4.87 (m, 1H), 4.91–4.85 (m, 1H), 4.42–4.37 (m, 2H), 3.80 (s, 3H), 3.71 (d, *J* = 13.0 Hz, 1H), 3.58–3.40 (m, 3H), 3.27 (t, *J* = 7.0 Hz, 1H), 1.89–1.77 (m, 1H), 1.77–1.71 (m, 1H), 1.70 (dd, *J* = 1.5, 1.0 Hz, 3H). **^{13}C NMR** (100 MHz, $CDCl_3$) δ 159.1, 145.7, 140.7, 130.5, 129.2, 128.2, 128.2, 126.7, 113.7, 113.0, 72.6, 67.9, 62.0, 55.2, 51.1, 34.1, 16.8. **IR** (neat, cm^{-1}) 3073, 3026, 2970, 2937, 2914, 2824, 2724, 1600, 1412, 1451, 1371, 1242, 1117, 1070, 1029, 896. **HRMS** (DART) Calc'd for $C_{21}H_{28}NO$ 326.21200, found 326.21245. $[\alpha]_D^{20}$ +4.7 (c = 1.0, $CHCl_3$).

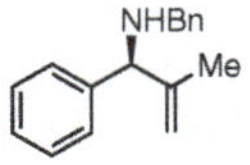

(*S*)-*N*-benzyl-2-methyl-1-phenylprop-2-en-1-amine (1.235h)[188]—Was synthesized according to **general procedure 8a** using **1.234 h** (1.23 g, 3.6 mmol, 0.24 M) and anhydrous HCl in 1,4-dioxane (1.8 mL, 7.2 mmol). The title compound was obtained as clear and light yellow oil (0.39 g, 1.6 mmol, 45%). **^{1}H NMR** (400 MHz, $CDCl_3$) δ 7.36–7.11 (m, 9H), 5.12 (dt, *J* = 2.0, 1.0 Hz, 1H), 4.84 (p, *J* = 1.5 Hz, 1H), 4.13 (s, 1H), 3.72–3.55 (m, 2H), 1.53 (s, 4H, overlapped with NH). **^{13}C NMR** (100 MHz, $CDCl_3$) δ 146.6, 142.3, 140.6, 128.3, 128.2, 128.1, 127.3, 127.0, 126.8, 111.3, 67.8, 51.5, 18.7. **IR** (neat, cm^{-1}) 3319, 3083, 3062, 3026, 2969, 2935, 2829, 1452, 1417, 1108. **HRMS** (ESI+) Calc'd for $C_{17}H_{20}N$ 238.1595, found 238.1590. $[\alpha]_D^{20}$ −23.8 (c = 1.0, $CHCl_3$).

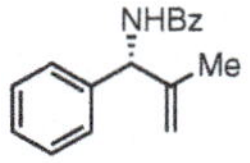

(*R*)-*N*-(2-methyl-1-phenylallyl)benzamide (1.235i)—Was synthesized according to **general procedure 8a** using *ent*-**1.234a** (1.0 g, 3.9 mmol, 0.39 M) and anhydrous HCl in 1,4-dioxane (2.0 mL, 7.96 mmol). The compound of the first step was obtained as clear and light yellow oil (503 mg, 3.42 mmol, 88%). Subsequently, a flame dried flask was charged with the oil (500 mg, 3.40 mmol, 0.26 M, 1 equiv) and DMAP (41.5 mg, 0.34 mmol, 0.1 equiv) and was purged with argon for 10 min. The amine was then dissolved in pyridine:DCM (16 mL, 3:1 v:v) and the resultant solution was cooled to 0 °C. At this time, benzoyl chloride (473 μL, 4.08 mmol, 1.2 equiv) was added dropwise over 10 min. The solution was then warmed to room temperature and stirred for 1 h until TLC analysis indicated full conversion of starting material. The resulting solution was concentrated *in vacuo*. The resulting yellow solid was dissolved in $CHCl_3$ and extracted with $NaHCO_3$ (2x) and brine. The combined organic layers were dried over $MgSO_4$, filtered and concentrated *in vacuo*. The pure benzoylated amine was obtained in analytically pure form by silica gel flash column chromatography using hexanes:EtOAc (4:1 v:

[188]Substrate prepared by Hyung Yoon.

v) as the mobile phase. **^{1}H NMR** (400 MHz, $CDCl_3$) δ 7.82–7.75 (m, 2H), 7.54–7.28 (m, 8H), 6.39 (d, J = 8.0 Hz, 1H), 5.69 (d, J = 8.0 Hz, 1H), 5.12–5.01 (m, 2H), 1.75 (s, 3H). **^{13}C NMR** (100 MHz, $CDCl_3$) δ 166.3, 143.9, 139.8, 134.4, 131.5, 128.8, 128.6, 127.8, 127.5, 126.9, 111.8, 58.8, 20.5. **IR** (neat, cm^{-1}) 3397, 3061, 3030, 1633, 1602, 1524, 1490, 1452. **HRMS** (DART) Calc'd for $C_{17}H_{18}NO$ 252.13909, found 252.13884. $[\alpha]_D^{20}$ +133.1 (c = 1.0, $CHCl_3$).

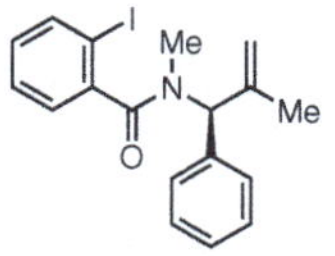

(*S*)-2-iodo-*N*-methyl-*N*-(2-methyl-1-phenylallyl)benzamide (1.237a)—Synthesized according to **general procedure 9** using 2-iodobenzoic acid (586 mg, 2.36 mmol) and **1.235a** (415 mg, 2.60 mmol). The crude amide was obtained in analytically pure form by silica gel flash column chromatography using hexanes:EtOAc (10:3 v:v) as the mobile phase. The title compound was obtained as an extremely viscous, clear and colourless oil (827 mg, 2.11 mmol, 89%). *NMR analysis showed the clean desired compound to be present as a complex mixture of 4 rotamers, and only approximate integration values are displayed.* **^{1}H NMR** (approximate integrations displayed) (500 MHz, $CDCl_3$) δ 7.92–7.75 (m, 1H), 7.44–7.27 (m, 6H), 7.25–6.62 (m, 2H), 6.39 (s, ~0.6H), 5.26–4.51 (m, ~2.4H), 3.25–2.36 (m, 3H), 1.93–1.39 (m, 3H). **^{13}C NMR** (125 MHz, $CDCl_3$) δ 171.9, 171.2 170.8 (br), 143.3, 143.2 overlapped with 143.1 (br), 142.8, 142.3, 141.9 overlapped with 141.86 (br), 139.6, 139.6 overlapped with 139.5, 139.2 (br), 139.0, 137.5 (br), 136.3 (br), 135.7, 130.3, 130.2, 130.0 (br), 129.0 (br), 128.6, 128.4 (br), 128.2, 127.8 (br), 127.7, 127.5, 127.5, 127.5, 127.1, 127.0, 126.6, 118.4, 115.2 (br), 114.2, 112.9 (br), 92.9, 92.8, 92.0 (br), 68.0, 67.9, 61.9 (br), 33.5 overlapped with 33.3 (br), 30.6, 30.4, 23.1, 22.2 (br), 21.4 (br), 21.2. **IR** (neat, cm^{-1}) 3502, 3442, 3419, 3086, 3060, 2971, 1635, 1633, 1419, 1392, 1073, 1031. **HRMS** (DART) Calc'd for $C_{18}H_{21}INO$ 392.05113, found 392.05104. Enantiomeric purity was determined by HPLC analysis in comparison with racemic material t_r = 10.95 (minor) t_r = 15.36 (major) (99:1 er shown; Chiralcel AD-H column, 10% ^{i}PrOH in hexanes 1.0 mL/min, 220 nm). $[\alpha]_D^{20}$ −64.2 (c = 0.5, $CHCl_3$).

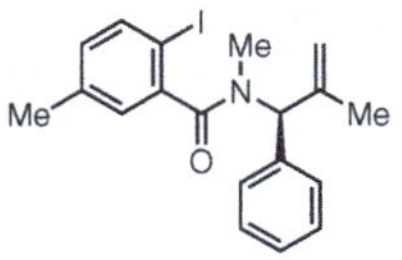

(*S*)-2-iodo-*N*,5-dimethyl-*N*-(2-methyl-1-phenylallyl)benzamide (1.237b)—Synthesized according to **general procedure 9** using 2-iodo-5-methylbenzoic acid (375 mg, 1.43 mmol) and **1.235a** (250 mg, 1.58 mmol). The crude amide was obtained in analytically pure form by silica gel flash column chromatography using hexanes:EtOAc:DCM (6:0.75:4 v:v:v) as the mobile phase. The title compound was obtained as a white solid (530 mg, 1.361 mmol, 95%, MP 63–65 °C). *NMR analysis showed the clean desired compound to be present as a complex mixture of 4 rotamers, and only approximate integration values are displayed.* **^{1}H NMR** (500 MHz, $CDCl_3$) δ 7.78–7.53 (m, 1H), 7.44–7.26 (m, 5H), 7.14–6.67 (m, 2H), 6.38 (s, ~0.5H), 5.34–4.48 (m, ~2.5H), 3.22–2.44 (m, 3H), 2.33–1.95 (m, 3H), 1.93–1.36 (m, 3H). **^{13}C NMR** (125 MHz, $CDCl_3$) δ 171.9 (br), 171.2, 170.8, 143.3, 142.9 (br), 142.8, 142.7 (br), 141.9, 141.78 (br), 141.5, 139.3, 139.1, 139.0, 138.8 (br), 138.6 (br), 138.5 (br), 137.7, 137.5 (br) overlapping 137.4, 136.2, 135.6, 131.1, 131.0, 130.0 (br), 129.9, 128.9 (br), 128.5, 128.3 (br), 128.2, 128.2, 128.0, 127.7 (br), 127.6, 127.5 (br), 127.4, 127.4, 127.3, 118.3, 115.1 (br), 114.0, 112.7 (br), 88.6, 88.3, 87.8 (br), 68.1, 67.8, 61.8 (br) overlapping 61.7 (br), 33.3 (br) overlapping 33.2 (br), 30.5, 30.3, 22.9, 22.2 (br), 21.3 (br), 21.0, 20.8, 20.8, 20.5. **IR** (neat, cm^{-1}) 3086, 3059, 3028, 2970, 2919, 1639, 1437, 1399, 1344, 1262, 1218, 1076, 1014. **HRMS** (DART) Calc'd for $C_{19}H_{21}INO$ 406.06678, found 406.06586. Enantiomeric purity was determined by HPLC analysis in comparison with racemic material t_r = 8.94 (minor) t_r = 11.1 (major) (>99:1 er shown; Chiralcel AD-H column, 10% ^{i}PrOH in hexanes 1.0 mL/min, 210 nm). $[\alpha]_D^{20}$ −52.8 (c = 0.5, $CHCl_3$).

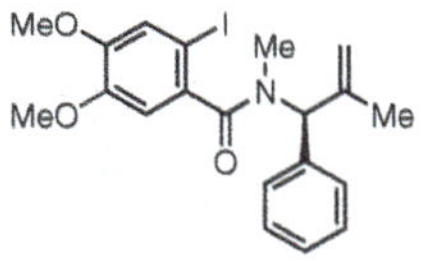

(*S*)-2-iodo-4,5-dimethoxy-*N*-methyl-*N*-(2-methyl-1-phenylallyl)benzamide (1.237c) —Synthesized according to **general procedure 9** using 2-iodo-4,5-dimethoxybenzoic acid (260 mg, 0.845 mmol) and **1.235a** (150 mg, 0.931 mmol). The crude amide was obtained in analytically pure form by silica gel flash column chromatography using hexanes:EtOAc (1:1 v:v) as the mobile phase. The title compound was obtained as a clear and colourless oil (240 mg, 0.532 mmol, 63%). *NMR analysis showed the clean desired compound to be present as a complex mixture of at least 3 rotamers, and only approximate integration values are displayed.* **^{1}H NMR** (500 MHz, $CDCl_3$) δ 7.42–7.26 (m, ~4.5H), 7.27–6.53 (m, 2H), 6.36–6.32 (m, ~0.5H), 6.07 (s, ~0.5H), 5.30–4.44 (m, 2.5H), 3.91–3.75 (m, 5H), 3.19 (s, ~1H), 3.13 (s, ~1H), 2.86–2.42 (m, ~2H), 2.03–1.42 (m, 3H). **^{13}C NMR** (125 MHz, $CDCl_3$) δ 171.8, 171.0, 170.7, 149.7, 149.6 (br), 149.5, 148.7, 148.2, 144.0, 143.2, 141.9, 139.9, 136.2 (br), 135.6, 135.3 (br), 134.5, 134.1, 130.1 (br) overlapping 129.9, 128.9 (br), 128.6, 128.4, 128.1, 127.7 (br), 127.4, 127.3, 121.6, 121.4, 121.3 (br), 118.4, 115.2 (br), 113.8, 112.6 (br), 110.1, 110.0, 109.i, 81.2, 80.7, 80.4, 68.2, 68.0, 62.1 (br), 56.2, 56.1, 56.1, 55.8, 55.0, 53.4, 33.4 (br), 30.5, 30.4, 23.2, 21.3. **IR** (neat, cm^{-1}) 3025, 3002, 2933, 2911, 1635, 1594, 1505, 1394, 1365, 1329, 1254, 1211, 1181, 1160, 1082, 1027. **HRMS** (DART) Calc'd for $C_{20}H_{21}INO_3$ 452.07226, found 452.07184. Enantiomeric purity was determined by HPLC analysis in comparison with racemic material t_r = 16.2 (minor) t_r = 21.2 (major) (>99:1 er shown; Chiralcel AD-H column, 10% ^{i}PrOH in hexanes 1.0 mL/min, 210 nm). $[\alpha]_D^{20}$ −49.2 (c = 1.0, $CHCl_3$).

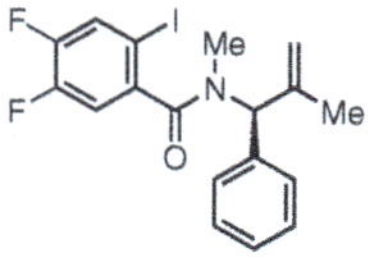

(*S*)-4,5-difluoro-2-iodo-*N*-methyl-*N*-(2-methyl-1-phenylallyl)benzamide (1.237d)[189] –Synthesized according to **general procedure 9** using 2-iodo-4,5-difluorobenzoic acid (240 mg, 0.845 mmol) and **1.235a** (150 mg, 0.931 mmol). The crude amide was obtained in analytically pure form by silica gel flash column chromatography using hexanes:EtOAc:DCM (6:0.75:4 v:v:v) as the mobile phase. The title compound was obtained as a clear and colourless oil (198 mg, 0.462 mmol, 55%). *NMR analysis showed the clean desired compound to be present as a complex mixture of at least 3 rotamers, and only approximate integration values are displayed.* **^{1}H NMR** (500 MHz, $CDCl_3$) δ 7.73–7.29 (m, 5H), 7.19–6.34 (m, 1H), 6.32 (s, ~0.5H), 5.42–4.53 (m, ~2.5H), 3.17 (s, ~0.5H), 2.83 (s, ~0.75H), 2.64–2.45 (m, ~1.5H), 1.92 (s, ~0.5H), 1.86–1.78 (m, ~1.5H), 1.51 (s, ~0.75H). **^{13}C NMR** (125 MHz, $CDCl_3$) δ 169.9, 169.2, 168.9, 168.90 151.7, 151.6, 151.2, 151.0, 150.9, 150.9, 150.8, 150.7, 149.7, 149.6, 149.1, 149.0, 149.0, 148.9, 148.9, 148.8, 148.7, 143.2, 142.4, 141.6, 139.8 (br), 139.1, 138.8, 138.7, 137.0 (br), 135.9 (br), 135.2, 123.0 (br), 128.7, 128.7, 128.6 (br), 128.4, 128.3, 128.2, 128.0 (br), 127.9, 127.2, 118.9, 116.4, 116.3, 116.2, 116.1, 115.8, 115.6, 115.2 (br), 114.4, 112.9 (br), 85.2, 85.2, 85.1, 85.0, 84.2, 84.2, 84.2, 84.1, 68.2, 68.1, 62.2, 33.3 (br), 30.8, 30.5, 23.9, 23.0, 21.4 (br), 21.1. **^{19}F NMR** (375 MHz, $CDCl_3$) δ–133.94 (dt, *J* = 20.5, 8.5 Hz), −134.09 (dt, *J* = 20.5, 8.5 Hz), −134.35 (ddd, *J* = 20.5, 9.5, 8.08 Hz), −136.02 (br), −136.20 (br), −136.71 to −136.89 (m). **IR** (neat, cm^{-1}) 3087, 3061, 3029, 2973, 2942, 2918, 1641, 1604, 1595, 1491, 1476, 1440, 1399, 1373, 1326, 1283, 1225, 1182, 1148, 1072. **HRMS** (DART) Calc'd for $C_{18}H_{17}F_2INO$ 428.03229, found 428.03107. Enantiomeric purity was determined by HPLC analysis in comparison with racemic material t_r = 9.63 (minor) t_r = 12.52 (major) (99:1 er shown; Chiralcel AD-H column, 5% ^{i}PrOH in hexanes 1.0 mL/min, 210 nm). $[\alpha]_D^{20}$ −64.8 (c = 0.5, $CHCl_3$).

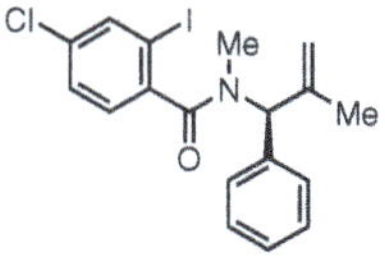

(*S*)-4-chloro-2-iodo-*N*-methyl-*N*-(2-methyl-1-phenylallyl)benzamide (1.237e)— Synthesized according to **general procedure 9** using 2-iodo-4-chlorobenzoic acid

[189]Substrate prepared by Hyung Yoon.

(260 mg, 0.845 mmol) and **1.235a** (150 mg, 0.931 mmol). The crude amide was obtained in analytically pure form by silica gel flash column chromatography using hexanes:EtOAc (1:1 v:v) as the mobile phase. The title compound was obtained as an extremely viscous and pale yellow oil (230 mg, 0.541 mmol, 64%). *NMR analysis showed the clean desired compound to be present as a complex mixture of 4 rotamers, and only approximate integration values are displayed.* **^{1}H NMR** (500 MHz, $CDCl_3$) δ 7.92–7.75 (m, ~1H), 7.41–7.18 (m, 6H), 7.08–6.53 (m, 1H), 6.35 (s, ~0.5H), 5.28–4.50 (m, ~2.5H), 3.16 (s, ~0.5H), 2.84 (s, ~0.8H), 2.63–2.45 (m, ~1.7H), 1.95–1.75 (m, ~2.2H), 1.50–1.46 (m, ~0.8H). **^{13}C NMR** (125 MHz, $CDCl_3$) δ 171.0, 170.3, 170.0 (br), 143.2, 142.6, 141.7 (br), 141.5 (br), 140.8, 140.4, 139.0, 138.9, 138.8, 138.6 (br), 137.3 (br), 136.1 (br), 135.4, 135.2, 135.0, 134.9, 123.0 (br), 128.8 (br), 128.7, 128.5, 128.3, 128.0, 127.9 (br), 127.8, 127.7, 127.4, 127.2, 118.6, 115.1 (br), 114.3, 112.9 (br), 93.2, 93.1, 92.2, 68.0, 68.0, 62.1 (br), 33.4 overlapped with 33.3 (br), 30.7, 30.4, 23.1, 22.2 (br), 21.4 (br), 21.2. **IR** (neat, cm^{-1}) 3085, 3028, 2969, 1635, 1596, 1327, 1248, 1097, 1070, 1025. **HRMS** (DART) Calc'd for $C_{18}H_{18}ClINO$ 426.01216, found 426.01379. Enantiomeric purity was determined by HPLC analysis in comparison with racemic material t_r = 10.6 (minor) t_r = 16.7 (major) (>99:1 er shown; Chiralcel AD-H column, 15% ^{i}PrOH in hexanes 1.0 mL/min, 210 nm). $[\alpha]_D^{20}$ −62.6 (c = 0.5, $CHCl_3$).

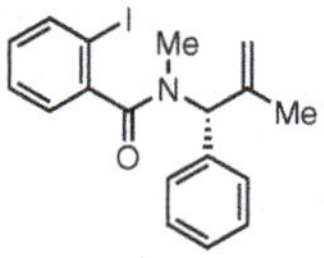

(*R*)-2-iodo-*N*-methyl-*N*-(2-methyl-1-phenylallyl)benzamide (1.237f)—Synthesized in accordance to the entire route of (*S*)-2-iodo-*N*-methyl-*N*-(2-methyl-1-phenylallyl) benzamide. **All analytic data was in accordance with that reported for the (*S*) analog shown above**. Enantiomeric purity was determined by HPLC analysis in comparison with racemic material t_r = 10.95 (major) t_r = 15.36 (minor) (>99:1 er. shown; Chiralcel AD-H column, 10% ^{i}PrOH in hexanes 1.0 mL/min, 210 nm). $[\alpha]_D^{20}$ +67.9 (c = 1.0, $CHCl_3$).

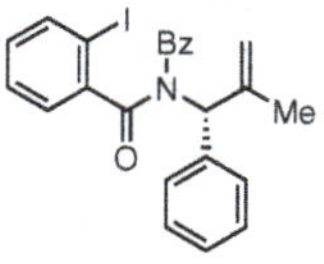

(*R*)-*N*-benzoyl-2-iodo-*N*-(2-methyl-1-phenylallyl)benzamide (1.237g)—The 2-iodobenzoyl chloride was synthesized according to **general procedure 9** using 2-iodobenzoic acid (327 mg, 1.32 mmol, 1.5 equiv, 0.15 M). A flame dried flask was charged with **1.235i** (222 mg, 0.884 mmol, 1 equiv, 0.18 M) and was purged with argon for 10 min. The amine was dissolved in anhydrous DMF and the resultant solution was cooled to 0 °C. At this time, LiHMDS (1 M in THF, 0.928 mL, 0.928 mmol, 1.05 equiv) was added drop wise over 10 min. The solution was allowed to stir for 30 min at this temperature before warming to room temperature. The solution was re-cooled to 0 °C and the 2-iodobenzoyl chloride in DCM was added over 5 min. At this time, the resulting solution was warmed to room temperature and was stirred until TLC analysis indicated no further conversion of starting material. At this time the reaction was quenched by doubling the reaction volume with a concentrated aqueous solution of $NaHCO_3$. The layers were separated and the aqueous layer was extracted with DCM (3×). The combined organic layers were washed with brine, dried over Na_2SO_4, filtered, and concentrated *in vacuo*. The crude amide was obtained in analytically pure form by silica gel flash column chromatography using hexanes:EtOAc:DCM (40:1:8 v:v:v) as the mobile phase. The title compound was obtained as a white solid (249 mg, 0.518 mmol, 59%, MP = 86–87 °C). **^{1}H NMR** (500 MHz, $CDCl_3$) δ 7.64 (d, J = 8.0 Hz, 2H), 7.53–7.50 (m, 1H), 7.41–7.32 (m, 4H), 7.29–7.17 (m, 4H), 7.13–7.07 (m, 2H), 6.77 (ddd, J = 8.0, 6.5, 2.5 Hz, 1H), 6.40 (s (br), 1H), 5.23 (s, 1H), 5.08 (s (br), 1H), 1.93 (s (br), 3H). **^{13}C NMR** (partial, severe peak broadening) (126 MHz, $CDCl_3$) δ 173.8, 170.7, 142.3, 140.9, 140.6, 137.4 (2, br), 131.8, 131.3, 129.9, 128.7, 128.5, 128.4, 127.9, 127.4, 115.6, 96.1, 66.5, 22.1. **IR** (neat, cm^{-1}) 3062, 3028, 2919, 1704, 1653, 1319, 1308, 1231, 1146. **HRMS** (DART) Calc'd for $C_{24}H_{21}INO_2$ 482.06113, found 482.05179. Enantiomeric purity was determined by HPLC analysis in comparison with racemic material t_r = 6.8 (major) t_r = 8.8 (minor) (>99:1 er shown; Chiralcel AD-H column, 15% ^{i}PrOH in hexanes 1.0 mL/min, 210 nm). $[\alpha]_D^{20}$ +20.7 (c = 0.73, $CHCl_3$).

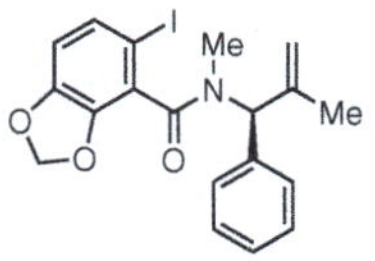

(*S*)-5-iodo-*N*-methyl-*N*-(2-methyl-1-phenylallyl)benzo[*d*][1, 3]dioxole-4-carboxamide (**1.235h**)–Synthesized according to **general procedure 9** using 5-iodobenzo[*d*][1, 3] dioxole-4-carboxylic acid (247 mg, 0.845 mmol) and **1.235a** (150 mg, 0.931 mmol). The crude amide was obtained in analytically pure form by silica gel flash column chromatography using hexanes:EtOAc (3:1 v:v) as the mobile phase. The title compound was obtained as a white solid (277 mg, 0.637 mmol, 68%, MP = 112–114 °C). *NMR analysis showed the clean desired compound to be present as a complex mixture of 4 rotamers, and only approximate integration values are displayed.* ^{1}H NMR (500 MHz, $CDCl_3$) δ 7.42–7.04 (m, 6H), 6.58–6.48 (m, ~1H), 6.35 (d, J = 6.7 Hz, ~0.7H), 6.07–5.66 (m, ~2H), 5.22–4.58 (m, ~2.3H), 3.11 (s, ~0.4H), 2.96 (s, ~0.3H), 2.66 (s, ~0.9H), 2.59 (s, ~1.4H), 1.93–1.85 (m, ~1.3H), 1.85–1.76 (m, ~1.4H), 1.65–1.64 (m, ~0.3H). **^{13}C NMR** (125 MHz, $CDCl_3$) δ 167.6, 167.4, 166.6, 166.5, 166.5, 148.1, 148.1, 148.0, 147.8, 144.7, 144.5, 144.3, 142.3, 141.6, 141.6, 141.4, 138.6, 136.9, 136.6, 136.3, 132.2, 132.2, 132.2, 132.0, 129.7, 129.1,

128.9, 128.9, 128.4, 128.3, 128.3, 128.3, 128.1, 127.9, 127.8, 127.7, 127.4, 127.2, 124.2, 124.0, 123.7, 123.5, 118.1, 115.5, 115.0, 113.4, 110.4, 110.4, 110.4, 110.3, 102.1, 102.0, 101.7, 81.7, 812, 80.9, 80.9, 67.7, 67.6, 62.0, 61.9, 32.6, 32.5, 31.0, 30.9, 22.7, 22.0, 21.3, 21.0. **IR** (neat, cm^{-1}) 3086, 3061, 3004, 2913, 1635, 1444, 1390, 1332, 1240, 1101, 1084, 1036. HRMS (DART) Calc'd for $C_{19}H_{19}INO_3$ 436.04096, found 436.04085. Enantiomeric purity was determined by HPLC analysis in comparison with racemic material t_r = 10.0 (minor) t_r = 14.92 (major) (>99:1 er shown; Chiralcel AD-H column, 10% ^{i}PrOH in hexanes 1.0 mL/min, 210 nm). $[\alpha]_D^{20}$ −32.5 (c = 1.0, $CHCl_3$).

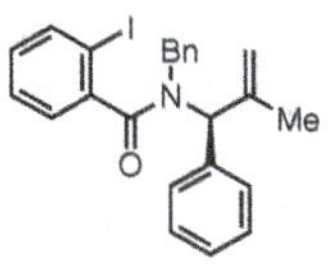

(*S*)-*N*-benzyl-2-iodo-*N*-(2-methyl-1-phenylallyl)benzamide (1.237i)[190]—Synthesized according to **general procedure 9** using 2-iodobenzoic acid (143 mg, 0.577 mmol) and **1.235h** (150 mg, 0.635 mmol). The crude amide was obtained in analytically pure form by silica gel flash column chromatography using hexanes:EtOAc:DCM (12:0.75:8 v:v:v) as the mobile phase. The title compound was obtained as a clear and light yellow oil (153 mg, 0.327 mmol, 57%). *NMR analysis showed the clean desired compound to be present as a complex mixture of 4 rotamers, and only approximate integration values are displayed.* **^{1}H NMR** (500 MHz, $CDCl_3$) δ 7.92–7.64 (m, 1H), 7.50–6.80 (m, 13H), 6.53 (d, *J* = 7.5 Hz, ~0.3H), 6.13–5.94 (m, ~0.15H), 5.46–3.92 (m, ~4.5H), 1.82 (s, ~0.5H), 1.61 (dd, *J* = 1.5, 1.0 Hz, ~0.3H), 1.60–1.58 (m, ~0.5H), 1.34 (dt, *J* = 1.5, 1.0 Hz, ~1.7H). **^{13}C NMR** (125 MHz, $CDCl_3$) δ 172.4, 171.7, 144.0, 143.3 142.2, 141.9, 139.8, 139.4, 138.5, 138.3, 137.9, 134.9, 130.8, 130.4, 130.1, 130.1, 129.8, 128.7, 128.5, 128.5, 128.3, 128.3, 128.3, 128.2, 128.0, 127.9, 127.8, 127.6, 127.6, 127.4, 127.3, 127.1, 127.0, 126.7, 126.2, 118.7, 113.9, 92.6, 92.3, 68.5, 68.4, 47.6, 47.4, 23.1, 21.3. **IR** (neat, cm^{-1}) 3085, 3062, 3028, 2969, 1641, 1602, 1584, 1495, 1467, 1452, 1429, 1404, 1362, 1324, 1311, 1294, 1159, 1077. **HRMS** (ESI+) Calc'd for $C_{24}H_{23}INO$ 468.0816, found 468.0816. Enantiomeric purity was determined by HPLC analysis in comparison with racemic material t_r = 12.3 (major) t_r = 14.76 (minor) (>99:1 er shown; Chirapak AS column, 7.5% ^{i}PrOH in hexanes 1.0 mL/min, 30 °C, 210 nm). $[\alpha]_D^{20}$ −33.8 (c = 1.0, $CHCl_3$).

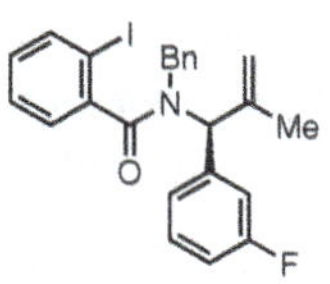

[190]Substrate prepared by Hyung Yoon.

(*S*)-*N*-benzyl-*N*-(1-(3-fluorophenyl)-2-methylallyl)-2-iodobenzamide (1.237j) —Synthesized according to **general procedure 9** using 2-iodobenzoic acid (451 mg, 1.82 mmol) and **1.235b** (450 mg, 2.00 mmol). The crude amide was obtained in analytically pure form by silica gel flash column chromatography using hexanes:EtOAc:DCM (24:1:16 v:v:v) as the mobile phase. The title compound was obtained a white solid (654 mg, 1.34 mmol, 74%, MP = 134–136 °C). The title compound was obtained as a clear and light yellow oil (153 mg, 0.327 mmol, 57%). *NMR analysis showed the clean desired compound to be present as a complex mixture of at least 3 rotamers, and only approximate integration values are displayed.* **^{1}H NMR** (500 MHz, $CDCl_3$) δ 7.95–7.70 (m, 1H), 7.42–6.55 (m, 12H), 6.00–5.77 (m, ~0.1H), 5.41–5.29 (m, ~1H), 5.23–5.03 (m, ~1.5H), 4.94–4.35 (m, ~1.3H), 4.07 (d, J = 15.0 Hz, ~0.5H), 1.84 (s, 0.5H), 1.61 (s, ~0.7H), 1.37–1.34 (m, ~1.7H). **^{13}C NMR** (125 MHz, $CDCl_3$) δ 172.4, 171.7, 163.8, 163.7, 163.7, 161.8, 161.7, 161.7, 143.4, 142.9, 142.0, 141.7, 141.4, 141.4, 140.0, 139.8, 139.4, 138.1, 137.8, 137.4, 137.4, 130.3, 130.2, 129.9, 129.9, 129.9, 129.8, 129.8, 129.8, 129.7, 128.7, 128.2, 128.1, 128.0, 127.7, 127.5, 127.4, 127.0, 126.9, 126.9, 126.5, 126.4, 126.4, 123.7, 123.7, 119.1, 117.7, 117.5, 115.3, 115.3, 115.2, 115.1, 114.9, 114.8, 114.6, 114.4, 114.4, 114.1, 92.5, 92.3, 68.1, 67.7, 67.7, 64.0, 47.5, 47.4, 23.2, 22.1, 21.2. **IR** (film, cm^{-1}) 3088, 3062, 3031, 3005, 2976, 1641, 1590, 1467, 1405, 1326, 1247, 1160, 1135. **HRMS** (ESI+) Calc'd for $C_{24}H_{22}FINO$ 486.06301, found 486.07371. Enantiomeric purity was determined by HPLC analysis in comparison with racemic material t_r = 40.7 (major) t_r = 43.9 (minor) (>99:1 er shown; Chiracel OD-H column, 1% ^{i}PrOH in hexanes 1.0 mL/min, 210 nm). $[\alpha]_D^{20}$ −37.6 (c = 1.0, $CHCl_3$).

(*S*)-*N*-benzyl-2-iodo-*N*-(1-(3-methoxyphenyl)-2-methylallyl)benzamide (1.237k) —Synthesized according to **general procedure 9** using 2-iodobenzoic acid (337 mg, 1.36 mmol) and **1.235c** (400 mg, 1.50 mmol). The crude amide was obtained in analytically pure by first triturating the crude using EtOAc/hexanes. The obtained precipitate was washed with distilled water to remove the remaining triethylammonium hydrochloride salt, and then drying to a constant weight under reduced pressure. The title compound was obtained as a free flowing white solid (552 mg, 1.11 mmol, 82%, MP = 141–143 °C). *NMR analysis showed the clean desired compound to be present as a complex mixture of 3 rotamers, and only approximate integration values are displayed.* **^{1}H NMR** (500 MHz, $CDCl_3$) δ 7.98–7.68 (m, 1H), 7.45–7.30 (m, 2H), 7.30–6.52 (m, 10H), 5.39–4.98 (m, ~3H), 4.90–4.77 (m, 1H), 4.49–4.36 (m, ~0.2H), 4.06 (d, J = 14.5, ~1H), 3.78 (s, ~0.5H), 3.67 (s, ~0.5H), 3.60 (s, 2H), 1.83 (s, ~0.5H), 1.64–1.59 (m, ~0.5H), 1.40–1.34 (m, 2H). **^{13}C NMR** (125 MHz, $CDCl_3$) δ 172.4, 171.6,

159.7, 159.6, 159.5, 143.9, 143.1, 142.2, 141.8, 140.2, 139.8, 139.3, 138.3, 138.2, 136.2, 130.1, 130.1, 129.4, 129.3, 129.2, 128.7, 128.6, 127.9, 127.9, 127.7, 127.4, 127.3, 127.1, 126.9, 126.7, 126.3, 122.7, 122.4, 120.5, 118.7, 116.4, 116.0, 114.7, 114.1, 113.9, 113.7, 112.9, 92.6, 92.3, 68.5, 68.3, 55.3, 55.1, 47.6, 47.5, 45.7, 23.1, 21.3, 8.6. **IR** (film, cm^{-1}) 3062, 3029, 3004, 2971, 2915, 1639, 1600, 1584, 1443, 1452, 1433, 1405, 1326, 1266, 1163, 1066. **HRMS** (ESI+) Calc'd for $C_{25}H_{25}INO_2$ 498.09300, found 498.09183. Enantiomeric purity was determined by HPLC analysis in comparison with racemic material t_r = 44.82 (major) t_r = 48.67 (minor) (>99:1 er shown; Chiracel OD-H column, 1% ^{i}PrOH in hexanes 1.0 mL/min, 210 nm). $[\alpha]_D^{20}$ −23.4 (c = 1.0, $CHCl_3$).

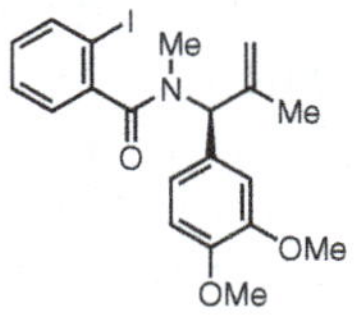

(*S*)-*N*-(1-(3,4-dimethoxyphenyl)-2-methylallyl)-2-iodo-*N*-methylbenzamide (1.237l)—Synthesized according to **general procedure 9** using 2-iodobenzoic acid (406 mg, 1.64 mmol) and **1.235d** (400 mg, 1.81 mmol). The crude amide was obtained in analytically pure form by silica gel flash column chromatography using hexanes:EtOAc (2:1 v:v) as the mobile phase. The title compound was obtained as a hygroscopic white foam (633 mg, 1.40 mmol, 86%). *NMR analysis showed the clean desired compound to be present as a complex mixture of 4 rotamers, and only approximate integration values are displayed.* **^{1}H NMR** (500 MHz, $CDCl_3$) δ 7.92–7.70 (m, 1H), 7.46–7.16 (m, ~1.7H), 7.15–6.79 (m, 4H), 6.75–6.61 (m, 0.2H), 6.42 (d, *J* = 2.0 Hz, ~0.1H), 6.31 (s (b), ~0.6H), 5.26–4.60 (m, ~2.4H), 3.89–3.84 (m, ~5.7H), 3.77 (s, ~0.3H), 3.15 (s, 0.35H), 2.83 (s, 1H), 2.62–2.44 (m, 1.65H), 1.89 (s, 0.35H), 1.88–1.79 (m, 1.65H), 1.45–1.43 (m, 1H). **^{13}C NMR** (125 MHz, $CDCl_3$) δ 171.8, 1701.0, 170.7 (br), 149.1, 148.9 (br), 148.8, 148.7, 148.6 (br), 148.3, 143.5, 143.2 (br) overlappling 143.0 (br), 142.9, 142.3, 142.1 (br), 142.0, 141.8 (br), 139.6, 139.4, 139.4, 139.1 (br), 131.5, 130.3, 130.1, 130.0, 130.0, 129.8 (br), 128.6, 128.6, 128.3, 127.9, 127.8, 127.4, 127.1, 127.01, 126.6, 122.1, 122.0, 121.2 (br), 119.6, 118.2, 114.6, 113.8, 113.5, 113.4, 112.6, 112.4, 111.0, 110.8, 110.7, 92.8, 92.0, 67.7, 61.7, 56.3, 56.0, 55.9, 33.4 (br) overlapping 33.18 (br), 30.24, 22.99, 22.03 (br), 21.42 (br), 21.20. **IR** (neat, cm^{-1}) 3001, 2954, 2934, 2914, 1635, 1591, 1515, 1464, 1440, 1419, 1393, 1337, 1254, 1154, 1141, 1073, 1028. **HRMS** (ESI+) Calc'd for $C_{20}H_{23}INO_3$ 452.07226, found 452.07321. Enantiomeric purity was determined by HPLC analysis in comparison with racemic material t_r = 8.2 (minor) t_r = 11.9 (major) (99:1 er shown; Chiracel AD-H column, 25% ^{i}PrOH in hexanes 1.0 mL/min, 210 nm). $[\alpha]_D^{20}$ −79.7 (c = 1.0, $CHCl_3$).

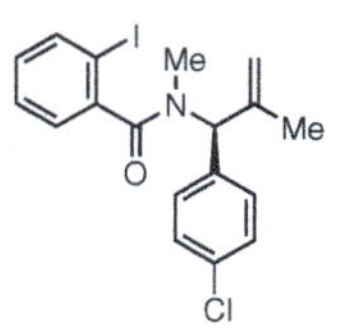

(*S*)-*N*-(1-(4-chlorophenyl)-2-methylallyl)-2-iodo-*N*-methylbenzamide (1.237m)[191]**–Synthesized according to general procedure 9** using 2-iodobenzoic acid (459 mg, 1.85 mmol) and **1.235e** (400 mg, 2.03 mmol). The crude amide was obtained in analytically pure form by silica gel flash column chromatography using hexanes: EtOAc:DCM (6:0.75:4 v:v:v) as the mobile phase. The title compound was obtained as an extremely viscous, clear and colourless oil (789 mg, 1.84 mmol, 99%). *NMR analysis showed the clean desired compound to be present as a complex mixture of 4 rotamers, and only approximate integration values are displayed.* **^{1}H NMR** (500 MHz, $CDCl_3$) δ 7.93–7.74 (m, 1H), 7.42–7.17 (m, 5H), 7.14–6.92 (m, ~1.8H), 6.66 (dd, J = 7.5, 1.5 Hz, ~0.2H), 6.36 (s, ~0.6H), 5.26–4.51 (m, ~2.4H), 3.15 (s, ~0.4H), 2.86 (s, ~0.7H), 2.69–2.50 (m, ~1.9H), 1.95–1.76 (m, ~2.3H), 1.50–1.40 (m, ~0.7H). **^{13}C NMR** (125 MHz, $CDCl_3$) δ 171.8, 171.1, 170.9, 142.8, 142.6, 142.1, 141.7, 141.5, 139.6, 139.5, 139.3 (br) overlappling 139.2 (br), 137.6, 136.2 (br), 135.0 (br), 134.1, 133.6 (br), 133.3, 131.3 overlapping 131.2 (br), 130.4, 130.3, 130.1, 128.8 overlappling 128.8, 128.6, 128.5 (br), 127.8, 127.5, 126.9, 126.9, 126.5, 118.6, 115.7 (br), 114.6, 113.6 (br), 92.8, 92.7, 92.0 (br), 67.4, 67.2, 61.2, 33.3 (br) overlapping 33.1 (br), 30.5, 30.3, 22.2 (br) overlapping 21.3 (br), 21.1. **IR** (neat, cm^{-1}) 3087, 3051, 2993, 2972, 2940, 2913, 2852, 1640, 1490, 1436, 1391, 1337, 1259, 1180, 1091, 1073, 1034. **HRMS** (ESI+) Calc'd for $C_{18}H_{18}ClINO$ 426.01071, found 426.01074. Enantiomeric purity was determined by HPLC analysis in comparison with racemic material t_r = 9.8 (minor) t_r = 12.3 (major) (>99:1 er shown; Chiracel AD-H column, 15% ^{i}PrOH in hexanes 1.0 mL/min, 210 nm). $[\alpha]_D^{20}$ −81.5 (c = 1.0, $CHCl_3$).

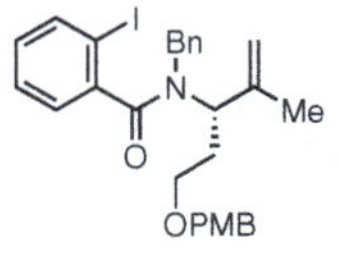

(*S*)-*N*-benzyl-2-iodo-*N*-(5-((4-methoxybenzyl)oxy)-2-methylpent-1-en-3-yl)benzamide (1.237n)—Synthesized according to general procedure **5** using 2-iodobenzoic acid (186 mg, 0.750 mmol) and **S4h** (256 mg, 0.788 mmol, 1.05 equiv). The crude amide was obtained in analytically pure form by silica gel flash column chromatography using hexanes:EtOAc (3:1 v:v) as the mobile phase. The title compound was obtained as a clear and faintly yellow oil (355 mg, 0.640 mmol, 85%). *NMR analysis showed the clean desired compound to be present as a*

[191]Substrate prepared by Hyung Yoon.

complex mixture of 4 rotamers, and only approximate integration values are displayed. **^{1}H NMR** (500 MHz, $CDCl_3$) δ 7.92–7.39 (m, 2H), 7.36–7.11 (m, ~5.5H), 7.10–6.93 (m, ~3.5H), 6.91–6.74 (m, 2H), 5.28–4.78 (m, 3H), 4.53–4.04 (m, 4H), 3.81 (s, 2H), 3.79 (s, 1H), 3.69–3.46 (m, 1H), 3.32–3.21 (m, 0.5H), 3.13–2.98 (m, 0.5H), 2.26–2.11 (m, 0.5H), 2.10–1.76 (m, 3H), 1.72 (s, 0.5H), 1.66–1.62 (m, ~1H). **^{13}C NMR** (125 MHz, $CDCl_3$) δ 171.9, 171.6 overlapping 171.5, 159.2 overlapping 159.1, 159.0, 143.1 (br), 142.8 (br), 142.7 (br), 142.2, 142.0, 141.2, 141.1, 140.1, 139.7, 139.4 (br), 139.2 (br), 139.0, 138.8, 137.3 (br), 137.1 (br), 130.6 (br) overlapping 130.5 (br), 130.2, 130.1, 130.0, 129.8, 129.5, 129.1, 129.0, 128.9, 128.3, 128.2, 128.2 overlapping 128.1 (br), 127.9 (br) overlapping 127.8 (br), 127.6, 127.6, 127.6, 127.4 (br), 127.2, 127.1, 127.0, 115.0, 114.8, 114.5, 113.8, 113.6, 113.6, 93.5, 93.2 (br), 92.9 (br), 92.7, 72.7, 72.4, 72.2, 67.5, 67.1, 66.6, 61.2, 60.6, 56.4 (br), 55.7 (br), 55.3, 55.3, 50.1 (br), 49.7 (br), 45.9, 45.1, 32.5, 31.8, 31.4 (br), 30.7 (br), 23.1, 22.9, 21.7. **IR** (film, cm^{-1}) 2950, 2934, 2858, 1634, 1612, 1585, 1511, 1463, 1409, 1360, 1302, 1247, 1172, 1100, 1079, 1033, 1016. **HRMS** (DART) Calc'd for $C_{28}H_{31}INO_3$ 556.13486, found 556.13404. Enantiomeric purity was determined by HPLC analysis in comparison with racemic material t_r = 22.4 (minor) t_r = 24.6 (major) (>99.5:0.5 er shown; Chiracel AD-H column, 10% ^{i}PrOH in hexanes 1.0 mL/min, 210 nm). $[\alpha]_D^{20}$ −64.3 (c = 0.54, $CHCl_3$).

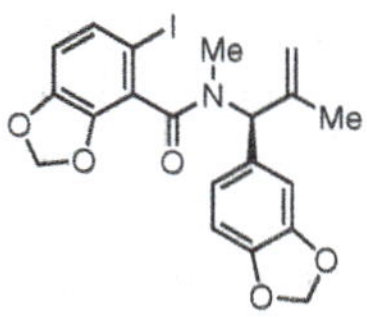

(*S*)-*N*-(1-(benzo[*d*][1, 3]dioxol-5-yl)-2-methylallyl)-5-iodo-*N*-methylbenzo[*d*][1, 3]dioxole-4-carboxamide (1.224)—Synthesized according to a modified version of **general procedure 9** using 5-iodobenzo[*d*][1, 3]dioxole-4-carboxylic acid (6.08 g, 20.9 mmol), oxalyl chloride (1 equiv), DMF (10 drops), and **1.226** (4.71 g, 23.0 mmol, 1.1 equiv). The crude amide was obtained in analytically pure form by silica gel flash column chromatography using hexanes:EtOAc (10:3 v:v) as the mobile phase. The title compound was obtained as a free flowing by hygroscopic white solid (8.37 g, 17.4 mmol, 84%, MP = 209–210 °C). *NMR analysis showed the clean desired compound to be present as a complex mixture of 4 rotamers, and only approximate integration values are displayed.* ^{1}H NMR (500 MHz, Chloroform-*d*) δ 7.30–7.24 (m, 1H), 6.89–6.85 (m, 1H), 6.81–6.76 (m, ~1.5H), 6.76–6.65 (m, ~0.5H), 6.27–6.22 (m, ~0.75H), 6.07–5.94 (m, ~3.5H), 5.92–5.56 (m, ~0.5H), 5.20–4.74 (m, ~2.25H), 3.04 (s, ~0.5H), 2.95 (s, ~0.5H), 2.66 (s, ~1H), 2.60 (s, ~1H), 1.86 (dt, *J* = 1.5, 1.0 Hz, ~1H), 1.81–1.79 (m, ~1.5H), 1.61 (dt, *J* = 1.5, 1.0 Hz, ~0.5H). **^{13}C NMR** (125 MHz, $CDCl_3$) δ 167.5, 167.4, 166.6, 166.5, 148.2, 148.2, 148.1, 147.9, 147.8, 147.8, 147.7, 147.6, 147.3, 147.2, 146.9, 146.9, 144.8, 144.4, 144.3, 142.5, 142.2, 141.7, 141.5, 132.6, 132.3, 132.2,

132.1, 131.9, 130.8, 130.3, 130.1, 124.3, 124.1, 123.8, 123.7, 123.5, 123.1, 122.6, 121.6, 117.6, 115.2, 114.7, 113.1, 110.5, 110.5, 110.5, 110.2, 109.7 109.5, 108.7, 108.2, 108.2, 108.1, 107.9, 102.1, 102.1, 101.9, 101.8, 101.2, 101.1, 101.1, 101.1, 81.8, 81.6, 81.1, 81.0, 67.4, 61.9, 61.8, 32.6, 32.5, 30.9, 30.8, 22.4, 22., 21.3, 21.0. **IR** (film, cm^{-1}) 2972, 2904, 1639, 1502, 1489, 1444, 1394, 1373, 1336, 1242, 1099, 1037. **HRMS** (DART+) Calc'd for $C_{20}H_{19}INO_5$ 480.03079, found 480.03057. Enantiomeric purity was determined by HPLC analysis in comparison with racemic material t_r = 8.2 (minor), t_r = 11.2 (major) (>99:1 er shown; Chiracel AD-H column, 20% ^{i}PrOH in hexanes 1.0 mL/min, 210 nm). $[\alpha]_D^{20}$ −32.2 (c = 0.8, $CHCl_3$).

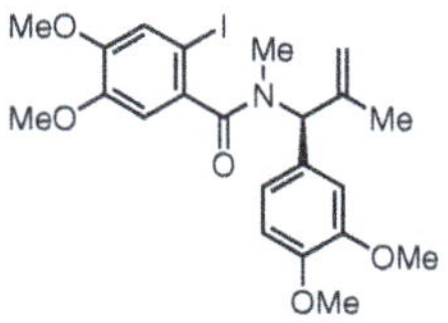

(*S*)-*N*-(1-(3,4-dimethoxyphenyl)-2-methylallyl)-2-iodo-4,5-dimethoxy-*N*-methyl benzamide (1.237o)—Synthesized according to **general procedure 9** using 2-iodo-4,5-dimethoxybenzoic acid (505 mg, 1.64 mmol) and **1.235d** (400 mg, 1.81 mmol). The crude amide was obtained in analytically pure form by silica gel flash column chromatography using hexanes:EtOAc (1:1 v:v) as the mobile phase. The title compound was obtained as a hygroscopic white solid (808 mg, 1.58 mmol, 97%, MP = 56–58 °C). *NMR analysis showed the clean desired compound to be present as a complex mixture of at least 3 rotamers, and only approximate integration values are displayed.* **^{1}H NMR** (500 MHz, $CDCl_3$) δ 7.32–7.09 (m, ~1.3H), 6.96–6.43 (m, ~3.5H), 6.27 (s, ~0.5H), 6.17 (s, ~0.2H), 5.29–4.53 (m, ~2.5H), 3.91–3.74 (m, 11H), 3.24 (s, 0.5H), 3.16 (s, 0.5H), 2.80 (s, ~1H), 2.64–2.48 (m, ~1.5H), 1.92 (s, ~0.5H), 1.87–1.78 (m, ~1.1H), 1.69 (s, ~0.2H), 1.50 (s, ~1.1H), 1.17 (s, ~0.6H). **^{13}C NMR** (125 MHz, $CDCl_3$) δ 171.8, 170.9, 170.6, 149.7, 149.6, 149.5, 149.1, 148.9, 148.8, 148.8, 148.5 (br), 148.3 (br), 148.3 (br), 144.3, 143.2, 142.7, 142.1 (br), 137.1, 135.3 (br), 134.6, 134.1, 133.3, 132.2, 130.0, 128.5, 127.9, 122.1 (br), 121.8, 121.6, 121.3 (br), 119.5, 118.2, 116.2, 113.5, 113.5, 113.4, 112.1 (br), 110.9, 110.9, 110.7, 110.2, 110.0, 109.8, 81.1, 80.9, 80.4, 67.9, 67.8, 61.9, 58.6, 56.3, 56.1, 56.1, 56.1, 56.0, 56.0, 55.9, 55.8, 55.8, 55.1, 33.3 (br), 30.5, 30.3, 23.8, 23.1, 21.3, 21.3, 20.9. **IR** (film, cm^{-1}) 2956, 2932, 2916, 1633, 1593, 1515, 1507, 1463, 1394, 1253, 1211, 1159, 1140, 1027. **HRMS** (ESI+) Calc'd for $C_{22}H_{27}INO_5$ 512.09339, found 512.09278. Enantiomeric purity was determined by HPLC analysis in comparison with racemic material t_r = 11.1 (minor) t_r = 15.0 (major) (>99:1 er shown; Chiracel AD-H column, 25% ^{i}PrOH in hexanes 1.0 mL/min, 210 nm). $[\alpha]_D^{20}$ −73.4 (c = 1.0, $CHCl_3$).

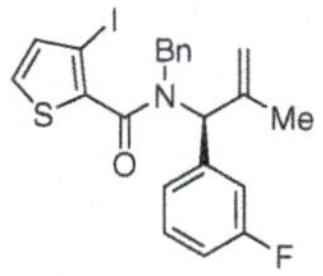

(S)-N-benzyl-N-(1-(3-fluorophenyl)-2-methylallyl)-3-iodothiophene-2-carboxamide (1.237)[192]—Synthesized according to **general procedure 9** using 3-iodothiophene-2-carboxylic acid (127.5 mg, 0.5 mmol) and **1.235b** (140.4 mg, 0.55 mmol). The crude amide was obtained in analytically pure form by silica gel flash column chromatography using hexanes:EtOAc (3:1 v:v) as the mobile phase. The title compound was obtained as a white crystalline solid (195 mg, 0.41 mmol, 82%, MP 100–101 °C). *NMR analysis showed the clean desired compound to be present as a complex mixture of 3 rotamers, and only approximate integration values are displayed.* **^{1}H NMR** (500 MHz, $CDCl_3$) δ 7.24–6.90 (m, 10H), 6.86 (tdd, J = 8.5, 2.5, 1.0 Hz, 1H), 5.68–4.86 (m (br), 4H), 4.34 (d, J = 15.5 Hz, 1H), 1.74–1.45 (m(br), 3H). **^{13}C NMR** (125 MHz, $CDCl_3$) δ 166.2, 163.7, 161.7, 142.8, 138.4, 137.4, 136.1, 135.1, 129.8, 129.8, 127.8, 127.7, 127.5, 126.6, 125.7, 125.8, 79.4, 21.6. **^{19}F NMR** (564 MHz, $CDCl_3$) δ-112.80 to -113.10 (m). **IR** (neat, cm^{-1}) 3107, 3080, 3063, 3034, 2940, 2928, 1628, 1616, 1589, 1489, 1441, 1427, 1395, 1316, 1283, 1262, 1246, 1234, 1152, 1138, 910, 874, 862. **HRMS** (DART) Calc'd for $C_{22}H_{20}FINOS$ 492.02943, found 492.02949. $[\alpha]_D^{20}$ −37.731 (c = 0.58, $CHCl_3$).

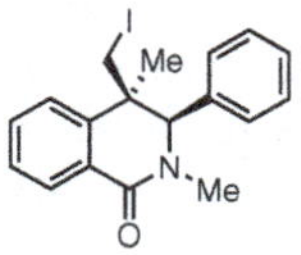

(3*R*,4*R*)-4-(iodomethyl)-2,4-dimethyl-3-phenyl-3,4-dihydroisoquinolin-1(2*H*)-one (1.244a)—The title compound was synthesized from **1a** (78.3 mg, 0.2 mmol) using **general procedure 10a**, and was isolated via silica gel flash column chromatography using hexanes:EtOAc (3:1 v:v) as the mobile phase (69.1 mg, 0.176 mmol, 88%, >95:5 dr (*cis:trans*)).

Major (*cis*-**1.244a**): Isolated as a white solid (MP = 136–137 °C). **^{1}H NMR** (500 MHz, $CDCl_3$) δ 8.22 (ddd, J = 7.5, 2.0, 0.5 Hz, 1H), 7.49 (td, J = 7.5, 2.0 Hz, 1H), 7.44 (td, J = 7.5, 2.0 Hz, 1H), 7.26–7.21 (m, 2H), 7.16–7.12 (m, 2H), 7.09–7.05 (m, 2H), 4.34 (s, 1H), 3.93 (d, J = 10.0 Hz, 1H), 3.06 (s, 3H), 2.97–2.93 (m, 1H), 1.63 (d, J = 1.0 Hz, 3H). **^{13}C NMR** (125 MHz, $CDCl_3$) δ 163.5, 140.1, 136.2, 132.3, 128.8, 128.8, 128.6, 128.6, 128.5, 127.6, 123.2, 72.0, 40.3, 34.3, 31.4, 16.2. **IR** (cm^{-1}, film) 3064, 3031, 2966, 2929, 2875, 1645, 1600, 1575, 1473, 1457, 1395, 1297, 1258, 1197, 1158. **HRMS** (ESI+) Calc'd for $C_{18}H_{19}INO$,

[192]Substrate prepared by Hyung Yoon.

392.0506; found, 392.0511. Enantiomeric purity was determined by HPLC analysis in comparison with racemic material t_r = 10.9 (minor) t_r = 11.9 (major) (>99:1 er shown; Chiracel AD-H column, 10% ^{i}PrOH in hexanes 1.0 mL/min, 210 nm). $[\alpha]_D^{20}$ +261.6 (c = 1.0, $CHCl_3$).

Minor (*trans*-**1.244a**): ^{1}**H NMR** (500 MHz, $CDCl_3$) δ 8.27–8.23 (m, 1H), 7.50–7.43 (m, 2H), 7.24–7.14 (m, 4H), 6.92–6.89 (m, 2H), 4.61 (s, 1H), 3.78 (dd, *J* = 10.5, 1.0 Hz, 1H), 3.34 (d, *J* = 10.5 Hz, 1H), 3.07 (s, 3H), 1.22 (d, *J* = 1.0 Hz, 3H). ^{13}**C NMR** (125 MHz, $CDCl_3$) δ 163.3, 138.7, 136.5, 132.5, 129.0, 128.6, 128.4, 128.3, 128.2, 128.0, 125.1, 70.7, 40.9, 34.3, 22.8, 22.4. **HRMS** (ESI+) Calc'd for $C_{18}H_{19}INO$, 392.0506; found, 392.0510. Enantiomeric purity was determined by HPLC analysis in comparison with racemic material t_r = 9.2 (minor) t_r = 16.5 (major) (99:1 er. shown; Chiracel AD-H column, 10% ^{i}PrOH in hexanes 1.0 mL/min, 210 nm).

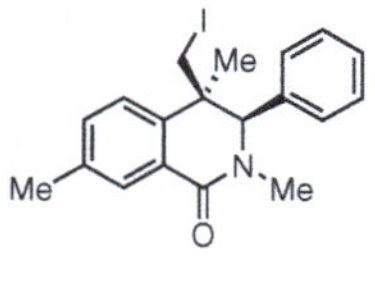

(3*R*,4*R*)-4-(iodomethyl)-2,4,7-trimethyl-3-phenyl-3,4-dihydroisoquinolin-1(2*H*)-one (1.244b)[193]—The title compound was synthesized from **1.237b** (81.0 mg, 0.2 mmol) using **general procedure 10b**, and was isolated via silica gel flash column chromatography using hexanes:EtOAc (3:1 v:v) as the mobile phase (71.6 mg, 0.177 mmol, 88%, 94:6 dr (*cis:trans*)). **Major**: Isolated as a white solid (MP = 47–49 °C). ^{1}**H NMR** (500 MHz, $CDCl_3$) δ 8.04–8.03 (m, 1H), 7.30–7.27 (m, 1H), 7.25–7.21 (m, 1H), 7.16–7.11 (m, 2H), 7.13–7.09 (m, 1H), 7.09–7.05 (m, 2H), 4.31 (s, 1H), 3.91 (d, *J* = 10.0 Hz, 0H), 3.05 (s, 3H), 2.94–2.90 (m, 1H), 2.43 (s, 3H), 1.60 (d, 1.0 Hz, 3H). ^{13}**C NMR** (125 MHz, $CDCl_3$) δ 163.7, 137.48, 137.1, 136.3, 132.9, 129.3, 128.6, 128.5, 128.4, 128.3, 123.1, 72.1, 40.0, 34.3, 31.4, 21.0, 16.4. **IR** (cm^{-1}, film) 3070, 3028, 2966, 2861, 1647, 1608, 1573, 1482, 1452, 1394, 1374, 1353, 1298, 1261, 1195, 1087. **HRMS** (DART) Calc'd for $C_{19}H_{21}INO$, 406.06678; found, 406.06645. Enantiomeric purity was determined by HPLC analysis in comparison with racemic material t_r = 19.2 (minor) t_r = 20.6 (major) (>99:1 er shown; Chiracel AD-H column, 5% ^{i}PrOH in hexanes 1.0 mL/min, 225 nm). $[\alpha]_D^{20}$ +216.8 (c = 0.5, $CHCl_3$).

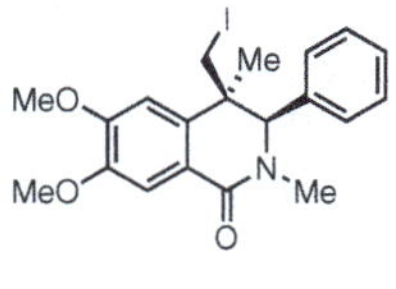

[193]Reaction was run by Hyung Yoon.

(3*R*,4*R*)-4-(iodomethyl)-6,7-dimethoxy-2,4-dimethyl-3-phenyl-3,4-dihydroiso quinolin-1(2*H*)-one (1.244c)—The title compound was synthesized from **1.237c** (90.2 mg, 0.2 mmol) using **general procedure 10a**, and was isolated via silica gel flash column chromatography using hexanes:EtOAc (1:1 v:v) as the mobile phase (73.6 mg, 0.163 mmol, 82%, 93:7 dr (*cis:trans*)). **Major**: Isolated as a white solid (MP = 50–52 °C). **^{1}H NMR** (500 MHz, $CDCl_3$) δ 7.74 (s, 1H), 7.27–7.23 (m, 1H), 7.18–7.14 (m, 2H), 7.12–7.07 (m, 2H), 6.72 (s, 1H), 4.28 (s, 1H), 3.98 (s, 3H), 3.90 (s, 3H), 3.82 (d, *J* = 10.0 Hz, 1H), 3.03 (s, 3H), 2.97 (dd, *J* = 10.0, 1.0 Hz, 1H), 1.63 (d, *J* = 1.0 Hz, 3H). **^{13}C NMR** (125 MHz, $CDCl_3$) δ 163.5, 151.9, 148.0, 136.3, 133.7, 128.7, 128.6, 128.4, 121.6, 111.3, 106.2, 72.3, 56.1, 56.1, 39.9, 34.2, 31.3, 16.2. **IR** (cm^{-1}, film) 3025, 3000, 2962, 1641, 1581, 1511, 1482, 1454, 1436, 1406, 13989, 1346, 1286, 1241, 1217, 1207, 1184, 1163. **HRMS** (DART) Calc'd for $C_{20}H_{23}INO_3$, 452.07226; found, 452.07256. Enantiomeric purity was determined by HPLC analysis in comparison with racemic material t_r = 14.7 (major) t_r = 16.0 (minor) (>99:1 er shown; Chiracel AD-H column, 10% ^{i}PrOH in hexanes 1.0 mL/min, 210 nm). $[\alpha]_D^{20}$ +223.0 (c = 1.0, $CHCl_3$).

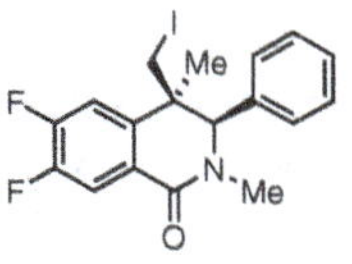

(3*R*,4*R*)-6,7-difluoro-4-(iodomethyl)-2,4-dimethyl-3-phenyl-3,4-dihydroiso-quinolin-1(2*H*)-one (1.244d)—The title compound was synthesized from **1.237d** (81.0 mg, 0.2 mmol), Pd(QPhos)$_2$ (23.0 mg, 0.075 mmol, 7.5 mol% [Pd]), and 1,2,2,6,6-pentamethylpiperidine (127 μL, 0.7 mmol, 3.5 equiv) using **general procedure 10a**, and was isolated via silica gel flash column chromatography using hexanes:EtOAc (3:1 v:v) as the mobile phase (43.5 mg, 0.107 mmol, 54%, 97:3 dr (*cis:trans*)). **Major**: Isolated as a white solid (MP = 111–113 °C). **^{1}H NMR** (500 MHz, $CDCl_3$) δ 8.05 (dd, *J* = 10.5, 8.5 Hz, 1H), 7.30–7.26 (m, 1H), 7.21–7.16 (m, 2H), 7.09–7.04 (m, 3H), 4.34 (s, 1H), 3.74 (d, *J* = 10.0 Hz, 1H), 3.04 (s, 3H), 2.91 (dd, *J* = 10.0, 1.0 Hz, 1H), 1.62 (d, *J* = 1.0 Hz, 3H). **^{13}C NMR** (125 MHz, $CDCl_3$) δ 161.8 (s), 152.6 (dd, *J* = 255.0, 13.0 Hz), 149.5 (dd, *J* = 250.0, 13.0 Hz), 137.6 (dd, *J* = 6.0, 4.0 Hz), 135.6, 128.9, 128.6, 128.4, 126.1 (dd, *J* = 5.5, 3.5 Hz), 118.4 (dd, *J* = 19.0, 1.5 Hz), 113.3 (d, *J* = 19.5 Hz), 72.0, 40.1 (d, *J* = 1.0 Hz), 34.4, 31.3 (d, *J* = 1.0 Hz), 14.6. **^{19}F NMR** (564 MHz, $CDCl_3$) δ-130.57 to -130.68 (m), -137.94 to -138.03 (m). **IR** (cm^{-1}, film) 3059, 3030, 3004, 2968, 2926, 2861, 1652, 1616, 1608, 1505, 1487, 1452, 1394, 1355, 1314, 1291, 1245, 1208, 1190, 1173, 1080. **HRMS** (DART) Calc'd for $C_{18}H_{17}F_2INO_3$, 428.03229; found, 428.03243. Enantiomeric purity was determined by HPLC analysis in comparison with racemic material t_r = 14.4 (major) t_r = 22.3 (minor) (>99:1 er shown; Chiracel AD-H column, 5% ^{i}PrOH in hexanes 1.0 mL/min, 210 nm). $[\alpha]_D^{20}$ +246.0 (c = 1.0, $CHCl_3$).

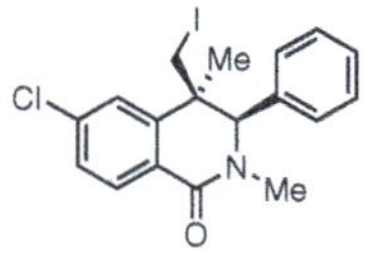

(3*R*,4*R*)-6-chloro-4-(iodomethyl)-2,4-dimethyl-3-phenyl-3,4-dihydroisoquinolin-1(2*H*)-one (1.244e)[194]—The title compound was synthesized from **1.237e** (85.0 mg, 0.2 mmol) using **general procedure 10a**, and was isolated via silica gel flash column chromatography using hexanes:EtOAc:DCM (6:0.75:4 v:v:v) as the mobile phase (64.3 mg, 0.151 mmol, 76%, 97:3 dr (*cis:trans*)). **Major**: Isolated as a white crystalline solid (MP = 59–61 °C). **^{1}H NMR** (500 MHz, $CDCl_3$) δ 8.16 (dd, *J* = 8.5, 0.5 Hz, 1H), 7.41 (dd, *J* = 8.5 2.0 Hz, 1H), 7.28–7.24 (m, 1H), 7.22 (d, *J* = 2.0 Hz, 1H), 7.19–7.14 (m, 2H), 7.10–7.05 (m, 2H), 4.35 (s, 1H), 3.85 (d, *J* = 10.0 Hz, 0H), 3.05 (s, 3H), 2.94–2.90 (m, 1H), 1.63 (d, *J* = 1.0 Hz, 3H). **^{13}C NMR** (125 MHz, $CDCl_3$) δ 162.6, 141.9, 138.4, 135.7, 130.3, 128.6, 128.4, 128.4, 127.8, 127.2, 123.8, 71.8, 40.3, 34.2, 31.2, 15.0. **IR** (cm^{-1}, film) 3063, 3028, 3001, 2966, 2926, 2855, 1647, 1593, 1450, 1402, 1392, 1308, 1257, 1197, 1160, 1101, 1079, 1052, 1031. **HRMS** (DART) Calc'd for $C_{18}H_{18}ClINO$, 426.01216; found, 426.01178. Enantiomeric purity was determined by HPLC analysis in comparison with racemic material t_r = 10.7 (major) t_r = 11.44 (minor) (>99:1 er shown; Chiracel AD-H column, 15% ^{i}PrOH in hexanes 1.0 mL/min, 210 nm). $[\alpha]_D^{20}$ +272.0 (c = 0.5, $CHCl_3$).

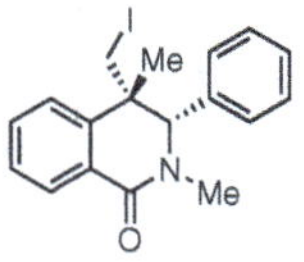

(3S,4*S*)-4-(iodomethyl)-2,4-dimethyl-3-phenyl-3,4-dihydroisoquinolin-1(2*H*)-one (*ent*-1.244)—The title compound was synthesized from ***ent*-1.237a** (78.3 mg, 0.2 mmol) using **general procedure 10a**, and was isolated via silica gel flash column chromatography using hexanes:EtOAc (3:1 v:v) as the mobile phase (70.3 mg, 0.178 mmol, 90%, >95:5 dr (*cis:trans*)). **All analytic data was in accordance with that reported for the (3*R*,4*R*) analog shown above** Enantiomeric purity was determined by HPLC analysis in comparison with racemic material t_r = 10.9 (major) t_r = 11.9 (major) (>99:1 er shown; Chiracel AD-H column, 10% ^{i}PrOH in hexanes 1.0 mL/min, 210 nm). $[\alpha]_D^{20}$ −264.0 (c = 1.0, $CHCl_3$).

[194]Reaction was run by Hyung Yoon.

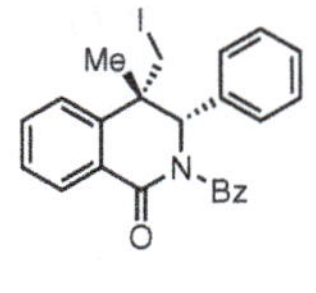

(3*S*,4*S*)-2-benzoyl-4-(iodomethyl)-40-methyl-3-phenyl-3,4-dihydroisoquinolin-1(2*H*)-one (1.244 g)—The title compound was synthesized from **1.237 g** (96.4 mg, 0.2 mmol) using **general procedure 10b** heating for 48 h, and was isolated as a white foam via silica gel flash column chromatography using hexanes:EtOAc (5:1 v:v) as the mobile phase (58.9 mg, 0.122 mmol, 61%, 65:35 dr (*cis:trans*)). The compounds were characterized as a ~63:37 (major:minor) mixture of diastereomers. **^{1}H NMR** (500 MHz, $CDCl_3$) δ 8.23–8.17 (m, 1H), 7.68–7.58 (m, 1H), 7.55–7.44 (m, 4H), 7.44–7.31 (m, 3H), 7.24–7.03 (m, 5H), 5.71 (s, 0.37H, minor), 5.68 (s, 0.65H, major), 4.01 (d, *J* = 10.0 Hz, ~0.63H, major), 3.87 (d, *J* = 10.5 Hz, ~0.37H, minor), 3.52 (d, *J* = 10.5 Hz, ~0.37H, minor), 3.08 (dd, *J* = 10.0, 1.0 Hz, 0.63H, major), 1.79 (d, *J* = 1.0 Hz, ~1.89H, major), 1.37 (s, ~1.11H, minor). **^{13}C NMR** (125 MHz, $CDCl_3$) δ 173.6, 173.4, 164.8, 164.3, 142.0, 140.8, 137.7, 136.9, 136.5, 136.4, 134.4, 134.3, 131.5, 131.5, 130.0, 129.7, 129.1, 128.6, 128.5, 128.4, 128.4, 128.4, 128.4, 128.2, 128.2, 128.2, 128.1, 128.1, 127.7, 127.6, 125.9, 124.1, 66.9, 66.4, 41.5, 40.9, 31.7, 23.2, 20.7, 15.4. **IR** (cm^{-1}, film) 3064, 3031, 2959, 2927, 1687, 1599, 1457, 1388, 1262. **HRMS** (DART) Calc'd for $C_{124}H_{21}INO_2$, 482.06170; found, 482.06134. Enantiomeric purity was determined by HPLC analysis in comparison with racemic material Major diastereomer: t_r = 8.0 (major enantiomer) t_r = 11.0 (minor enantiomer), minor diastereomer: T_R = 9.2 (major enantiomer) T_R = 23.6 (minor enantiomer) (Major diastereomer: >99:1 er, Minor diastereomer: >99:1 er shown; Chiracel AD-H column, 30% ^{i}PrOH in hexanes 1.0 mL/min, 210 nm). *Note*: In the absence of base the product was obtained in 94% NMR yield as a 52:48 mixture of diastereomers after 24 h.

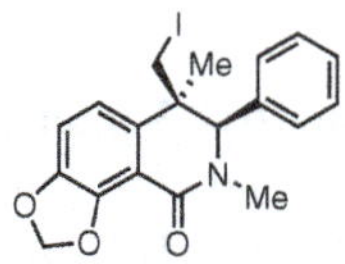

(6*R*,7*R*)-6-(iodomethyl)-6,8-dimethyl-7-phenyl-7,8-dihydro-[1, 3]dioxolo[4,5-*h*] isoquinolin-9(6*H*)-one (1.244 h)—The title compound was synthesized from **1.237 g** (87.1 mg, 0.2 mmol) using **general procedure 10b**, and was isolated via silica gel flash column chromatography using hexanes:EtOAc (2:3 v:v) as the mobile phase (70.9 mg, 0.163 mmol, 82%, 94:6 dr (*cis:trans*)). **Major:** Isolated as a white solid (MP = 155–157 °C). **^{1}H NMR** (500 MHz, $CDCl_3$) δ 7.27–7.22 (m, 1H), 7.18–7.14 (m, 2H), 7.14–7.11 (m, 2H), 6.85 (d, *J* = 8.0 Hz, 1H), 6.63 (dd, *J* = 8.0, 0.5 Hz, 1H), 6.18 (d, *J* = 1.5 Hz, 1H), 6.16 (d, *J* = 1.5 Hz, 1H), 4.28 (s, 1H), 3.83 (d, *J* = 10.0 Hz, 0H), 3.02 (s, 3H), 2.90–2.86 (m, 1H), 1.59 (d, *J* = 1.0 Hz, 3H). **^{13}C NMR** (125 MHz, $CDCl_3$) δ 161.5, 148.4, 148.1, 136.1,

133.4, 128.6, 128.5, 128.3, 116.1, 112.3, 110.5, 102.5, 72.1, 40.3, 33.7, 31.3, 16.5. **IR** (cm^{-1}, film) 3006, 2967, 2903, 1646, 1602, 1456, 1437, 1374, 1255, 1220, 1207, 1133, 1083, 1048, 1028. **HRMS** (DART) Calc'd for $C_{19}H_{19}INO_3$, 436.04096; found, 436.04015. Enantiomeric purity was determined by HPLC analysis in comparison with racemic material t_r = 46.4 (minor) t_r = 50.20 (major) (>99:1 er shown; Chiracel AD-H column, 5% ^{i}PrOH in hexanes 1.0 mL/min, 210 nm). $[\alpha]_D^{20}$ + 153.2 (c = 1.0, $CHCl_3$).

(3*R*,4*R*)-2-benzyl-4-(iodomethyl)-4-methyl-3-phenyl-3,4-dihydroisoquinolin-1 (2*H*)-one (1.244i)[195]—The title compound was synthesized from **1.237i** (93.4 mg, 0.2 mmol) using **general procedure 10b**, and was isolated via silica gel flash column chromatography using hexanes:EtOAc:DCM (6:0.75:4 v:v) as the mobile phase (88.6 mg, 0.189 mmol, 95%, 98:2 dr (*cis:trans*)). The major and minor enantiomers were not separated and are herein characterized as a 98:2 ratio, specific was not obtained and is not provided. Isolated as a white foam. **^{1}H NMR** (500 MHz, $CDCl_3$) δ 8.32–8.26 (m, 1H), 7.53–7.43 (m, 2H), 7.43–7.29 (m, 5H), 7.29–7.21 (m, 1H), 7.23–7.17 (m, 1H), 7.15 (ddd, *J* = 8.0, 7.0, 1.0 Hz, 2H), 7.09–7.06 (m, 2H), 5.64 (d, *J* = 14.0 Hz, 1H), 4.34 (s, 1H), 3.82 (d, *J* = 10.0 Hz, 1H), 3.54 (d, *J* = 14.5 Hz, 1H), 2.86 (dd, *J* = 10.0, 1.0 Hz, 1H), 1.25 (d, *J* = 1.0 Hz, 3H). **^{13}C NMR** (125 MHz, $CDCl_3$) δ 163.4, 140.2, 136.5, 136.2, 132.5, 129.3, 129.1, 128.9, 128.9, 128.6, 128.6, 128.4, 127.8, 127.7, 123.3, 67.3, 48.2, 39.8, 31.6, 16.0. **IR** (cm^{-1}, film) 3064, 3028, 2978, 2925, 1647, 1600, 1578, 1495, 1467, 1454, 1445, 1433, 1353, 1297, 1261, 1153, 1080. **HRMS** (ESI+) Calc'd for $C_{24}H_{23}INO$, 468.08243; found, 468.08175. Enantiomeric purity was determined by HPLC analysis in comparison with racemic material t_r = 9.3 (major) t_r = 12.3 (minor) (>99:1 er shown; Chiracel AD-H column, 10% ^{i}PrOH in hexanes 1.0 mL/min, 210 nm).

(3*R*,4*R*)-2-benzyl-3-(3-fluorophenyl)-4-(iodomethyl)-4-methyl-3,4-dihydroiso-quinolin-1(2*H*)-one (1.244j)—The title compound was synthesized from **1.237j** (97.0 mg, 0.2 mmol) using **general procedure 10b**, and was isolated via silica gel flash column chromatography using hexanes:EtOAc:DCM (2:1:1 v:v) as the

[195]Reaction was run by Hyung Yoon.

mobile phase (two repetitions) (83.7 mg, 0.173 mmol, 87%, 98:2 dr (*cis:trans*)). **Major**: Isolated as an off-white foam. **^{1}H NMR** (500 MHz, $CDCl_3$) δ 8.30 (ddd, *J* = 7.5, 1.5, 1.0 Hz, 1H), 7.55–7.43 (m, 2H), 7.42–7.29 (m, 5H), 7.23–7.17 (m, 1H), 7.13 (td, *J* = 8.0, 6.0 Hz, 1H), 6.94 (tdd, *J* = 8.5, 2.5, 1.0 Hz, 1H), 6.91 (d, *J* = 7.0 Hz, 1H), 6.76 (d, *J* = 10.5 Hz, 1H), 5.62 (d, *J* = 14.5 Hz, 1H), 4.35 (s, 1H), 3.84 (d, *J* = 10.0 Hz, 1H), 3.60 (d, *J* = 14.5 Hz, 1H), 2.86 (d, *J* = 10.0 Hz, 0H), 1.25 (s, 3H). **^{13}C NMR** (125 MHz, $CDCl_3$) δ 163.3, 162.3 (d, *J* = 247.0 Hz), 139.8, 138.9 (d, *J* = 6.5 Hz), 136.3, 132.7, 130.0 (d, *J* = 8.0 Hz), 129.3, 129.2, 128.7, 127.9 (d, *J* = 2.0 Hz), 124.8 (d, *J* = 3.0 Hz), 123.3, 115.9, 115.7, 115.5, 66.9 (d, *J* = 2.0 Hz), 48.4, 39.9, 31.5 15.7. **^{19}F NMR** (376 MHz, $CDCl_3$) δ-111.8 (q, *J* = 8.5 Hz). **IR** (cm^{-1}, film) 3019, 3010, 2925, 1647, 1599, 1591, 1488, 1467, 1445, 1296, 1264, 1216, 1152, 1080. **HRMS** (ESI +) Calc'd for $C_{24}H_{22}FINO$, 486.0725; found, 486.0738. Enantiomeric purity was determined by HPLC analysis in comparison with racemic material t_r = 6.6 (major) t_r = 7.8 (minor) (>99:1 er shown; Chiracel AD-H column, 20% ^{i}PrOH in hexanes 1.0 mL/min, 210 nm). $[\alpha]_D^{20}$ + 162.0 (c = 0.5, $CHCl_3$).

(3*R*,4*R*)-2-benzyl-4-(iodomethyl)-3-(3-methoxyphenyl)-4-methyl-3,4-dihydroiso-quinolin-1(2*H*)-one (1.244k)—The title compound was synthesized from **1.237k** (99.4 mg, 0.2 mmol) using **general procedure 10b**, and was isolated via silica gel flash column chromatography using hexanes:EtOAc (3:1 v:v) as the mobile phase (89.9 mg, 0.181 mmol, 90%, >98:2 dr (*cis:trans*)). **Major:** Isolated as a tackey light yellow oil. **^{1}H NMR** (500 MHz, $CDCl_3$) δ 8.28 (dd, *J* = 7.5, 2.0 Hz, 1H), 7.51–7.30 (m, 7H), 7.22–7.18 (m, 1H), 7.06 (t, *J* = 8.0 Hz, 1H), 6.78 (dd, *J* = 8.0, 2.0 Hz, 2H), 6.66 (d, *J* = 8.0 Hz, 1H), 6.60 (s, 1H), 5.64 (d, *J* = 14.5 Hz, 1H), 4.31 (s, 1H), 3.82 (d, *J* = 10.0 Hz, 1H), 3.61–3.53 (m, 4H), 2.89 (d, *J* = 10.0 Hz, 1H), 1.25 (s, 3H). **^{13}C NMR** (125 MHz, $CDCl_3$) δ 163.4, 159.2, 140.1, 137.7, 136.5, 132.4, 129.3, 129.3, 129.3, 128.9, 128.9, 128.5, 127.7, 127.6, 123.2, 121.2, 114.2, 67.2, 54.9, 48.2, 39.8, 31.5, 16.0. **IR** (cm^{-1}, film) 3064, 3027, 2003, 2964, 2927, 2867, 1647, 1600, 1466, 1260, 1151, 1050. **HRMS** (DART) Calc'd for $C_{25}H_{25}INO_2$, 498.09300; found, 498.09418. Enantiomeric purity was determined by HPLC analysis in comparison with racemic material t_r = 18.4 (major) t_r = 24.3 (minor) (>99:1 er shown; Chiracel AD-H column, 5% ^{i}PrOH in hexanes 1.0 mL/min, 210 nm). $[\alpha]_D^{20}$ + 130.5 (c = 0.5, $CHCl_3$).

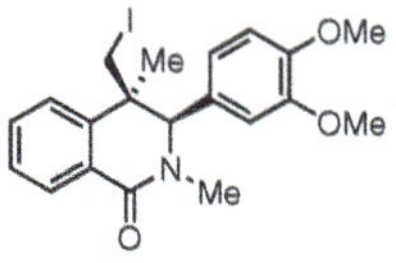

(3*R*,4*R*)-3-(3,4-dimethoxyphenyl)-4-(iodomethyl)-2,4-dimethyl-3,4-dihydroiso quinolin-1(2*H*)-one (1.244l)—The title compound was synthesized from **1.237l** (92.0 mg, 0.2 mmol) using **general procedure 10b**, and was isolated via silica gel flash column chromatography using hexanes:EtOAc (2:1 v:v) as the mobile phase (68.2 mg, 0.151 mmol, 76%, 93:7 dr (*cis:trans*)). An analytically pure sample of the major diastereomer was obtained by recrystallization in THF/hexanes at −20 °C. **Major**: Isolated as a white solid after recrystallization (MP = 223–223 °C). **^{1}H NMR** (500 MHz, $CDCl_3$) δ 8.26–8.19 (m, 1H), 7.48 (td, J = 7.5, 1.5 Hz, 1H), 7.42 (td, J = 7.5, 1.0 Hz, 1H), 7.23–7.19 (m, 1H), 6.73–6.70 (m, 1H), 6.64 (d, J = 8.5 Hz, 1H), 6.47 (s, 1H), 4.29 (s, 1H), 3.93 (d, J = 10.0 Hz, 1H), 3.79 (s, 3H), 3.50 (s, 3H), 3.08 (s, 3H), 2.94 (dd, J = 10.0, 1.0 Hz, 1H), 1.62 (d, J = 1.0 Hz, 3H). **^{13}C NMR** (125 MHz, $CDCl_3$) δ 163.6, 149.0, 148.4, 140.2, 132.3, 129.1, 128.6, 128.4, 127.6, 123.3, 121.5, 110.8, 110.5, 71.6, 55.7, 55.3, 40.5, 34.3, 31.1, 16.6. **IR** (cm^{-1}, film) 2960, 2929, 2905, 1646, 1600, 1523, 1260, 1241, 1081, 1026. **HRMS** (DART) Calc'd for $C_{20}H_{23}INO_3$, 452.07226; found, 452.07117. Enantiomeric purity was determined by HPLC analysis in comparison with racemic material t_r = 6.6 (major) t_r = 8.7 (minor) (>99:1 er shown; Chiracel AD-H column, 25% ^{i}PrOH in hexanes 1.0 mL/min, 210 nm). $[\alpha]_D^{20}$ + 168.9 (c = 1.0, $CHCl_3$).

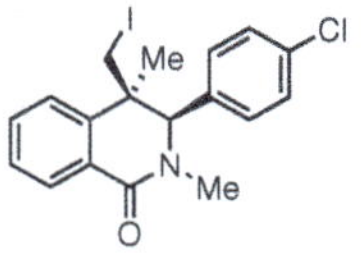

(3*R*,4*R*)-3-(4-chlorophenyl)-4-(iodomethyl)-2,4-dimethyl-3,4-dihydroisoquinolin-1(2*H*)-one (1.244 m)[196]—The title compound was synthesized from (*S*)-*N*-(1-(4-chlorophenyl)-2-methylallyl)-2-iodo-*N*-methylbenzamide **1.237 m** (85.3 mg, 0.2 mmol) using **general procedure 10a**, and was isolated via silica gel flash column chromatography using hexanes:EtOAc:DCM (12:1:8 v:v:v) as the mobile phase (69.8 mg, 0.164 mmol, 82%, 98:2 dr (*cis:trans*)). **Major**: Isolated as a white powder (MP = 117–118 °C). **^{1}H NMR** (500 MHz, $CDCl_3$) δ 8.21 (dd, J = 7.5, 1.5 Hz, 1H), 7.52–7.47 (m, 1H), 7.47–7.41 (m, 1H), 7.21 (d, J = 8.0 Hz, 1H), 7.12 (d, J = 8.5 Hz, 2H), 7.02 (d, J = 8.5 Hz, 2H), 4.33 (s, 1H), 3.94 (d, J = 10.0 Hz, 1H), 3.05 (s, 3H), 2.89 (d, J = 10.5 Hz, 1H), 1.62 (s, 3H). **^{13}C NMR** (125 MHz, $CDCl_3$) δ 163.4, 139.8, 134.8, 134.5, 132.5, 129.9, 128.9, 128.6, 128.6, 127.0, 123.2, 71.4, 40.3, 34.3, 31.3, 15.9. **IR** (cm^{-1}, film) 3070, 3000, 2968, 2925, 2863, 1652, 1643, 1601, 1576, 1492, 1473, 1397, 1375, 1216, 1197, 1111, 1094, 1082. **HRMS** (DART) Calc'd for $C_{18}H_{18}ClINO$, 426.01216; found, 426.01119. Enantiomeric purity was determined by HPLC analysis in comparison with racemic material t_r = 12.4 (major) t_r = 14.0 (minor) (>99:1 er shown; Chiracel AD-H column, 20% ^{i}PrOH in hexanes 0.5 mL/min, 210 nm). $[\alpha]_D^{20}$ + 287.2 (c = 1.0, $CHCl_3$).

[196]Reaction was run by Hyung Yoon.

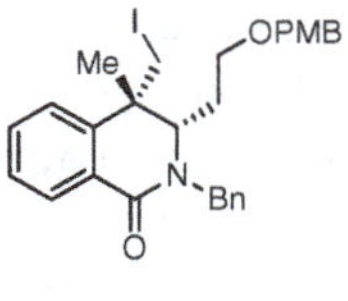

(3*S*,4*S*)-2-benzyl-4-(iodomethyl)-3-(2-((4-methoxybenzyl)oxy)ethyl)-4-methyl-3,4-dihydroisoquinolin-1(2*H*)-one (1.244n)—The title compound was synthesized from **1.237n** (222 mg, 0.4 mmol) using **general procedure 10a**, and was isolated via silica gel flash column chromatography using hexanes:EtOAc:DCM (4:1:4 v:v:v) as the mobile phase (201 mg, 0.362 mmol, 91%, 80:20 dr (*cis:trans*). **Major**: Isolated as a viscous, colourless, and translucent oil. **^{1}H NMR** (500 MHz, Chloroform-*d*) δ 8.10 (dd, *J* = 7.5, 1.5 Hz, 1H), 7.46–7.39 (m, 3H), 7.37–7.33 (m, 1H), 7.32–7.25 (m, 5H), 7.14 (d, *J* = 7.5 Hz, 1H), 6.91–6.88 (m, 2H), 5.73 (d, *J* = 14.0 Hz, 1H), 4.50 (d, *J* = 11.5 Hz, 1H), 4.41 (d, *J* = 11.5 Hz, 1H), 3.85 (d, *J* = 14.0 Hz, 1H), 3.82–3.79 (m, 4H), 3.77 (dd, *J* = 10.5, 3.5 Hz, 1H), 3.56 (td, *J* = 10.0, 4.5 Hz, 1H), 3.44 (ddd, *J* = 10.0, 5.5, 4.0 Hz, 1H), 3.38 (d, *J* = 10.0 Hz, 1H), 1.96 (dddd, *J* = 14.0, 10.0, 5.5, 3.5 Hz, 1H), 1.43–1.35 (m, 1H), 0.76 (s, 3H). **^{13}C NMR** (125 MHz, $CDCl_3$) δ 163.5, 159.2, 140.6, 136.8, 132.2, 130.1, 130.0, 129.3, 128.9, 128.4, 128.4, 127.7, 127.2, 123.1, 113.7, 72.4, 65.8, 60.6, 55.2, 49.8, 39.5, 29.8, 28.3, 16.2. **IR** (cm^{-1}, film) 2959, 2929, 2861, 1639, 1600, 1495, 1470, 1321, 1248, 1102, 1080, 1034. **HRMS** (DART): Calc'd for $C_{28}H_{31}INO_3$, 556.13486; found, 556.13416. Enantiomeric purity was determined by HPLC analysis in comparison with racemic material t_r = 14.1 (major) t_r = 17.3 (minor) (> 99:1 er shown; Chiracel AD-H column, 15% ^{i}PrOH in hexanes 1.0 mL/min, 210 nm). $[\alpha]_D^{20}$ −130.2 (c = 0.55, $CHCl_3$).

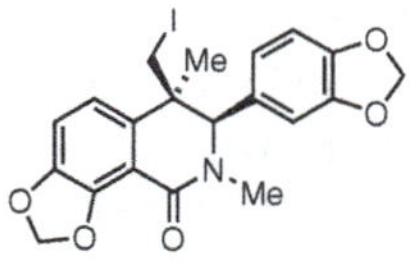

(6*R*,7*R*)-7-(benzo[d][1, 3]dioxol-5-yl)-6-(iodomethyl)-6,8-dimethyl-7,8-dihydro-[1, 3]dioxolo[4,5-h]isoquinolin-9(6H)-one (1.223)—The title compound was synthesized from **1.224** (95.8 mg, 0.2 mmol), $Pd(QPhos)_2$ (23.0 mg, 0.075 mmol, 7.5 mol% [Pd]), and 1,2,2,6,6-pentamethylpiperidine (127 μL, 0.7 mmol, 3.5 equiv) using **general procedure 10b**, and was isolated via silica gel flash column chromatography using hexanes:EtOAc (2:1 v:v) as the mobile phase (80.4 mg, 0.168 mmol, 84%, 92:8 dr (*cis:trans*)).

Major (*cis*-**1.223**): Isolated as a white solid (MP = 209–210 °C). *Note*: X-ray quality crystals of the major diastereomer could be obtained by recrystallization of the purified material via the slow diffusion of hexanes into EtOAc. **^{1}H NMR** (600 MHz, $CDCl_3$) δ 6.86 (d, *J* = 8.0 Hz, 1H), 6.75 (dd, *J* = 8.0, 2.0 Hz, 1H), 6.65 (d, *J* = 6.5 Hz, 1H), 6.64 (d, *J* = 6.5 Hz, 1H), 6.44 (d, *J* = 2.0 Hz, 1H), 6.19 (d, *J* = 1.5 Hz, 1H), 6.16 (d, *J* = 1.5 Hz, 1H), 5.90 (d, *J* = 1.5 Hz, 1H), 5.89

(d, J = 1.5 Hz, 1H), 4.20 (s, 1H), 3.84 (d, J = 10.0 Hz, 1H), 3.03 (s, 3H), 2.95 (d, J = 10.0 Hz, 1H), 1.57–1.57 (m, 3H). **^{13}C NMR** (175 MHz, $CDCl_3$) δ 161.3, 148.5, 148.2, 147.7, 147.5, 133.3, 129.8, 123.2, 116.2, 112.1, 110.6, 108.0, 107.8, 102.5, 101.1, 72.0, 40.2, 33.6, 31.3, 16.5. **IR** (neat, cm^{-1}) 3007, 2901, 1643, 1602, 1460, 1253, 1220, 1047. **HRMS** (DART) Calc'd for $C_{20}H_{19}INO_5$ 480.03079, found 480.02961. Enantiomeric purity was determined by HPLC analysis in comparison with racemic material t_r = 13.9 (minor) t_r = 15.4 (major) (> 99:1 er shown; Chiracel AD-H column, 20% ^{i}PrOH in hexanes 1.0 mL/min, 210 nm). $[\alpha]_D^{20}$ + 146.6 (c = 0.525, $CHCl_3$).

Minor (*trans*-**1.223**): Isolated as an off-white foam. **^{1}H NMR** (600 MHz, $CDCl_3$) δ 6.85 (d, J = 8.0 Hz, 1H), 6.66 (d, J = 5.1 Hz, 1H), 6.65 (d, J = 5.0 Hz, 1H), 6.52 (dd, J = 8.0, 1.8 Hz, 1H), 6.36 (d, J = 1.8 Hz, 1H), 6.19 (d, J = 1.3 Hz, 1H), 6.17 (d, J = 1.3 Hz, 1H), 5.90 (d, J = 1.4 Hz, 1H), 5.88 (d, J = 1.4 Hz, 1H), 4.46 (s, 1H), 3.72 (d, J = 10.1 Hz, 1H), 3.28 (d, J = 10.2 Hz, 1H), 3.03 (s, 3H), 1.18 (s, 3H). **^{13}C NMR** (151 MHz, $CDCl_3$) δ 160.9, 149.0, 148.1, 147.7, 147.6, 131.9, 130.4, 122.1, 118.3, 112.2, 111.0, 107.9, 107.6, 102.6, 101.1, 70.5, 41.0, 33.6, 23.3, 22.5. **IR** (neat, cm^{-1}) 3009, 2964, 2916, 1645, 1600, 1504, 1485, 1456, 1255, 1236, 1039. **HRMS** (DART+) Calc'd for $C_{20}H_{19}INO_5$ 480.02811, found 480.02938. $[\alpha]_D^{20}$ + 73.69 (c = 1.00, $CHCl_3$).

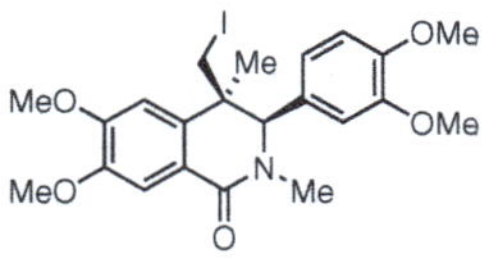

(3*R*,4*R*)-3-(3,4-dimethoxyphenyl)-4-(iodomethyl)-6,7-dimethoxy-2,4-dimethyl-3,4-dihydroisoquinolin-1(2*H*)-one (1.244o)—5 mol% [Pd]: The title compound was synthesized from **1.237o** (102 mg, 0.2 mmol) using **general procedure 10b**, and was isolated via silica gel flash column chromatography using hexanes:EtOAc (1:2 v:v) as the mobile phase (69.5 mg, 0.136 mmol, 68%, 90:10 dr (*cis:trans*)). 7.5 mol% [Pd]: The title compound was synthesized from **1.237o** (102 mg, 0.2 mmol), $Pd(QPhos)_2$ (22.9 mg, 0.015 mmol, 7.5 mol% [Pd]), and 1,2,2,6,6-pentamethylpiperidine (127 μL, 0.7 mmol, 3.5 equiv) using **general procedure 10b**, and was isolated via silica gel flash column chromatography using hexanes:EtOAc (1:2 v:v) as the mobile phase (93.4 mg, 0.182 mmol, 91%, 87:13 dr (*cis:trans*)). **Major:** Isolated as a white solid (MP = 133–134 °C). **^{1}H NMR** (500 MHz, $CDCl_3$) δ 7.73 (s, 1H), 6.69 (s, 1H), 6.67–6.58 (m, 3H), 4.22 (s, 1H), 3.95 (s, 3H), 3.88 (s, 3H), 3.83 (d, J = 10.0 Hz, 1H), 3.79 (s, 3H), 3.60 (s, 3H), 3.04 (s, 3H), 2.95 (d, J = 10.0 Hz, 1H), 1.60 (s, 3H). **^{13}C NMR** (125 MHz, $CDCl_3$) δ 163.5, 151.9, 148.9, 148.3, 148.0, 133.6, 128.5, 121.8, 121.1, 111.5, 111.1, 110.6, 106.3, 71.8, 56.1, 56.1, 55.7, 55.5, 40.1, 34.2, 31.2, 16.6. **IR** (cm^{-1}, film) 3002, 2962, 2934, 1641, 1600, 1516, 1240, 1287, 1259, 1240, 1217, 1185, 1144, 1048, 1027. **HRMS** (DART) Calc'd for $C_{22}H_{27}INO_5$, 512.09339; found, 512.09190. Enantiomeric purity was determined by HPLC analysis in comparison with racemic material t_r = 20.2 (major) t_r = 22.5 (minor) (>99:1 er shown; Chiracel AD-H column, 20% ^{i}PrOH in hexanes 0.5 mL/min, 210 nm). $[\alpha]_D^{20}$ +248.5 (c = 1.0, $CHCl_3$).

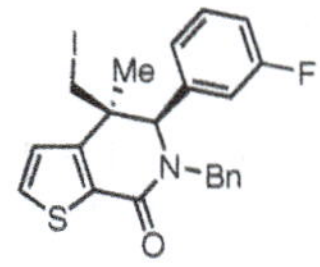

(*R*,*R*)-6-benzyl-5-(3-fluorophenyl)-4-(iodomethyl)-4-methyl-5,6-dihydrothieno [2,3-c]pyridin-7(4H)-one (1.244p)[197]—The title compound was synthesized from **1.237p** 78.3 mg, 0.2 mmol) using **general procedure 10a**, and was isolated via silica gel flash column chromatography using hexanes:EtOAc (3:1 v:v) as the mobile phase (69.1 mg, 0.176 mmol, 88%, >95:5 dr (*cis:trans*)). Major: Isolated as a white solid (MP = 136–137 °C). **H NMR1** (500 MHz, $CDCl_3$) δ 8.22 (ddd, *J* = 7.5, 2.0, 0.5 Hz, 1H), 7.49 (td, *J* = 7.5, 2.0 Hz, 1H), 7.44 (td, *J* = 7.5, 2.0 Hz, 1H), 7.26–7.21 (m, 2H), 7.16–7.12 (m, 2H), 7.09–7.05 (m, 2H), 4.34 (s, 1H), 3.93 (d, *J* = 10.0 Hz, 1H), 3.06 (s, 3H), 2.97–2.93 (m, 1H), 1.63 (d, *J* = 1.0 Hz, 3H). **C NMR13** (125 MHz, $CDCl_3$) δ 163.5, 140.14, 136.2, 132.3, 128.8, 128.8, 128.6, 128.6, 128.4, 127.6, 123.2, 72.0, 40.3, 34.3, 31.4, 16.2. IR (cm^{-1}, film) 3064, 3031, 2966, 2929, 2875, 1645, 1600, 1575, 1473, 1457, 1395, 1297, 1258, 1197, 1158. HRMS (ESI+) Calc'd for $C_{18}H_{19}INO$, 392.0506; found, 392.0511. Enantiomeric purity was determined by HPLC analysis in comparison with racemic material t_r = 10.9 (minor) t_r = 11.9 (major) (>99:1 er shown; Chiracel AD-H column, 10% ^{i}PrOH in hexanes 1.0 mL/min, 210 nm). $[\alpha]_D^{20}$ +261.6 (c = 1.0, $CHCl_3$).

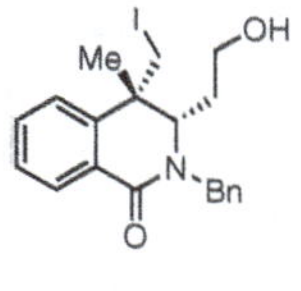

(3*S*,4*S*)-2-benzyl-3-(2-hydroxyethyl)-4-(iodomethyl)-4-methyl-3,4-dihydroiso-quinolin-1(2*H*)-one [(*S*,*S*)-**1.250**]—DDQ (51 mg, 0.223 mmol, 1.18 equiv) was added in a single portion to a 4.4 mL solution (4:1 DCM:pH 7 buffer) of (*S*,*S*)-**1.244 m** (105 mg, 0.189 mmol, 1 equiv). This was allowed to stir for 3 h at which time the reaction was poured into 20 mL of 10:1 H_2O:saturated aqueous $NaHCO_3$. The aqueous layer was extracted with EtOAc (3 × 15 mL), and the combined organic layers were washed with brine, dried over $MgSO_4$, filtered and

[197]Reaction was run by Hyung Yoon.

concentrated *in vacuo*. The crude carbinol was purified via silica gel flash column chromatography using hexanes:EtOAc (3:2 v:v) as the mobile phase to afford the title compound was a free-flowing white solid (80.5 mg, 0.185 mmol, 98%, MP = 136-138 °C). **^{1}H NMR** (600 MHz, $CDCl_3$) δ 8.13–8.08 (m, 1H), 7.51–7.48 (m, 2H), 7.45 (td, *J* = 7.5, 1.5 Hz, 2H), 7.38–7.32 (m, 3H), 7.31–7.28 (m, 1H), 7.16–7.14 (m, 1H), 5.78 (d, *J* = 14.0 Hz, 1H), 3.99 (d, *J* = 14.0 Hz, 1H), 3.84–3.79 (m, 2H), 3.78–3.69 (m, 2H), 3.38 (dd, *J* = 10.0, 1.0 Hz, 1H), 1.82 (dddd, *J* = 14.0, 10.0, 6.5, 3.5 Hz, 1H), 1.46–1.39 (m, 2H), 0.78 (d, *J* = 1.0 Hz, 3H). **^{13}C NMR** (150 MHz, $CDCl_3$) δ 163.6, 140.6, 136.8, 132.3, 130.0, 129.0, 128.5, 128.5, 127.9, 127.4, 123.2, 60.4, 58.9, 50.3, 39.4, 32.6, 28.2, 16.6. **IR** (cm^{-1}, film) 3391, 3004, 2984, 1623, 1599, 1464, 1318, 1299, 1265, 1154, 1080, 1061. **HRMS** (DART) Calc'd for $C_{20}H_{23}INO_2$, 436.07735; found, 436.07767. Enantiomeric purity was determined by HPLC analysis in comparison with racemic material t_r = 8.26 (minor) t_r = 11.4 (major) (>99:1 er shown; Chiracel AD-H column, 20% ^{i}PrOH in hexanes 1.0 mL/min, 210 nm). $[\alpha]_D^{20}$ −132.8 (c = 0.54, $CHCl_3$).

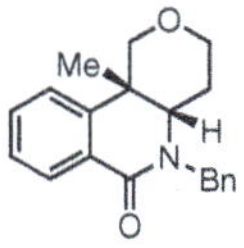

(4a*R*,10b*R*)-5-benzyl-10b-methyl-3,4,4a,5-tetrahydro-1*H*-pyrano[4,3-*c*]isoquinolin-6(10b*H*)-one [(*S*,*S*)-**1.251**]—A dry 2 dram vial was charged with (*S*,*S*)-**1.250** (40.0 mg, 0.092 mmol 1 equiv) and TBAI (33.9 mg, 0.092, 1 equiv), and its contents were purged with argon for 10 min, before being taken up in dry THF (3 mL). To this suspension was added NaH (7.3 mg, 0.183 mmol, 2.0 equiv) and the vial was capped with a Teflon line screw cap, sealed further with Teflon tape, and placed in a pre-heated oil bath at 65 °C where it was stirred for 18 h. At this time TLC analysis indicated full and clean conversion of starting material, and the reaction was quenched with saturated NH_4Cl (3 mL). The aqueous layer was extracted with EtOAc (3 × 3 mL) and the combined organic layers were washed with brine, dried over Na_2SO_4, filtered, and concentrated. The crude material was purified via silica gel flash column chromatography using hexanes:EtOAc (3:2 v:v) as the mobile phase to afford the title compound as a hygroscopic white solid (26.4 mg, 0.086 mmol, 93%, MP = 34–35 °C) **^{1}H NMR** (600 MHz, $CDCl_3$) δ 8.18 (dd, *J* = 8.0, 1.0 Hz, 1H), 7.50 (td, *J* = 7.5, 1.5 Hz, 1H), 7.39–7.25 (m, 7H), 5.43 (d, *J* = 14.5 Hz, 1H), 4.44 (d, *J* = 13.0 Hz, 1H), 4.07 (d, *J* = 14.5 Hz, 1H), 3.85 (dd, *J* = 11.5, 5.0 Hz, 1H), 3.39–3.33 (m, 1H), 3.21–3.15 (m, 2H), 1.69 (ddt, *J* = 13.5, 4.0, 2.0 Hz, 1H), 1.54 (dddd, *J* = 13.5, 13.0, 12.0, 5.0 Hz, 1H), 0.81 (s, 3H). **^{13}C NMR** (150 MHz, $CDCl_3$) δ 163.1, 141.8, 137.2, 132.5, 129.1, 128.9, 128.6, 128.5, 127.6, 126.8, 124.1, 72.3, 67.6, 60.9, 48.4, 38.6, 28.2, 24.9. **IR** (cm^{-1}, neat) 2959, 2923, 2847, 1645, 1602, 1456, 1287, 1146, 1100. **HRMS** (ESI+)

Calc'd for $C_{20}H_{22}O_2$, 308.16505; found, 308.16536. Enantiomeric purity was determined by HPLC analysis in comparison with racemic material T_R = 19.1 (minor) T_R = 22.0 (major) (99:1 er shown; Chiracel AD-H column, 10% ^{i}PrOH in hexanes 1.0 mL/min, 210 nm). $[\alpha]_D^{20}$ −109.7 (c = 0.34, $CHCl_3$).

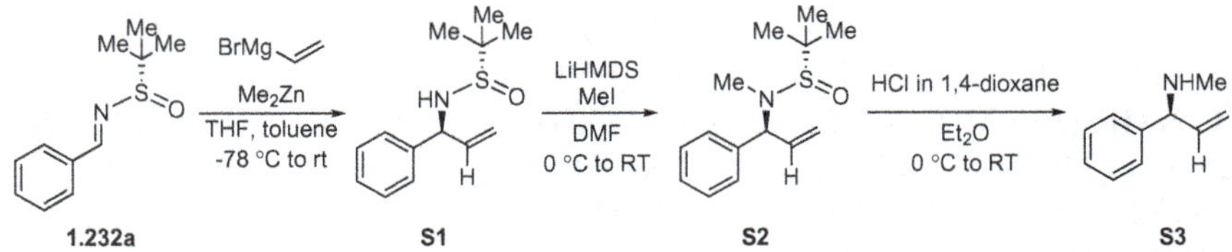

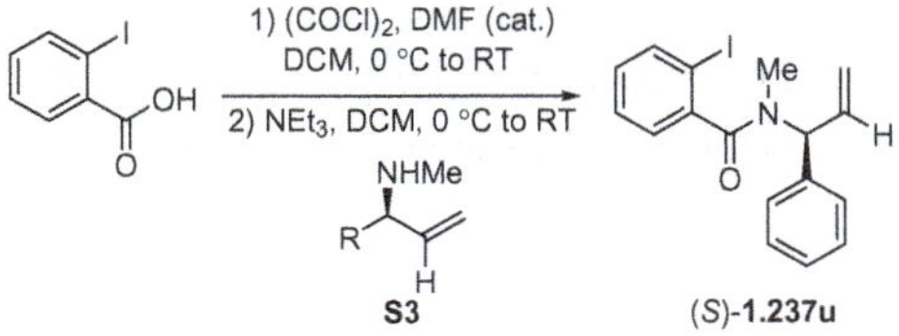

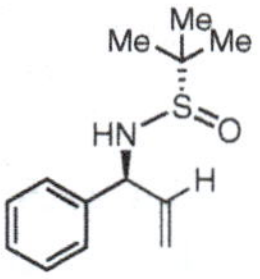

(*S*)-2-methyl-*N*-((*S*)-1-phenylallyl)propane-2-sulfinamide (S1)—Was synthesized according to **general procedure 6** using **1.232a** (1.2 g, 5.7 mmol, 0.13 M). ^{1}H NMR analysis of the crude product determined the diastereoselectivity of the addition to be 95:5. The crude allylic sulfinamide was obtained in analytically pure form by silica gel flash column chromatography using hexanes:EtOAc (1:1 v:v) as the mobile phase. The title compound was obtained as a clear and yellow oil (823 mg, 3.47 mmol, 61%). **^{1}H NMR** (400 MHz, $CDCl_3$) δ 7.38–7.27 (m, 5H), 5.98–5.88 (m, 1H), 5.38 (m, 1H), 5.24 (m, 1H), 5.01–4.95 (m, 1H), 3.45 (s, 1H), 1.25 (s, 9H). **^{13}C NMR** (100 MHz, $CDCl_3$) δ 141.5, 138.3, 128.8, 128.0, 117.4, 117.4, 61.4, 55.6, 22.7. **IR** (neat, cm^{-1}) 3200, 3062, 2979, 2957, 2925, 2902, 2867, 1491, 1473, 1363, 1012. **HRMS** (DART) Calc'd for $C_{13}H_{20}NOS$ 238.12626 found 238.12656. $[\alpha]_D^{20}$ + 129.31 (c = 0.53, $CHCl_3$).

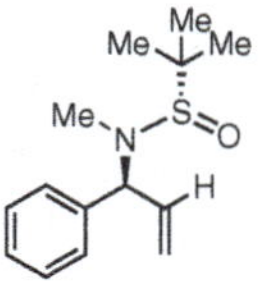

(*S*)-*N*,2-dimethyl-*N*-((*S*)-1-phenylallyl)propane-2-sulfinamide (S2)—Was synthesized according to **general procedure** 7 using **S1** (776 mg, 3.27 mmol, 0.2 M) and iodomethane (408 μL, 6.54 mmol). The crude alkylated amine was obtained in analytically pure form by silica gel flash column chromatography using hexanes: EtOAc (10:2 v:v) as the mobile phase. The title compound was obtained as clear and light yellow oil (685 mg, 2.73 mmol, 84%). **^{1}H NMR** (400 MHz, Chloroform-*d*) δ 7.38–7.24 (m, 5H), 6.10–5.99 (m, 1H), 5.34–5.31 (m, 1H), 5.30–5.26 (m, 1H), 4.89 (d, *J* = 7.5 Hz, 1H), 2.48 (s, 3H), 1.21 (s, 9H). **^{13}C NMR** (100 MHz, $CDCl_3$) δ 139.6, 136.6, 128.5, 127.9, 127.6, 117.9, 70.3, 58.5, 28.8, 23.8. **IR** (neat, cm^{-1}) 2977, 2956, 2925, 1493, 1475, 1452, 1360, 1188, 1075. **HRMS** (DART) Calc'd for $C_{14}H_{22}NOS$ 252.14201, found 252.14221. $[\alpha]_D^{20}$ +12.8 (c = 0.5, $CHCl_3$).

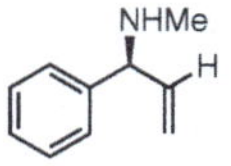

(*S*)-*N*-methyl-1-phenylprop-2-en-1-amine (S3)—Synthesized according to **general procedure 8a** using **S2** (632 mg, 2.5 mmol, 0.36 M) and anhydrous HCl in 1,4-dioxane (1.26 mL, 5 mmol). The title compound was obtained as clear and light yellow oil (275 mg, 1.87 mmol, 75%). **^{1}H NMR** (400 MHz, $CDCl_3$ δ 7.36–7.30 (m, 4H), 7.29–7.21 (m, 1H), 5.97–5.87 (m, 1H), 5.25–5.19 (m, 1H), 5.14–5.09 (m, 1H), 4.05 (d, *J* = 7.0 Hz, 1H), 2.38 (s, 3H), 1.38 (s, 1H). **^{13}C NMR** (100 MHz, $CDCl_3$) δ 142.7, 140.8, 128.5, 127.2, 127.2, 115.0, 68.1, 34.4. **IR** (neat, cm^{-1}) 3331, 3081, 3062, 3026, 2846, 2788, 1490, 1475, 1452, 1133, 1109. **HRMS** (DART) Calc'd for $C_{10}H_{14}N$ 148.11299, found 148.11262. $[\alpha]_D^{20}$ +28.2 (c = 0.5, $CHCl_3$).

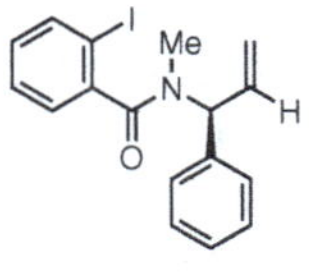

(*S*)-2-iodo-*N*-methyl-*N*-(1-phenylallyl)benzamide [(*S*)-**1.237u**]–Synthesized according to **general procedure 9** using 2-iodobenzoic acid (367 mg, 1.48 mmol) and **S3** (241 mg, 1.63 mmol). The crude amide was obtained in analytically pure form by silica gel flash column chromatography using hexanes:EtOAc:DCM (6:0.75:4 v:v:v) as the mobile phase. The title compound was obtained as an extremely viscous, clear and colourless oil (186 mg, 1.29 mmol, 87%). *NMR analysis showed the clean desired compound to be present as a mixture of 3 rotamers, and only approximate integration values are displayed.* **^{1}H NMR** (600 MHz, $CDCl_3$) δ 7.87 (dd, *J* = 8.0, 1.0 Hz, ~0.2H), 7.84 (dd, *J* = 8.0, 1.0 Hz, ~0.2H), 7.80 (d, *J* = 8.0 Hz, ~0.6H), 7.52–7.12 (m, ~7H), 7.10–7.00 (m, ~1H), 6.62 (d, *J* = 6.0 Hz, ~0.6H), 6.17 (ddd, *J* = 17.5, 10.5, 5.5 Hz, ~0.6H), 6.09 (ddd, *J* = 17.5, 10.5, 6.6 Hz, ~0.2H), 6.02 (ddd, *J* = 17.2, 10.5, 5.5 Hz, ~0.2H), 5.65–5.30 (m, ~1.6H), 5.24–5.10 (m, ~0.8H), 3.04 (s, ~0.6H), 2.83 (s, ~0.6H), 2.59–2.43 (m, ~1.8H). **^{13}C NMR** (125 MHz, $CDCl_3$)

δ 171.2, 171.0, 170.7 (br), 142.7, 142.2, 141.8, 139.7, 139.4, 139.2, 138.4 (br), 137.9, 137.7 (br), 137.1, 134.3, 134.2, 133.5 (br), 133.1 (br), 130.4, 130.1, 130.0, 128.8 (br) overlapping 128.8, 128.6, 128.5, 128.3, 127.9, 127.9, 127.6, 127.5, 127.0 (br) overlapping 126.9, 126.9, 126.7, 120.8, 119.4 (br), 118.0, 117.9 (br), 93.0, 92.8, 92.0 (br), 64.4, 64.3, 58.0, 32.1, 29.4, 29.1. **IR** (neat, cm^{-1}) 3061, 3029, 2927, 1654, 1622, 1470, 1427, 1398, 1324, 1090. **HRMS** (DART) Calc'd for $C_{17}H_{17}INO$ 378.03510, found 378.03548. Enantiomeric purity was determined by HPLC analysis in comparison with racemic material $t_r = 9.9$ (minor) $t_r = 20.9$ (major) (>99:1 er shown; Chiralcel AD-H column, 10% ^{i}PrOH in hexanes 1.0 mL/min, 210 nm). $[\alpha]_D^{20}$ −81.3 (c = 0.52, $CHCl_3$).

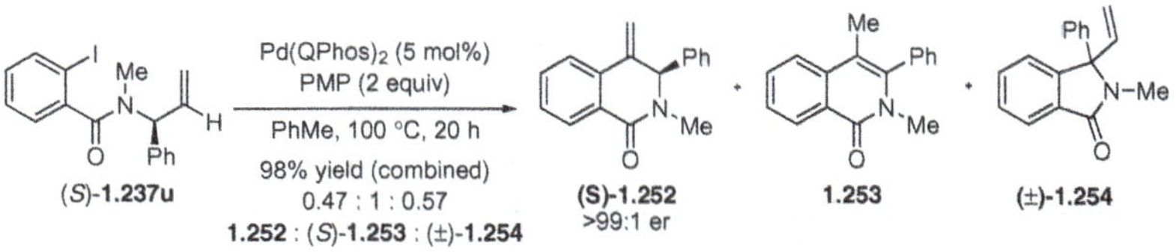

One oven-dried 2 dram vial was cooled under argon, charged with (*S*)-**1.237u** (75.4 mg, 0.2 mmol, 1 equiv), and purged with argon for 10 min. A second oven-dried vial was cooled under argon, charged with $Pd(QPhos)_2$ (15.2 mg, 0.01 mmol, 5 mol% [Pd]), and purged with argon for 10 min. (*S*)-**1.237u** was dissolved in dry and degassed PhMe (2 mL, 0.1 M) and 1,2,2,6,6-pentamethylpiperidine (72.4 μL, 0.4 mmol, 2 equiv) was added via a dry microsyringe. This well-mixed solution of amide and base was added into the vial containing the catalyst. The vial was fitted with a Teflon® lined screw cap under a stream of argon passed through a large inverted glass funnel, sealed using Teflon® tape, and placed in a pre-heated oil bath at 100 °C for 20 h. At this time ^{1}H NMR analysis of a small aliquot indicated full conversion of starting material, and the reaction vial was cooled, and its contents were filtered through a 2 cm plug of silica gel in a pipette eluting with 100% EtOAc. The crude reaction mixture was purified via silica gel flash column chromatography using hexanes:EtOAc (4:1 v:v) as the mobile phase then by silica gel flash column chromatography using DCM:hexanes: EtOAc (12:3:8 v:v:v) as the mobile phase.

(**1.253**) Isolated as a white solid with a melting point of 95–97 °C (lit. 95–96 °C)[198] and all spectral data was in accordance with the reported literature.[199] ^{1}**H NMR** (400 MHz, $CDCl_3$) δ 8.56–8.53 (m, 1H), 7.73–7.68 (m, 2H), 7.56–7.45 (m, 4H), 7.30–7.25 (m, 2H), 3.27 (s, 3H), 2.03 (s, 3H). ^{13}**C NMR** (100 MHz, $CDCl_3$) δ 162.6, 140.2, 137.1, 135.9, 132.1, 129.4, 129.0, 128.7, 128.1, 126.4, 125.3, 123.2, 110.4, 34.2, 14.8.

[198]Reference [173].

[199]Reference [173]

((*S*)-**1.252**, proposed) Isolated as a tacky and colourless film. **^{1}H NMR** (500 MHz, $CDCl_3$) δ 8.23–8.19 (m, 1H), 7.44–7.40 (m, 3H), 7.26–7.20 (m, 3H), 7.16–7.13 (m, 2H), 5.57 (s, 1H), 5.41 (d, *J* = 1.0 Hz, 1H), 5.18 (d (br), *J* = 1.0 Hz, 1H), 3.13 (s, 3H). **^{13}C NMR** (150 MHz, $CDCl_3$) δ 163.7, 141.1, 139.7, 134.2, 132.1, 128.9, 128.8, 128.1, 127.9, 127.4, 125.7, 123.9, 113.2, 68.8, 34.0. **IR** (cm^{-1}, film) 2956, 2923, 1645, 1600, 1484, 1450, 1396, 1265. **HRMS** (DART) Calc'd for $C_{17}H_{16}NO$, 250.12319; found, 250.12381. Enantiomeric purity was determined by HPLC analysis in comparison with racemic material t_r = 9.9 (major) t_r = 11.1 (minor) (> 99:1 er shown; Chiracel AD-H column, 10% ^{i}PrOH in hexanes 1.0 mL/min, 210 nm). *Note: This compound has been only partially characterized in the literature,*[200] *and we have found slight discrepancies in the ^{1}H NMR data.*

[(±)-**1.254**]. Isolated as a light yellow oil. ^{1}H NMR (500 MHz, $CDCl_3$) δ 7.89 (ddd, *J* = 7.0, 1.5, 1.0 Hz, 1H), 7.50–7.41 (m, 2H), 7.36–7.28 (m, 3H), 6.45 (dd, *J* = 17.5, 10.5 Hz, 1H), 5.38 (dd, *J* = 10.5, 1.0 Hz, 1H), 5.16 (dd, *J* = 17.5, 1.0 Hz, 1H), 2.89 (s, 3H). **^{13}C NMR** (150 MHz, $CDCl_3$) δ 168.5, 149.2, 139.0, 135.9, 131.7, 130.8, 128.9, 128.2, 128.2, 127.1, 123.8, 123.1, 117.4, 72.3, 25.8. **IR** (cm^{-1}, film) 2955, 2952, 2923, 1683, 1470, 1372. **HRMS** (DART) Calc'd for $C_{17}H_{16}NO$, 250.12319; found, 250.12344. Enantiomeric ratio was determined to be 50:50 by HPLC analysis, Chiracel AD-H column, 1.5% ^{i}PrOH in hexanes 1.0 mL/min, 235 nm).

Scale-up of Pd-Catalyzed Carboidination of chiral linear *N*-allyl carboxamides:

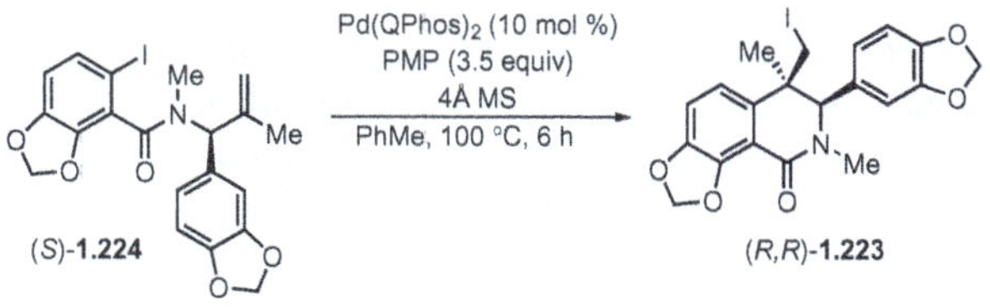

(*S*)-**1.224** (*R*,*R*)-**1.223**

An oven dried 500 mL round bottom flask was charged with (*S*)-**1.224** (5.0 g, 10.4 mmol, 1 equiv), $Pd(QPhos)_2$ (1.59 g, 1.04 mmol, 10 mol%), and activated 4Å molecular sieves (1.5 g) and purged with argon for 30 min. The solid components were taken up in dry and degassed PhMe (208 mL, 5 FPT cycles, transferred via cannula) and dry and degassed 1,2,2,6,6-pentamethylpiperidine (6.61 mL, 35.6 mmol, 3.5 equiv, 5 FTP cycles). The reaction vessel was fitted with an oven dried reflux condenser which had been cooled under a flow of argon for 10 min. The reaction was heated for 6 h at 100 °C. At this time, TLC indicated the full consumption of starting material, the reaction was cooled and filtered through a plug of silica gel eluting with 100% EtOAc, and the solution concentrated under reduced pressure. ^{1}H NMR analysis of the crude reaction indicated a 91:9 mixture

[200]Reference [174].

of diastereomers (*cis*:*trans*). The crude mixture of dihydroquinolines was carefully purified using silica gel flash column chromatography using hexanes:EtOAc (1:1 v:v) as the mobile phase to yield the separate diastereomers (4.18 g, 8.72 mmol, 84% (combined values)). Characterization data for both the major and minor diastereomers is identical to that reported above.

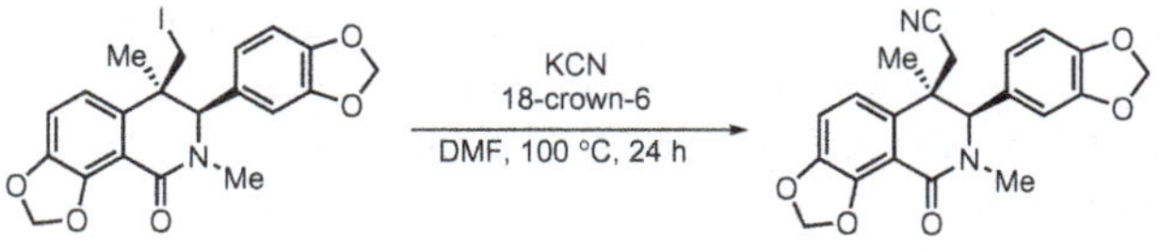

(*R*,*R*)-**1.223** (*R*,*R*)-**1.222**

A dry 100 mL flask containing a stirbar was charged with (*R*,*R*)-**1.223** (1.2 g, 2.5 mmol, 1 equiv), powdered KCN (815 mg, 12.5 mmol, 5 equiv), and 18-crown-6 (3.30 g, 12.5 mmol, 5 equiv), and the solid mixture was dissolved in anhydrous DMF (30 mL). The reaction vessel was fitted with a reflux condenser placed into an oil bath pre-heated to 100 °C. After 24 h the reaction had another 2.5 equivalents of both powdered KCN and 18-crown-6. After an additional 24 h, TLC indicated full conversion of starting material. The deep brown/red solution was diluted with saturated aqueous $NaHCO_3$ (20 mL) and water (20 mL). The aqueous layer was extracted with EtOAc (4 × 50 mL), and the combined organic layers were sequentially washed with water (3 × 20 mL) and brine (20 mL), dried over Na_2SO_4, filtered, and concentrated (*Note*: ***appropriate measures were taken to handle all resulting reaction waste with care at basic pH, segregated from all other waste contamination, in an appropriately labeled vessel, stored in a well operating fume hood***). The crude nitrile was purified via silica gel flash column chromatography using DCM:MeOH (80:1 v:v) as the mobile phase and the pure title compound was afforded as an off-white solid (632 mg, 1.67 mmol, 67%, MP = 242–244 °C). *Note*: X-ray quality crystals of the nitrile could be obtained by recrystallization of the purified material via the slow diffusion of hexanes into EtOAc. **^{1}H NMR** (500 MHz, $CDCl_3$) δ 6.86 (d, *J* = 8.0 Hz, 1H), 6.69 (d, *J* = 1.0 Hz, 2H), 6.46 (d, *J* = 8.0 Hz, 1H), 6.38 (s, 1H), 6.20 (d, *J* = 1.5 Hz, 1H), 6.18 (d, *J* = 1.0 Hz, 1H), 5.91 (d, *J* = 1.5 Hz, 1H), 5.90 (d, *J* = 1.5 Hz, 1H), 4.27 (s, 1H), 3.04 (s, 3H), 2.80 (d, *J* = 16.5 Hz, 1H), 2.23 (dd, *J* = 16.5, 1.0 Hz, 1H), 1.65 (d, *J* = 1.0 Hz, 3H). **^{13}C NMR** (125 MHz, $CDCl_3$) δ 160.7, 148.9, 148.6, 148.2, 148.0, 133.9, 129.4, 122.5, 117.6, 115.5, 111.7, 110.9, 108.3, 107.3, 102.7, 101.3, 71.5, 40.5, 33.8, 29.8, 26.4. **IR** (neat, cm^{-1}) 3010, 1645, 1602, 1504, 1485, 1458, 1240, 1232. **HRMS** (DART+) Calc'd for $C_{21}H_{19}INO_5$ 379.12940, found 379.13017. $[\alpha]_D^{20}$ +94.7 (c = 0.47, $CHCl_3$).

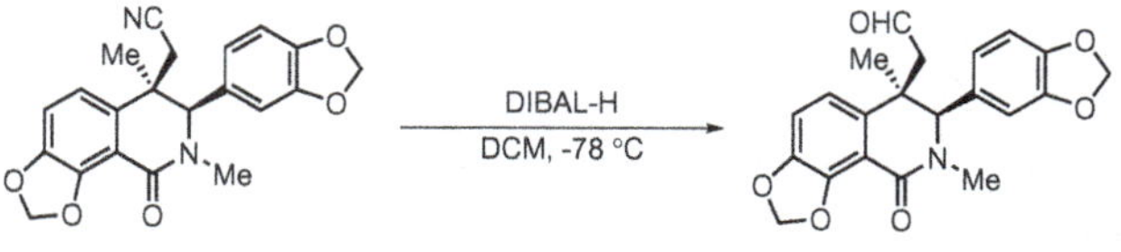

(*R*,*R*)-**1.222** (*R*,*R*)-**1.257**

A dry 50 mL Schlenk flask was charged with (*R*,*R*)-**1.222** (400 mg, 1.06 mmol, 1 equiv) and purged with argon for 10 min. The contents of the flask were dissolved in freshly distilled DCM (15 mL) and the resulting clear and colourless solution was cooled to −78 °C. At this temperature the solubility of the nitrile is decreased, and the solution becomes slightly opaque but colourless. At this time DIBAL-H (1 M in PhMe, 2.11 m, 2.11 mmol, 2 equiv) was added drop wise over 15 min. After 90 min at this temperature additional DIBAL-H (2.11 mL, 2.11 mmol, 2 equiv) was added drop wise over 15 min. After stirring for an additional 30 min, TLC analysis indicated full conversion of starting material. The reaction was slowly quenched by the addition of MeOH (1 mL) such that the internal reaction temperature never exceeded −75 °C. Once effervescence ceased, more MeOH (1 mL) was added using the same care, and the reaction was warmed to room temperature, diluted with DCM (10 mL), and a saturated aqueous solution of Rochelle's salt (25 mL) was added. This bi-phasic mixture was vigorously stirred at room temperature for 12 h. At this time the phases were separated, and the aqueous layer was extracted with DCM (2 × 15 mL). The combined organic layers were washed with brine, dried over Na_2SO_4, filtered, and concentrated in vacuo. The crude material was purified using silica gel flash column chromatography using hexanes:EtOAc (1:2 v:v) as the mobile phase. The pure aldehydes (*R*,*R*)-**1.257** was obtained as a white foam (207.1 mg, 0.544 mmol, 51%). **^{1}H NMR** (500 MHz, $CDCl_3$) δ 9.39 (t, *J* = 1.4 Hz, 1H), 6.86 (d, *J* = 8.0 Hz, 1H), 6.62 (d, *J* = 8.0 Hz, 1H), 6.54 (d, *J* = 8.0 Hz, 1H), 6.47 (dd, *J* = 8.0, 2.0 Hz, 1H), 6.33 (d, *J* = 2.0 Hz, 1H), 6.19 (d, *J* = 1.5 Hz, 1H), 6.16 (d, *J* = 1.5 Hz, 1H), 5.89 (d, *J* = 1.5 Hz, 1H), 5.88 (d, *J* = 1.5 Hz, 1H), 4.50 (s, 1H), 3.02 (s, 3H), 2.83 (dd, *J* = 18, 2.0 Hz, 1H), 2.57 (d, *J* = 18 Hz, 1H), 1.59 (s, 3H). **^{13}C NMR** (125 MHz, $CDCl_3$) δ 200.0, 160.8, 148.4, 148.3, 147.9, 147.9, 136.3, 130.9, 122.3, 116.3, 111.9, 110.9, 108.1, 107.5, 102.5, 101.2, 71.0, 49.5, 40.9, 33.7, 30.5. **IR** (neat, cm^{-1}) 3007, 2907, 2847, 1700, 1653, 1487, 1448, 1240, 1228. **HRMS** (DART+) Calc'd for $C_{21}H_{22}NO_5$ 368.14980, found 368.14874. $[\alpha]_D^{20}$ +130.98 (c = 0.6, $CHCl_3$).

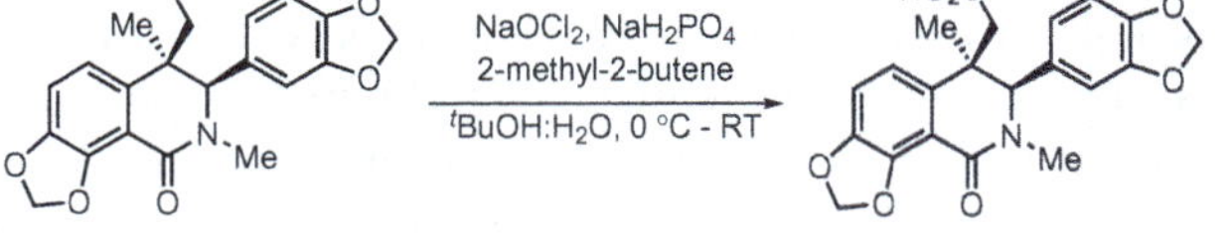

(*R*,*R*)-**1.157** (*R*,*R*)-**1.155**

(*R*,*R*)-**1.257** (450 mg, 1.18 mmol, 1 equiv) was suspended in a solution of ^{t}BuOH (15 mL) and 2-methyl-2-butene (1.87 mL, 17.7 mmol, 15 equiv) and the resulting mixture was cooled to 0 °C. To this was added an aqueous solution (15 mL) of $NaClO_2$ (1.07 g, 11.8 mmol, 10 equiv) and NaH_2PO_4 (1.07 g, 9.91 mmol, 8.4 equiv) (*Note*: to dissolve $NaOCl_2$ and NaH_2PO_4 in water, sonicate in a well-operative fume hood, and handle with care). The mixture was allowed to warm to room temperature, and after stirring at this temperature for 90 min TLC analysis showed full and clean conversion of starting material. The reaction was cooled to 0 °C and was quenched with 50 mL of saturated $NaSO_3$ and was diluted with water (25 mL). The aqueous layer was extracted with DCM (3 × 50 mL), washed with brine (50 mL), dried over Na_2SO_4, filtered, and concentrated *in vacuo.* The crude material was purified using silica gel flash column chromatography using DCM:MeOH (15:1 v:v) as the mobile phase and the title compound was afforded (*R*,*R*)-**1.155** as a white foam (370 mg, 0.93 mmol, 79%). **^{1}H NMR** (500 MHz, $CDCl_3$) δ 6.85 (d, *J* = 8.0 Hz, 1H), 6.62 (d, *J* = 8.0 Hz, 1H), 6.59 (d, *J* = 8.0 Hz, 1H), 6.53 (dd, *J* = 8.0, 2.0 Hz, 1H), 6.37 (d, *J* = 2.0 Hz, 1H), 6.17 (d, *J* = 2.0 Hz, 1H), 6.15 (d, *J* = 2.0 Hz, 1H), 5.89 (d, *J* = 1.5 Hz, 1H), 5.88 (d, *J* = 1.5 Hz, 1H), 4.80 (s, 1H), 3.04 (s, 3H), 2.94 (d, *J* = 17.0 Hz, 1H), 2.34 (d, *J* = 17.0 Hz, 1H), 1.62 (s, 3H). **^{13}C NMR** (125 MHz, $CDCl_3$) δ 175.4, 161.2, 148.3, 148.2, 147.7, 147.6, 136.9, 130.9, 122.2, 115.9, 111.8, 110.7, 108.0, 107.7, 102.5, 101.1, 69.5, 40.0, 39.5, 33.9, 28.8. **IR** (neat, cm^{-1}) 2973, 2909, 1725, 1625, 1597, 1505, 1489, 1447, 1254, 1241, 1228, 1043. **HRMS** (DART+) Calc'd for $C_{21}H_{20}NO_7$ 398.12398, found 398.12356. $[\alpha]_D^{20}$ +125.73 (c = 0.47, $CHCl_3$).

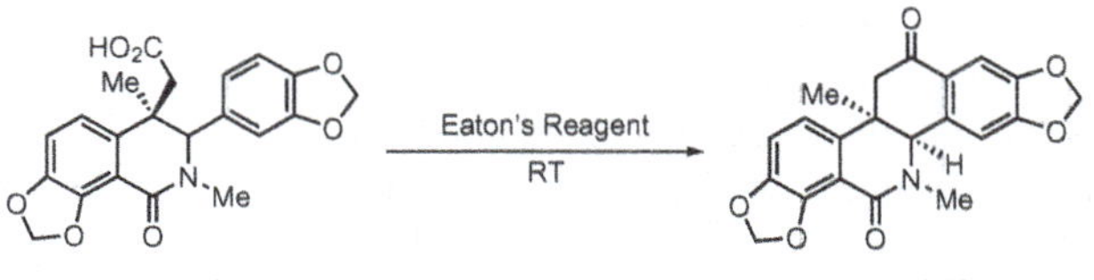

A flask was charged with a solution of Eaton's reagent (18 mL, large excess/solvent) and cooled to 0 °C in an ice bath. To this was added solid (*R*,*R*)-**1.255** (368 mg, 0.928 mmol, 1 equiv) and the reaction mixture was allowed to warm to room temperature where it was stirred for 2 h. A small aliquot of the reaction mixture was carefully worked up and the resulting TLC analysis indicated full conversion of starting material. The reaction mixture was poured into 50 mL of an ice cooled saturated aqueous solution of $NaHCO_3$. The resulting aqueous solution was extracted with DCM (3 × 50 mL). The combined organic layers were washed with brine (50 mL), dried over Na_2SO_4, filtered, and concentrated in vacuo. The crude material was purified via silica gel flash column chromatography using DCM:MeOH (20:1 v:v) was the mobile phase. The title compound was afforded as a white solid (230 mg, 0.61 mmol, 66%, MP = >300°C). *Note*: X-ray quality crystals of the ketone could be obtained by recrystallization of the purified material via the slow diffusion of hexanes into a saturated acetone solution. **^{1}H NMR** (600 MHz, $CDCl_3$) δ 7.29 (s, 1H), 6.72–6.68 (m, 2H), 6.60 (d, *J* = 8.0 Hz, 1H), 6.08 (d, *J* = 1.5 Hz, 1H), 5.98 (d, *J* = 1.5 Hz, 1H), 5.96–5.94 (m, 2H), 4.49 (d, *J* = 1.0 Hz, 1H), 3.53 (s, 3H), 3.35 (d, *J* = 17.0 Hz, 1H), 2.79 (d, *J* = 17.0 Hz, 1H), 1.46 (s, 3H). **^{13}C NMR** (150 MHz, $CDCl_3$) δ 192.6, 161.7, 153.1, 148.3, 148.0, 147.7, 139.4, 135.3, 127.0, 116.5, 111.6, 110.8, 105.9, 105.8, 102.2, 102.0, 67.9, 47.7, 42.4, 37.5, 28.9. **IR** (neat, cm^{-1}) 3018, 2916, 1672, 1647, 1482, 1281. **HRMS** (DART+) Calc'd for $C_{21}H_{18}NO_6$ 381.11341, found 381.11429. $[\alpha]_D^{20}$ −88.31 (c = 0.42, $CHCl_3$).

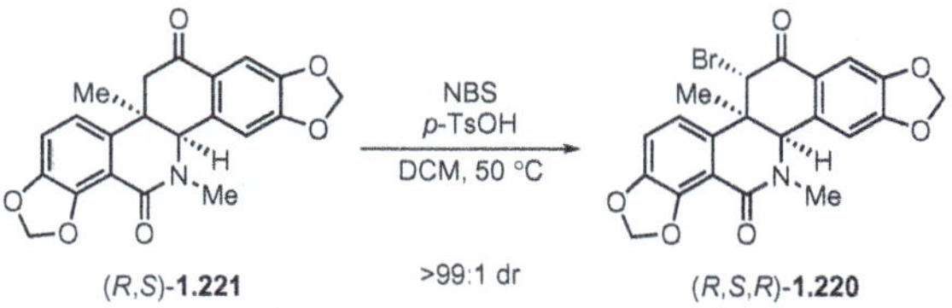

A flame-dried 10 mL round bottom flask equipped with a stir bar was chared with NBS (49.2 mg, 0.276 mmol, 2 equiv) and *p*-TsOH (4.6 mg, 0.0267, 20 mol%) and purged with nitrogen for 10 min. At this time a DCM solution (5 mL) of (*R*,*S*)-**1.221** (52.4 mg, 0.138 mmol, 1 equiv) was added via syringe. The flask was fitted with a reflux condenser and the reaction was heated at 50 °C for 6 h at which time TLC analysis indicated full conversion of the ketone. The reaction was quenched with saturated aqueous NH_4Cl (5 mL) after separating the layers, the aqueous phase was extracted with DCM (3×). The combined organic layers were washed with brine, dried over Na_2SO_4, filtered, and concentrated in vacuo to afford the crude α-bromo ketone which was present in >99:1 dr by crude ^{1}H NMR analysis of a homogeneous aliquot. The crude material was purified by silica gel flash column chromatography using hexanes:EtOAc (1:2 v:v) followed by DCM:MeOH (20:1 v:v) as the mobile phases. The pure α-bromo ketone was obtained as a colorless solid (53.5 mg, 0.117 mmol, 85% yield). **Note**: X-ray quality crystals of the ketone could be obtained by recrystallization of the purified material via the slow diffusion of hexanes into a saturated acetone solution. **^{1}H NMR** (500 MHz, $CDCl_3$) δ 7.32 (s, 1H), 6.73–6.70 (m, 2H), 6.58 (d, *J* = 8.0 Hz, 1H), 6.09 (d, *J* = 1.5 Hz, 1H), 6.00 (d, *J* = 1.0 Hz, 1H), 5.98 (d, *J* = 1.0 Hz, 1H), 5.95 (d, *J* = 1.5 Hz, 1H), 5.07 (s, 1H), 4.72 (d, *J* = 0.5 Hz, 1H), 3.56 (s, 3H), 1.59 (s, 3H). **^{13}C NMR** (125 MHz, $CDCl_3$) δ 186.9, 161.9, 153.8, 148.8, 148.5, 148.1, 138.5, 132.8, 124.1, 116.7, 111.5, 110.9, 107.0, 106.0, 102.5, 102.3, 64.5, 57.5, 45.2, 37.9, 26.6. **IR** (thin film, cm^{-1}) 2973, 2952, 1676, 1656, 1504, 1484, 1446, 1376, 1275, 1254, 1225. **HRMS** (ESI+) Calc'd for $C_{21}H_{17}BrNO_6$ 458.02392, found 458.02404. $[\alpha]_D^{20}$ −94.58 (c = 0.5, $CHCl_3$).

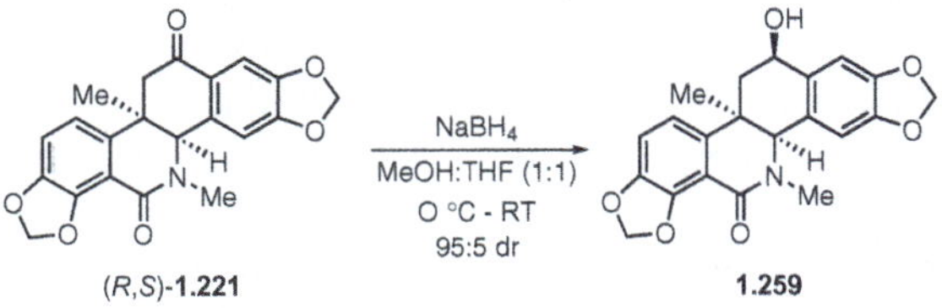

A dry 2 dram vial was charged with (*R*,*S*)-**1.221** (25 mg, 0.0659 mmol, 1 equiv), which was taken up in a 1:1 solution of dry THF:MeOH (2 mL). The resulting solution was colled to 0 °C before adding solid $NaBH_4$ (25 mg, 0.659 mg, 10 equiv). The mixture was aged at this temperature until all gas evolution ceased, and the reaction was warmed to room temperature and stirred for 15 h. At this time TLC analysis indicated full conversion of starting material, and the reaction was quenched by the addition of a saturated aqueous solution of NH_4Cl (2 mL), and the aqueous phase was extracted with DCM (3 × 3 mL). The combined organic layers were washed with brine, dried over Na_2SO_4, filtered, and concentrated in vacuo to afford to the desired carbinol as a sufficiently pure off-white solid. ^{1}H NMR analysis of the product showed the product to be present as a 95:5 mixture of diasteromers. The crude diastereomeric mixture of carbinols was purified via silica gel flash column chromatrography using DCM:MeOH (20:1 v:v) as the mobile phase. The product was obtained as a white solid (22.7 mg, 0.0596 mmol, 91%). **^{1}H NMR** (Major) (500 MHz, $CDCl_3$) δ 6.91–6.85 (m, 2H), 6.78 (d, *J* = 8.0 Hz, 1H), 6.59 (d, *J* = 1.0 Hz, 1H), 6.09 (d, *J* = 1.5 Hz, 1H), 5.96 (d, *J* = 1.5 Hz, 1H), 5.89 (d, *J* = 1.5 Hz, 1H), 5.87 (d, *J* = 1.5 Hz, 1H), 4.68 (m (br), 1H), 4.07 (s, 1H), 3.37 (s, 3H), 2.72 (dd, *J* = 15.0, 4.0 Hz, 1H), 2.25 (dd, *J* = 15.0, 6.5 Hz, 1H), 1.38 (s, 3H). **^{13}C NMR** (Major) (125 MHz, $CDCl_3$) δ 161.9, 148.3, 147.8, 147.7, 147.4, 137.6, 132.5, 128.8, 118.2, 112.3, 110.7, 107.9, 105.6, 102.2, 101.2, 67.5, 66.2, 41.6, 37.6, 36.8, 30.2. **HRMS** (ESI+) Calc'd for $C_{21}H_{20}NO_6$ 384.1344, found 384.1346.

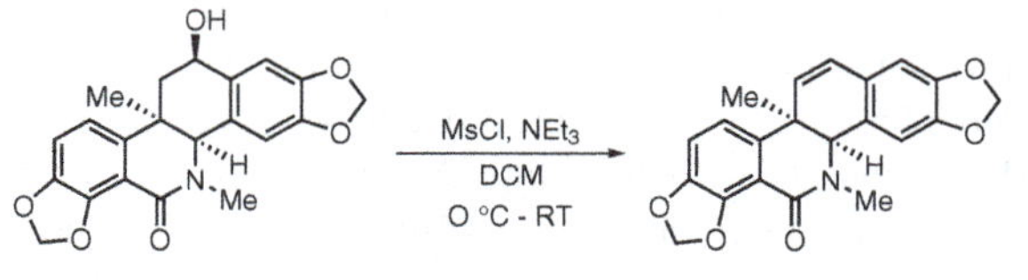

A dry 2 dram vial was charged with a 95:5 diasteromeric mixture of carbinols (9.8 mg, 0.0257 mmol, 1 equiv) and purged with argon. Freshly distilled DCM (1.5 mL) was added via syringe and the resulting solution was cooled to 0 °C. Freshly distilled NEt_3 (108 μL, 0.771 mmol, 30 equiv) and MsCl (30 μL, 0.386 mmol, 15 equiv) were added by micro syringe in that order. The resulting yellow solution was stirred for 30 min at this temperature before warming to room temperature where it was aged for an additional 1 h. At this time TLC analysis indicated full conversion of the starting material, and 1 mL of a 5% aqueous HCl solution was added, and the aqueous phase was extracted with DCM (3 × 3 mL). The combined organic layers were washed with brine, dried over Na_2SO_4, filtered and concentrated in vacuo. The crude alkene was purified using silica gel flash column chromatography using DCM:MeOH (60:1 v:v) as the mobile phase. The desired alkene (*R,R*)-**1.201** was isolated as a white solid (9.2 mg, 0.0253 mmol, 99%). **^{1}H NMR** (500 MHz, $CDCl_3$) δ 6.69 (d, *J* = 8.0 Hz, 1H), 6.61 (s, 1H), 6.58 (d, *J* = 8.0 Hz, 1H), 6.51 (s, 1H), 6.42 (d, *J* = 9.5 Hz, 1H), 6.11–6.06 (m, 2H), 5.99–5.95 (m, 1H), 5.90–5.85 (m, 2H), 4.37 (s, 1H), 3.38 (s, 3H), 1.47 (s, 3H). **^{13}C NMR** (125 MHz, $CDCl_3$) δ 161.8, 147.7, 147.5, 147.4, 146.8, 137.3, 132.7, 129.1, 127.3, 127.2, 116.9, 112.5, 110.7, 107.2, 105.8, 102.1, 101.1 68.0, 38.8, 36.4, 28.6.

IR (film, cm^{-1}) 2960, 2902, 2869, 1645, 1483, 1502, 1483, 1262, 1253, 1236, 1220, 1053. **HRMS** (DART+) Calc'd for $C_{21}H_{18}NO_5$ 364.11850, found 364.11843. $[\alpha]_D^{20}$ −37.88 (c = 0.26, $CHCl_3$).

References

1. Petrone, D.A., Malik, H.A., Clemenceau, A., Lautens, M.: Org. Lett. **14**, 4806 (2012)
2. Petrone, D.A., Yoon, H., Weinstabl, H., Lautens, M.: Angew. Chem. Int. Ed. **53**, 7908 (2014)
3. Petrone, D.A., Le, C.M., Newman, S.G., Lautens, M.: Pd-catalyzed carboiodination: early developments to recent advancements (Chap. 7). In: Colacot, T. (ed.) New trends in cross-coupling. The Royal Society of Chemistry, Cambridge (2015)
4. Petrone, D.A., Ye, J., Lautens, M.: Chem. Rev. **116**, 8003–8004 (2016)
5. Johansson Seechurn, C.C.C., Kitching, M.O., Colacot, T.J., Snieckus, V.: Angew. Chem. Int. Ed. **51**, 5062 (2012)
6. Negishi, E.: In Negishi E, de Meijere, A. (eds.) Handbook of Organopalladium Chemistry for Organic Synthesis, vol. 1. 1st edn., p. 213. Wiley, New York (2002)
7. Stille, J.K., Lau, K.S.: Acc. Chem. Res. **10**, 434 (1977)
8. Amatore, C., Jutand, A.: J. Organomet. Chem. **576**, 254 (1999)
9. Cadado, A.L., Espinet, P.: Organometallics **17**, 954 (1998)
10. Amii, H., Uneyama, K.: Chem. Rev. **109**, 2119 (2009)
11. Clot, E., Eisenstein, O., Jasim, N., MacGregor, S.A., McGrady, J.E., Perutz, R.N.: Acc. Chem. Res. **44**, 333 (2011)
12. Ohashi, M., Shibata, M., Saijo, H., Kambara, T., Ogoshi, S.: Organometallics **32**, 3631 (2013)
13. de Jong, G.T., Bickelhaupt, F.M.: J. Phys. Chem. A **109**, 9685 (2005)
14. Collman, J.P., Hegedus, L.S., Norton, J.R., Finke, R.G.: Principles and applications of organotransition metal chemistry, 2nd edn, pp. 176–258. University Science Books, Mill Valley (1987)
15. Spessard, G.O., Meissler, G.L.: Organometallic chemistry, pp. 171–175. Prentice Hall, Upper Saddle River, New Jersey (1996)
16. Christmann, U., Vilar, R.: Angew. Chem. Int. E. **44**, 366 (2005)
17. Hartwig, J.F.: Organotransition metal chemistry: from bonding to catalysis, 1st edn, pp. 895–898. University Science Books, Mill Valley (2010)
18. Mateo, C., Fernández-Rivas, C., Echavarren, A.M., Cárdenes, D.J.: J Organometallics **16**, 1997 (1997)
19. Denmark, S.E., Sweise, R.F.: Acc. Chem. Res. **35**, 835 (2002)
20. Miyaura, N., Yamada, K., Sufinome, H., Suzuki, A.: J. Am. Chem. Soc. **107**, 972 (1985)
21. Lennox, A.J.J., Lloyd-Jones, G.C.: Angew. Chem. Int. Ed. **52**, 7362 (2013)
22. Ridgway, B.H., Woerpel, K.A.: J. Org. Chem. **63**, 458 (1998)
23. Matos, K., Soderquist, J.A.: J. Org. Chem. **63**, 461 (1998)
24. Miyaura, N.: Top. Curr. Chem. **219**, 11 (2002)
25. Cárdenes, D.J., Mateo, C., Echavarren, A.M.: Angew. Chem. Int. Ed. **33**, 2445 (1996)
26. Casado, A.L., Espinet, P.: J. Am. Chem. Soc. **120**, 8978 (1998)
27. Icci, A., Angelucci, F., Bassetti, L., Lo Sterzo, C.: J. Am. Chem. Soc. **124**, 1060 (2002)
28. Cotter, W.D., Barbour, L., McNamera, K.L., Hechter, R., Lachicotte, R.J.: J. Am. Chem. Soc. **120**, 11016 (1998)
29. Hartwig, J.F.: Inorg. Chem. **46**, 1936 (2007)
30. Heck, R.F.: J. Am. Chem. Soc. **90**, 5518 (1968)
31. Heck, R.F.: J. Am. Chem. Soc. **90**, 5526 (1968)

32. Heck, R.F.: J. Am. Chem. Soc. **90**, 5531 (1968)
33. Heck, R.F.: J. Am. Chem. Soc. **90**, 5535 (1968)
34. Heck, R.F.: J. Am. Chem. Soc. **90**, 5538 (1968)
35. Heck, R.F.: J. Am. Chem. Soc. **90**, 5542 (1968)
36. Heck, R.F.: J. Am. Chem. Soc. **90**, 5546 (1968)
37. Fujiwara, Yuzo, Moritani, I., Danno, S., Asano, R., Teranishi, S.: J. Am. Chem. Soc. **91**, 7166 (1968)
38. Mizoroki, T., Mori, K., Ozaki, A.: Bull. Chem. Soc. Jpn **44**, 581 (1971)
39. Heck, R.F., Nolley, J.P.: J. Org. Chem. **14**, 2320 (1972)
40. Mizoroki, T., Mori, K., Ozaki, A.: Bull. Chem. Soc. Jpn **46**, 1505 (1973)
41. Dieck, H.A., Heck, R.F.: J. Am. Chem. Soc. **96**, 1133 (1974)
42. Coulson, D.R.: Chem. Commun. 1530 (1968)
43. Dieck, H.A., Heck, R.F.: J. Org. Chem. **40**, 1083 (1975)
44. Kim, J.-I.I., Patel, B.A., Heck, R.F.: J. Org. Chem. **46**, 1067 (1981)
45. Ziegler, C.B., Heck, R.F.: J. Org. Chem. **43**, 2941 (1978)
46. Julia, M., Duteil, M., Grard, C., Kuntz, E.: Bull. Soc. Chim. Fr. 2791 (1973)
47. Spencer, A.: J. Organomet. Chem. **270**, 115 (1984)
48. Davison, J.B., Simon, N.M., Sojka, S.A.: J. Mol. Catal. **22**, 349 (1984)
49. Ben-David, Y., Portney, M., Milstein, D.: J. Am. Chem. Soc. **111**, 8742 (1989)
50. Ben-David, Y., Portney, M., Milstein, D.: J. Chem. Soc. Chem. Commun. 1816 (1989)
51. Ben-David, Y., Portney, M., Gozin, M., Milstein, D.: Organometallics **11**, 1995 (1992)
52. Portney, M., Ben-David, Y., Milstein, D.: Organometallics **12**, 4734 (1993)
53. Herrmann, W.A., Brossner, C., Öfele, K., Reisinger, C.-P., Priermeier, T., Beller, M., Fischer, H.: Angew. Chem. Int. Ed. **34**, 1844 (1995)
54. Herrmann, W.A., Elison, M., Fischer, J., Köcher, C., Artus, G.R.J.: Angew. Chem. Int. Ed. **34**, 2371 (1995)
55. Reetz, M.T., Lohmer, G., Schwickardi, R.: Angew. Chem. Int. Ed. **37**, 481 (1998)
56. Ehrentraut, A., Zapf, A., Beller, M.: Synlett **11**, 1589 (2000)
57. Kauffmann, D.E., Nouroozian, M., Henze, H., Synlett, 1091 (1996)
58. Beller, M., Zapf, A., Synlett, 792 (1998)
59. Littke, A.F., Fu, G.C.: J. Org. Chem. **64**, 10 (1999)
60. Littke, A.F., Fu, G.C.: J. Am. Chem. Soc. **123**, 6989 (2001)
61. Shaughnessy, K.H., Kim, P., Hartwig, J.F.: J. Am. Chem. Soc. **126**, 1184 (2004)
62. Heck, R.F.: Org. React. **27**, 345 (1982)
63. Davies, G.D., Hallberg, A.: Chem. Rev. **89**, 1433 (1989)
64. Heck, R.F.: In: Trost, B.M., Flemming, I. (eds.), Comprehensive Organic Synthesis, vol. 4. Permagon Press, Oxford (1991)
65. De Meijere, A., Meyer, F.E.: Angew. Chem. Int. Ed. **33**, 2739 (1994)
66. Brase, S., deMeijere, A.: In: Diederich, F., Stang, P.J. (eds.) Metal-Catalyzed Cross-Coupling Reactions. Wiley, New York (1998)
67. Beletskaya, I.P., Cheprakov, A.V.: Chem. Rev. **100**, 3009 (2000)
68. Beletskaya, I.P., Cheprakov, A.V.: In T. Colacot, (ed.) New Trends in Cross-Coupling: Theory and Application, pp. 355–478 . The Royal Society of Chemistry, Cambridge (2015)
69. DeAngelis, A., Colacot, T.J.: In Colacot, T. (ed.) New Trends in Cross-Coupling: Theory and Application, pp. 20–90. The Royal Society of Chemistry, Cambridge (2015)
70. Chartoire, A., Nolan, S.P.: In Colacot, T. (ed.), New Trends in Cross-Coupling: Theory and Application, pp. 139–227. The Royal Society of Chemistry, Cambridge (2015)
71. Negishi, E.-I., Coperet, C., Ma, S., Liou, S.-Y., Liu, F.: Chem. Rev. **96**, 365 (1996)
72. Negishi, E.-i, Copéret, C., Ma, S., Liou, S.-Y., Liu, F.: Chem. Rev. **95**, 365 (1996)
73. Mori, M., Chiba, K., Ban, Y.: Tetrahedron Lett. **18**, 1037 (1977)
74. Cortese, N.A., Ziegler, C.B., Hrnjez, B.J., Heck, R.F.: J. Org. Chem. **43**, 2952 (1978)
75. Odle, R., Belvins, B., Ratcliff, M., Hegedus, L.S.: J. Org. Chem. **45**, 2790 (1980)
76. Narula, C.K., Mak, K.T., Heck, R.F.: J. Org. Chem. **48**, 2792 (1983)
77. Grigg, R., Stevenson, P., Worakun, T.: J. Chem. Soc. Chem. Commun. 1073 (1984)

78. Tour, J.M., Negishi, E.-I.: J. Am. Chem. Soc. **107**, 8289 (1985)
79. Abelman, M.M., Oh, T., Overman, L.E.: J. Org. Chem. **52**, 4130 (1987)
80. Larock, R.C., Song, H., Baker, B.E., Gong, W.H.: Tetrahedron Lett. **29**, 2919 (1988)
81. Negishi, E.-I., Zhang, Y., O'Connor, B.: Tetrahedron Lett. **29**, 2915 (1988)
82. Sato, Y., Sodeoka, M., Shibasaki, M.: Chem. Lett. **19**, 1953 (1990)
83. Nicolaou, K.C., Bulger, P.G., Sarlah, D.: Angew. Chem. Int. Ed. **44**, 4442 (2005)
84. Jeffery, T.: Tetrahedron **52**, 10113 (1996)
85. Danishefsky, S.J., Masters, J.J., Young, W.B., Link, J.T., Synder, L.B., Magee, T.V., Kung, D.K., Isaacs, R.C.A., Bornmann, W.G., Alaimo, C.A., Coburn, C.A., Di Grandi, M.J.: J. Am. Chem. Soc. **118**, 2843 (1996)
86. Kagan, H.B., Diter, P., Gref, A., Guillaneux, D., Masson-Szymcak, A., Rebriére, F., Riant, O., Samuel, O., Taidien, S.: Pure Appl. Chem. **68**, 29 (1996)
87. Shibasaki, M., Boden, C.D.J., Kojima, A.: Tetrahedron **53**, 7371 (1997)
88. Shibasaki, M., Vogl, E.M., Ohshima, T.: Adv. Synth. Catal. **346**, 1533 (2004)
89. Sato, Y., Sodeoka, M., Shibasaki, M.: J. Org. Chem. **54**, 4738 (1989)
90. Carpenter, N.E., Kucera, D.J., Overman, L.E.: J. Org. Chem. **54**, 5846 (1989)
91. Ozawa, F., Kubo, A., Hayashi, T.: J. Am. Chem. Soc. **113**, 1417 (1991)
92. Cabri, W., Candiani, I., DeBernardinis, S., Francalanci, F., Penco, S.: J. Org. Chem. **56**, 5796 (1991)
93. Brown, J.M., Pérez-Torrente, J., Alcock, N.W., Clase, H.J.: Organometallics **14**, 207 (1995)
94. Brown, J.M., Hii, K.K.: Angew. Chem. Int. Ed. **35**, 657 (1996)
95. Hii, K.K., Claridge, T.D.W., Brown, J.M., Smith, A., Deeth, R.J.: Helv. Chim. Acta **84**, 3043 (2001)
96. Ashimori, A., Overman, L.E.: J. Org. Chem. **57**, 4571 (1992)
97. Overman, L.E., Poon, D.J.: Angew. Chem. Int. Ed. **36**, 519 (1997)
98. Thorn, D.L., Hoffman, R.: J. Am. Chem. Soc. **100**, 2079 (1978)
99. Samsel, E.G., Norton, J.R.: J. Am. Chem. Soc. **106**, 5505 (1984)
100. Ashimori, A., Bachand, B., Calter, M.A., Govek, S.P., Overman, L.E., Poon, D.J.: J. Am. Chem. Soc. **120**, 6488 (1998)
101. Lapierre, A.J.B., Geib, S.J., Curran, D.P.: J. Am. Chem. Soc. **129**, 494 (2007)
102. Tsuji, J.: Palladium Reagents and Catalysts—New Perspectives for the 21st Century, 2nd edn, pp. 1–26. Wiley, West Sussex (2004)
103. Vigalok, A.: Chem. Eur. J. **14**, 5102 (2008)
104. Casado, A.L., Espinet, P.: J. Am. Chem. Soc. **120**, 9878 (1998)
105. Xue, L., Lin, Z.: Chem. Soc. Rev. **39**, 1692 (2010)
106. Ettorre, R.: Inorg. Nuc. Chem. Lett. **5**, 45 (1968)
107. Fahey, D.R.: J. Chem. Soc. Chem. Commun. 417 (1970)
108. Fahey, D.R.: J. Organomet. Chem. **27**, 283 (1971)
109. Roy, A.H., Hartwig, J.F.: J. Am. Chem. Soc. **123**, 1232 (2001)
110. Roy, A.H., Hartwig, J.F.: Organometallics **23**, 1533 (2004)
111. Littke, A.F., Dai, C., Fu, G.C.: J. Am. Chem. Soc. **122**, 4020 (2000)
112. Alcazar-Roman, L.M., Hartwig, J.F.: J. Am. Chem. Soc. **123**, 12905 (2000)
113. Stambuli, J.P., Bühl, M., Hartwig, J.F.: J. Am. Chem. Soc. **124**, 9346 (2002)
114. Roy, A.H., Hartwig, J.F.: J. Am. Chem. Soc. **125**, 13944 (2003)
115. Grushin, V.V., Alper, H.: Chem. Rev. **94**, 1047 (1994)
116. Lam, K.C., Marder, T.B., Lin, Z.: Organometallics **26**, 758 (2007)
117. Yahav-Levi, A., Goldberg, I., Vigalok, A.: J. Am. Chem. Soc. **128**, 8710 (2006)
118. Ruddick, J.D., Shaw, B.L.: J. Chem. Soc. (A) 2969 (1969)
119. Goldberg, K.I., Yan, J.Y., Winter, E.L.: J. Am. Chem. Soc. **116**, 1573 (1994)
120. Goldberg, K.I., Yan, J.Y., Breitung, E.M.: J. Am. Chem. Soc. **117**, 6889 (1995)
121. Maitlis, P.M., Haynes, A., Sunley, G.J., Howard, M.J.: J. Chem. Soc. Dalton Trans. 2187 (1996)
122. Kampmeier, J.A., Rodehorst, R.M., Phillip, J.B.: J. Am. Chem. Soc. **103**, 1847 (1981)

123. Gonsalvi, L., Gaunt, J.A., Adams, H., Castro, A., Sunley, G.J., Haynes, A.: Organometallics **22**, 1047 (2003)
124. Frech, C.M., Milstein, D.: J. Am. Chem. Soc. **128**, 12434 (2006)
125. Newman, S.G., Lautens, M.: J. Am. Chem. Soc. **132**, 11416 (2010)
126. Newman, S.G., Lautens, M.: J. Am. Chem. Soc. **133**, 1778 (2011)
127. Shelby, Q., Kataoka, N., Mann, G., Hartwig, J.F.: J. Am. Chem. Soc. **122**, 10718 (2000)
128. Tollman, C.A.: Chem. Rev. **77**, 313 (1977)
129. Lan, Y., Liu, P., Newman, S.G., Lautens, M., Houk, K.N.: Chem. Sci. **3**, 1987 (2012)
130. Liu, H., Li, C., Qiu, D., Tong, X.: J. Am. Chem. Soc. **133**, 6187 (2011)
131. Newman, S.G., Howell, J.K., Nicolaus, N., Lautens, M.: J. Am. Chem. Soc. **133**, 14916 (2011)
132. Klapars, A., Buchwald, S.L.: J. Am. Chem. Soc. **124**, 14844 (2002)
133. Jia, X., Petrone, D.A., Lautens, M.: Angew. Chem. Int. Ed. **51**, 9870 (2012)
134. Ellis, G.P., Lockhart, I.M.: In Ellis, G.P. (eds.) The Chemistry of Heterocyclic Compounds, Chromenes, Chromanones, and Chromones. Wiley-VCH, New York (1977)
135. Kamigauchu, M., Noda, Y., Nishijo, J., Iwasaki, K., Tobetto, K., In, Y., Tomoo, K., Ishita, T.: Bioorg, Med. Chem. Let. 13, 1867 (2005)
136. Kim, D.K.: Arch. Pharmacal Res. **25**, 817 (2002)
137. Ma, W.G., Fukushi, Y., Tahara, S.: Fitoterapia **70**, 258 (1999)
138. Ninomiya, I., Yamamoto, O., Naito, T.: J. Chem. Soc. Chem. Commun. 437 (1976)
139. Cushman, M., Abbaspour, A., Gupta, Y.P.: J. Am. Chem. Soc. **105**, 2873 (1983)
140. Cushman, M., Abbaspour, A., Gupta, Y.P.: J. Am. Chem. Soc. **112**, 5898 (1990)
141. Hanaoka, M., Yoshida, S., Mukai, C.: Tetrahedron Lett. **29**, 6621 (1988)
142. Cushman, M., Chen, J.-K.: J. Org. Chem. **52**, 1517 (1987)
143. Tani, C., Takao, N.: Yakugaku Zasshi **82**, 755 (1962)
144. Nielsen, L., Lindsay, K.B., Faber, J., Nielsen, N.C., Skrydstrup, T.: J. Org. Chem. **72**, 10035 (2007)
145. Low, D.W., Pattison, G., Wieczsysty, M.D., Churchill, G.H., Lam, H.W.: Org. Lett. **14**, 2548 (2012)
146. Weinstabl, H., Suhartono, M., Qureshi, Z., Lautens, M.: Angew. Chem. Int. Ed. **52**, 5305 (2012)
147. Blackmond, D.G.: Angew. Chem. Int. Ed. **44**, 4302 (2005)
148. Maresca, L., Natile, G.: Comments Inorg. Chem. **14**, 349 (1993)
149. Albano, V.G., Natile, G., Panunzi, A.: Cood. Chem. Rev. **133**, 67 (1994)
150. Hatano, M., Terada, M., Mikami, K.: Angew. Chem. Int. Ed. **40**, 249 (2001)
151. Greeves, N.: Reduction of C=O to CHOH by metal hydrides. In: Trost, T., Flemming, I. (eds.) Comprehensive Organic Synthesis, vol. 7. Permagon Press, Oxford (1991)
152. Gabbutt, C.D., Heron, B.M., Instone, A.C.: Tetrahedron **62**, 737 (2006)
153. Subramanian, V., Batchu, B.R., Barange, D., Pal, M.: J. Org. Chem. **70**, 4778 (2005)
154. Miura, K., Wang, D., Hosomi, A.: J. Am. Chem. Soc. **127**, 9366 (2005)
155. Mizuno, A., Kusama, H., Iwasawa, N.: Chem. Eur. J. **16**, 8248 (2010)
156. Hurada, T., Kurokawa, H., Kagamikhara, Y., Tanaka, S., Inoue, A., Oku, A.: J. Org. Chem. **57**, 1412 (1992)
157. Zhao, J.-F., Loh, T.-P.: Angew. Chem. Int. Ed. **48**, 7232 (2009)
158. Shchepin, R., Xu, C., Dussault, P.: Org. Lett. **12**, 4772 (2010)
159. Nakajima, M., Kotani, S., Ishizaka, T., Hashimoto, S.: Terahedron Lett. **46**, 157 (2005)
160. Hon, Y.-S., Hsu, T.-R., Chen, C.-Y., Lin, Y.-H., Chnag, F.-J., Hseish, C.-H., Szu, P.: Tetrahedron **59**, 1509 (2003)
161. Zheng, J.-S., Chang, H.-N., Liu, L., Wang, F.-L.: J. Am. Chem. Soc. **133**, 11080 (2011)
162. Joly, G.D., Jacobsen, E.N.: Org. Lett. **4**, 1795 (2002)
163. Druais, V., Hall, M.J., Corsi, C., Wandeborn, S.V., Meyer, C., Cossy, J.: Tetrahedron **66**, 6358 (2010)
164. He, H., Zaturska, D., Kim, J., Aguirre, J., Llauger, L., She, Y., Wu, N., Immormino, R.M., Gewirth, D.T., Chiosis, H.: J. Med. Chem. **49**, 381 (2006)

165. Weinstabl, H., Suhartono, M., Qureshi, Z., Lautens, M.: Angew. Chem. Int. Ed. **52**, 5305 (2013)
166. Mattson, R.J., Sloan, C.P., Lockhart, C.C., Catt, J.D., Gao, Q., Huang, S.: J. Org. Chem. **64**, 8004 (1999)
167. Lv, P., Huang, K., Xie, L., Xu, X.: Org. Biomol. Chem. **9**, 3133 (2011)
168. Faber, J., Lindsay, K.B., Nielson, L., Nielson, N.C.: Skrydstrup, Troels. J. Org. Chem. **72**, 10035 (2007)
169. Kosciolowicz, A., Rozwadowska, M.D.: Tetrahedron Asymmetry **17**, 1444 (2007)
170. Amin, S.R., Bean, C.J., Bettigeri, S.V., Forbes, D.C., Law, A.M., Stockman, R.A.: Synth. Commun. **39**, 2405 (2009)
171. Esumi, T., Fukino, M., Hatakeyama, S., Ishihara, J., Okamoto, N., Shibahara, S., Takahashi, K., Tashiro, Y.: Synthesis **17**, 2935 (2009)
172. Bjerglund, K., Friis, S., Hernandez, D., Lindsay, K.B., Mittag, T., Mose, R., Nielsen, L., Skrydstrup, T.: J. Org. Chem. **75**, 3283 (2010)
173. Ackerman, L., Lygin, L.V., Hofmann, N.: Angew. Chem. Int. Ed. **50**, 6379 (2011)
174. Grigg, R., Sansano, J.M., Santhakumar, V., Sridharan, V., Thangavelanthum, R., Thornton-Pett, M., Wilson, D.: Tetrahedron Lett. **53**, 11803 (1997)

Chapter 2
Pd-Catalyzed Diastereoselective Carbocyanation Reactions of Chiral *N*-Allyl Carboxamides and Indoles

2.1 Introduction

Although still in its early stages, the aryliodination reaction is a noteworthy transformation due to its ability to rapidly construct all-carbon quaternatry centers. Chapter 1 outlined key discoveries which facilitated a significant broadening of the reaction scope, and the first application of this methodology in the synthesis of a natural product. Therein, the nucleophilic cyanation which follows the key Pd-catalyzed aryliodination produces the product in moderate yields using harsh conditions and excess amounts of toxic and costly reagents. We saw this as an opportunity to develop an alternate route towards this key structure, and hypothesized that an aryl halide could be directly converted into the desired nitrile via an arylcyanation reaction using Pd catalysis and exogenous cyanide. Since this transformation has been underexplored, we thought that our efforts would have implications beyond a second generation synthetic route to corynoline. This chapter outlines our work in developing two arylcyanation methodologies which utilize either chiral *N*-allyl carboxamides or indoles as substrates.

2.1.1 Palladium-Catalyzed Cyanation of Aromatic Halides

The nitrile is a valuable synthetic handle which can be easily converted to many useful functional groups. The development of methodologies which incorporate this moiety into organic molecules has spanned across many synthetic subdisciplines, and

Portions of this chapter have appeared in print. See Refs. [1, 2].

D. A. Petrone, *Stereoselective Heterocycle Synthesis via Alkene Difunctionalization*, Springer Theses, https://doi.org/10.1007/978-3-319-77507-4_2

2.1 2.2 2.3 2.4

Scheme 2.1 Cyanation of aryl halides using CuCN

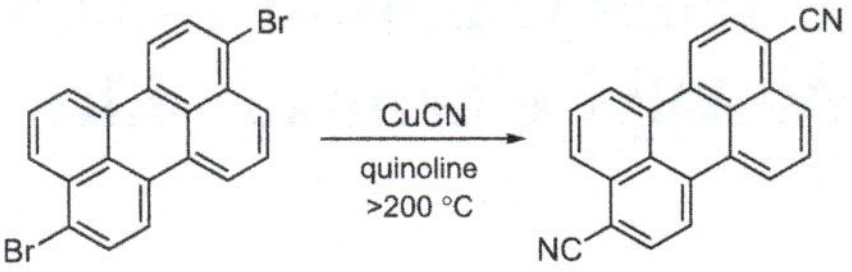

2.5 2.6

Scheme 2.2 Pongratz's synthesis of perylene-3,9-dicarbonitrile via cyanation

includes efforts from the Lewis acid catalysis and organocatalysis communities.[1] One of the most common ways in which nitriles can be incorporated is via the cyanation of aryl halides. This is traditionally carried out by reacting an aryl halide **2.1** with CuCN **2.2** to obtain the corresponding nitrile **2.3** and a stoichiometric equivalent of a copper halide byproduct **2.4** (Scheme 2.1).

An early report concerning a cyanation of an arylhalide using a transition metal-based reagent was by Pongratz in 1927, where it was shown that CuCN could be used to convert dibromide **2.5** into perylene-3,9-dicarbonitrile **2.6** in 95% yield.[2] This reaction is formally referred to as the Rosenmund-von Braun reaction (Scheme 2.2).[3]

One drawback to this method and Sandmeyer-type cyanations[4] is the need to employ a stoichiometric amount of CuCN, which leads to the production of equimolar amounts of heavy metal waste. Furthermore, reactions are usually run at high temperature, and aryl chlorides and bromides display decreased reactivity under the initial reaction conditions.[5] Although copper has been applied in countless examples of these reactions,[6] the introduction of Pd salts as catalysts has greatly enhanced the scope of this reaction (Scheme 2.3), and this approach has become an important alternative to such traditional methods. In addition, Pd-catalyzed cyanations employ inexpensive cyanide sources such as sodium, potassium or zinc cyanide.

Sakakibara reported the first Pd-catalyzed cyanation of an aryl halide which utilized KCN as a cyanide source (Scheme 2.4).[7] They showed that iodo and bromobenzene as well as 1-bromonaphthalene could be converted into the corresponding aryl nitriles using $Pd(CN)_2$ as the catalyst. This reaction failed to proceed

---

[1]References [3, 4].

[2]Reference [5].

[3]Although this reaction is referred to as the Rosenmund-von Braun reaction, Rosenmund contribution showed that ArX reacts with KCN, KOH and H_2O in a CuCN promoted reaction leading to the benzoic acid derivatives. von Braun used this reaction, yet did not study it in detail or expand the scope. See Refs. [6–9].

[4]Reference [10].

[5]Reference [11].

[6]References [9, 12].

[7]Reference [13].

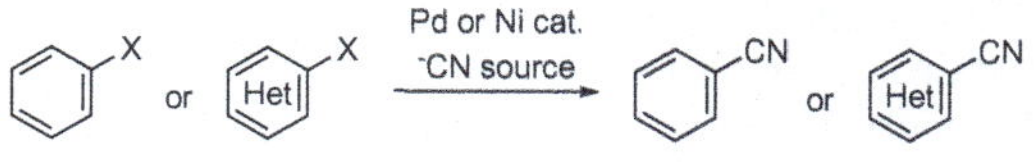

2.7 2.8 2.9 2.10

Scheme 2.3 Ni- and Pd-catalyzed cyanation of aromatic halides

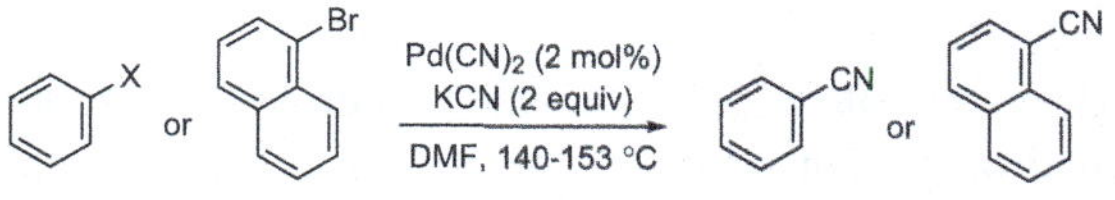

2.11 (X = Br, I) 2.12 2.13 2.14

Scheme 2.4 First example of a Pd-catalyzed cyanation of an aryl halide by Sakakibara

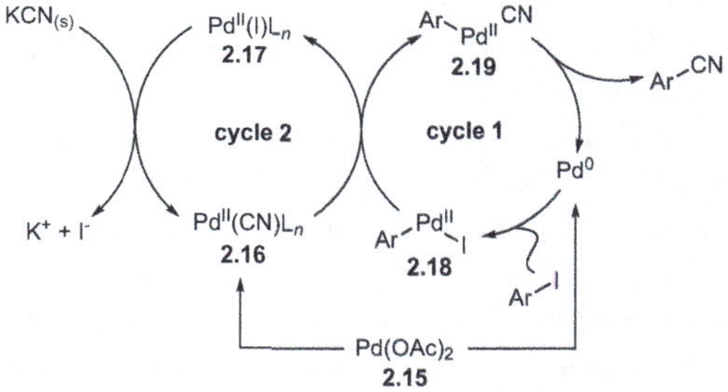

Scheme 2.5 Proposed synergistic mechanism for the Pd-catalyzed cyanation of aryl halides

in the absence of the Pd catalyst, or when run is the presence of phosphine or phosphite ligands. They also noted that the exclusion of air and moisture was necessary in order to obtain reproducible results.

Early mechanistic studies were carried out by Takagi and co-workers which led to the mechanistic proposal in Scheme 2.5.[8] This mechanism involves two synergistic catalytic cycles, wherein cycle 1 represents that of a standard cross coupling (i.e. oxidative addition, transmetallation, reductive elimination), and cycle 2 involves Pd acting as a cyanide carrier. Since these seminal reports, many incremental advances to the reaction conditions for cyanation of aromatic halides have been made. These include the use of phosphines[9] and phosphorous triamides as ligands,[10] crown ethers,[11] alumina-supported cyanide reagents,[12] and the use of alkyl nitriles[13] and dialkyl cyanoboronates[14] as cyanide sources.

Akin to the previously detailed Mizoroki-Heck reaction, efforts to render the Pd-catalyzed cyanation efficient with aryl halides came much later than the initial

[8]Reference [14].
[9]Reference [15].
[10]Reference [16].
[11]Reference [17].
[12]References [18, 19].
[13]Reference [20].
[14]Reference [21].

Scheme 2.6 First general Pd-catalyzed cyanation of aryl chlorides

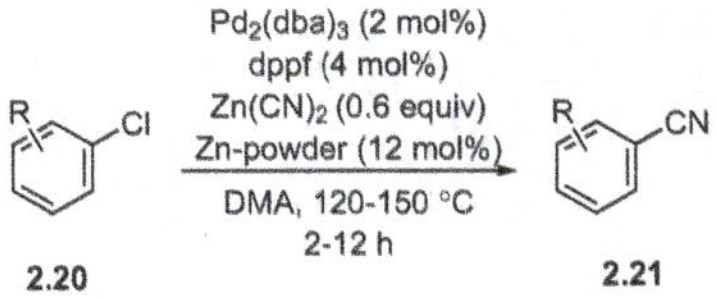

2.20 **2.21**

discovery of the reaction. Due to their low reactivity, the cyanation of aryl chlorides had only been achieved using Ni or using Pd with activated (hetero)aryl chlorides prior to the year 2000.[15] Jin and Confalone from DuPont Pharmaceuticals reported the first general Pd-catalyzed cyanation of aryl chlorides in 2000 (Scheme 2.6).[16] The authors showed that the use of a Pd/dppf catalyst combination in the presence of substoichiometric amounts of $Zn(CN)_2$ afforded the desired benzonitriles in high yields. The authors noted that the use of catalytic amounts of Zn powder was necessary to reduce Pd(II) which is formed to some extent during side reactions.

Based on their work concerning the use of aryl chlorides in the Mizoroki-Heck reaction,[17] Suzuki-Miyaura couplings[18] and aminations,[19] Beller and co-workers reported an efficient catalyst system for the cyanation of aryl chlorides in 1998 (Scheme 2.7a).[20] Early in the study the authors identified that a combination of an inorganic base (i.e. Na_2CO_3) and crown ether led to improved results. However, after further optimization they found that TMEDA could be used as a co-catalytic additive which takes the place of both these reagents. In 2007, Littke and Soumeillant reported that bulky phosphine ligands were effective in promoting the cyanation of aryl chlorides (Scheme 2.7b). Therein, two catalyst systems were developed where system 1 was most effective for electron-rich and neutral aryl chlorides, whereas system 2 worked best for electron poor and sulfur-containing aryl chlorides.[21] Beller discovered that potassium hexacyanoferrate(II) ($K_4[Fe(CN)_6]$) could be used in the cyanation of (hetero)aryl bromides and heteroaryl chlorides in 2004 (Scheme 2.7c).[22] $K_4[Fe(CN)_6]$ is a non-toxic compound with a higher LD_{50} value than NaCl, and is used in the food industry for metal precipitation. These authors reported improved conditions for the cyanation of aryl chlorides in 2007 which utilized their class of cataCXium® ligands (Scheme 2.7d). Under these conditions both electron-rich and -poor aryl chlorides could be effectively cyanated using low catalyst loadings.

[15]Reference [22].
[16]Reference [23].
[17]Reference [24].
[18]Reference [25].
[19]Reference [26].
[20]References [27, 28].
[21]Reference [29].
[22]Reference [30].

Scheme 2.7 Improvements to the Pd-catalyzed cyanation of aryl chlorides

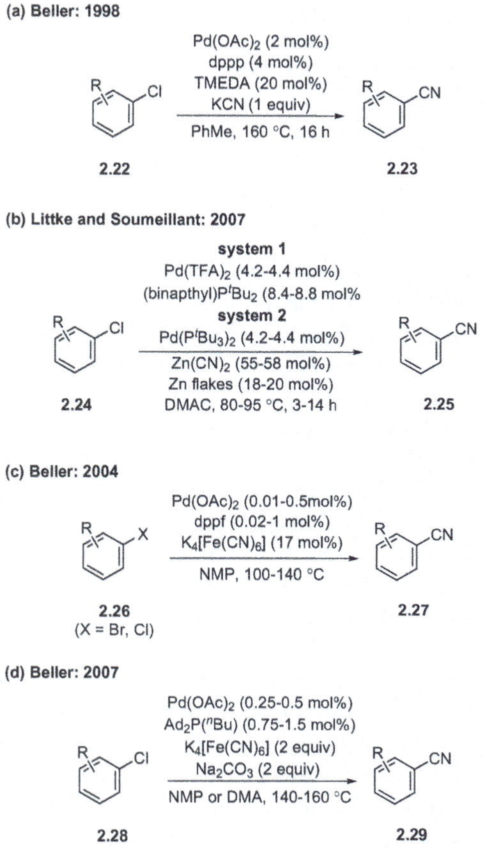

The success and industrial importance of these and other catalytic aryl halide cyanations [11] has inspired the development of other classes of transition-metal-catalyzed cyanation reactions which afford products possessing higher levels of complexity.

2.1.2 Ni-Catalyzed Carbocyanation Proceeding via C–C Bond Cleavage

A second valuable class of cyanation reactions is the Ni-catalyzed addition of an organonitrile **2.30** across an unsaturated carbon–carbon moiety **2.31** to generate a product containing two new carbon–carbon bonds **2.32**. This reaction proceeds via oxidative addition of a Ni(0) species into the carbon–CN bond to generate Ni(II)

Scheme 2.8 General representation of a Ni-catalyzed carbocyanation

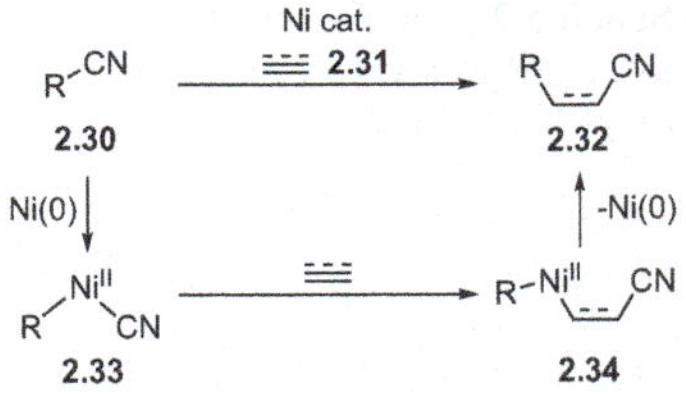

species **2.33**,[23] followed by insertion of the Ni–CN bond across an alkene or alkyne substrate to generate alkyl or vinyl Ni(II)-species **2.34**. A final carbon–carbon bond-forming reductive elimination generates the product (Scheme 2.8).

Nakao and Hiyama[24] reported the first Ni-catalyzed carbocyanation of alkynes **2.35** using benzonitiles **2.36** in 2004 which employed a $Ni(COD)_2/PMe_3$ catalyst combination (Scheme 2.9a).[25] This reaction produced *cis*-β-aryl-acrylonitriles **2.37** in good yields in examples using both symmetrical and unsymmetrical alkynes. The authors state that although insertion of the alkyne into the Ni–Ar bond instead of the Ni–CN bond cannot be ruled out, the cyanonickelation better explains the regioselectivity observed in certain examples. The authors later showed that the addition of Lewis acids to the reaction enabled a vast expansion of the scope which included alkyl, alkenyl, and aryl nitrile substrates.[26] Although the scope of these reactions only included alkynes, the reaction was later expanded to strained alkenes like norbornene.[27] However, the reaction conditions were not general even across the various classes of nitrile substrates, and required several combinations of Lewis acids, phosphine ligands and temperatures to proceed satisfactorally. In 2006, the same authors developed an efficient and general protocol for the Ni-catalyzed allylcyanation of alkynes using an electron deficient triarylphosphine ligand (Scheme 2.9c).[28] Alkyl nitriles have also been shown to undergo a Ni-catalyzed addition across both alkynes and allenes by using BPh_3 as a key Lewis acid additive.[29]

Interesting extensions of this chemistry were reported in 2009 by Nakao and Hiyama where cyanoformates **2.46** could be added across allenes **2.44**[30] and alkynes **2.45** (Scheme 2.9d).[31] These reactions provide access to interesting classes of molecules which contain two functional groups in the carboxylate oxidation state. Alkyl nitiles **2.49** containing a heteroatom directing group were shown to

[23]References [31–35].
[24]Reference [36].
[25]References [37, 38].
[26]Reference [39].
[27]Reference [40].
[28]References [41, 42].
[29]Reference [43].
[30]Reference [44].
[31]Reference [45].

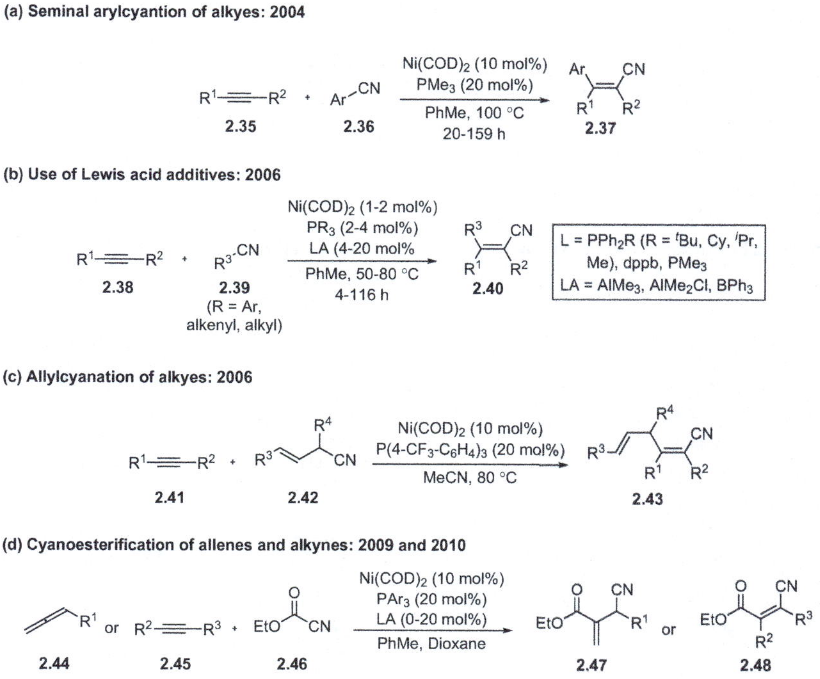

Scheme 2.9 Advances in Ni-catalyzed carbocyanation by Nakao and Hiyama

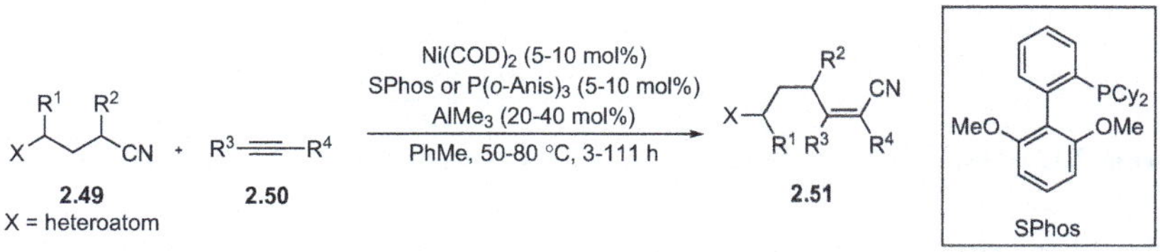

Scheme 2.10 Ni-catalyzed heteroatom-enabled alkylcyanation of alkynes

undergo a Ni-catalyzed alkylcyanation of alkynes **2.50** in 2010 (Scheme 2.10).[32] By using either SPhos or an anisole-based phosphine ligand, in addition to a catalytic amount of an aluminum Lewis acid co-catalyst, a series of products **2.51** could be obtained in good yields. The heteroatom group played an important role in suppressing β-hydrogen elimination from either the alkyl Ni(II)CN species arising from C–CN oxidative addition, or the alkyl(vinyl)Ni(II) species which is formed after migratory insertion of the alkyne substrate.

[32]Reference [46].

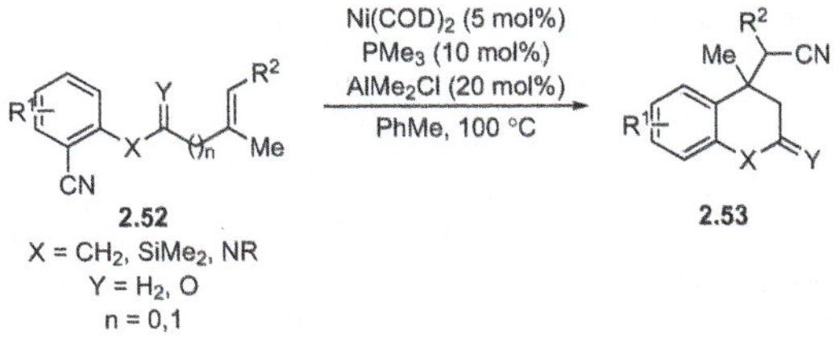

Scheme 2.11 Intramolecular Ni-catalyzed arylcyanation of alkenes

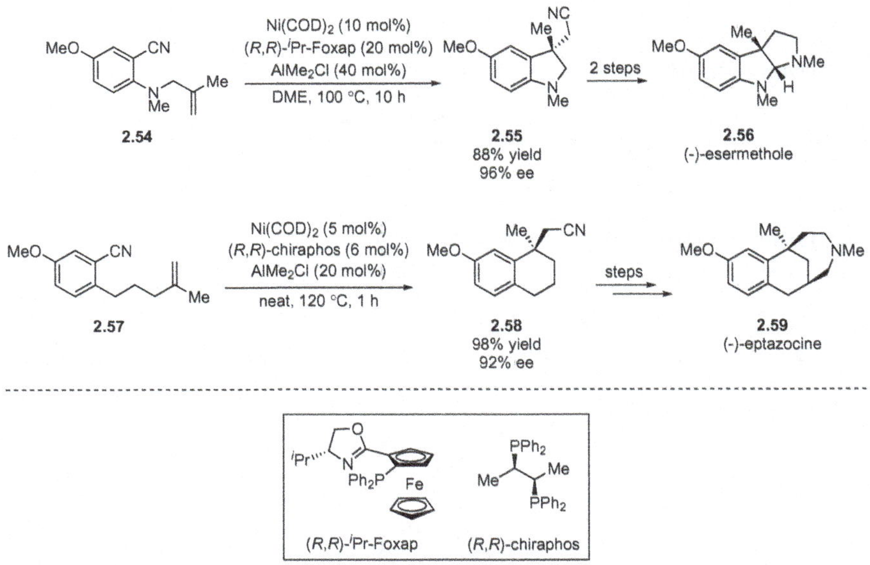

Scheme 2.12 Asymmetric intramolecular alkene arylcyanations by Nakao and Hiyama

In 2008, Hiyama and Nakao reported the intramolecular Ni-catalyzed arylcyantion of alkenes **2.52** (Scheme 2.11).[33] They showed that various 5-, 6- and 7-membered carbo-, sila- and azacycles **2.53** could be obtained in moderate to excellent yields typically using a $Ni(COD)_2/PMe_3$ combination in the presence of a catalytic amount of $AlMe_2Cl$.

The authors also reported two enantioselective examples which employed either (*R*,*R*)-^{i}Pr-Foxap or (*R*,*R*)-chiraphos as the chiral ligands. In addition, this methodology could be employed towards the synthesis of the alkaloid natural prodcuts (-)-esermethole **2.56** and (-)-eptazocine **2.59** (Scheme 2.12). The scope of the enantioselective process leading to indolines was subsequently expanded upon in 2010 by the same authors.[34]

[33]Reference [47].

[34]Reference [48].

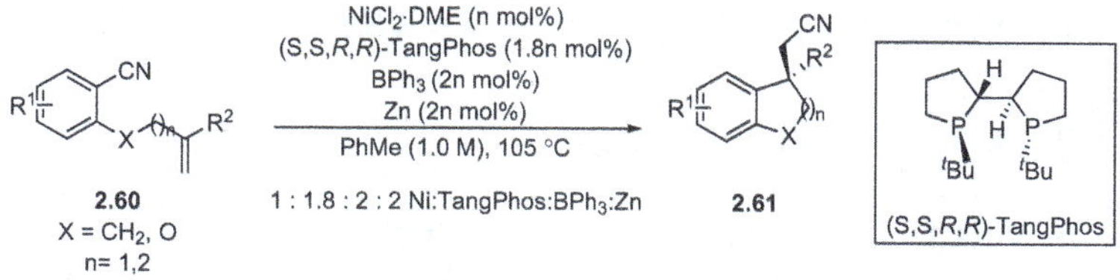

Scheme 2.13 Asymmetric intramolecular alkene arylcyanations by Watson and Jacobsen

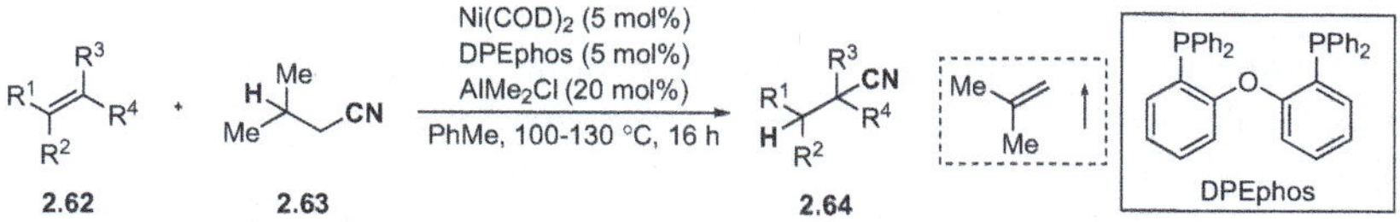

Scheme 2.14 Ni-catalyzed transfer hydrocyanation by Morandi

Almost simultaneously, Watson and Jacobsen reported an asymmetric intramolecular Ni-catalyzed alkene aryl cyanation where up to 97% ee could be obtained (Scheme 2.13).[35] The use of easy-to-handle $NiCl_2$·DME represents a major improvement to previous Ni-catalyzed arylcyanations which commonly utilize highly air-sensitive $Ni(COD)_2$ as the pre-catalyst. The addition of Zn to the reaction conditions helps to facilitate the formation of the active Ni(0) species from the Ni(II) pre-catalyst.

In 2016, Morandi reported a Ni-catalyzed alkene-nitrile interconversion which proceeds via hydrocyanation (Scheme 2.14).[36] This reaction utilized an C(sp^3)–CN oxidative addition in a substrate that bears a β-hydrogen for elimination to occurs at the stage of alkyl–Ni(II)–CN. This step generates the desired H–Ni(II)–CN intermediate which undergoes the subsequent hydrocyanation reaction. The authors showed that this reaction was reversible under certain conditions. However, using **2.63** as the optimal source of HCN allowed the reaction to drive off isobutene gas and proceed to completion. Under identical condition, the authors were able to perform a highly efficient dehydrocyanation when either norbornene or norbornadiene was added as a HCN acceptor.

[35]Reference [49].

[36]Reference [50].

2.1.3 Pd-Catalyzed Carbocyanation or Alkene and Alkynes

2.1.3.1 Heck-Type Domino Reactions

As discussed in Chap. 1, the Mizoroki-Heck reaction mechanism can divert through what is known as anion capture cascade or Heck-type domino reaction. These processes typically involve the trapping of an alkyl or vinyl Pd(II) intermediate (i.e. **2.66** or **2.69**) which is present after oxidative addition of an aryl or vinyl halide and subsequent alkene or alkyne insertion (Scheme 2.15).

Grigg was the first to study these transformations in great detail, and in 1988 they showed that formate could act as a hydride surrogate in the Pd-catalyzed intramolecular hydroarylation of alkynes (Scheme 2.16).[37,38] In these examples, hydride can be used as a nucleophile (generated from formic acid and piperidine) to trap a vinyl Pd(II) species which results from oxidative addition and subsequent intramolecular 5- or 6-exo-dig carbopalladation.[39]

This class of reactions allows the straightforward functionalization of otherwise difficult to access Pd(II) species. Since the seminal reports from Grigg, several interesting permutations of this class of reaction have been appeared, and terminating steps involving inter-[40] and intramolecular[41] C–H functionalizations, carbonylations [55], aminations,[42] vinylations,[43] and borylations[44] have all been reported (Scheme 2.17).

Another important variation to this reaction involves the termination of a species like **2.66** or **2.69** with a cyanide nucleophile.[45] The first intermolecular example of this reaction was reported by Torri in 1990 (Scheme 2.18a). They showed that a tandem assembly of vinyl halides, norbornene and cyanide could be accomplished in the presence of a Pd(0) catalyst. This reaction proceeds in a *syn* fashion, and since th Pd(II) species is never in the presence of suitable β-hydrogen atoms, no Mizoroki-Heck reaction products were observed. This reaction was extended to the intermolecular carbocyanation reaction of vinyl halides containing distal stereocenters.[46]

---

[37] Reference [51].

[38] Around the same time, an intermolecular variant was reported. See Ref. [52].

[39] References [53–55].

[40] Reference [56].

[41] References [57–60].

[42] References [54, 61].

[43] References [62, 63].

[44] References [64, 65].

[45] Reference [66].

[46] Reference [67].

Scheme 2.15 General representation of Heck-type domino reactions

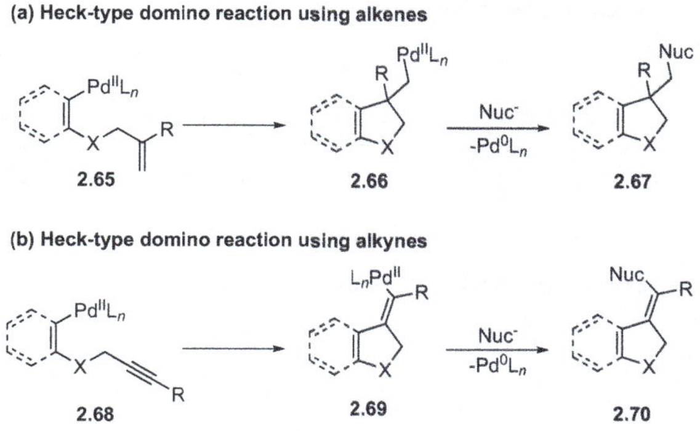

Scheme 2.16 Intramolecular hydroarylation of alkynes catalyzed by Pd

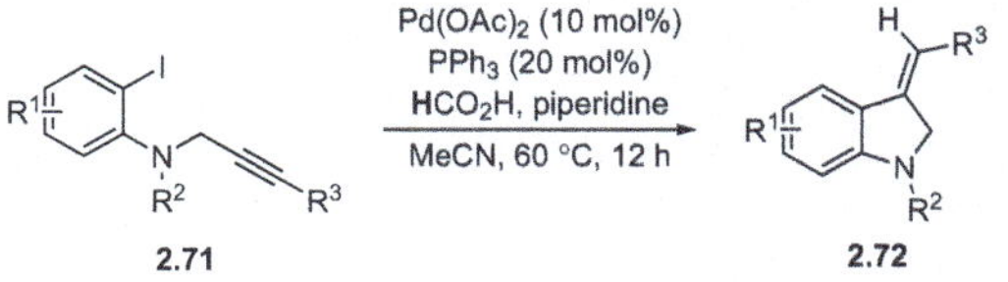

The first intramolecular arylcyanation reaction was reported by Grigg in 1993 (Scheme 2.18b).[47] They showed that both aryl and vinyl halides could be transformed into diverse nitrile-containing heterocycles via both 5- and 6-exo-trig cyclizations. In addition, they also described the first cyanide trapping of a 2° alkyl Pd(II) intermediate. Zhu and co-workers reported the intramolecular arylcyanation of electron-deficient alkenes en route to oxindole products in 2007 (Scheme 2.18c).[48] This report represented the first use of potassium ferro(II)cyanide as an arylcyanation reaction. Furthermore, the authors were also able to obtain moderate levels of enantioselectivity in this transformation by using the chiral biphosphine ligand (*S*)-Difluorphos (Scheme 2.18d). The same group has even reported the use of this reaction as the key step in the synthesis of various natural products.[49] This reaction was subsequently expanded to the use of non-electron-deficient alkenes in 2011 by Kim and co-workers who used nearly identical reaction conditions to those of Zhu.[50] Other than these sporadic examples, reports of the Pd-catalyzed arylcyanation reaction have been scarce.[51]

---

[47]Reference [68].

[48]Reference [69].

[49]References [70, 71].

[50]Reference [72].

[51]Reference [73].

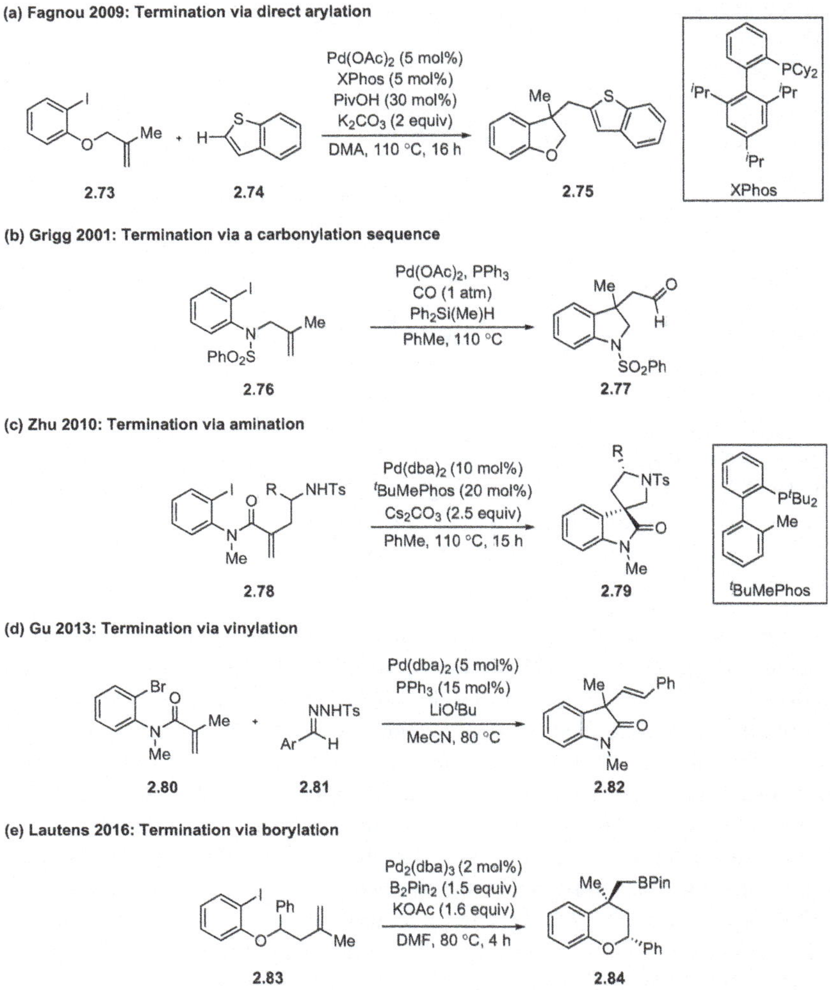

Scheme 2.17 Examples of various terminating steps in the Heck-type domino reaction

2.1.3.2 Other Classes of Pd-Catalyzed Carbocyanation[52]

Takaya reported the first intermolecular Pd-catalyzed *cis*-addition of aroyl cyanides **2.94** across alkynes **2.95** in 1994 (Scheme 2.19a).[53] Interestingly, this reaction was found to be most optimal when a PPh_3/dppb mixed-ligand system was employed. The authors propose that the operative mechanism could involve the formation of

[52]For an example of the addition of TMSCN across alkynes, see: Ref. [74].

[53]Reference [75].

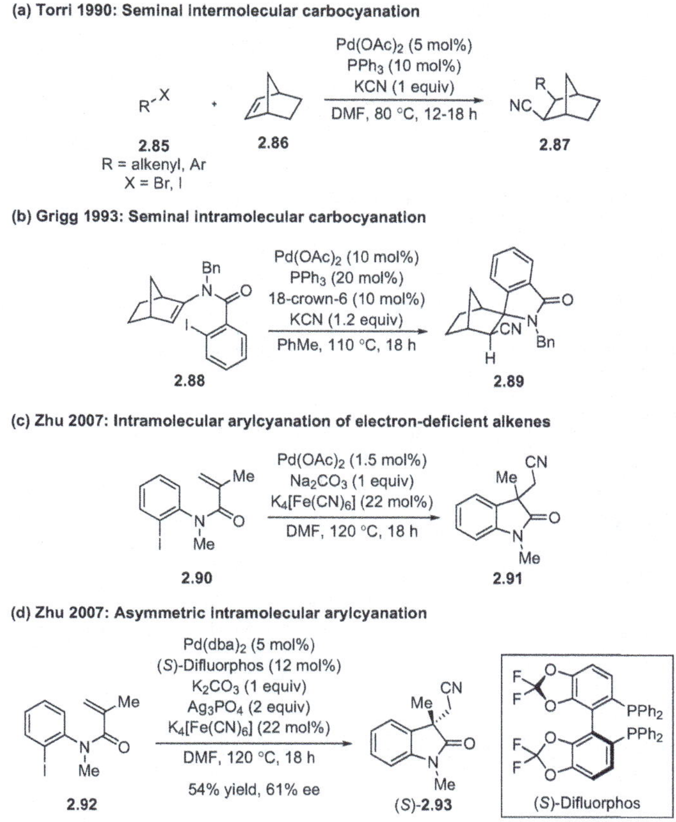

Scheme 2.18 Pd-catalyzed carbocyanation reactions

ynone **2.97** which undergoes a regioselective hydrocyanation by in situ generated H–Pd(II)–CN to generate the final product **2.96**.[54] Nishihara showed that ethyl cyanoformate **2.98** could be added across norbornene in 2005 (Scheme 2.19b).[55] The scope of this reaction was later expanded to other strained alkenes and cyanoformates in 2006.[56]

Alkynyl **2.100** and alkenyl cyanoformamides **2.101** were shown to undergo Pd-catalyzed intramolecular cyanoamidation reaction by Takemoto and co-workers in 2006 (Scheme 2.19c).[57] This methodology could be used to access 5-, 6- and 7-membered lactams in good yields using a simple catalyst system. The same authors reported that this the reaction of alkenyl cyanoformates could undergo a

---

[54]Reference [76].

[55]Reference [77].

[56]Reference [78].

[57]References [79, 80].

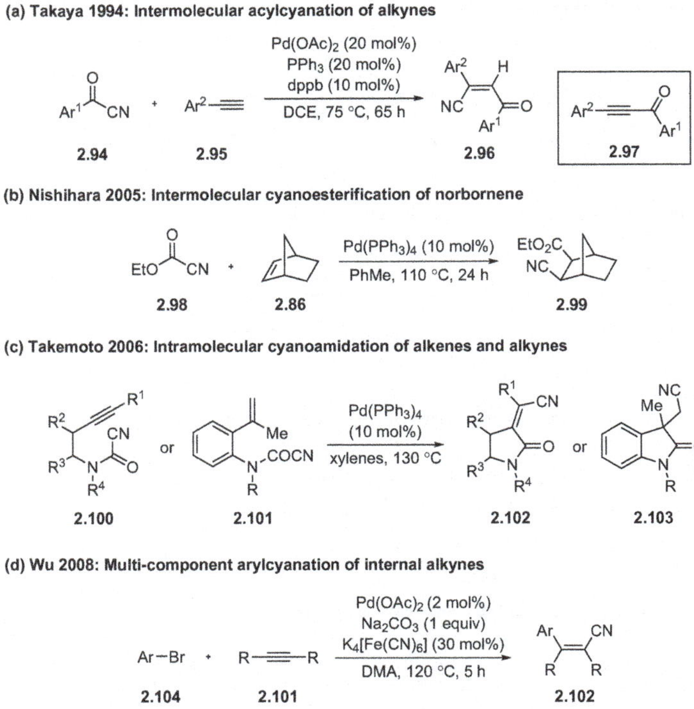

Scheme 2.19 Other classes of Pd-catalyzed carbocyanation reactions

moderately enantioselective amidation reaction in the presence of chiral phosphine ligands.[58] Wu reported a Pd-catalyzed three-component arylcyanation of internal alkynes using aryl bromides and $K_4[Fe(CN)_6]$ (Scheme 2.19d).[59] Using similar conditions to that of Zhu [69] the authors were able to obtain the β-arylalkenylnitrile products in good moderate to good yields. A mechanism involving oxidative addition to the aryl bromide followed by arylpalladation and subsequent carbon–CN reductive elimination was proposed.

2.1.4 Research Goals: Part 1

Our previous work on the synthesis of (+)-corynoline inspired us to develop a second generation route to this molecule which would circumvent some potential drawbacks. First, although the developed aryliodination of chiral *N*-allyl

[58]Reference [81].

[59]Reference [82].

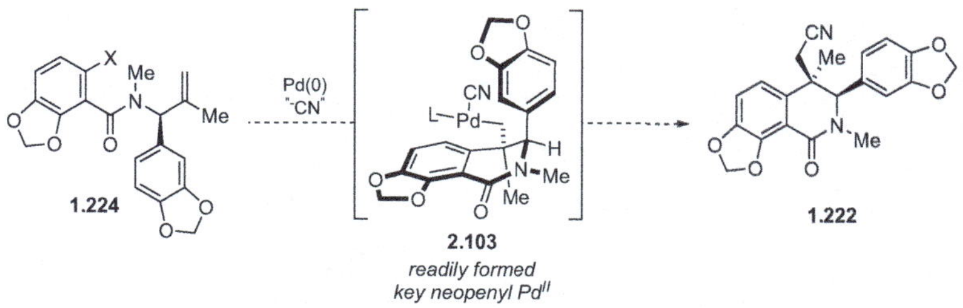

Scheme 2.20 Proposed diastereoselective Pd-catalyzed arylcyanation of chiral *N*-allyl carboxamides

carboxamides provided an efficient route to complex dihydroisoquinoline products, it utilized QPhos. This is one of the most costly achiral phosphine ligands on the market (~$151,736.51/mol),[60] a fact which can potentially cause problems with overall scalability. Furthermore, a moderate yielding nucleophilic cyanation of a sterically hindered neopentyl iodide was necessary in the subsequent step. This transformation makes use of high temperatures/long reaction times, in addition to excess amounts of KCN and 18-crown-6. In this regard, we envisioned that executing a diastereoselective arylcyanation reaction on a similar amide substrate would provide access to the required nitrile. Indeed, this reaction would increase overall step efficiency and ensure a more efficient route to synthesize derivatives of (+)-corynoline (Scheme 2.20). Furthermore, the paucity of stereoselective Pd-catalyzed arylcyanation methods in the literature further motivated us to study this transformation in a more complex setting.

2.1.5 *Results and Discussion: Diastereoselective Pd-Catalyzed Arylcyanation of Chiral* N*-Allyl Carboxamides*

2.1.5.1 Starting Material Preparation

The majority of the chiral substrates were prepared in a similar fashion to those used in Chap. 1. However, the corresponding 2-bromobenzoyl chloride derivatives **2.104** were used instead of the corresponding 2-iodobenzoyl chlorides (Scheme 2.21).

Some substrates were not obtained via the exact route shown in Chap. 1. For example substrates with vinylic groups other than CH_3 could be prepared via lithiation of vinyl bromides (i.e. **2.106** and **2.111**) at low temperature followed by

[60] Price taken from: http://www.strem.com/catalog/v/26-3575/32/iron_312959-24-3 on 04/18/2016.

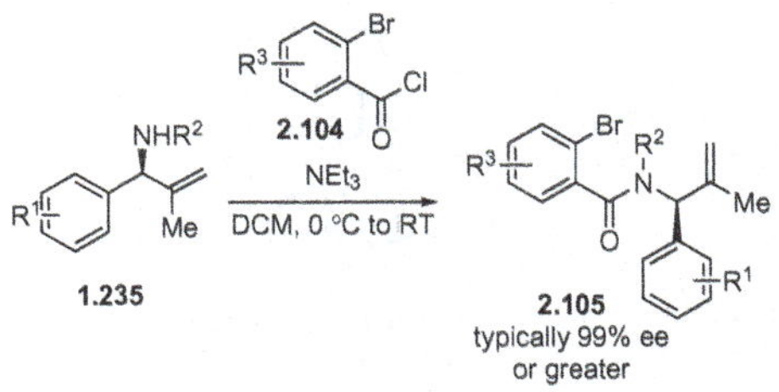

Scheme 2.21 General route to enantioenriched aryl bromide-containing *N*-allyl carboxamides

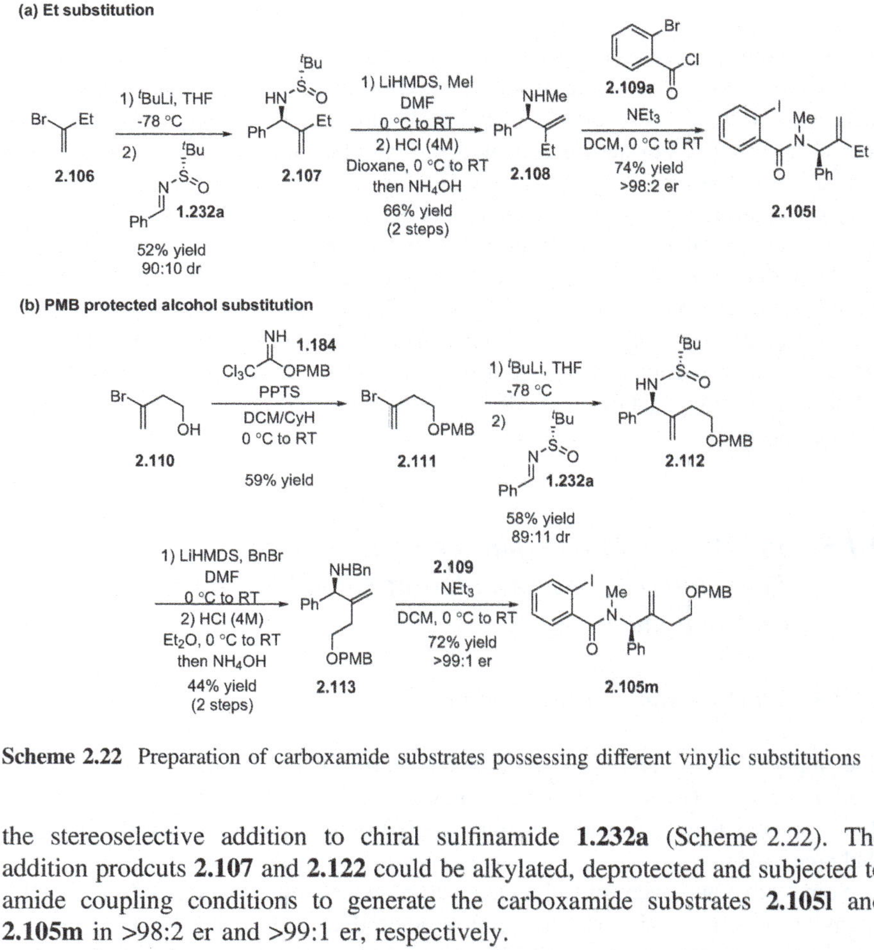

Scheme 2.22 Preparation of carboxamide substrates possessing different vinylic substitutions

the stereoselective addition to chiral sulfinamide **1.232a** (Scheme 2.22). The addition prodcuts **2.107** and **2.122** could be alkylated, deprotected and subjected to amide coupling conditions to generate the carboxamide substrates **2.105l** and **2.105m** in >98:2 er and >99:1 er, respectively.

Brominated indole substrates were also prepared in a straightforward manner from **2.114**.[61] *N*-benzylation affords **2.115** which could be saponified to generate

[61]Reference [83].

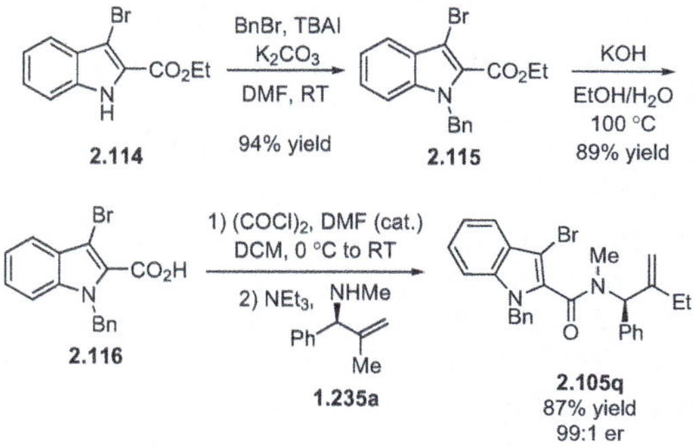

Scheme 2.23 Preparation of a 3-bromo indole substrate

the corresponding indole 2-carboxylate **2.116**. Acid **2.116** could be efficiently coupled to generate the carboxamide substrate **2.105q** (Scheme 2.23).

2.1.5.2 Optimization

We first began to optimize the arylcyanation reaction using chiral amide (*S*)-**1.237a**. Due to the fact that the Pd-catalyzed arylcyanation reaction is relatively underexplored, it semed like there was a more narrow pool of reaction conditions to initially test this reaction. Therefore, we looked into the direct arylhalide cyanation literature as a means to come to a consensus on where to begin (Table 2.1). Based on earlier reports, our initial reaction conditions utilized $Pd(dba)_2$ (5 mol%), dppf (5 mol%), $Zn(CN)_2$ (55 mol%) in DMF (0.1 M) at 120 °C for 2 h (entry 1). The reaction

Table 2.1 Optimization of the intramolecular arylcyanation of (*S*)-**1.237a**: Zn loading[a]

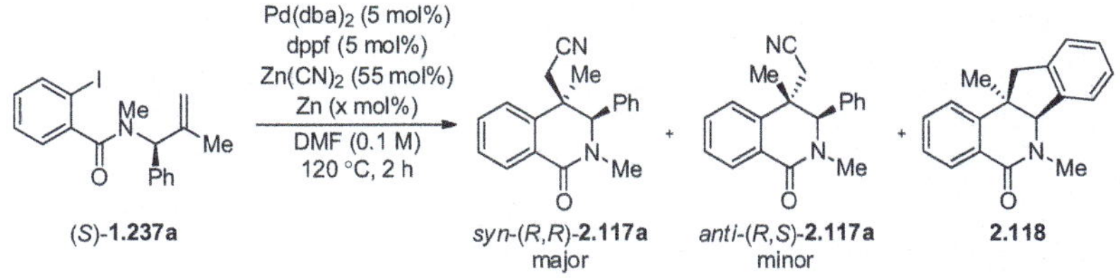

Entry	Zn (x mol%)	Conv (%)[b]	% yield **2.177a** (*syn*+*anti*)[b]	dr[c]	% yield **2.118** (%)[b]
1	0	51	49	>95:5	7
2[d]	10	>95	93	>95:5	<5
3	20	>95	87	>95:5	6

[a]Reactions run on 0.1 mmol scale
[b]Determined by ^{1}H NMR analysis of the crude reaction mixture using 1,3,5-trimethoxybenzene as an internal standard
[c]Determined by ^{1}H NMR analysis of the crude reaction mixture
[d]Reaction conducted by Hyung Yoon

proceeded with partial conversion and the desired diastereomeric products were obtained in 49% yield in >95:5 dr. Byproduct **2.118**, which results from direct intramolecular C–H functionalization, was obtained in 7% yield [55]. This byproduct originates from the Pd(II)-species resulting from the major alkene insertion pathway and artificially lowers the diastereoselectivity, in cases where it is formed to a larger extent. Zn powder was tested as an additive in order to increase the conversion of starting material (vide supra). When 10 mol% of purified Zn powder was added to the reaction, we observed full conversion of starting material and 93% yield of the desired products with only a trace amount of **2.118** (entry 2). Increasing the Zn loading to 20 mol% led to a decreased yield of the desired product (entry 3). The use of solvents such as PhMe, 1,4-dioxane and DMA lead to only trace conversions of the starting material with neither **2.117** or **2.118** being observed in any case.

Table 2.2 shows a selection of the various ligands which were screened in this reaction. The ^{i}Pr analogs of dppf led to unsatisfactory yields of **2.117** (entry 2),

Table 2.2 Optimization of the intramolecular arylcyanation of (*S*)-**1.237a**: ligand screen[a]

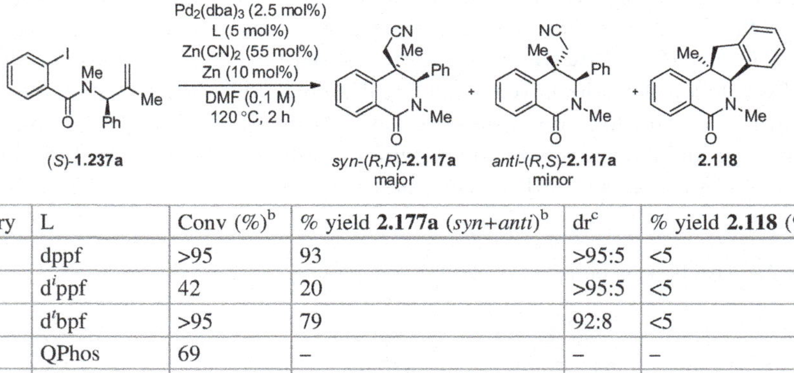

(*S*)-**1.237a** *syn*-(*R*,*R*)-**2.117a** major *anti*-(*R*,*S*)-**2.117a** minor **2.118**

Entry	L	Conv (%)[b]	% yield **2.177a** (*syn*+*anti*)[b]	dr[c]	% yield **2.118** (%)[b]
1[f]	dppf	>95	93	>95:5	<5
2	d^{i}ppf	42	20	>95:5	<5
3[f]	d^{t}bpf	>95	79	92:8	<5
4	QPhos	69	–	–	–
5[d]	P^{t}Bu$_3$	87	57	91:9	0
6[e,f]	Pd(P^{t}Bu$_3$)$_2$	>95	84	88:12	0
7[e,f]	Pd(d^{t}bpf)Cl$_2$	>95	89	92:8	0
8[e,g]	Pd(P^{t}Bu$_3$)$_2$	>95	84	92:8	0
9[e,f,h,i]	Pd(P^{t}Bu$_3$)$_2$	45	27	93:7	0

[a]Reactions run on 0.1 or 0.3 mmol scale
[b]Determined by ^{1}H NMR analysis of the crude reaction mixture using 1,3,5-trimethoxybenzene as an internal standard
[c]Determined by ^{1}H NMR analysis of the crude reaction mixture
[d]The HBF$_4$ salt of this ligand was used in combination to KO^{t}Bu (5 mol%)
[e]The reaction was run with Pd(P^{t}Bu$_3$)$_2$ (5 mol%)
[f]Reaction conducted by Hyung Yoon
[g]The reaction was run with the corresponding aryl bromide substrate
[h]The reaction was run with the corresponding aryl chloride substrate
[i]The reaction was run for 4 h

whereas the corresponding ^{t}Bu analog performed slightly better, yet not as well as dppf (entry 3). QPhos led to no observable quantities of the desired product, and an overall messy reaction profile as observed by ^{1}H NMR analysis of the crude reaction mixture (entry 4). Employing $P^tBu_3 \cdot HBF_4$ in the presence of an equimolar amount of KO^{t}Bu, afforded the desired product in 57% yield with 91:9 dr. Interestingly, byproduct **2.118** was not observed in this reaction. When $Pd(P^tBu_3)_2$ was used as the catalyst in the absence of additional ligands, the desired product could be obtained in 84% yield with a slightly diminished dr (entry 6). Slightly improved results could be obtained when using the Pd(II) pre-catalyst Pd(d^{t}bpf)Cl_2, however, we found that these results were not reproducible (entry 7). The corresponding aryl bromide substrate led to similar yields of the desired product, yet the dr increased slightly to 92:8, whereas the arylchloride analog provided poor yields of the desired product (entries 8 and 9).

The product was obtained with slightly better yield and dr when the reaction temperature was lowered to 110 °C (Table 2.3, entry 2), whereas decreasing the temperature to 100 °C led to decreased overall product yields (entry 3). Both the yield and dr of the reaction improved incrementally when the loading of Zn additive was decreased, (entries 4 and 5). In the absence of Zn metal, the product could be isolated in 90% yield with 95:5 dr.

The optimal reaction conditions were found to be $Pd(P^tBu_3)_2$ (5 mol%) in the presence of $Zn(CN)_2$ (55 mol%) in DMF at 110 °C for 2 h. The absolute and

Table 2.3 Optimization of the intramolecular arylcyanation of (*S*)-**2.105a**: temperature and Zn loading screen[a]

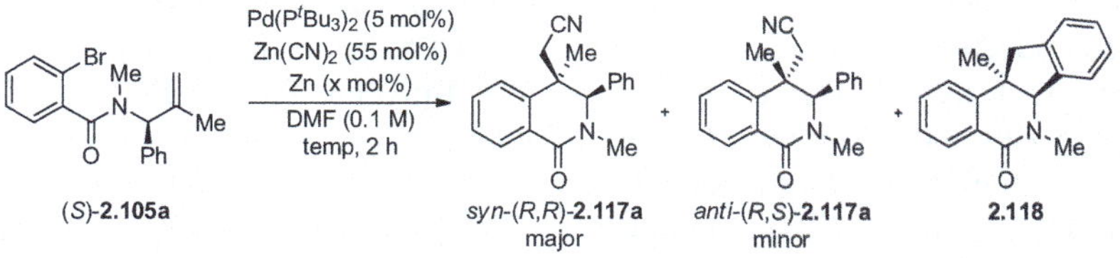

Entry	Temp (°C)	Zn (x mol%)	Conv (%)[b]	% yield **2.177a** (*syn*+*anti*)[b,c]	dr[d]	% yield **2.118** (%)[b]
1[e]	120	10	>95	84	92:8	0
2[e]	110	10	>95	88(86)	93:7	0
3[e]	100	10	>95	73	93:7	0
4[e]	110	5	>95	89(88)	94:6	0
5	110	0	>95	86(90)[f]	95:5	0

[a]Reactions run on 0.3 mmol scale
[b]Determined by ^{1}H NMR analysis of the crude reaction mixture using 1,3,5-trimethoxybenzene as an internal standard
[c]Value in parentheses represents an isolated yield
[d]Determined by ^{1}H NMR analysis of the crude reaction mixture
[e]Reaction conducted by Hyung Yoon
[f]Average value over three experiments

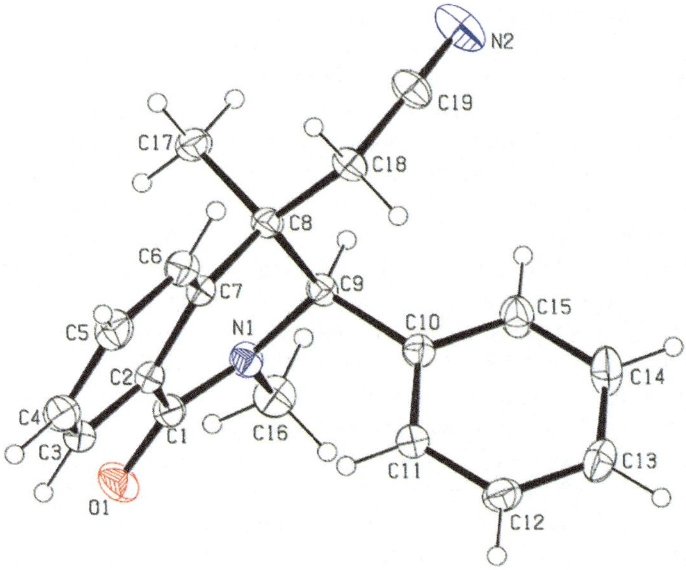

Fig. 2.1 Crystal structure of the *syn*-arylcyanation product *syn*-(*R*,*R*)-**2.117a**

relative stereochemistry of the major diastereomer was unambiguously determined to be *syn* by an X-ray diffraction study (Fig. 2.1).

We next sought to further examine the optimized reaction conditions by altering several individual parameters. This study would provide us with a better understanding of how each aspect of the reaction conditions affected the overall yield and diastereoselectivity (Table 2.4). The use of other Pd pre-catalysts containing bulky ligands, such as $Pd(QPhos)_2$ and Buchwald's ^{t}BuXPhos-Pd-G1 **[Pd2]**, produced inferior yields and selectivities (entries 2–3). Other solvents such as 1,4-dioxane or PhMe led to greatly reduced reactivity and yields (entries 4–5). When the reaction was run using the conditions reported by Zhu [69], formation of **2.117** did not occur. Instead, arylation product **2.118** was observed in 32% yield (entry 6). This result can be rationalized by the slower rate of cyanide transfer to the Pd catalyst with respect to that of $Zn(CN)_2$,[62] as well as the presence of carbonate base which can assist in the C–H metallation step of the mechanism leading to byproduct formation. $Zn(CN)_2$ appears to have the optimal reactivity profile, as both cyanide equivalents are transferred over the course of the reaction. Its highly covalent nature leads to a decreased amount of free cyanide in solution, which may deter catalyst deactivation.[63] Both aryl iodide and chloride derivatives led to inferior results

[62]Reference [84].

[63]Reference [85].

Table 2.4 Studying the effects of altering individual reaction parameters[a]

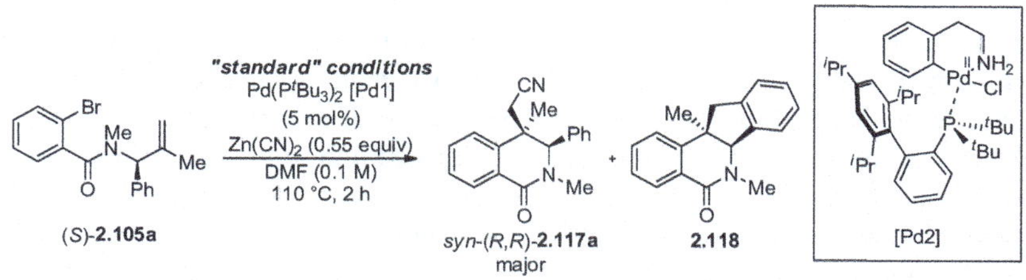

Entry	Variation from the ***"standard" conditions***	dr[b]	Yield **2.117a** (%)[c,d]	Yield **2.118** (%)[c]
1[e]	None	95:5	86 (90)[e]	0
2	Pd(QPhos)$_2$ instead of **[Pd1]**	90:10	69	0
3	**[Pd2]** instead of **[Pd1]**	94:6	4	0
4	PhMe instead of DMF	–	0	0
5	1,4-dioxane instead of DMF	75:25	11	0
6[f]	$K_4Fe(CN)_6$ instead of $Zn(CN)_2$	–	0	32
7	ArI instead of ArBr	93:7	55	0
8[g]	ArI instead of ArBr	88:12	77	0
9	ArCl instead of ArBr	94:6	17	0
10	90 °C	96:4	73	0

[a]Reactions run on 0.3 mmol scale
[b]Determined by ^{1}H NMR analysis of the crude reaction mixture
[c]Determined by ^{1}H NMR analysis of the crude reaction mixture using 1,3,5-trimethoxybenzene as an internal standard
[d]Values in brackets represent isolated yields
[e]Average value over three experiments
[f]Run in the presence of Na_2CO_3 (1 equiv)
[g]Reaction run in the presence of Zn dust (10 mol%)

(entries 7 and 9). However, the reactivity of the ArI derivative could be restored by adding a catalytic amount of Zn dust to the reaction (entry 8). The decreased yields of ArI substrates under the standard conditions may be a result of trace Ullman-type dimerization of the starting material, consequently generating Pd(II). In this reaction, zinc may act to maintain appropriate concentration levels of the active Pd(0) catalyst.[64] Finally, lowering the reaction temperature to 90 °C led to sluggish conversion of **2.105a**, while no improvement in diastereoselectivity was observed (entry 10).

[64]References [86–89].

2.1.5.3 Examination of the Substrate Scope

The generality of the optimized conditions was tested on a series of chiral *N*-allyl carboxamide substrates (Table 2.5). In all cases, the enantioenriched substrates could be cyclized to the desired products with no erosion of enantiomeric ratio (98:2 to >99:1 er) as determined by HPLC analysis. In addition, either enantiomer of the product could be accessed with nearly complete enantioenrichment (entries 1 and 7). Sterically hindered aryl bromide **2.105b** posessing an *o*-Me substituent was reacted under the standard reaction conditions, and yielded the desired product **2.117b** in 89% with a >95:5 dr. Dihalogenated carboxamide **2.105c** (entry 3) was efficiently cyclized to afford the desired chlorodihydroisoquinolinone **2.117c** in 89% yield with 94:6 dr. After converting the *N*-protecting group from methyl to benzyl **2.105d**, product **2.117d** was obtained in 85% yield with slightly diminished selectivity (91:9 dr, entry 4). This result contrasts those obtained in the aryliodination discussed in Chap. 1, where converting the *N*-Me to an *N*-Bn led to an overall increase in both

Table 2.5 Scope of the dihydroisoquinolinone synthesis by arylcyanation[a]

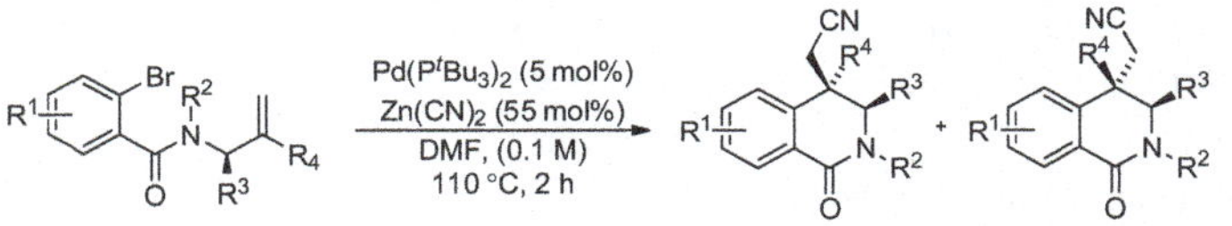

(*S*)-**2.105** *syn*-(*R*,*R*)-**2.117** major *anti*-(*R*,*S*)-**2.117** minor

Entry	Substrate		Product		Yield[b]	dr (*syn* +*anti*)[c] (er)[d]
1	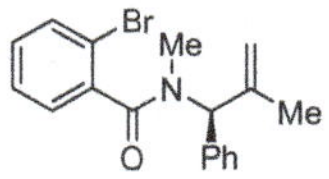	(*S*)-**2.105a**	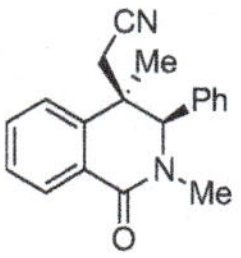	(*R*,*R*)-**2.117a**	90%	95:5 (>99:1)
2	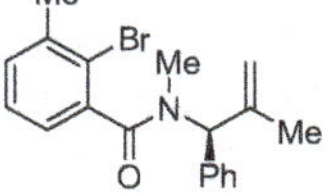	(*S*)-**2.105b**[h]	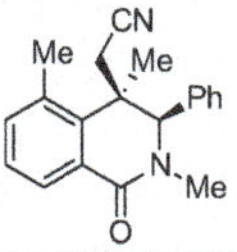	(*R*,*R*)-**2.117b**[h]	89%	>95:5 (>99:1)
3	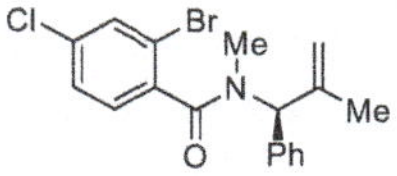	(*S*)-**2.105c**[h]	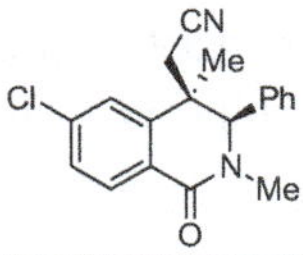	(*R*,*R*)-**2.117c**[h]	89%	94:6 (99:1)
4[f]		(*S*)-**2.105d**[h]	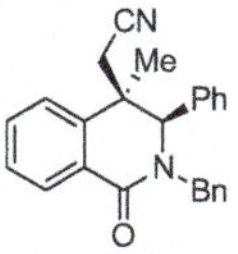	(*R*,*R*)-**2.117d**[h]	85%	91:9 (98:2)

(continued)

Table 2.5 (continued)

Entry	Substrate		Product		Yield[b]	dr (*syn* +*anti*)[c] (er)[d]
5		(*S*)-**2.105e**[h]	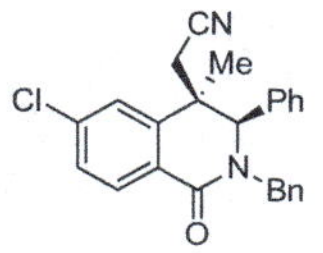	(*R*,*R*)-**2.117e**[h]	70%	91:9 (>99:1)
6		(*S*)-**2.105f**[h]	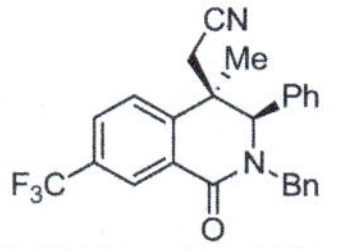	(*R*,*R*)-**2.117f**[h]	78%	89:11 (99:1)
7		(*R*)-**2.105a**	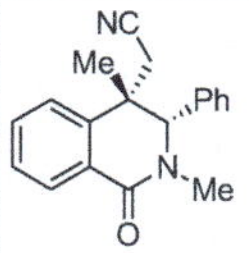	(*S*,*S*)-**2.117a**[f]	95%	95:5 (>99:1)
8	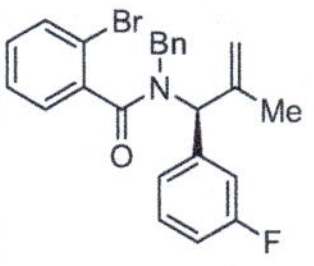	(*S*)-**2.105g**	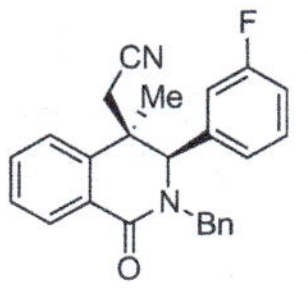	(*R*,*R*)-**2.177g**	93%	90:10 (99:1)
9	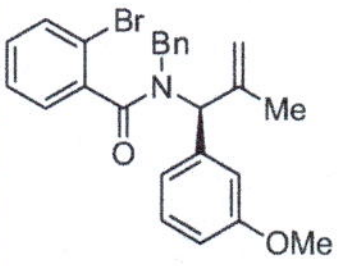	(*S*)-**2.105h**[h]	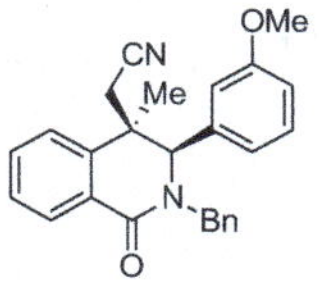	(*S*,*S*)-**2.117h**	86%	90:10 (>99:1)
10	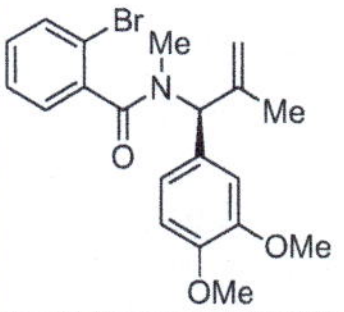	(*S*)-**2.105i**	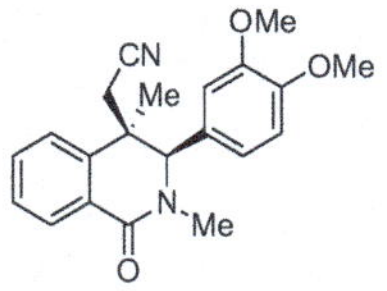	(*R*,*R*)-**2.117i**	94%	94:6 (99:1)
11[e]	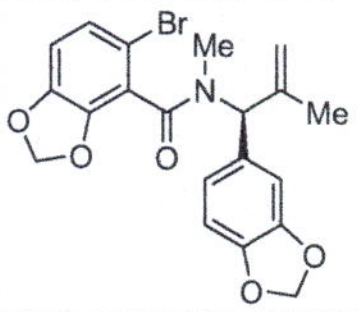	(*S*)-**2.105j**	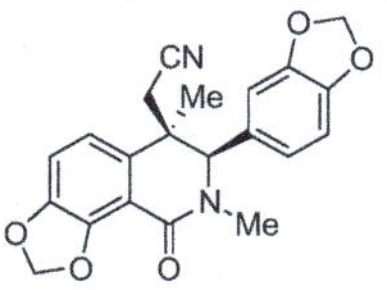	(*R*,*R*)-**1.222**	79%	>95:5 (>99:1)
12	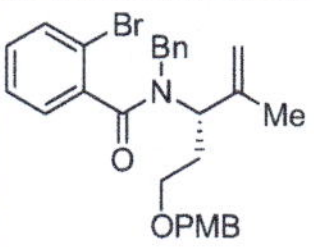	(*S*)-**2.105k**[h]	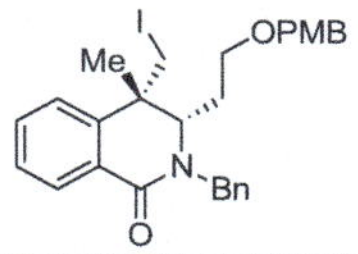	(*R*,*R*)-**2.117k**[h]	66%	53:47 (>99:1)
13	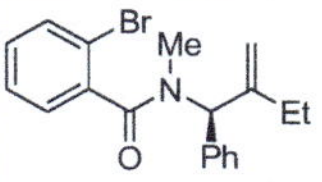	(*S*)-**2.105l**	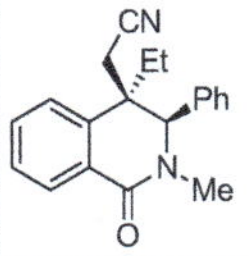	(*R*,*R*)-**2.117l**	73%	95:5 (99:1)

(continued)

Table 2.5 (continued)

Entry	Substrate		Product		Yield[b]	dr (*syn* +*anti*)[c] (er)[d]
14	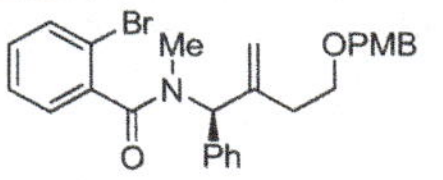	(*S*)-**2.105m**	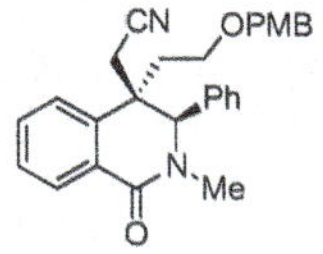	(*R*,*R*)-**2.117m**	77%	92:8 (99:1)
15[f]	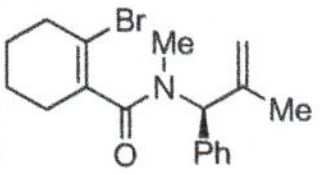	(*S*)-**2.105n**	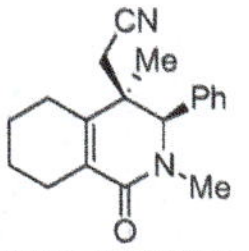	(*R*,*R*)-**2.117n**	67%	87:13 (99:1)
16		(*S*)-**2.105o**[h]		(*R*,*R*)-**2.117o**[h]	70%	93:7 (98:2)
17	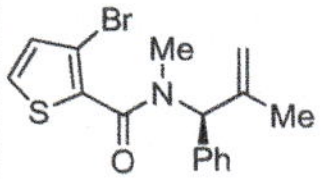	(*S*)-**2.105p**[h]	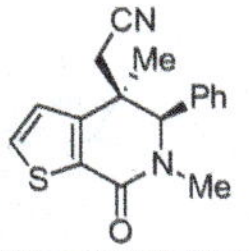	(*R*,*R*)-**2.117p**	87%	93:7 (>99:1)
18	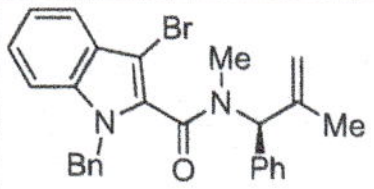	(*S*)-**2.105q**[h]	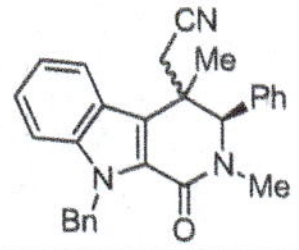	(*R*,*R*)-**2.117q**[h]	85%	50:50 (99:1)[g]
19	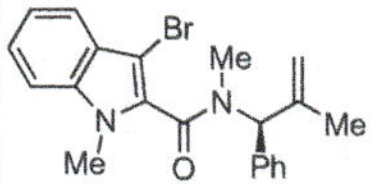	(*S*)-**2.105r**[h]	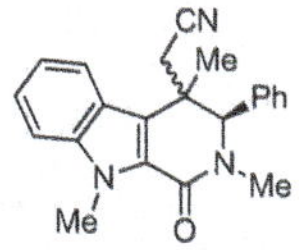	(*R*,*R*)-**2.117r**[h]	77%	40:60 (>99:1)[g]
20	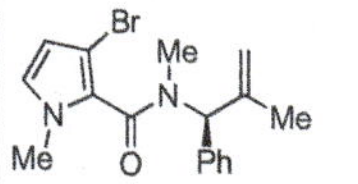	(*S*)-**2.105s**[h]	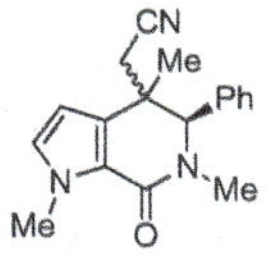	(*R*,*R*)-**2.117s**[h]	92%	50:50 (99:1)[g]

[a]Reactions were run on a 0.3 mmol scale

[b]Combined isolated yield of both the *cis* and *trans* diastereomers

[c]Determined by ^{1}H NMR analysis of the crude reaction mixture

[d]Determined by HPLC using a chiral stationary phase

[e]Reaction was run with 2.5 mol% $Pd(P^{t}Bu_3)_2$

[f]Reaction run in the presence of 50 mol% of PMP

[g]er value for both diastereoemers

[h]This compound was prepared by Hyung Yoon

yield and diastereoselectivty. Reacting chloro- and trifluoromethyl-substituted aryl bromides **2.105e** and **2.105f** under the optimized conditions afforded the desired products in 70 and 78% yields with 91:9 and 89:11 dr, respectively (entries 5 and 6). Substitution on the allylic aromatic group (**2.105g–i**) shows no deleterious effects and the corresponding products (**2.117g–i**) were obtained in high yields with good to excellent levels of selectivity (entries 8–10).

(+)-Corynoline precursor **1.222** could be obtained in 79% yield with >95:5 dr when the catalyst loading was lowered to 2.5 mol% (entry 11). We were able to increase the efficiency of the previously reported route from Chap. 1 from 56% over two steps (Pd-catalyzed carboiodination followed by nucleophilic cyanation) to 79% in a single Pd-catalyzed transformation. Alkyl PMB ether **2.105k** was transformed to the desired product **2.117k** in 66% yield with almost no diastereoselectivity (53:47 dr). This finding can be rationalized by the decreased steric demand of the alkyl group compared to the Ar group. Other vinylic alkyl groups (R^4) such as Et (**2.105l**) and an alkyl PMB ether (**2.105m**) were tolerated leading to the desired products **2.117l** and **2.117m** in 73 and 77% yield and 95:5 and 92:8 dr, respectively. Cyclohexenyl bromide substrate **2.105n** could be cyclized to afford product **2.117n** in 67% yield with 87:13 dr with the aid of additional PMP. In this example, trace amounts of cyclopropanation products were observed, which are thought to arise from cyclopropanative carbopalladation of the activated indole olefin by the neopentyl Pd(II) intermediate.

Heteroaromatic halides also shown to be reactive under the optimized conditions. 3-Bromopicolinic acid derivative **2.105o** and thiophene **2.105p** were transformed to the corresponding products **2.117o** and **2.117p** in 70 and 87% yield, respectively, with 93:7 dr in both cases. Notably, 3-bromoindoles **2.105q** and **2.105r** successfully underwent heteroarylcyanation, affording **2.117q** and **2.117r** in 77 and 85% yield, albeit as almost 1:1 mixtures of diastereomers. In the case of *N*-Me indole **2.105q**, a switch in stereochemistry to the *anti* product was observed. We became curious about the observation wheret the thiophene-containing substrate led to a significantly higher dr than the corresponding indole substrates. Therefore, pyrrole substrate **2.105s** was prepared in order to probe whether it was extended aromatic structure of the indole substrates that was leading to this sharp decrease in observed selectivity with respect to the thiophene. This substrate underwent efficient conversion to product **2.117s** in 92% yield with no diastereoselectivity. This finding echoes the results of the indole substrates, which suggests that the extended aromatic structure of the indole substrates is not causing the observed decrease in selectivity. Although it is not clear what the exact origin of the decreased selectivity is, the electron rich nature of such hetereoaryl Pd(II) intermediates resulting from carbon–halogen oxidative addition could be playing a role. Perhaps, when highly electron-rich substrates are employed, the rate of olefin insertion in enchaned. This may make the facial insertion step leading to the normal minor diasteromer competitive with respect to that of the opposite olefin face, thus leading to a more evan ratio or diastereomers. Nevertheless, examples **2.117o–s** represent the first Pd-catalyzed heteroarylcyanation reactions to the best of our knowledge.

2.1.5.4 Product Derivatization

The reaction of chlorine-containing amide (*S*)-**2.105c** with twice the loading of Zn $(CN)_2$ (110 mol%) led to clean formation of the double cyanation product **2.119** in 67% yield with 95:5 dr (Scheme 2.24). Simply monitoring this reaction by ^{1}H NMR and TLC using established compound reference standards shows that the arylcyanation of the aryl bromide is proceeding faster than the direct cyanation of the aryl chloride. Furthermore, no direct aryl bromide cyanation products were observed which suggests that alkene coordination/insertion may be a faster process than cyanaide transmetallation.

These functionalized dihydroisoquinoline scaffolds were shown to be amenable to further modification. The individual diastereomers **2.117a** and **2.117c** were chosen to examine the reactivity of the key functional groups. A Mizoroki-Heck reaction of **2.117c** was accomplished using a variation of Fu's conditions [90], and *trans* alkene product **2.220** was obtained in 91% yield (Scheme 2.25a). Primary amide **2.221** could be obtained in 83% yield via nitrile hydrolysis of **2.117a** under basic conditions (Scheme 2.25b). The respective acid derivative was not obtained, again possibly owing to the severe steric crowding of this functional group. Finally, the global reduction of cyclized product **2.117a** was achieved using an excess of $LiAlH_4$ in refluxing THF and diamine **2.222** was obtained in 85% yield (Scheme 2.25c).

Inspired by a challenging nucleophilic cyanation in our synthesis of (+)-corynoline, we have developed a diastereoselective Pd-catalyzed arylcyanation of enantioenriched carboxamides which yields neopentyl nitrile containing dihydroisoquinolinones with full preseveration of enantioenrichment and with yields and dr's up to 95% and >95:5, respectively. We have developed reaction conditions which tolerate various functionality and substrate classes, most notably vinyl and heteroaromatic bromide substrates. The use of heteroaromatic bromides represents the first examples of a Pd-catalyzed heteroarylcyanation reaction. In contrast to the corresponding Pd-catalyzed aryliodination, this reaction seemed to be less sensitive to alkene substitutions. Since various alkyl groups can be used at this position, it opens up more possibilities for product derivatization as well as utility in complex molecule synthesis.

Scheme 2.24 Tandem Pd-catalyzed arylcyanation/direct cyanation reaction of (*S*)-**2.105c**

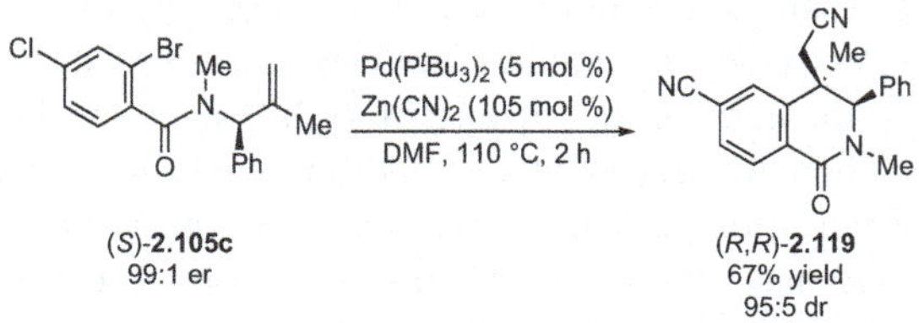

Scheme 2.25 Derivatization of the alkyl nitrile-containing dihydroisoquinolinones

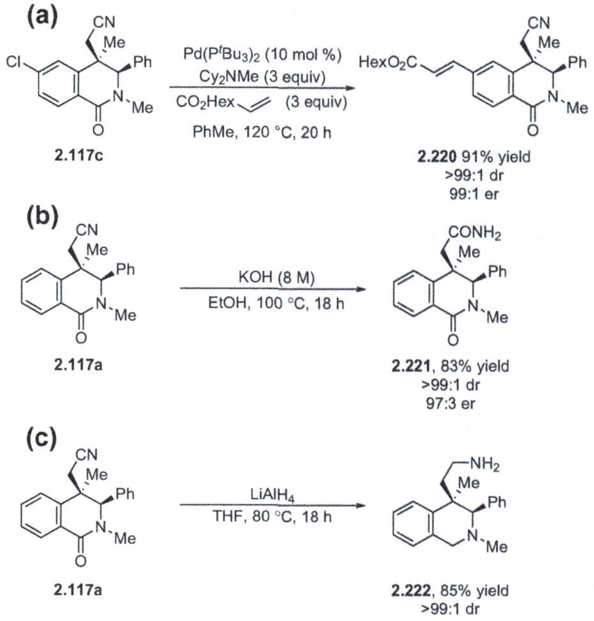

2.1.6 *Research Goal: Part 2*

With a robust procedure for a diastereoselective Pd-catalyzed alkene difunctionalization in hand, we next thought to extend this methodology to a less explored class of substrate which could be used to access other heterocyclic classes. We became inspired by a 2001 report from Grigg which included a single example of *N*-*o*-iodobenzoyl indole **2.223** reacting with 2-furyltributylstannane **2.224** in a Pd-catalyzed carbonylative Stille reaction (Scheme 2.26a).[65] This reaction proceeds via a dearomatization of the indole heterocycle, and provides the desired product **2.225** in 32% yield as a single diastereoisomer. Byproduct **2.226** was also observed which presumably results from the direct carbonylative Still-type coupling of the aryl halide. This transformation represented the first Pd-catalyzed Heck-type dearomatization of an indole. In 2012, Yao and Wu reported that *N*-*o*-halobenzoyl indoles **2.227** could undergo a Pd-catalyzed dearomative Mizoroki Heck reaction under basic conditions (Scheme 2.26b).[66] This reaction produced indoline products **2.228** containing an exocyclic methylene which were obtained with various levels

[65]Reference [55].

[66]Reference [91].

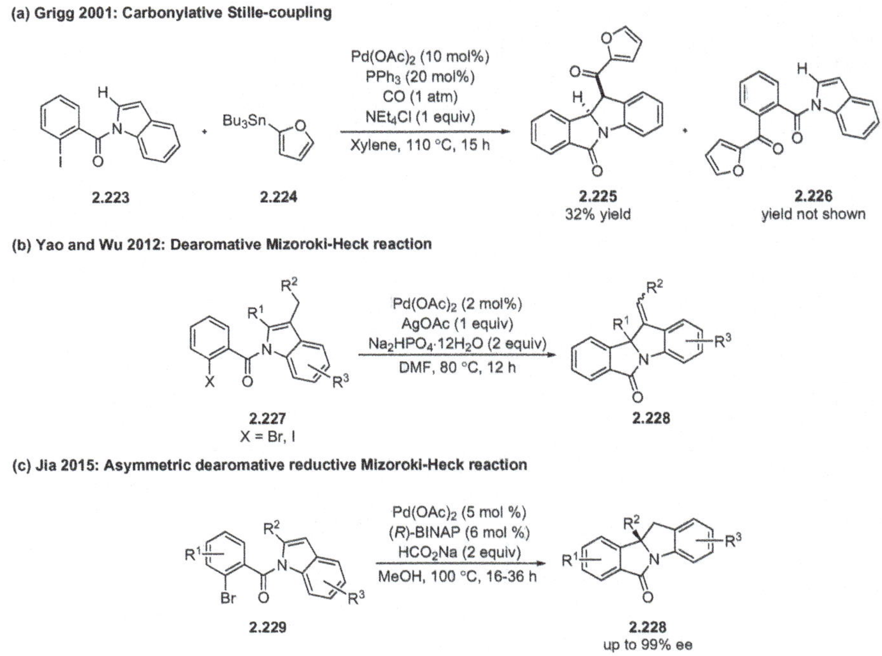

Scheme 2.26 Pd-catalyzed Heck-type indole dearomatizations

of *E*/*Z* selectivity. In 2015, Jia utilized *N*-*o*-bromobenzoyl indoles **2.229** as substrates in an enantioselective intramolecular Pd-catalyzed reductive Mizoroki-Heck reaction (Scheme 2.26c).[67] This transformation used sodium formate as the hydride source, and generated fused indoline cores **2.230** with high levels of enantioselectivity. This transformation represented the first enantioselective indole dearomatization reaction via a domino Heck-type mechanistic manifold. Chapter 3 will contain a formal introduction and discussion of dearomatization reaction and their implications in synthesis.

We invisioned that the development of a diastereoselective dearomative arylcyanation employing this substrate class **2.229** would facilitate the formation of similar fused indolines **2.231** which posess a tetrasubstituted tertiary stereocenter, a functionalizable benzylic carbon in addition to a nitrile which can be further derivatized (Scheme 2.27).

---

[67]Reference [92].

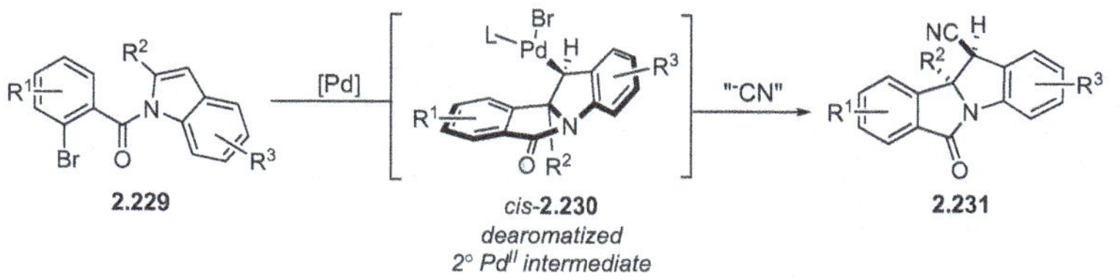

Scheme 2.27 Proposed Pd-catalyzed dearomative arylcyanation

2.1.7 Results and Discussion: Diastereoselective Pd-Catalyzed Dearomative Arylcyanation of Indoles

2.1.7.1 Starting Material Preparation

The starting *N*-o-bromobenzoyl indoles **2.229** can be easily prepared by reacting 2-bromobenzoyl chlorides **2.104** with free indoles **2.232** under two sets of reaction conditions. The first involves the deprotonation of the indole nitrogen using NaH and reaction with the benzoyl chloride derivative in THF at elevated temperatures [92]. The second involves the use of NaOH as a base in conjunction with TBAB which acts as a phase transfer catalyst (Scheme 2.28a).[68] Although many of the required indole derivatives **2.232** are commercially available, more complex indoles could easily be obtained using chemistry developed in our laboratory (Scheme 2.28b).[69] Dibromoolefins **2.233** can be reacted with various alkyl (**2.234/2.235**) or aryl (**2.236**) boron reagents in a tandem Pd-catalyzed intramolecular Buchwald-Hartwig/intermolecular Suzuki-Miyaura cross coupling sequence. Boron reagents like **2.234** can be easily obtained via an alkene hydroboration using 9-BBN dimer in THF. Unlike standard methods such as the Fischer indole synthesis, this route by Fang and Lautens leads to the clean formation of products as single regioisomers.

2.1.7.2 Optimization

We initiated the reaction discovery process by employing substrate **2.229a** under the optimized reaction conditions from part 1 of this chapter (Scheme 2.29). Bromo aromatic **2.229a** was fully consumed after 17 h and the desired arylcyanation product **2.237** was obtained in 71% yield with 1.5:1 dr in combination with 9% of the direct aryl cyanation product **2.238**. We were able to confirm the major product to be *syn*-**2.237a** by X-ray crystallographic analysis (Fig. 2.2). We were curious

[68]Reference [93].

[69]Reference [94].

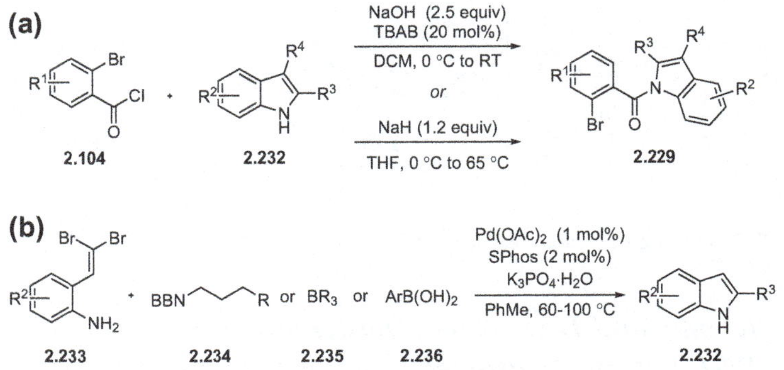

Scheme 2.28 Synthesis of indoles and *N*-benzoyl indoles

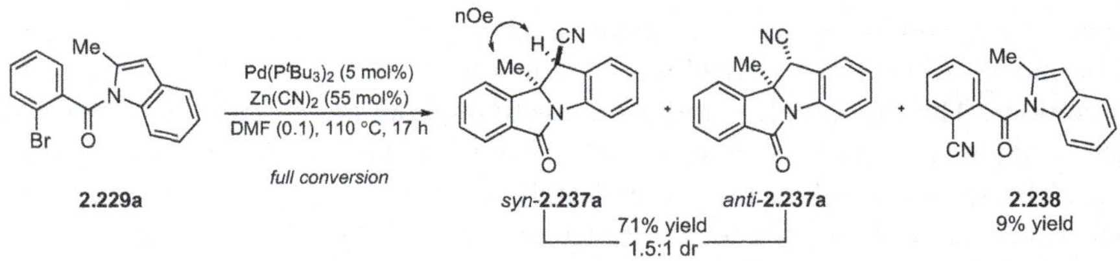

Scheme 2.29 Initial attempt at the dearomative arylcyanation reaction

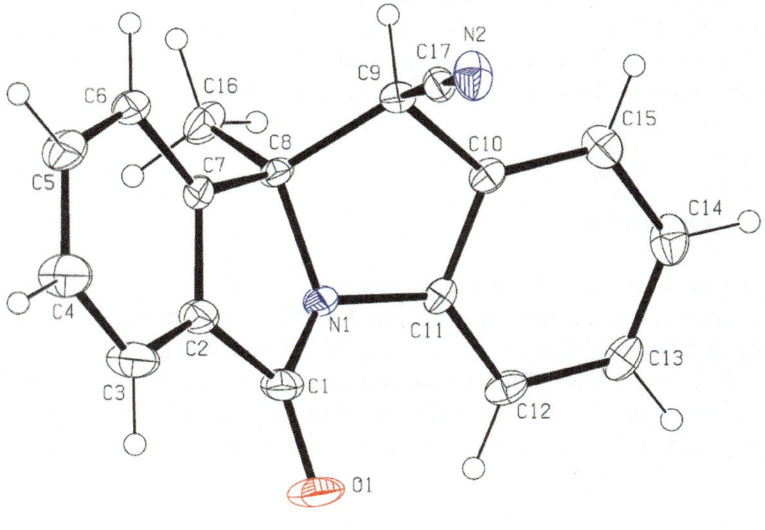

Fig. 2.2 Crystal structure of the *syn*-arylcyanation product *syn*-**2.237a**

about the formation of the minor diastereomer, and hypothesized that it could arise either via an isomerization of the proposed dearomatized benzylic Pd(II) intermediate *cis*-**2.230** or by isomerization of product *syn*-**2.237** under the reaction conditions (vide infra).

With the proof-of-concept in hand, we carried forward with the optimization of the reaction conditions (Table 2.6). Other Pd(0) pre-catalysts containing bulky phosphine ligands such as P^tBu_2Ph and $P(o\text{-Tol})_3$ were tested and the desired product was obtained with 1.6:1 dr and in 67 and 26% yield, respectively (entries 2 and 3). The bulky bidentate phosphine ligand d^tbpf was tested by employing Pd(dtbpf)Cl_2 in the presence of 10 mol% purified Zn dust. This reaction led to the product in 74% yield with 1.6:1 dr and the formation of **2.238** was inhibited (entry 4). Zn dust was added to this reaction in order to initially reduce Pd(II) to Pd(0), and no product is observed when the reaction is run in the absence of Zn (entry 5). We found that the use of Pd (d^tbpf)Cl_2 was not reproducible over several runs, and thus we returned to using Pd $(P^tBu_3)_2$. The effect of using different solvents in the dearomative arylcyanation was studied next. MeOH afforded 76% yield of **2.237** with 2.3:1 dr (entry 6), whereas PhMe only afforded trace amounts of the desired product with 39% of the direct aryl halide cyanation (entry 7). **2.237** was obtained in 76% yield and 2.3:1 dr when the reaction was run in 1,4-dioxane (entry 8), while MeCN afforded **2.237** in 68% yield as a single diastereoisomer (entry 9). Since we were not able to improve the yield further using this catalyst system, we sought to explore the use of the d^tbpf ligand in combination with a different Pd source. When 10 mol% of both $Pd(OAc)_2$ and d^tbpf **2.239** was used, the product was obtained in 84% yield with 19:1 dr (entry 10). Increasing the scale of the reaction to 0.2 mmol scale under these conditions led to an increase in product yield to 95% with no change in dr (entry 11). Lowering the Pd and ligand loading to 5 mol% led to nearly identical results (entry 12). We found that when the concentration was decreased to 0.067 M the product was obtained in nearly 99% yield as a single diastereomer (entry 13). This reaction was found not to function well even at 100 °C and the product was only obtained in 37% yield with >20:1 dr (entry 14). The less toxic cyanide source $K_4[Fe(CN)_6]$ was not compatible under the reaction conditions, and none of the desired product was observed (entry 15). It should be noted that other bulky bidentate ligands such as d^ippf **2.240** and ^{t}BuXantphos **2.241** led to no formation of product. The optimized conditions were found to be $Pd(OAc)_2$ (5 mol%) and d^tbpf (5 mol%) with $Zn(CN)_2$ (55 mol%) in DMF (0.067 M) at 110 °C for 18 h. It should be noted that any efforts to develop an asymmetric variant of this reaction using chiral bidentate phosphine ligands failed and led to complete recovery of the starting material.

At this point we were still perplexed as to why solvents such as DMF and dioxane led to the product being obtained with lower levels of diastereoselectivity. We are under the assumption that a *syn* dearomative carbopallation via transition state **2.242** is occurring to generate *syn*-**2.230** (Scheme 2.30).[70] We hypothesized that the stereochemistry of this benzylic Pd(II) intermediate could be converting to

[70]Reference [95].

Table 2.6 Optimization of the intramolecular dearomative arylcyanation of **2.229a**[a]

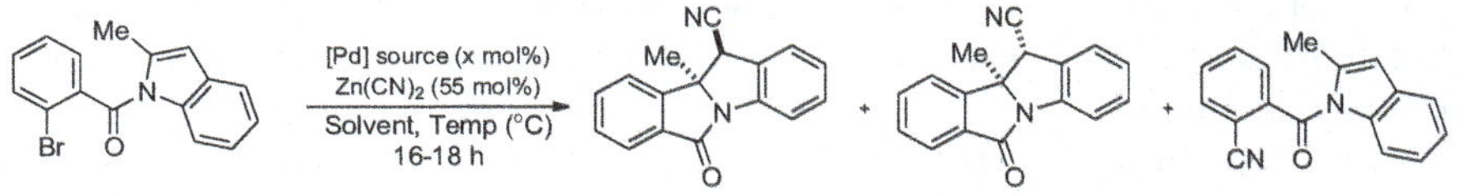

2.229a *syn*-**2.237a** *anti*-**2.237a** **2.238**

Entry	[Pd] source (x mol%)	Solvent (M)	Temp. (h)	% yield **2.237**[b] (*syn*+*anti*)	dr[c]	% yield **2.238**[b]
1	$Pd(P^tBu_3)_2$ (10)	DMF (0.1)	110 (17)	71	1.5:1	9[d]
2	$Pd(P^tBu_2Ph)_2$ (10)	DMF (0.1)	110 (17)	67	1.6:1	0
3	$Pd[P(o\text{-}Tol)_3]_2$ (10)	DMF (0.1)	110 (17)	26	1.6:1	2
4[e]	$Pd(d^tbpf)Cl_2$ (10)	DMF (0.1)	110 (18)	74	1.6:1	0
5[f]	$Pd(d^tbpf)Cl_2$ (10)	DMF (0.1)	110 (16)	0	–	0
6[g]	$Pd(P^tBu_3)_2$ (10)	MeOH (0.1)	100 (16)	76	2.3:1	6
7[g]	$Pd(P^tBu_3)_2$ (10)	PhMe (0.1)	110 (16)	4	>20:1	39
8[g]	$Pd(P^tBu_3)_2$ (10)	Dioxane (0.1)	110 (16)	76	2.3:1	6
9	$Pd(P^tBu_3)_2$ (10)	MeCN (0.1)	110 (16)	68	>20:1	0[h]
10	$Pd(OAc)_2/d^tbpf$ (10)	MeCN (0.1)	110 (20)	84	19:1	<5
11[i]	$Pd(OAc)_2/d^tbpf$ (10)	MeCN (0.1)	110 (17)	95	19:1	<5
12[i,j]	$Pd(OAc)_2/d^tbpf$ (5)	MeCN (0.1)	110 (17)	94	19:1	0
13[i,j,h]	**$Pd(OAc)_2/d^tbpf$ (5)**	**MeCN (0.067)**	**110 (18)**	**99(98)**[k]	**>20:1**	**0**
14[i,j]	$Pd(OAc)_2/d^tbpf$ (5)	MeCN (0.067)	100 (18)	37	>20:1	0
15[j,d]	$Pd(OAc)_2/d^tbpf$ (5)	MeCN (0.1)	100 (17)	<5	–	0

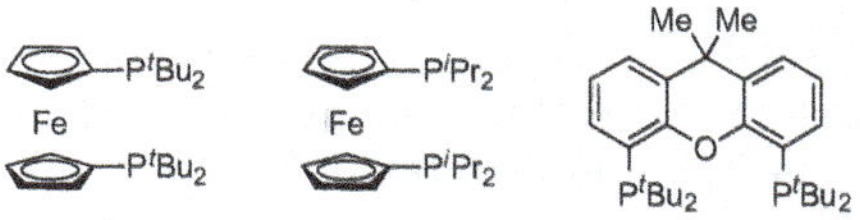

2.239 **2.240** **2.241**

[a]Reactions run on a 0.1 mmol scale

[b]Determined by ^{1}H NMR analysis of the crude reaction mixture using 1,3,5-trimethoxybenzene as an internal standard

[c]Determined by ^{1}H NMR analysis of the crude reaction mixture

[d]0.22 equivalents of $K_4[Fe(CN)_6]$ was used instead of $Zn(CN)_2$

[e]Reaction run in the presence of 10 mol% Zn dust

[f]Reaction run in with absence of Zn dust

[g]Reaction run by Andy Yen

[h]Average values over two runs

[i]Reaction run on 0.2 mmol scale

[j]Catalyst solution transferred with cannula

[k]Yield in brackets represents the isolated yield averaged over two runs

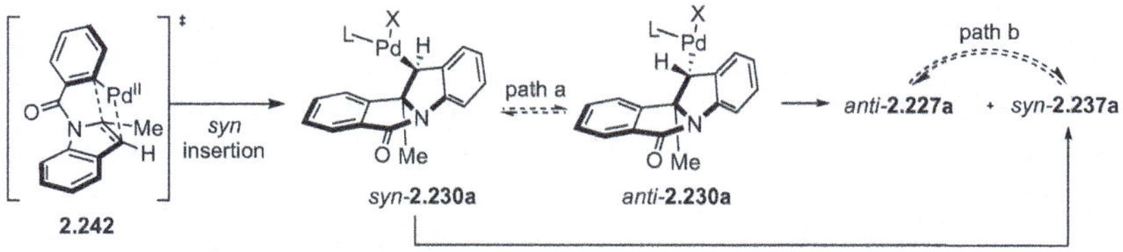

2.242

Scheme 2.30 Possible avenues for the formation of *trans*-**2.230a**

Table 2.7 Control studies to probe the possible *syn*-to-*anti* epimerization **2.237a**

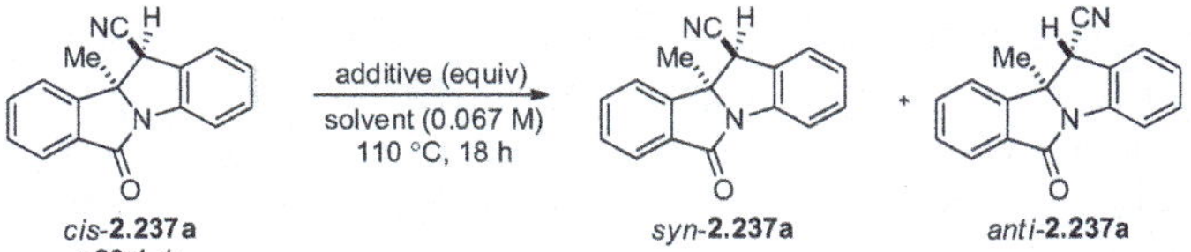

Entry	Solvent	Additive	% yield **2.137a** (*syn*+*anti*)[a,b]	*syn*: *anti*[a]
1	MeCN	–	92	>20:1
2	1,2-DCE	–	91	>20:1
3	PhMe	–	96	>20:1
4	Dioxane	–	99	>20:1
5	DMF	–	90	1.8:1
6	PhMe	$HNBu_2$ (0.5)	89	1:3
7	DMF[c]	N_2 (sparge)	99	>20:1
8	Dioxane	$Zn(CN)_2$ (0.55 equiv)	96	>20:1
9	Dioxane	$ZnBr_2$ (0.55 equiv)	78	>20:1
10	Dioxane	$Pd(OAc)_2$/d^{t}bpf (5 mol%), $Zn(CN_2)$ (0.55)[d]	91	1.2:1
11[d]	Dioxane	$Pd(OAc)_2$/d^{t}bpf (5 mol%)	93	1:3
12[d]	Dioxane	$Pd(OAc)_2$ (0.05)	83	4.7:1

[a]Determined by ^{1}H NMR analysis of the crude reaction mixture
[b]Isolated yields
[c]Reaction sparged with N_2 for 18 h
[d]Reaction run for 4 h

anti-**2.230** under certain reaction conditions. Upon transmetallation with $Zn(CN)_2$ and final carbon-carbon bond-forming reductive elimination, *anti*-**2.230** leads to *anti*-**2.237a** (path a). A second possibility was that *syn*-**2.237a** was simply epimerizing under the reaction conditions (path b). We thought that the latter scenario could be probed in a more practically straightforward manner by subjected purified sample of *syn*-**2.230a** to various reaction conditions (Table 2.7).

When *cis*-**2.237a** was heated in MeCN, 1,2-dichloroethane, PhMe, or 1,4-dioxane, it was recovered in >90% yield as a single stereoisomer in all cases (entries 1–4). Heating *syn*-**2.237a** in DMF led to a significant erosion in the *syn*:*anti* ratio, and slightly reduced recovery yields. Since DMF is known to thermally decompose as well as hydrolyze in the presence of moisture to CO and $HNMe_2$,[71] we considered the possibility of this byproduct facilitating the epimerization via a deprotonation/reprotonation mechanism. To test the prospects of this base-mediated epimerization process, *syn*-**2.237a** was heated in PhMe in the presence of $HNBu_2$ (0.5 equiv). These conditions were judiciously chosen because PhMe led to no epimerization of *syn*-**2.237a** (entry 3), while $HNBu_2$ is a less volatile 2° amine analog of $HNMe_2$.[72] This experiment showed significant epimerization of *syn*-**2.237a**, and *anti*-**2.237a** was actually obtained as the major isomer (*syn*:*anti* = 1:3, entry 6). To further support the prospect of $HNMe_2$ promoting the unwanted epimerization process, *syn*-**2.237a** was heated in DMF for 18 h with constant nitrogen sparging to remove traces of $HNMe_2$ from the reaction environment (entry 7). This resulted in *syn*-**2.237a** being recovered in 99% yield as a single diasterioisomer, thus providing further evidence for a base mediated process.

Reduced dr's were also observed during the optimization when dioxane was used as the solvent. However, the same argument as before can clearly not be made to rationalize this observation. Although neither $Zn(CN)_2$ nor $ZnBr_2$ (0.55 equiv) caused any epimerization (entries 8 and 9), when *syn*-**2.237a** was reacted with Pd $(OAc)_2$ and D^{t}BPF with or without $Zn(CN)_2$ in dioxane, 1.2:1 and 1:3 mixtures of *syn*-**2.237a** and *anti*-**2.237a** were obtained in 91 and 93% yields, respectively (entries 10 and 11). Even $Pd(OAc)_2$ alone was found to cause epimerize in dioxane to some extent (entry 12). Although the exact origin for epimerization in dioxane remains unclear, one possibility involves $Pd(OAc)_2$ coordinating to the lone pair of the nitrile and metallating the benzylic C–H bond. The stereochemistry of the resulting α-pallado nitrile could invert, thus leading to the observed mixture. A second possibility involves the Pd-catalyed autooxidation of dioxane which leads to the formation of acidic byproducts such as glycolic acid, glyoxylic acid and formic acid.[73] These acids can potentially lead to epimerization of this sensitive stereocenter at high temperatures.

2.1.7.3 Examination of the Substrate Scope

The conditions optimized found for the dearomative arylcyanation of **2.229a** were tested on a series of indole substrates possessing sterically and electronically diverse *o*-bromobenzoyl groups (Table 2.8). Indoline **2.237b** could be obtained in

[71]Reference [96].

[72]Dimethyl amine is a gas at ambient temperature and pressure with a boiling point of 7 °C. Dibutyl amine has a boiling point of 159 °C.

[73]Reference [97].

51% yield as a single diastereomer, suggesting that the reaction is sensitive to steric hindrance in close proximity to the C–Br bond (entry2). F- and CF_3-containing indolines **2.237c** and **2.237e** could be obtained in 94 and 96% yields 20:1 and 12:1 dr (entries 3 and 5). Furthermore, Cl-containing **2.237d** was afforded in 93% yield with >20:1 dr (entry 4). Substrates bearing electron-donating groups were also well tolerated, and **2.237f** bearing an OMe moiety was afforded in 94% yield with >20:1 dr (Entry 6).

An ethyl group (**2.237g**, 91%), alkyl chlorides (**2.237j**, 64%) as well as alkyl alcohols and amines protected as benzyl ethers (**2.237h**, 83%) or phthalimides (**2.237i**, 98%) could be incorporated into the final products, which were obtained with >20:1 dr (entries 7–10). Substrates bearing aromatic groups at the indole

Table 2.8 Scope of the fused indoline synthesis by arylcyanation[a]

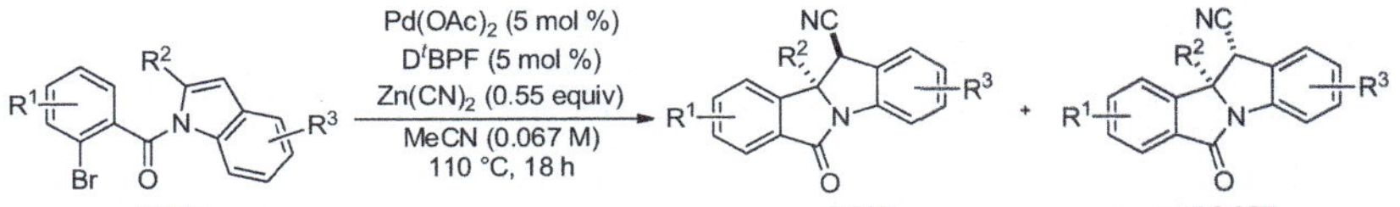

2.229 syn-2.237 anti-2.237

Entry	Substrate		Product		Yield[b]	dr (*syn*: *anti*)[c]
1	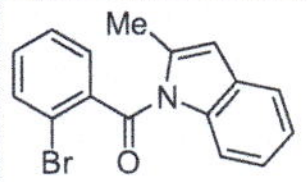	**2.229a**	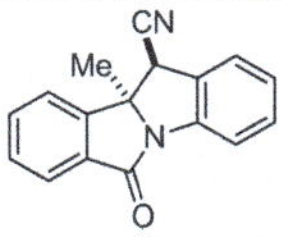	**2.237a**	98% (88%)[d]	>20:1
2	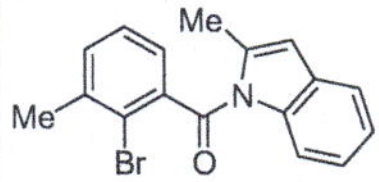	**2.229b**[e]	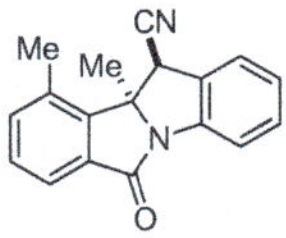	**2.237b**[e]	51%	>20:1
3		**2.229c**[e]	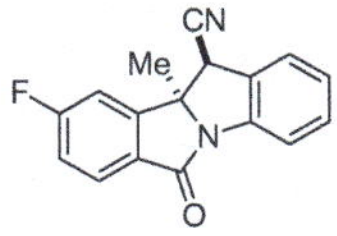	**2.237c**[e]	94%	20:1
4[f]		**2.229d**[e]	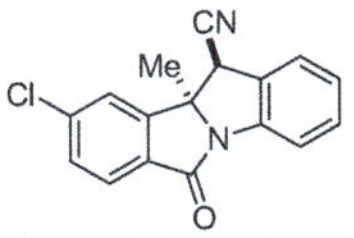	**2.237d**[e]	93%	>20:1
5	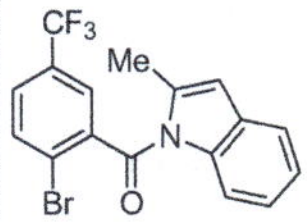	**2.229e**[e]	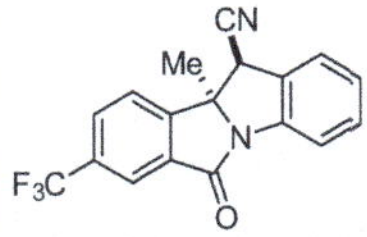	**2.237e**[e]	96%	12:1
6	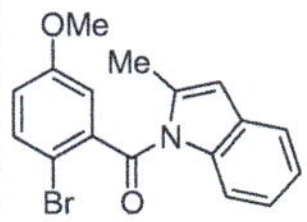	**2.229f**[e]	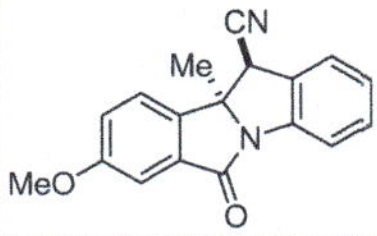	**2.237f**[g]	94%	>20:1

(continued)

Table 2.8 (continued)

Entry	Substrate		Product		Yield[b]	dr (*syn*: *anti*)[c]
7	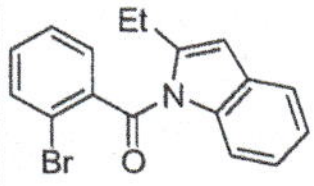	**2.229g**	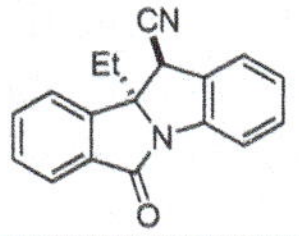	**2.237g**	91%	>20:1
8	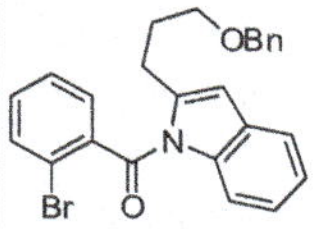	**2.229h**	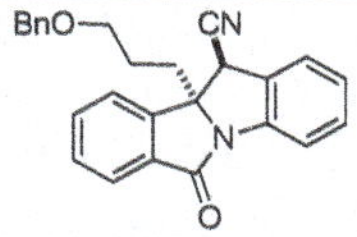	**2.237h**	83%	>20:1
9	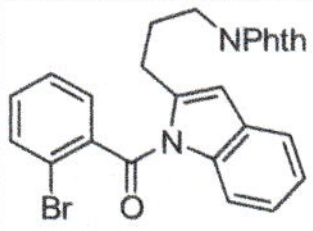	**2.229i**	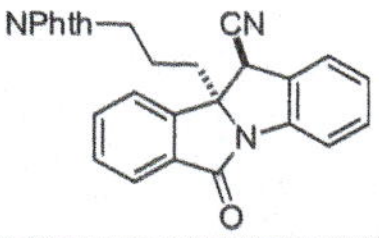	**2.237i**	98%	>20:1
10	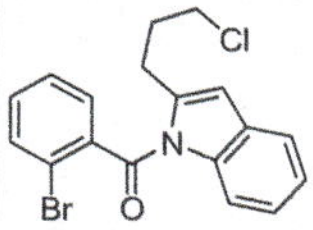	**2.229j**	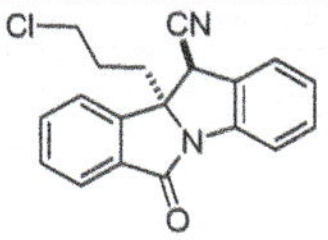	**2.237j**	64%	>20:1
11[h]	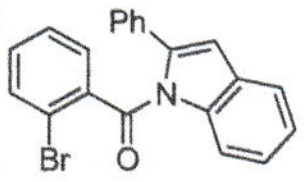	**2.229k**	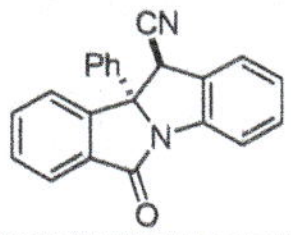	**2.237k**	91%	>20:1
12	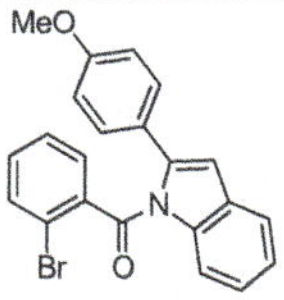	**2.229l**[g]	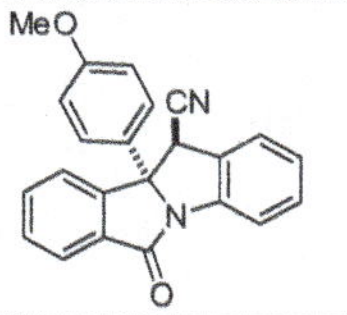	**2.237l**	85%	>20:1
13	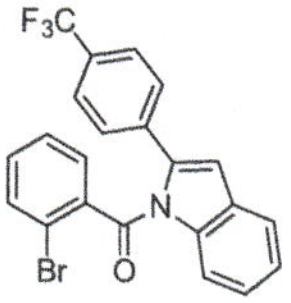	**2.229m**[g]	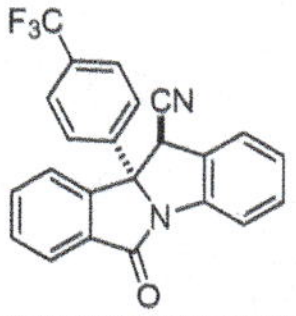	**2.237m**[e]	89%	>20:1
14		**2.229n**[e]	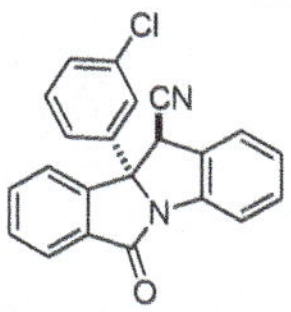	**2.237n**[e]	96%	>20:1
15[f]	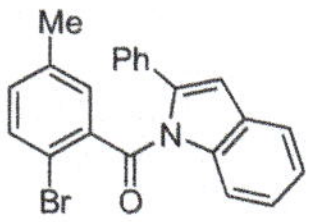	**2.229o**[e]	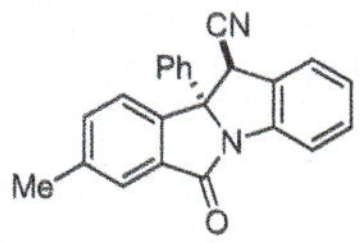	**2.237o**	89%	>20:1

(continued)

Table 2.8 (continued)

Entry	Substrate		Product		Yield[b]	dr (*syn*:*anti*)[c]
16	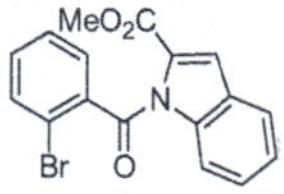	**2.229p**	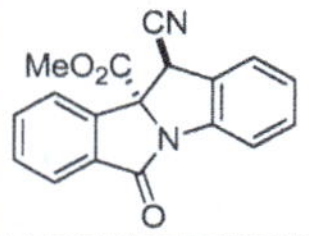	**2.237p**	72%	11:1
17[h]	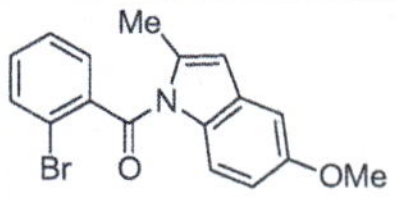	**2.229q**[e]	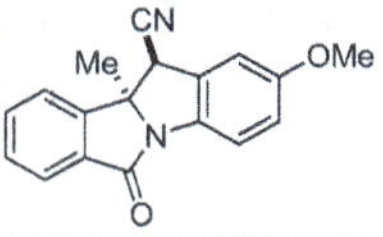	**2.237q**[e]	93%	>20:1
18[h]	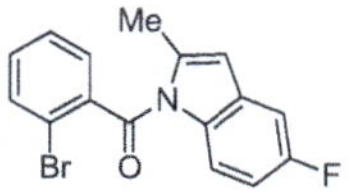	**2.229r**[e]	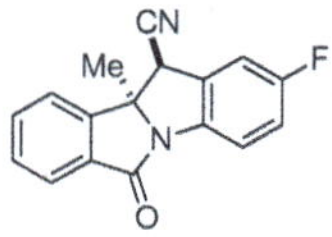	**2.237r**[e]	86%	>20:1
19[f]	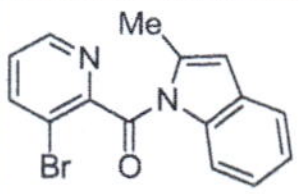	**2.229s**[e]	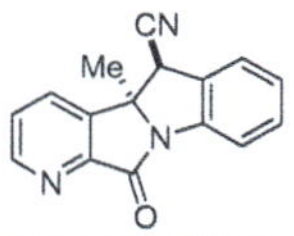	**2.237s**[e]	40%	>20:1
20	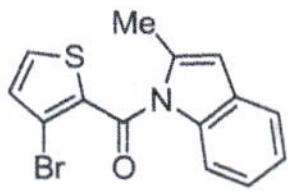	**2.229t**	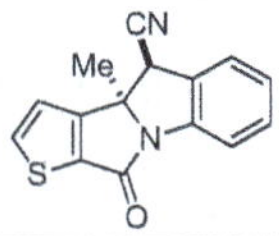	**2.237t**	78%	>20:1
21	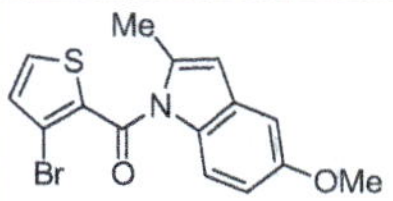	**2.229u**	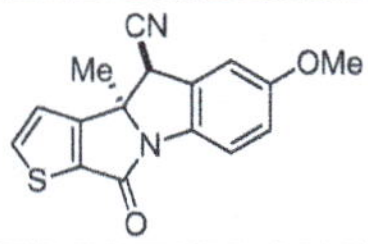	**2.237u**	69%	>20:1

[a]Reactions were run on a 0.2 mmol scale unless otherwise stated
[b]All yields shown are combined isolated yields of the diastereomers
[c]dr's were determined by ^{1}H NMR analysis of the crude reaction mixture
[d]Reaction was run on a 1 g (3.17 mmol) scale
[e]This compound was prepared by Andy Yen
[f]$Pd(P^tBu_3)_2$ (10 mol%)
[g]This compound was prepared by Nicolas Zeidan
[h]Reaction was run for 5 h

2-position were tolerated, and it was found that electron neutral, rich, and poor substrates were all converted to the desired products **2.237k–2.237o** with excellent yields and selectivities (entries 11–15). Methyl-indole-2-carboxylate **2.229p** was also reactive, and was converted to the corresponding indoline **2.237p** in 72% yield with 11:1 dr (entry 16). At this time we do not know the exact origin of the minor diastereomer. Substrates **2.229q** and **2.229r** were synthesized and subjected to the standard conditions in order to test the effects of perturbing the electronics of the indole moiety. Electron rich indoline **2.237q** was obtained in 93% yield in >20:1 dr, whereas electron deficient F-containing **2.237r** was obtained in 86% with >20:1 dr after only 5 h (entries 17 and 18). Pyridine-**2.229s** and thiophene-containing **2.229t** and **2.229u** were tested as a means to incorporate heteroaromatic groups into the indoline products. They were found to provide the desired products **2.237s–2.237u** with >20:1 dr in 40, 78 and 69% yield, respectively (entries 19–21).

2.1.7.4 Limitations

In all cases, this reaction presumably proceeds via a dearomatized Pd(II) intermediate (i.e. *syn*-**2.230**) which lacks any suitable β-hydrogen atoms. Although all of the above substrates are substituted at the indole 2-position, substrates which are unsubstituted at this position could still undergo the reaction. This is because the Pd atom in *cis*-**2.230** would be on the opposite face to the only β-hydrogen atom. However, a competing cyclization via C–H functionalization could also occur over the dearomative carbopalladation step.[74] The concomitant production of H–Pd(II)–X would inevidably halt the catalytic cycle, since it cannot be converted back to active Pd(0) under the base-free reaction conditions. When **2.229v** was subjected to the reaction conditions, **2.237v** was obtained in 32% yield in >20:1 dr. The majority of the remaining mass balance was recovered **2.229v**, and a trace amount of **2.243** was observed in the crude ^{1}H NMR spectrum (Scheme 2.31a).

Substrate **2.229w** which posesses a 2,3-dimethylindole motif was also tested. This reaction failed to produce any of the desired indoline **2.237w**, and trace amount of the product arising from carbopalladation/β-H elimination **2.244** was observed (Scheme 2.31b) [91]. There were several other substrates which failed to provide satisfactory yields of the desired products (Scheme 2.32). Dioxolane **2.229x** only afforded the desired product in low yield albeit as a single diastereomer. Aza-indole **2.229y** in addition to vinyl bromide **2.229z** afforded complex mixtures of products. Electron deficient **2.229aa** and 3-bromo indoles **2.229ab** and **2.229ac** were hardly consumed under the reaction conditions.

[74]Reference [98].

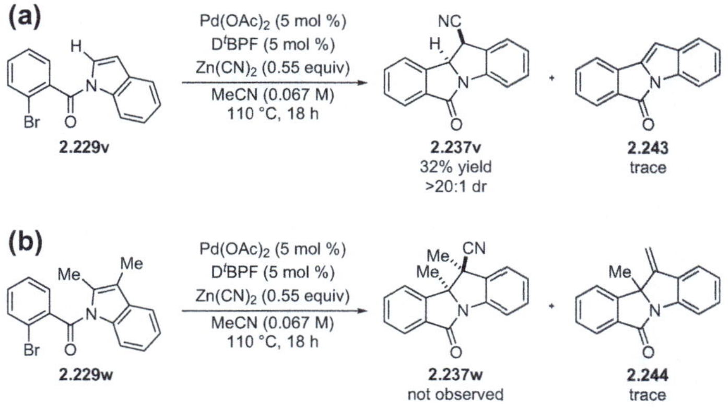

Scheme 2.31 Reaction of substrates containing unsubstituted and 2,3-disubstituted indole motif

Scheme 2.32 Substrates met with challenges in the dearomative arylcyanation

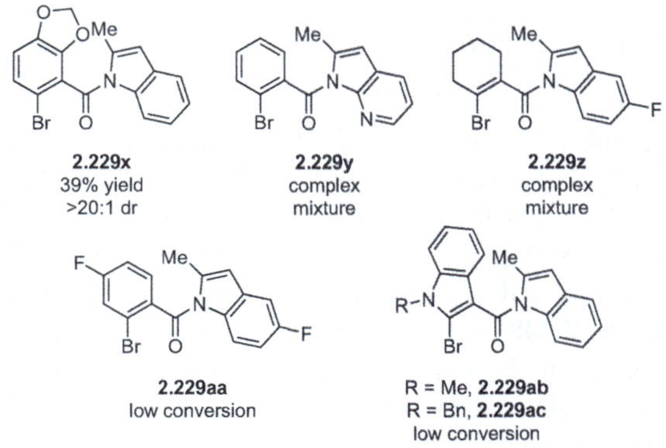

2.1.7.5 Derivativation of Indoline Products

We carried out a series of derivatization experiments to examine the synthetic utility of these indoline products (Scheme 2.33). We reasoned that trapping of an α-cyano anion[75] with Davis' oxaziridine[76] would generate the desired cyanoalkoxide, which could spontaneously eliminate cyanide under the reaction conditions to yield a cyclic aryl ketone. This reaction proceeded using NaHMDS as base to generate ketone **2.245** in 74% yield. It was found that **2.237a** could be alkylated in >20:1 dr using NaH and either *tert*-butyl bromoacetate or bromoacetaldehyde dimethyl

[75]Reference [99].

[76]Reference [100].

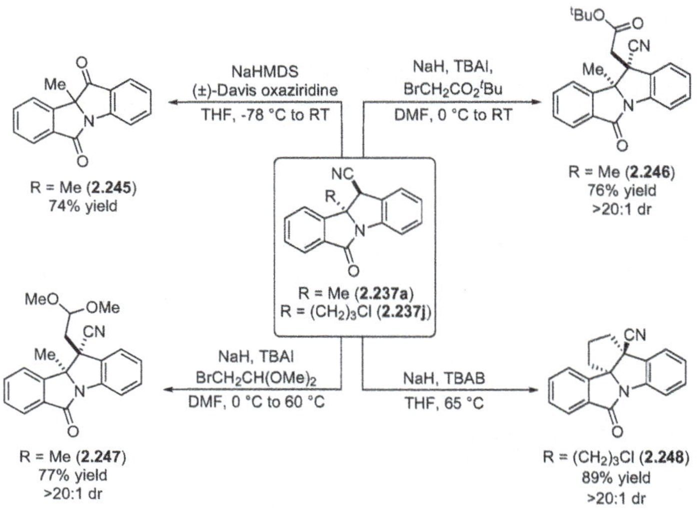

Scheme 2.33 Derivitization of the fused indoline products

acetal to generate ester **2.246** or acetal **2.247** in 76 and 77% yields, respectively. The relative stereochemistry of **2.246** and **2.247** were initially determined by NOE studies, and then ultimately unambiguously confirmed by X-ray crystallographic analysis (Fig. 2.2). We thought that indoline **2.237j** was primed to undergo an intramolecular alkylation, and in practice, afforded the *syn*-angularly fused carbocycle **2.248** in 89% yield. We were never able to selectively reduce either the lactam or the nitrile functionalities using standard reagents (i.e. $LiAlH_4$, DIBAL, BH_3). These reactions typically led to complex mixtures of compounds consistent with an unselective reduction process and/or epimerization of the benzylic stereocenter (Fig. 2.3).

2.1.8 Summary

Based on the previously outlined synthesis of (+)-corynoline, we were inspired to develop a second generation route to a key nitrile intermediate which involved a diastereoselective intramolecular Pd-catalyzed arylcyanation reaction of an enantioenriched *N*-allyl carboxamide. A broad array of enantioenriched dihydroisoquinoline products could be obtained in high yields with good to excellent levels of diastereoselectivity using $Pd(P^tBu_3)_2$ as the catalyst in the presence of a substoichiometric amount of $Zn(CN)_2$. Most importantly, this transformation was ammenble to the highly oxygenated precursor required for the synthesis of (+)-corynoline, and the reaction proceeds on gram-scale with better yields and

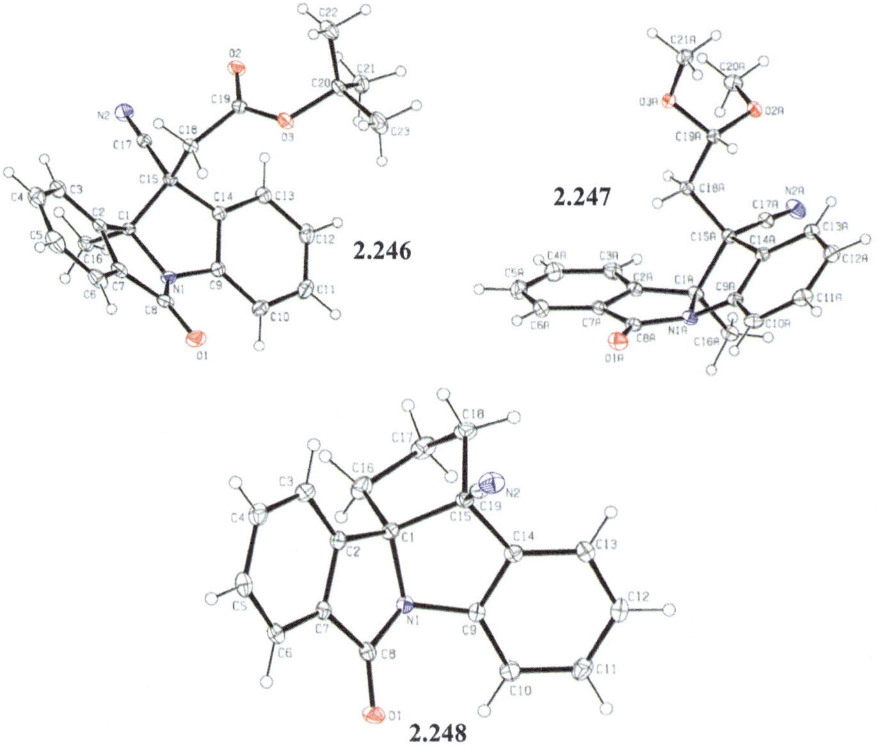

Fig. 2.3 Crystal structures of alkylation products **2.246**, **2.247** and **2.248**

diastereoselectivity than the previously studied 2-step Pd-catalyzed aryliodination/nucleophilic cyanation sequence.

Due to the recent interest in metal-catalyzed dearomatization reactions, and their ability to transform simple aromatic substrates into complex molecular architectures, we sought to explore *N*-benzoyl indoles as substrates in the arylcyanation reaction. We achieved the first Pd-catalyzed dearomative bisfunctionalization of indoles that proceeds via an intramolecular arylcyanation mechanism. This also represents the first arylcyanation reaction which proceedes via a transmetallation to a 2° benzylic Pd(II) species. Optimizaiton of the reaction conditions provided a protocol which afforded a broad variety of functionalized fused indoline cores in high yields as single diastereomers in most cases. DMF was found to not be a suitable solvent in this reaction, presumably due to its propensity to slightly decompose to $HNMe_2$ and CO under the reaction conditions. The presence of this basic byproduct can lead to the epimerization of the sensitive benzylic stereocenter.

The products generated in this reaction were shown to be amenable to several functionalization involving stereoselective alkylations which afford all-carbon quaternary centers, in addition to oxidative dehydrocyanation leading to ketones.

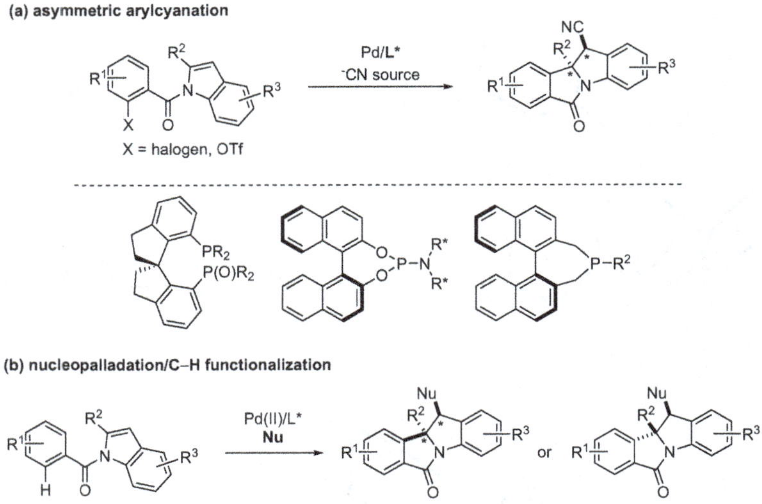

Scheme 2.34 Furture directions for the intramolecular dearomative indole bisfunctionalization

Although all preliminary effords toward developing enantioselective variant have failed, further studies should revolve around studying the new classes of ligands which are affording high ee's in previously non-selective Pd-catalyzed transformation (Scheme 2.34a). Furthermore, one could imagine that by using an electrophilic chiral Pd(II) catalyst, reaction involving *cis* or *trans* nucleopalladation of the indole C2–C3 double bond, followed by C–H fuctionalization could occur (Scheme 2.34b).

2.1.9 Experimental

General Considerations. Unless otherwise stated, all catalytic reactions were carried out under an inert atmosphere of dry argon or dry nitrogen utilizing glassware that was oven (120 °C) or flame dried and cooled under argon, whereas the work-up and isolation of the products from the catalytic reactions were conducted on the bench-top using standard techniques. Reactions were monitored using thin-layer chromatography (TLC) on EMD Silica Gel 60 F254 plates. Visualization of the developed plates was performed under UV light (254 nm) or using $KMnO_4$, *p*-anisaldehyde, or ceric ammonium molybdate (CAM) stain. Organic solutions were concentrated under reduced pressure on a Büchi rotary evaporator. Tetrahydrofuran was distilled over sodium, toluene was distilled over sodium and degassed via 5 freeze-pump-thaw cycles and stored over activated 4 Å molecular sieves, triethyl amine was distilled over potassium hydroxide, DCM was distilled over calcium hydride, and anhydrous *N,N*-dimethylformamide and diethyl ether were purchased from Aldrich and used as recieved. silica gel flash column chromatography was performed on Silicycle 230–400 mesh silica gel. All standard reagents including benzaldehyde derivatives, (*S*)-2-methylpropane-2-sulfinamide and (*R*)-2-methylpropane-2-sulfinamide, isopropenyl magnesium bromide, and dimethylzinc, were purchased from Sigma Aldrich, Alfa Aesar, Combi-Blocks, or Oakwood and were used without further purification. The 'racemic' assays for HPLC determination of enantiomeric excess for the final dihydroquinolinone products were synthesized using **general procedures 5–8** from Chap. 1 followed by **general procedures 1 and 2** from this chapter, except the 'racemic' 2-methylpropane-2-sulfinamide was employed. NOTE: it was later found that 2-methylpropane-2-sulfinamide from Combi-Blocks product # **QB-2211** and batch number **L77452** was present with a slight enantiomeric excess of the (*S*) enantiomer (confirmed by the company, and our independent analysis). This explains the slight enantiomeric excess in our 'racemic' HPLC assays of some linear starting materials and the corresponding dihydroisoquinolinone products. We have simultaneously confirmed that this is not due to an overlapped impurity causing an apparent ee. All chiral *N*-alkyl allylic amines were synthesized via the full synthetic route previously described in Chap. 1.[77] Indole, 2-methylindole, 2,3-dimethlyindole, 5-fluoro-2-methylindole, 5-methoxy-2-methylindole, 2-phenylindole, and methyl-2-indolecarboxylate were also obtained from commercial sources, while others were prepared by the indicated procedures. All 2-bromobenzoic acid derivatives were purchased from commercial sources. When the corresponding 2-bromobenzoyl chloride derivates were not commercially available, they were prepared using $(COCl)_2$ and DMF (cat.) in DCM at 0 °C to room temperature. 2-bromobenzoyl chloride was purchased from Sigma-Aldrich and used as received. 2-(4-methoxyphenyl)indole, 2-(4-trifluoromethylphenyl)indole, 2-(3-chlor ophenyl)indole, and 2-ethylindole were prepared following the procedure of Fang and Lautens, and the data was in accordance with that of the reported literature.[78] Methyl-2-indolecarboxylate was synthesized in accordance with the procedure of Pujol,[79] and the data was in accordance with the reported literature.[80] ***$Zn(CN)_2$ is a toxic substance and appropriate measures were taken to handle all resulting reaction waste with care at basic pH, segregated from all other waste contamination, in an appropriately labeled vessel, stored in a well operating fume hood.***

[77]Reference [101].

[78]Reference [102].

[79]Reference [103].

[80]Reference [104].

5-bromobenzo[*d*][1,3]dioxole-4-carboxylic acid was synthesized from the corresponding aldehydes[81] via an adapted literature procedure for the Pinnick oxidation of the 5-iodo derivative. **2-bromocyclohex-1-enecarboxylic acid** was synthesized in two steps from cyclohexanone according to the following known literature procedures, respectively.[82,83] **1-methyl-3-bromo-1H-indole-2-carboxylic acid** was synthesized according to the following known literature procedures along a synthetic route similar to that of the *N*-Bn derivative [83].[84,85] **3-bromo-1-methyl-1H-pyrrole-2-carboxylic acid** was synthesized from methyl 3-bromo-1-methyl-1H-pyrrole-2-carboxylate[86] by a procedure outlined below.

Instrumentation. NMR characterization data was collected at 296 K on a Varian Mercury 300, Varian Mercury 400, Varian 600, or a Bruker Avance III spectrometer operating at 300, 400, 500, or 600 MHz for ^{1}H NMR, and 75, 100, 125, for 150 MHz or ^{13}C NMR. ^{1}H NMR spectra were internally referenced to the residual solvent signal ($CDCl_3$ = 7.26 ppm, Toluene-d_8 = 2.3 ppm) or TMS. ^{13}C NMR spectra were internally referenced to the residual solvent signal ($CDCl_3$ = 77.0 ppm) and are reported as observed. Data for ^{1}H NMR are reported as follows: chemical shift (δ ppm), multiplicity (s = singlet, d = doublet, t = triplet, q = quartet, m = multiplet, br = broad), coupling constant (Hz), integration. Coupling constants have been rounded to the nearest 0.5 Hz. Melting point ranges were determined on a Fisher-Johns Melting Point Apparatus and are reported uncorrected. All reported diasteromeric ratios in data section are those obtained from ^{1}H NMR analysis of the crude reaction mixtures using 5 s delay. NMR yields for the optimization section were obtained by ^{1}H NMR analysis of the crude reaction mixture using 5 s delay and 1,3,5-trimethoxybenzene as an internal standard. Enantiomertic ratios were determined using HPLC on a chiral stationary phase using a complete Agilent HPLC system. All retention times have been rounded to the nearest 0.1 min. High resolution mass spectra (HRMS) were obtained on a micromass 70S-250 spectrometer (EI) or an ABI/Sciex QStar Mass Spectrometer (ESI) or a JEOL AccuTOF medel JMS-T1000LC mass spectrometer equipped with an IONICS® Direct Analysis in real Time (DART) ion source at Advanced Instrumentation for Molecular Structure (AIMS) in the Department of Chemistry at the University of Toronto.

[81]Reference [105].

[82]Reference [106].

[83]Reference [107].

[84]Reference [108].

[85]Reference [109].

[86]Reference [110].

General Procedures—Diastereoselective Synthesis of Dihydroisoquinolinones via Pd-Catalyzed Arylcyantion

General procedure 1) Synthesis of Linear Enantioenriched *N*-allyl Carboxamides

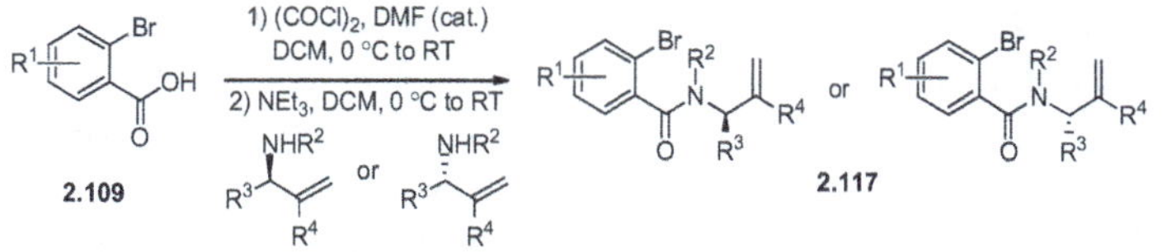

2.109 1.235 2.117

A flame dried flask was charged with the *o*-bromobenzoic acid derivative (**2.109**, 1 equiv) and was purged with argon for 10 min. The contents of the flask were taken up in dry DCM (10 mL), and anhydrous DMF (2 drops) was added. The resulting suspension was cooled to 0 °C and oxalyl chloride (2 equiv) was added drop wise over 5 min. Once gas evolution had ceased, the reaction mixture was warmed to room temperature, and stirred vigorously. Once the slow gas evolution ceased at this temperature, the reaction mixture was concentrated to dryness on a rotary evaporator, followed by high-vacuum to remove any excess oxalyl chloride. The flask containing the crude acid chloride was fitted with a septum and purged with argon for 10 min. The contents of the flask were taken up in dry DCM (10 mL) and cooled to 0 °C. To this solution was added dry NEt_3 (2 equiv) which immediately resulted in the appearance of a persistent orange-to-red colour. This then had a DCM solution (2.5 mL) of chiral allylic amine (**1.235**, **2.108** or **2.113**, 1.1 equiv) and dry NEt_3 (2 equiv) added drop wise over 10 min. The reaction was stirred at this temperature for 10 min and then warmed to room temperature where it was stirred for an additional 10 min. At this time the reaction was quenched by doubling the reaction volume with a concentrated aqueous solution of $NaHCO_3$. The layers were separated and the aqueous layer was extracted with DCM (3×). The combined organic layers were washed with brine, dried over Na_2SO_4, filtered, and concentrated in vacuo. The pure chiral *N*-allyl carboxamides **2.117** were purified using silica gel flash column chromatography using the indicated mobile phase.

General Procedure 2) Pd-Catalyzed Diastereoselective Arylcyanation of Chiral *N*-Allyl Cabroxamide

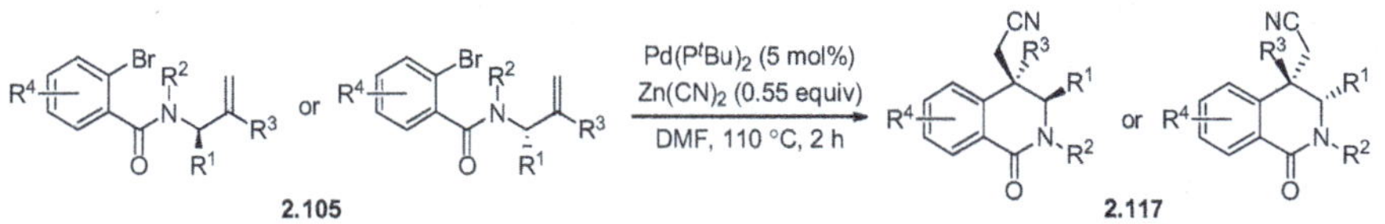

2.105 2.117

10a—Using liquid amides: One oven-dried 2 dram vial was cooled under argon, charged with the liquid chiral *N*-ally carboxamides (0.3 mmol, 1 equiv, **2.105**), and purged with argon for 10 min. A second oven-dried vial was cooled under argon, charged with $Pd(P^tBu_3)_2$ (0.015 mmol, 5 mol% [Pd]) and $Zn(CN)_2$ (0.165 mmol, 0.55 equiv) and purged with argon for 10 min. The amide was dissolved in distilled, dry and degassed DMF (3 mL, 0.1 M) was added via a dry microsyringe. This solution of amide was added into the vial containing the catalyst and cyanide source. The vial was fitted with a Teflon® lined screw cap under a stream of argon passed through a large inverted glass funnel, sealed using Teflon® tape, and placed in a pre-heated oil bath at 110 °C for 2 h. At this time either TLC analysis or ^1H NMR analysis of a small aliquot indicated full conversion of starting material, and the reaction vial was cooled and H_2O and EtOAc or DCM was added. The combined layers were washed with water (4×) and brine (2×), dried over Na_2SO_4, filtered, and concentrated in vacuo. The combined organic layers were dried filtered through a 2 cm plug of silica gel in a pipette eluting with 100% EtOAc. The crude dihydroisoquinolinones were purified via silica gel flash column chromatography using the indicated mobile phase.

2b—Using Solid amides: An oven-dried 2 dram vial was cooled under argon, charged with the solid chiral linear amide (0.3 mmol, 1 equiv, **2.105**), $Pd(P^tBu_3)_2$ (0.015 mmol, 5 mol% [Pd]), and $Zn(CN)_2$ (0.165 mmol, 0.55 equiv) and purged with argon for 10 min. The contents of the vial were taken up in distilled, dry and degassed DMF (3 mL, 0.1 M). The vial was fitted with a Teflon® lined screw cap under a stream of argon passed through a large inverted glass funnel, sealed using Teflon® tape, and placed in a pre-heated oil bath at 110 °C for 2 h. At this time either TLC analysis or ^1H NMR analysis of a small aliquot indicated full conversion of starting material, and the reaction vial was cooled and H_2O and EtOAc or DCM was added. The combined layers were washed with water (4×) and brine (2×), dried over Na_2SO_4, filtered, and concentrated in vacuo. The combined organic layers were dried filtered through a 2 cm plug of silica gel in a pipette eluting with 100% EtOAc. The crude dihydroisoquinolinones were purified via silica gel flash column chromatography using the indicated mobile phase.

General Procedure 3) Pd-catalyzed indole synthesis using *gem*-dibromoolefins

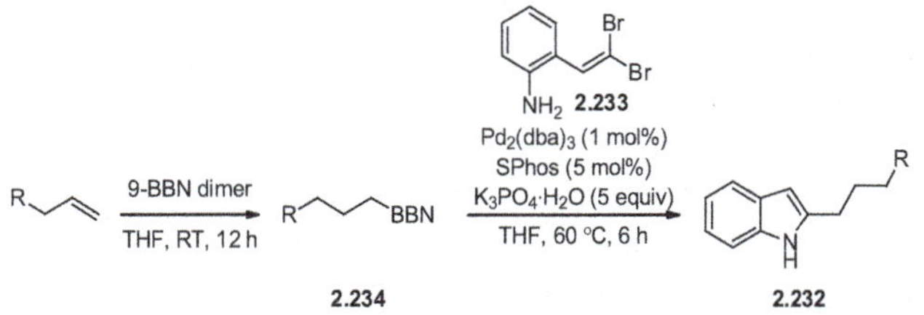

Synthesized in accordance to a modified procedure of Fang and Lautens [94]. Inside of an argon filled glovebox, a flame dried round bottom flask contained a magnetic stir bar was charged with 9-BBN dimer (0.825 equiv). The flask was sealed with a rubber septum, removed from the glovebox, and attached to an argon manifold. The solid 9-BBN dimer was taken up in dry THF to produce a 0.45 M solution (calculated based on moles of dimer), and the suspension was stirred at room temperature for approximately 10 min until a homogeneous solution was obtained. A second flame-dried round bottom flask was charged with 2-(2,2-dibromovinyl)aniline (1 equiv), $Pd_2(dba)_3$ (1 mol%), SPhos (5 mol%) and potassium phosphate monohydrate (5 equiv), and the mixture was purged with argon for 15 min. At this time, the solution of the alkyl 9-BBN reagent was added via cannula to the solid mixture. The reaction was placed into a pre-heated oil bath at 60 °C where it was stirred (800 rpm) for 6 h. At this time the reaction was cooled to room temperature and 30% aqueous H_2O_2 ($\sim$1 mL/mmol of bromoolefin) was added carefully, and the resulting solution was stirred for 30 min. The mixture was diluted with water and was extracted with EtOAc (3×) and the combined organic layers were washed with 1 M aqueous HCl (3×) and then sequentially with water and brine, dried over Na_2SO_4, filtered and concentrated under reduced pressure. The crude indole derivative was purified by flash column silica gel chromatography using the indicated solvent system.

General Procedure 4) *N*-(2-bromobenzoyl)indole synthesis using NaH.

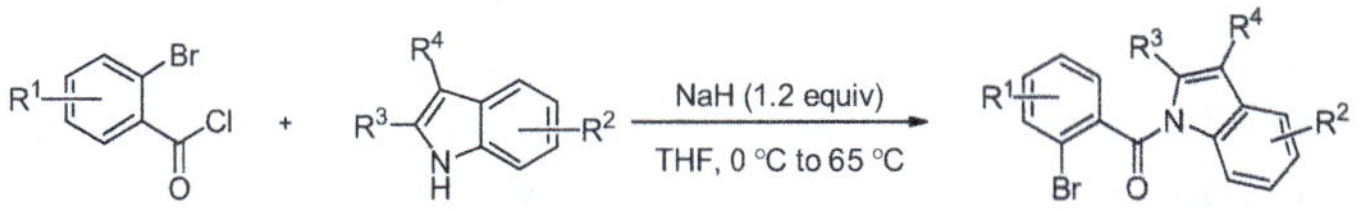

2.104 2.232 2.229

Synthesized in accordance to a modified procedure of Jia [92]. A 60% dispersion of NaH in mineral oil (1.2 equiv) was added to a stirred solution of the appropriate indole derivative **2.232** (1 equiv, $\sim$0.5 M) in THF at 0 °C and the corresponding solution was stirred for 5 min before warming to room temperature where it was stirred for 30 min. The solution of the sodium indolate was re-cooled to 0 °C at which time a solution of appropriate 2-bromobenzoyl chloride derivative **2.104** (2 equiv or 2.2 equiv, $\sim$1 M) in THF was added dropwise. Once the addition was complete, the reaction was allowed to warm to room temperature and then was stirred at 65 °C for 30 min. At this time the extent of completion of the reaction was determined by conversion of the indole derivate by TLC analysis. The reaction was cooled to room temperature and quenched with a saturated solution of NH_4Cl. The reaction mixture was then diluted with water and EtOAc, and after separating the layers, the aqueous layer was extracted with EtOAc (3×). The combined organic layers were washed sequentially with water and brine, dried over sodium sulfate, filtered and concentrated under reduced pressure. The crude *N*-(2-bromobenzoyl) indole derivative **2.229** was purified by flash column silica gel chromatography using the indicated solvent system.

General Procedure 5) *N*-(2-bromobenzoyl)indole synthesis using phase transfer conditions.

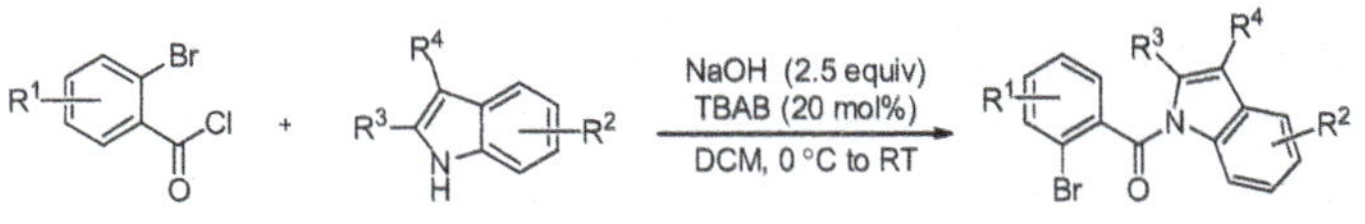

2.104 2.232 2.229

Synthesized in accordance to a modified procedure of Hastings and Weedon [93]. A DCM solution of the appropriate 2-bromobenzoyl chloride derivative **2.104** (1.5 equiv, 2 M) was added dropwise to a suspension of the appropriate indole derivative **2.232** (1 equiv), sodium hydroxide (2.5 equiv), and TBAB (20 mol%) in DCM (0.5 M with respect to indole) at 0 °C. The reaction was stirred at this temperature for 30 min before warming to room temperature (occasionally, a slight exotherm was observed leading to mild warming of the reaction mixture). At this time another 0.5 equivalent of the 2-bromobenzoyl chloride derivative in DCM (2 M) was added dropwise and the reaction was stirred for 30 min. The volume of the reaction was doubled with water and the layers were separated. The aqueous layer was extracted with DCM (3×) and the combined organic layers were washed with brine, dried over Na_2SO_4, filtered, and concentrated under reduced pressure. The crude *N*-(2-bromobenzoyl)indole derivate was purified by flash column silica gel chromatography using the indicated solvent system.

General Procedure 6) Pd-Catalyzed dearomative indole bisfunctionalization via arylcyanation.

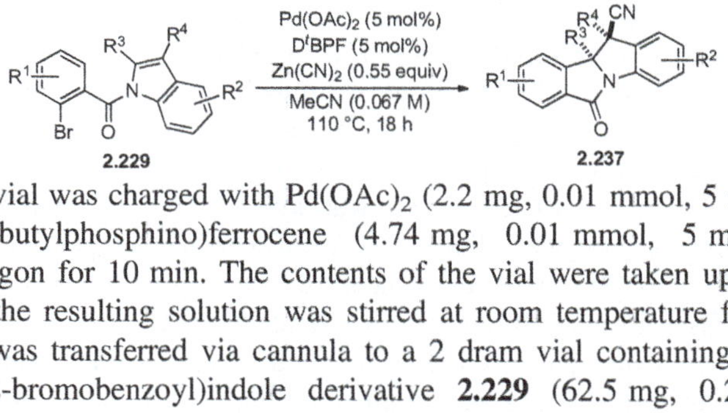

2.229 2.237

A dry 2 dram vial was charged with $Pd(OAc)_2$ (2.2 mg, 0.01 mmol, 5 mol%) and 1,1′-bis(di-*tert*-butylphosphino)ferrocene (4.74 mg, 0.01 mmol, 5 mol%) was purged with argon for 10 min. The contents of the vial were taken up in MeCN (1.5 mL) and the resulting solution was stirred at room temperature for 20 min. This solution was transferred via cannula to a 2 dram vial containing the corresponding *N*-(2-bromobenzoyl)indole derivative **2.229** (62.5 mg, 0.2 mmol, 1 equiv) and $Zn(CN)_2$ (12.9 mg, 0.11 mmol, 55 mol%) under argon. The vial which contained the catalyst/ligand mixture was rinsed with fresh MeCN (1.5 mL), and this too was transferred via cannula. The vial was sealed with a Teflon® lined screw-cap and placed in an oil bath pre-heated to 110 °C and was stirred at 700 rpm for 18 h. At this time the reaction was cooled and passed through a short pad of silica gel eluting with EtOAc, and concentrated. The dr was determined by ^1H NMR analysis of the crude reaction mixture, and was subsequently purified via silica gel flash column chromatography using hexanes/EtOAc as the mobile phase to give the corresponding indoline **2.237**.

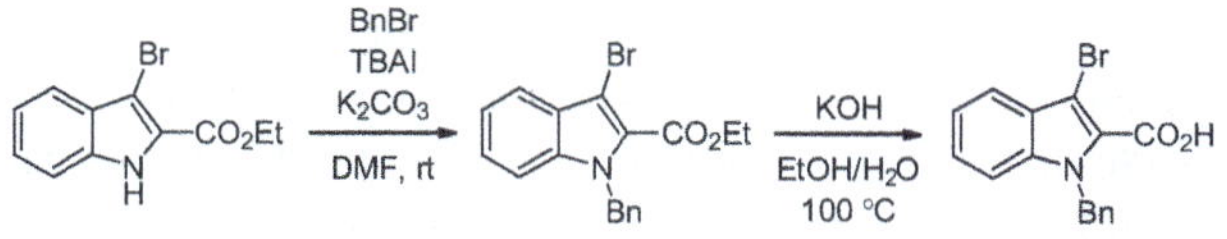

2.114 2.115 2.116

1-benzyl-3-bromo-1H-indole-2-carboxylic acid (2.116) A 25 mL round bottom flask was charged with **2.114** (1.0 g, 3.74 mmol, 1 equiv), TBAI (121 mg, 0.374 mmol, 10 mol%), K_2CO_3 (1.03 g, 7.48 mmol, 2 equiv), and DMF (12.5 mL). The resulting suspension had neat BnBr added (888 μL, 7.48 mmol, 2 equiv), and the solution was allowed to stir for 2 h, before having the reaction doubled with water. This solution was then extracted with EtOAc (3×) and the combined organic layers were sequentially washed with water (4×) and brine, dried over Na_2SO_4, filtered, and concentrated under reduced pressure. The crude alkylated indole was purified using silica gel flash column chromatography using hexanes:EtOAc (18:1 → 9:1 v:v). The title compound was afforded as a clear and light yellow oil (1.26 g, 3.53 mmol, 94%). **^{1}H NMR** (400 MHz, $CDCl_3$) δ 7.71 (dt, J = 8.0, 1.0 Hz, 1H), 7.36–7.32 (m, 2H), 7.28–7.18 (m, 4H), 7.05–7.01 (m, 2H), 5.80 (s, 2H), 4.38 (q, J = 7.0 Hz, 2H), 1.38 (t, J = 7.0 Hz, 3H). **^{13}C NMR** (100 MHz, $CDCl_3$) δ 161.1, 138.0, 137.7, 128.6, 127.3, 126.9, 126.4, 126.1, 125.3, 121.5, 121.5, 110.8, 99.6, 61.2, 48.7, 14.1. **IR** (neat, cm^{-1}) 3063, 3032, 2982, 1699, 1606, 1506, 1452, 1255, 1186, 1124, 1028. **HRMS** (ESI+) Calc'd for $C_{18}H_{17}BrNO_2$ 358.04427, found 358.04342.

An EtOH solution (11 mL) of **2.115** (1.17 g, 3.27 mmol, 1 equiv) was heated to reflux, and an aqueous solution (66 mL) of KOH (550 mg, 9.82 mmol, 3 equiv) was added. The resulting homogenous solution was refluxed for 14 h. At which time TLC analysis indicated full consumption of starting material. The reaction was allowed to cool to room temperature and then to 0 °C in an ice bath where it was acidified to pH = 2 using concentrated HCl. The resulting white precipitate was filtered off and dried to a constant weight in air, then under high vacuum overnight. Acid **2.116** was obtained as a white amorphous powder (952 mg, 2.89 mmol, 89%). **^{1}H NMR** (400 MHz, DMSO-d_6) δ 13.7 (s (br), 1H), 7.70–7.66 (m, 1H), 7.64 (dt, J = 8.0, 1.0 Hz, 1H), 7.42 (ddd, J = 8.0, 7.0, 1.0 Hz, 1H), 7.34–7.22 (m, 4H), 7.08–7.04 (m, 2H), 5.90 (s, 2H). **^{13}C NMR** (100 MHz, DMSO-d_6) δ 162.87, 139.13, 138.32, 129.45, 128.09, 127.16, 127.04, 126.98, 122.58, 121.47, 112.64, 98.56, 48.81. **IR** (neat, cm^{-1}) 3080, 1672, 1502, 1442, 1427, 1352, 1276, 1128. **HRMS** (ESI+) Calc'd for $C_{16}H_{13}BrNO_2$ 330.01297, found 330.01278.

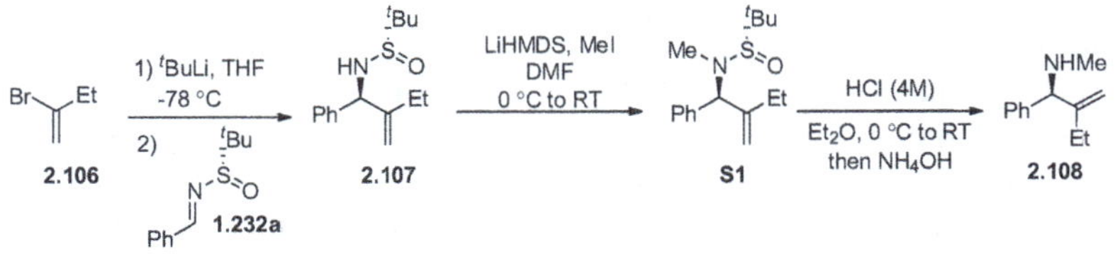

2.106 2.107 S1 2.108

(*S*)-2-methyl-*N*-((*S*)-2-methylene-1-phenylbutyl)propane-2-sulfinamide (2.108) *Note*: The reaction was carried out using an electronic thermocouple to ensure a proper and accurate temperature profile. A flame dried Schlenk flask was charged with **2.106** (0.756 g, 5.71 mmol, 1.2 equiv) and freshly distilled THF (15.5 mL), and the resulting solution was cooled to −78 °C. To this was added ^{t}BuLi (6.72 mL, 11.4 mmol, 2.4 equiv, 1.7 M in pentanes) at a rate such that the internal reaction temperature did not exceed −70 °C. The bright yellow solution was allowed to stir at this temperature for 30 min, before warming to −10 °C. Once this temperature was achieved the solution of organolithium was re-cooled to −78 °C and added via cannula to a THF solution (15.5 mL) of the **1.232a** (1.0 g, 4.76 mmol, 1 equiv) at a rate such that the internal temperature did not exceed −75 °C. The resulting solution was stirred at this temperature for 30 min before being warmed to room temperature. The reaction was slowly quenched at 0 °C with a saturated aqueous solution of NH_4Cl, and extracted with EtOAc (3×). The combined organic layers were washed with brine, dried over Na_2SO_4, filtered, and concentrated in vacuo. ^{1}H NMR analysis of the crude reaction mixture indicated that the product was present as a 90:10 mixture of diastereomers. Also present was a trace amount (~5%) of the direct *t*BuLi addition product. The crude material was purified by silica gel flash column chromatography using hexanes:EtOAc (10:7 v:v) to afford pure **2.107** as a clear and colourless oil (0.789 g, 2.96 mmol, 52%, dr = >98:2). This material was taken up in dry DMF (10 mL), and the resulting solution was cooled to 0 °C, at which time LiHMDS (3.55 mL, 3.55 mmL, 1.2 equiv, 1.0 M in THF) was added dropwise. The resulting deep orange/red solution was allowed to stir at this temperature for 30 min before warming to room temperature. MeI (369 μL, 5.92 mmol, 2 equiv) was added dropwise and the mixture was stirred for 30 min. The reaction was quenched with a saturated aqueous solution of NH_4Cl, and extracted with EtOAc (3×). The combined organic layers were sequentially washed with water (4×) and brine, dried over Na_2SO_4, filtered, and concentrated in vacuo. The crude material was purified by silica gel flash column chromatography using hexanes:EtOAc (10:3 v:v) to afford pure **S1** as a clear and light yellow oil (0.630 g, 2.25 mmol, 76%). **^{1}H NMR** (300 MHz, $CDCl_3$) δ 7.36–7.22 (m, 5H), 5.17–5.13 (m (br), 1H), 5.13–5.10 (m (br), 1H), 4.98 (s, 1H), 2.53 (s, 3H), 2.14–1.82 (m, 2H), 1.17 (s, 9H), 1.03 (t, J = 7.4 Hz, 3H). **^{13}C NMR** (100 MHz, $CDCl_3$) δ 148.8, 138.7, 128.7, 128.2, 127.4, 113.2, 71.8, 58.5, 29.6, 26.7, 23.7, 12.1. **IR** (neat, cm^{-1}) 3086, 2964, 1645, 1494, 1456, 1361, 1074. **HRMS** (ESI+) Calc'd for $C_{16}H_{26}NOS$ 280.17351, found 280.17451. $[\alpha]_D^{20}$ −53.1 (c = 0.67, $CHCl_3$).

(*S*)-*N*-methyl-2-methylene-1-phenylbutan-1-amine (**2.108**)—A flame dried flask was charged with **S1** (630 mg, 2.25 mmol, 1 equiv) was purged with argon for 10 min. The contents of the flask were then taken up in anhydrous Et_2O (8 mL) and the resulting solution was cooled to 0 °C. At this time, an anhydrous HCl solution (1.13 mL, 4.5 mmol, 2 equiv, 4 M in 1,4-dioxane) was added drop wise via syringe, and the resulting mixture was stirred at this temperature for 30 min. At this time the reaction mixture was warmed to room temperature and the white precipitate was filtered through a fritted funnel using a water aspirator and the filter cake was washed with a small amount of ice cooled Et_2O. The filter cake was dissolved through the frit into a new flask using distilled water. The aqueous filtrate had its volume doubled with Et_2O and with stirring, was slowly basified to pH 10 using 1 M aqueous NH_4OH. The layers were separated, and the aqueous layer was extracted with Et_2O (2×). The combined organic layers were washed with brine, dried over Na_2SO_4, filtered and concentrated in vacuo. The title compound was obtained as clear and light yellow oil (342 mg, 1.95 mmol, 87%). **^{1}H NMR** (300 MHz, $CDCl_3$) δ 7.36–7.19 (m, 1H), 5.22–5.20 (m, 1H), 4.93–4.90 (m, 1H), 4.02 (s, 1H), 2.34 (s, 3H), 2.03–1.73 (m, 2H), 0.96 (t, J = 7.5 Hz, 3H). **^{13}C NMR** (100 MHz, $CDCl_3$) δ 152.2, 142.4, 128.2, 127.4, 127.0, 108.3, 70.3, 34.7, 25.2, 12.1. **IR** (neat, cm^{-1}) 3335, 2964, 2787, 1643, 1435, 1130, 1074. **HRMS** (ESI+) Calc'd for $C_{12}H_{18}N$ 176.14392, found 176.14350. $[\alpha]_D^{20}$ −82.0 (c = 0.59, $CHCl_3$).

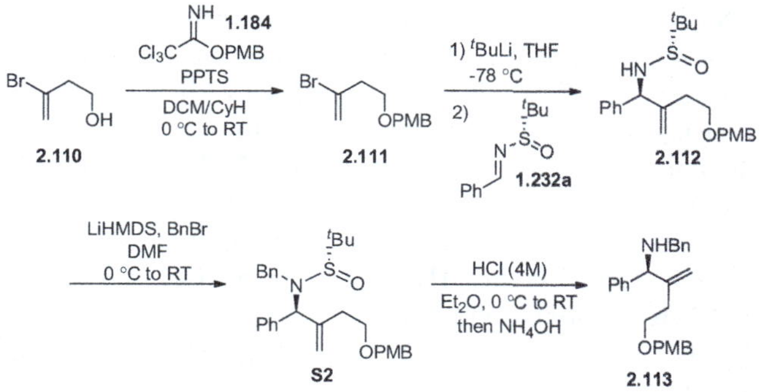

(*S*)-*N*-((*S*)-4-((4-methoxybenzyl)oxy)-2-methylene-1-phenylbutyl)-2-methylpropane-2-sulfinamide (2.113)—3-bromobut-3-en-1-ol **2.110** was synthesized according to the procedure by Gordon and co-workers and spectra data was in accordance with that reported therein.[87] A dry flask was charged with **2.110** (2.2 g, 14.6 mmol, 1 equiv) which was dissolved in DCM (3.3 mL) and cyclohexane (17 mL), and the resulting solution was cooled to 0 °C. To this was added a DCM solution (2.6 mL) of **1.184** (5.44 g, 19.2 mmol, 1.32 equiv) was added over 5 min, followed by the addition of solid PPTS (257 mg, 1.02 mmol, 7 mol%) at once. The resulting solution was stirred at this temperature for 45 min at which time the reaction was warmed to room temperature and stirred for 18 h. The reaction mixture was then filtered over a short pad of silica gel eluting with hexanes:EtOAc (4:1 v:v). The collected fraction was washed with $NaHCO_3$ and brine, dried over Na_2SO_4, filtered and concentrated in vacuo, and the crude PMB ether was purified via silica gel flash column chromatography using 5% EtOAc in hexanes as the mobile phase to afford the **2.111** as a clear and light yellow oil (2.32 g, 8.59 mmol, 59%). Spectral data for this compound were in accordance with that reported in the literature.[88]

[87]Reference [111].

[88]Reference [112].

(*S*)-N-((*S*)-4-((4-methoxybenzyl)oxy)-2-methylene-1-phenylbutyl)-2-methylpropane-2-sulfinamide (2.112b)—A flame dried Schlenk flask was charged with **2.111** (1.08 g, 3.99 mmol, 1.5 equiv) and the flask was ran through three vacuum-argon purge cycles, the vinyl bromide was dissolved in freshly distilled THF (8.6 mL), and the resulting solution was cooled to −78 °C. To this was added ^{t}BuLi (5.17 mL, 8.8 mmol, 3.3 equiv, 1.7 M in pentanes) over a period of 10 min. The bright yellow solution was allowed to stir at this temperature for 30 min, at which time a THF solution (2.2 mL) of the phenyl sulfinamide (558 mg, 2.67 mmol, 1 equiv) was added over 15 min. The resulting solution was stirred at this temperature for 2 h at which time TLC analysis of a small worked-up aliquot indicated full conversion of starting material. The cooling bath was removed and the reaction was slowly quenched with a saturated aqueous solution of NH_4Cl, and extracted with EtOAc (3×). The combined organic layers were washed with brine, dried over Na_2SO_4, filtered, and concentrated in vacuo. ^{1}H NMR analysis of the crude reaction mixture indicated the product to be present as an 89:11 mixture of diastereomers. The crude material was purified by silica gel flash column chromatography using hexanes:EtOAc (1:1 v:v) to afford **2.112** (dr = >95:5) as a clear and colourless oil (622 mg, 1.55 mol, 58%). **^{1}H NMR (Major)** (400 MHz, $CDCl_3$) δ 7.37–7.26 (m, 5H), 7.22–7.18 (m, 2H), 6.88–6.82 (m, 2H), 5.38–5.36 (m, 1H), 5.11–5.08 (m, 1H), 5.01–4.98 (m, 1H), 4.35 (s, 2H), 3.80 (s, 3H), 3.53 (d, *J* = 3.0 Hz, 1H), 3.50–3.36 (m, 2H), 2.34–2.24 (m, 1H), 2.23–2.11 (m, 1H), 1.25 (s, 9H). **^{13}C NMR (Major)** (100 MHz, $CDCl_3$) δ 159.1, 145.4, 140.6, 130.3, 129.2, 128.6, 127.9, 127.3, 114.3, 113.7, 72.5, 68.6, 63.3, 55.7, 55.2, 32.0, 22.7.

IR (neat, cm^{-1}) 3209, 3062, 3030, 2965, 2863, 1645, 1612, 1586, 1516, 1453, 1443, 1363, 1302, 1208, 1095. **HRMS** (ESI+) Calc'd for $C_{23}H_{31}NO_3NaS$ 424.1902, found 424.1916. $[\alpha]_D^{20}$ +28.2 (c = 1.0, $CHCl_3$).

(*R*)-*N*-benzyl-*N*-((*S*)-4-((4-methoxybenzyl)oxy)-2-methylene-1-phenylbutyl)-2-methylpropane-2-sulfinamide (S2)—In a dry flask **2.112** was taken up in dry DMF (4.4 mL), and the resulting solution was cooled to 0 °C, at which time LiHMDS (1.59 mL, 1.59 mmol, 1.2 equiv, 1.0 M in THF) was added dropwise. The resulting deep orange/red solution was allowed to stir at this temperature for 30 min before warming to room temperature. BnBr (314 μL, 2.64 mmol, 2 equiv) was added dropwise and the mixture was stirred for 30 min. The reaction was quenched with a saturated aqueous solution of NH_4Cl, and extracted with EtOAc (3×). The combined organic layers were sequentially washed with water (4×) and brine, dried over Na_2SO_4, filtered, and concentrated in vacuo. The crude material was purified by silica gel flash column chromatography using hexanes:DCM:EtOAc (5:1:1 v:v:v) to afford pure **S2** as a clear and colourless oil (543 mg, 1.10 mmol, 84%) **^{1}H NMR** (500 MHz, $CDCl_3$) δ 7.43–7.35 (m, 4H), 7.33–7.23 (m, 6H), 7.13–7.06 (m, 2H), 6.84–6.78 (m, 2H), 5.89 (s, 1H), 5.42–5.34 (m, 1H), 4.83 (s, 1H), 4.71 (dd, *J* = 16.5, 1.0 Hz, 1H), 4.22 (s, 2H), 3.97 (d, *J* = 16.5 Hz, 1H), 3.79 (s, 3H), 3.38–3.27 (m, 2H), 2.20 (dt, *J* = 14.5, 7.0 Hz, 0H), 1.97–1.87 (m, 1H), 0.88 (s, 9H). **^{13}C NMR** (125 MHz, $CDCCl_3$) δ 159.0, 141.5, 138.3, 138.3, 130.8, 130.3, 129.1, 128.6, 128.0, 127.9, 127.8, 127.1, 114.5, 113.6, 72.4, 68.9

(br) overlapping 68.8, 58.2, 55.2, 47.0, 35.1, 23.0. **IR** (neat, cm^{-1}) 3085, 2954, 1652, 1611, 1586, 1514, 1496, 1454, 1443, 1363, 1248, 1173, 1099, 1064, 1037. **HRMS** (DART) Calc'd for $C_{30}H_{38}NO_3S$ 492.25724, found 492.25795. $[\alpha]_D^{20}$ −162.6 (c = 0.67, $CHCl_3$).

(*S*)-*N*-benzyl-4-((4-methoxybenzyl)oxy)-2-methylene-1-phenylbutan-1-amine (2.113)—A flame dried flask was charged with **S2** (543 mg, 1.1 mmol, 1 equiv) was purged with argon for 10 min. The contents of the flask were then taken up in anhydrous Et_2O (4 mL) and the resulting solution was cooled to 0 °C. At this time, an anhydrous HCl solution (0.55 mL, 2.2 mmol, 2 equiv, 4 M in 1,4-dioxane) was added drop wise via syringe, and the resulting mixture was stirred at this temperature for 30 min. At this time the reaction mixture was warmed to room temperature and the volume of the reaction mixture was doubled with distilled water and with stirring, basified to pH 10 with a 1 M aqueous solution of NH_4OH. The layers were separated, and the aqueous layer was extracted with Et_2O (2×). The combined organic layers were washed with brine, dried over Na_2SO_4, filtered and concentrated in vacuo. The crude alkylated amine was obtained in analytically pure form by silica gel flash column chromatography using hexanes:EtOAc:DCM (5:1:1 v:v) as the mobile phase. The title compound was obtained as clear and light yellow oil (226 mg, 0.584 mmol, 53%). **^{1}H NMR** (500 MHz, $CDCl_3$) δ 7.42–7.37 (m, 2H), 7.37–7.30 (m, 6H), 7.29–7.24 (m, 2H), 7.23–7.19 (m, 2H), 6.90–6.84 (m, 2H), 5.39 (s, 1H), 5.05 (d, *J* = 2.0 Hz, 1H), 4.38–4.32 (m, 2H), 4.25 (s, 1H), 3.80 (s, 3H), 3.77–3.69 (m, 2H), 3.49–3.43 (m, 2H), 2.39–2.27 (m, 1H), 2.24–2.14 (m, 1H). **^{13}C NMR** (125 MHz, $CDCl_3$) δ 159.1, 147.7, 142.1, 140.5, 130.5, 130.5, 129.3, 128.3, 128.3, 128.2, 127.6, 127.2, 126.9, 113.7, 111.6, 72.5, 69.0, 67.3, 55.3, 51.6, 32.6. **IR** (neat, cm^{-1}) 3084, 3062, 3029, 3003, 2953, 2930, 1611, 1586, 1506, 1428, 1363, 1303, 1248, 1173, 1096, 1032. **HRMS** (ESI+) Calc'd for $C_{26}H_{29}NO_2$ 388.2271, found 388.2272. $[\alpha]_D^{20}$ − 26.7 (c = 0.9, $CHCl_3$).

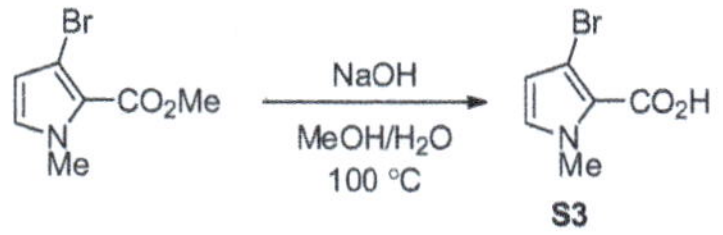

To a MeOH solution (30 mL) of 3-bromo-1-methyl-1H-pyrrole-2-carboxylate (427 mg, 1.96 mmol, 1 equiv) was added a 2 M aqueous solution of NaOH (10 mL). The resulting solution was heated to reflux for 3 h. At this time TLC analysis indicated full consumption of starting material and the reaction was acidified to pH = 2 using 1 M aqueous HCl. The resulting white precipitate was filtered off and dried to a constant weight in air, then under high vacuum overnight. **S3** was obtained as a white amorphous powder (357 mg, 1.75 mmol, 89%). **^{1}H NMR** (400 MHz, DMSO-d_6) δ 12.73 (s, 1H), 7.12 (d, *J* = 2.7 Hz, 1H), 6.27 (d, *J* = 2.7 Hz, 1H), 3.86 (s, 3H). **^{13}C NMR** (100 MHz, DMSO-d_6) δ 162.0, 130.3, 121.2, 112.7, 104.2, 39.0. **IR** (neat, cm^{-1}) 2956, 2922, 2848, 1684, 1433, 1271. **HRMS** (ESI+) Calc'd for $C_6H_7BrNO_2$ 203.96602, found 203.96557.

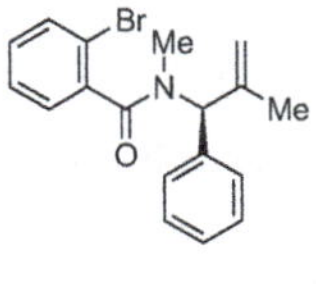

(*S*)-2-bromo-*N*-methyl-*N*-(2-methyl-1-phenylallyl)benzamide (2.105a)—Synthesized according to **general procedure 1** using 2-bromobenzoyl chloride (388 mg, 1.93 mmol) and **1.235a** (342 mg, 2.13 mmol). The crude amide was obtained in analytically pure form by silica gel flash column chromatography using hexanes: EtOAc (3:1 v:v) as the mobile phase. The title compound was obtained as an extremely viscous, clear and colourless oil (586 mg, 1.7 mmol, 88%). *NMR analysis showed the clean desired compound to be present as a complex mixture of 4 rotamers, and only approximate integration values are displayed.* ^{1}H NMR (500 MHz, $CDCl_3$) δ 7.64–7.50 (m, 1H), 7.40–7.13 (m, ~7.5H), 7.07 (td, *J* = 7.5, 1.0 Hz, ~0.2H), 7.04–7.00 (m, ~0.3H), 6.75 (dd, *J* = 7.5, 1.5 Hz, ~0.2H), 6.39 (s, ~0.5H), 5.24–4.58 (m, ~2.5H), 3.19–2.48 (m, 3H), 1.92–1.44 (m, 3H). **^{13}C NMR** (125 MHz, $CDCl_3$) δ 170.4, 169.7, 169.4 (br), 143.1, 142.7, 141.8, 138.9, 138.8, 138.7, 138.2, 137.9, 137.4, 136.4, 135.6, 133.0, 132.9, 132.8, 132.7, 130.2, 130.1, 130.1, 129.7, 129.0, 128.6, 128.5 (br), 128.4 (br), 128.4, 128.1, 127.8, 127.7, 127.7, 127.6, 127.5, 127.5, 127.1, 127.1, 126.8, 119.2, 119.1, 118.8 (br) overlapping 118.8 (br), 118.0, 114.8, 114.1, 113.0, 67.8, 67.8, 61.9 (br) overlapping 61.8 (br), 33.1, 33.0, 30.4, 30.2, 22.4, 21.9, 21.4, 21.1. **IR** (neat, cm^{-1}) 3086, 3059, 3028, 2972, 2943, 2918, 2852, 1616, 1476, 1387, 1327, 1250, 1180, 1121, 1076, 1032. **HRMS** (DART) Calc'd for $C_{18}H_{19}BrNO$ 344.06500, found 344.06504. Enantiomeric purity was determined by HPLC analysis in comparison with racemic material t_r = 10.0 (minor) t_r = 14.8 (major) (>99:1 er shown; Chiralcel AD-H column, 10% ^{i}PrOH in hexanes 1.0 mL/min, 220 nm). $[\alpha]_D^{20}$ −81.3 (c = 0.74, $CHCl_3$).

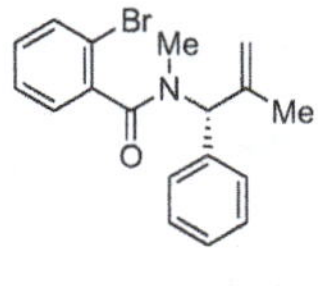

(*R*)-2-bromo-*N*-methyl-*N*-(2-methyl-1-phenylallyl)benzamide (*ent*-2.105a)—Synthesized **general procedure 1** using 2-bromobenzoyl chloride and *ent*-**1.235a**. All analytic data was in accordance with that reported for the (*S*) analog shown above. Enantiomeric purity was determined by HPLC analysis in comparison with racemic material t_r = 9.9 (major) t_r = 14.8 (minor) (>99:1 er. shown; Chiralcel AD-H column, 10% ^{i}PrOH in hexanes 1.0 mL/min, 210 nm). $[\alpha]_D^{20}$ +82.8 (c = 0.7, $CHCl_3$).

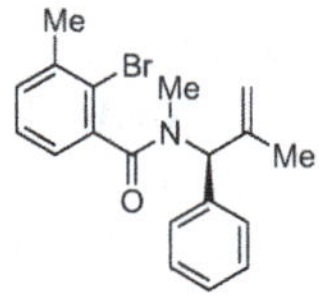

(*S*)-2-bromo-*N*,3-dimethyl-*N*-(2-methyl-1-phenylallyl)benzamide (2.105b)[89]—Synthesized according to **general procedure 1** using 2-bromo-3-methylbenzoic acid (215 mg, 1 mmol) and **1.235a** (177 mg, 1.1 mmol). The crude amide was obtained in analytically pure form by silica gel flash column chromatography using hexanes:EtOAc (3:1 v:v) as the mobile phase. The title compound was obtained as clear and colourless oil (268 mg, 0.75 mmol, 75%). *NMR analysis showed the clean desired compound to be present as a complex mixture of 4 rotamers, and only approximate integration values are displayed.* **^{1}H NMR** (600 MHz, $CDCl_3$) δ 7.40–7.13 (m, 7H), 7.11–7.08 (m, ~0.3H), 7.04–6.93 (m, ~0.7H), 6.56 (ddd, J = 7.5, 1.5, 0.5 Hz, ~0.1H), 6.44–6.38 (m, ~0.6H), 5.25–5.12 (m, 1H), 5.05–4.91 (m, ~1.2H), 4.59–4.56 (m, ~0.2H), 3.17 (s, ~0.4H), 2.85 (s, ~0.8H), 2.61 (s, ~0.8H), 2.53 (s, ~0.9H), 2.48–2.36 (m, 3H), 1.89–1.85 (m, ~1.2H), 1.85–1.81 (m, 1H), 1.48–1.45 (m, ~0.8H). **^{13}C NMR** (125 MHz, $CDCl_3$) δ 170.9, 170.1, 169.9, 169.8, 143.1, 142.8, 141.9, 141.8, 139.5, 139.3, 139.1 overlapping 139.0, 138.9, 138.8, 138.7, 138.4, 137.4, 136.5, 135.7, 130.9, 130.9, 130.7, 129.7, 129.7, 128.9, 128.5, 128.5, 128.3, 128.3, 128.0, 127.7, 127.5 overlapping 127.5, 127.5, 127.4, 126.9, 126.7, 124.9, 124.7 overlapping 124.7, 124.5, 121.4, 121.3, 121.1, 121.0, 118.0, 114.7, 114.0, 113.0, 67.8, 67.7, 61.8, 61.7, 33.1, 32.9, 30.3, 30.2, 23.3 overlapping 23.3, 23.2, 23.1, 22.50 21.9, 21.4, 21.0. **IR** (neat, cm^{-1}) 3086, 3059, 3025, 2974, 2945, 2920, 2855, 1635, 1435, 1393, 1327, 1110, 1080, 1028, 1005. **HRMS** (DART) Calc'd for $C_{19}H_{21}BrNO$ 358.08039, found 358.08065. Enantiomeric purity was determined by HPLC analysis in comparison with racemic material t_r = 14.9 (minor) t_r = 21.1 (major) (>99:1 er shown; Chiralcel AD-H column, 10% ^{i}PrOH in hexanes 0.75 mL/min, 210 nm). $[\alpha]_D^{20}$ −82.0 (c = 0.59, $CHCl_3$).

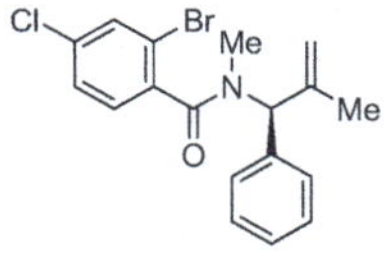

(*S*)-2-bromo-4-chloro-*N*-methyl-*N*-(2-methyl-1-phenylallyl)benzamide (2.105c)[90]—Synthesized according to **general procedure 1** using 2-bromo-4-chlorobenzoic acid (353 mg, 1.5 mmol) and **1.235a** (266 mg, 1.65 mmol). The crude amide was obtained in analytically pure form by silica gel flash column chromatography using hexanes:EtOAc (4:1 v:v) as the mobile phase. The title compound was obtained as a clear and colourless oil (503 mg, 1.3 mmol, 89%). *NMR analysis showed the clean desired compound to be present as a complex mixture of at least 4 rotamers, and only approximate integration values are displayed.* **^{1}H NMR** (500 MHz, $CDCl_3$) δ 7.69–7.53 (m, ~0.8H), 7.42–7.09 (m, 6H), 7.09–6.97 (m, ~0.5H), 6.67 (d, J = 8.2 Hz, ~0.2H), 6.36 (s, ~0.5H), 5.26–5.13 (m, 1H), 5.07–4.85 (m, ~1.25H), 4.59 (s, ~0.2H), 3.23–2.47 (m, 3H), 1.93–1.78 (m, 2H), 1.49 (s, 1H). **^{13}C NMR** (125 MHz, $CDCl_3$) δ 169.5, 168.8, 168.6, 143.0, 142.5,

[89]This compound was prepared by Hyung Yoon.

[90]This compound was prepared by Hyung Yoon.

141.7 (br), 138.5, 137.3 (br), 137.1 (br), 136.7, 136.4, 136.2 (br), 135.4, 135.4, 135.3, 135.2, 132.7, 132.7, 132.6 (br) overlapping 132.5 (br), 129.7, 129.0, 128.6, 128.5 (br), 128.4, 128.2, 128.2 (br) overlapping 128.1 (br), 127.9, 127.8 (br), 127.6, 127.4, 127.4, 127.2, 119.9, 119.8, 119.4, 118.3, 114.8, 114.2, 113.0, 67.9, 67.9, 62.1, 33.1 (br) overlapping 33.0 (br), 30.4, 30.2, 29.7, 22.4, 21.8, 21.4, 21.1. **IR** (neat, cm^{-1}) 3086, 3028, 2972, 2918, 1636, 1586, 1549, 1450, 1373, 1248, 1099, 1074, 1030, 912. **HRMS** (DART) Calc'd for $C_{18}H_{18}BrClNO$ 378.02603, found 378.02738. Enantiomeric purity was determined by HPLC analysis in comparison with racemic material t_r = 12.8 (minor) t_r = 22.5 (major) (99:1 er shown; Chiralcel AD-H column, 10% ^{i}PrOH in hexanes 1.0 mL/min, 210 nm). $[\alpha]_D^{20}$ −82.4 (c = 1.0, $CHCl_3$).

(*S*)-*N*-benzyl-2-bromo-*N*-(2-methyl-1-phenylallyl)benzamide (2.105d)[91]—Synthesized according to **general procedure 1** using 2-bromobenzoyl chloride (258 mg, 0.96 mmol) and **1.235h** (250 mg, 1.05 mmol). The crude amide was obtained in analytically pure form by silica gel flash column chromatography using hexanes: EtOAc (7:2 v:v) as the mobile phase. The title compound was obtained as a white solid (290 mg, 0.7 mmol, 72%). *NMR analysis showed the clean desired compound to be present as a complex mixture of 4 rotamers, and only approximate integration values are displayed.* **^{1}H NMR** (500 MHz, $CDCl_3$) δ 7.67–7.55 (m, ~0.8H), 7.52–6.80 (m, ~12.5H), 6.72–5.97 (m, ~0.5H), 5.42–4.00 (m, ~4.7H), 1.84–1.78 (m, ~0.5H), 1.61 (s, ~0.3H), 1.57–1.54 (m, ~0.6H), 1.36–1.32 (m, ~1.7H). **^{13}C NMR** (125 MHz, $CDCl_3$) δ 171.1, 170.4, 143.9, 143.3, 142.5, 138.6, 138.3, 138.1 overlapping 138.0, 137.2, 134.7, 133.1, 132.8, 132.4, 130.6 overlapping 130.5 (br), 130.2 overlapping 130.1 (br), 129.8 (br), 128.8 (br), 128.5, 128.4, 128.3 overlapping 128.3 overlapping 128.2, 128.1, 127.9, 127.8 (br), 127.5 overlapping 127.5, 127.4, 127.1, 126.9, 126.7, 126.1, 119.3, 118.8, 118.5, 113.7, 113.2, 68.4, 68.3, 64.5, 64.0 51.3, 50.3, 47.54 47.3, 22.7, 22.1, 21.2. **IR** (film, cm^{-1}). **HRMS** (DART) Calc'd for $C_{24}H_{23}BrNO$ 420.09630, found 420.09643. Enantiomeric purity was determined by HPLC analysis in comparison with racemic material t_r = 12.4 (minor) t_r = 16.6 (major) (>99:1 er shown; Chiralpak AS column, 7.5% ^{i}PrOH in hexanes 1.0 mL/min, 210 nm). $[\alpha]_D^{20}$ −41.0 (c = 1.0, $CHCl_3$).

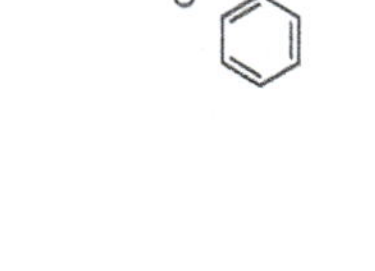

[91]This compound was prepared by Hyung Yoon.

(*S*)-*N*-benzyl-2-bromo-4-chloro-*N*-(2-methyl-1-phenylallyl)benzamide (2.105e)[92]—Synthesized according to **general procedure 1** using 2-bromo-4-chlorobenzoic acid (193 mg, 0.82 mmol) and **1.235h** (210 mg, 0.9 mmol). The crude amide was obtained in analytically pure form by silica gel flash column chromatography using hexanes:EtOAc (7:2 v:v) as the mobile phase. The title compound was obtained as a clear crystalline solid (273 mg, 0.6 mmol, 73%, MP = 111 – 112 °C). *NMR analysis showed the clean desired compound to be present as a complex mixture of 4 rotamers, and only approximate integration values are displayed.* **^{1}H NMR** (600 MHz, $CDCl_3$) δ 7.68–7.63 (m, ~0.5H), 7.60 (d, J = 2.0 Hz, ~0.2H), 7.53–6.81 (m, ~11.5H), 6.76 (d, J = 8.0 Hz, ~0.1H), 6.74–6.01 (m, ~0.6H), 5.32 (d, J = 14.5 Hz, ~0.5H), 5.29–5.27 (m, ~0.6H), 5.19–5.12 (m, ~1.2H), 5.08 (s, ~0.2H), 4.90 (s, ~0.6H), 4.81 (s, ~0.2H), 4.67 (d, J = 14.5 Hz, ~0.2H), 4.53 (d, J = 16.0 Hz, ~0.1H), 4.39 (d, J = 16.0 Hz, ~0.2H), 4.05 (d, J = 14.5 Hz, ~0.6H), 1.84–1.32 (m, 3H). **^{13}C NMR** (125 MHz, $CDCl_3$) δ 170.4, 169.6, 143.9, 143.2, 138.41 138.1, 137.9, 137.3, 137.00 136.8, 136.6, 135.4, 135.3, 135.0, 134.4, 132.9, 132.5, 132.0, 130.6, 129.6, 128.7, 128.5, 128.4, 128.4, 128.3, 128.2, 128.0, 128.0, 127.9, 127.7, 127.5, 127.5, 127.3, 126.8, 126.2, 119.9, 119.4, 118.8, 113.9, 113.3, 68.5, 68.4, 47.7, 47.4, 25.4, 22.8, 22.1, 21.2. **IR** (neat, cm^{-1}) 3086, 3063, 3030, 3007, 2974, 2918, 1643, 1586, 1559, 1495, 1472, 1452, 1431, 1408, 1327, 1099. **HRMS** (DART) Calc'd for $C_{24}H_{22}BrClNO$ 454.05733, found 454.05810. Enantiomeric purity was determined by HPLC analysis in comparison with racemic material t_r = 12.1 (major) t_r = 14.8 (minor) (>99:1 er shown; Chiralcel AD-H column, 10% ^{i}PrOH in hexanes 1.0 mL/min, 210 nm). $[\alpha]_D^{20}$ −41.1 (c = 0.515, $CHCl_3$).

(*S*)-*N*-benzyl-2-bromo-*N*-(2-methyl-1-phenylallyl)-5-(trifluoromethyl)benzamide (2.105f)[93]—Synthesized according to **general procedure 1** using 2-bromo-5-trifluoromethyl-benzoic acid (258.24 mg, 0.96 mmol) and **1.235h** (250 mg, 1.05 mmol). The crude amide was obtained in analytically pure form by a series of two silica gel flash column chromatography using hexanes:EtOAc (15:1 v:v) and DCM respectively as the mobile phase. The title compound was obtained as white powder (314 mg, 0.64 mmol, 67%, MP = 89–91 °C). *NMR analysis showed the clean desired compound to be present as a complex mixture of 4 rotamers, and only approximate integration values are displayed.* **^{1}H NMR** (500 MHz, $CDCl_3$) δ 7.78 (d, J = 8.5 Hz, ~0.5H), 7.70 (d, J = 8.5 Hz, ~0.2H), 7.67 (d, J = 2.0 Hz, ~0.5H), 7.61 (d, J = 8.5 Hz, ~0.2H), 7.56–7.42 (m, 1H), 7.42–7.19 (m, 4H), 7.19–7.11 (m, 1H), 7.07–6.96 (m, 2H), 6.96–6.83 (m, 2H), 6.63–6.14 (m, ~0.5H), 5.38–5.14 (m, ~2.5H), 4.96 (s, ~0.2H), 4.82 (s, ~0.6H), 4.80 (s, ~0.2H), 4.69 (d, J = 14.5 Hz, ~0.1H), 4.61 (d, J = 17.0 Hz, ~0.2H), 4.45–4.30 (m, ~0.2H), 4.05 (d, J = 15.0 Hz, ~0.6H), 1.90–

[92]This compound was prepared by Hyung Yoon.

[93]This compound was prepared by Hyung Yoon.

1.76 (m, ~0.6H), 1.55 (s, ~0.5H), 1.35–1.31 (m, ~1.8H). **^{13}C NMR** (125 MHz, $CDCl_3$) δ 169.9, 169.0, 144.0, 143.1, 142.3, 139.6, 139.1, 138.9, 138.5, 138.0, 137.8, 137.4, 136.9, 134.1, 133.6, 133.4, 132.9, 130.5, 129.7 (q, *J* = 33.5 Hz), 128.9, 128.6 (d, *J* = 1.0 Hz), 128.5, 128.4, 128.1, 127.9, 127.8, 127.6, 127.5, 127.1, 127.0, 126.7 (q, *J* = 4.0 Hz), 126.3, 124.8–124.6 (m), 124.5, 123.2–123.1 (m), 123.0–122.9 (m), 122.3, 122.0, 120.2, 119.3, 114.1, 113.3, 68.8, 68.5, 63.8, 51.1, 49.9, 47.9 47.4, 22.8, 22.1, 20.9. **^{19}F NMR** (564 MHz, $CDCl_3$) δ-62.8, -63.0, -63.2, -63.3. **IR** (neat, cm^{-1}) 3088, 3065, 3030, 2918, 1645, 1605, 1497, 1454, 1418, 1393, 1335, 1173, 1132, 1080, 1034. **HRMS** (DART) Calc'd for $C_{25}H_{22}BrF_3NO$ 488.08369, found 488.08439. Enantiomeric purity was determined by HPLC analysis in comparison with racemic material t_r = 10.05 (major) t_r = 11.0 (minor) (99:1 er shown; Chiralcel AD-H column, 5% ^{i}PrOH in hexanes 1.0 mL/min, 210 nm). $[\alpha]_D^{20}$ −26.403 (c = 0.515, $CHCl_3$).

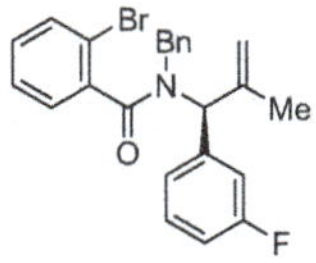

(*S*)-*N*-benzyl-2-bromo-*N*-(1-(3-fluorophenyl)-2-methylallyl)benzamide (2.105g)— Synthesized according to **general procedure 1** using 2-bromobenzoyl chloride (145 mg, 0.722 mmol) and **1.235b** (179 mg, 0.795 mmol). The crude amide was obtained in analytically pure form by silica gel flash column chromatography using hexanes:EtOAc (6:1 v:v) as the mobile phase. The title compound was obtained as a white solid (134 mg, 0.307 mmol, 43%, MP = 116–118 °C). *NMR analysis showed the clean desired compound to be present as a complex mixture of at least 3 rotamers, and only approximate integration values are displayed.* **^{1}H NMR** (500 MHz, $CDCl_3$) δ 7.66–7.61 (m, ~0.6H), 7.59 (dd, *J* = 8.0, 1.0 Hz, ~0.2H), 7.48 (d, *J* = 12.5 Hz, ~0.1H), 7.38–7.31 (m, ~1.7H), 7.30–7.27 (m, ~0.6H), 7.25–7.16 (m, ~1.4H), 7.16–6.90 (m, ~7H), 6.87–6.57 (m, ~1.3H), 6.06–5.84 (m, ~0.1H), 5.41 (dd, *J* = 15.0, 1.0 Hz, ~0.6H), 5.30 (d, *J* = 1.0 Hz, ~0.6H), 5.21–5.19 (m, ~0.6H), 5.18–5.13 (m, ~0.6H), 5.11 (s, ~0.3H), 4.92 (s, ~0.6), 4.80 (s, ~0.2H), 4.70 (d, *J* = 15.0 Hz, ~0.2H), 4.51–4.38 (m, ~0.3H), 4.01 (d, *J* = 15.0 Hz, ~0.6H), 1.86–1.76 (m, ~0.6H), 1.55 (s, 1H), 1.37–1.31 (m, 2H). **^{13}C NMR** (125 MHz, $CDCl_3$) δ 171.2, 170.4, 163.7, 161.7, 143.3, 142.8, 138.1–137.8 (m), 137.2 (d, *J* = 7.0 Hz), 133.2, 132.8, 130.3, 130.3, 129.9 (d, *J* = 8.0 Hz), 129.7 (d, *J* = 8.0 Hz), 128.4, 128.0, 127.7, 127.6, 127.5, 127.4, 127.2, 126.9 (d, *J* = 14.0 Hz), 126.4–126.2 (m), 123.6 (d, *J* = 3.0 Hz), 119.2, 119.0, 118.7, 117.4 (d, *J* = 22.0 Hz), 115.2 (d, *J* = 21.0 Hz), 114.5 (d, *J* = 21.0 Hz), 114.2, 68.0, 67.6 (d, *J* = 2.0 Hz), 47.5, 47.3, 22.7, 21.1. **^{19}F NMR** (470 MHz, $CDCl_3$) δ-112.5 to -112.6 (m), -113.2 to -113.2 (m). **IR** (film, cm^{-1}) 3032, 1643, 1589, 1473, 1406, 1309, 1219. **HRMS** (DART) Calc'd for $C_{24}H_{22}BrFNO$ 438.08688, found 438.08791. Enantiomeric purity was determined by HPLC analysis in comparison with racemic material t_r = 11.3 (minor) t_r = 13.2 (major) (>99:1 er shown; Chiralcel AD-H column, 10% ^{i}PrOH in hexanes 1.0 mL/min, 230 nm). $[\alpha]_D^{20}$ −47.7 (c = 0.52, $CHCl_3$).

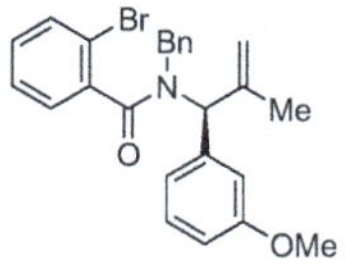

(S)-N-benzyl-2-bromo-N-(1-(3-methoxyphenyl)-2-methylallyl)benzamide (2.105h)[94]—Synthesized according to **general procedure 1** using 2-bromobenzoyl chloride (124 mg, 0.62 mmol) and **1.235c** (184 mg, 0.68 mmol). The crude amide was obtained in analytically pure form by silica gel flash column chromatography using hexanes: DCM:EtOac (10:2:1 v:v:v) as the mobile phase. The title compound was obtained as a white solid (192 mg, 0.43 mmol, 70%, MP = 122–123 °C). *NMR analysis showed the clean desired compound to be present as a complex mixture of at least 3 rotamers, and only approximate integration values are displayed.* **^{1}H NMR** (500 MHz, $CDCl_3$) δ 7.63 (dd, J = 8.0, 1.0 Hz, ~0.6H), 7.62–7.56 (m, ~0.2H), 7.37 (dd, J = 7.5, 2.0 Hz, ~0.6H), 7.35–7.30 (m, 1H), 7.29–7.25 (m, 1H), 7.25–7.10 (m, ~1.2H), 7.08 (t, J = 8.0 Hz, ~0.8H), 7.05–7.00 (m, 2H), 6.95 (dd, J = 7.0, 2.5 Hz, ~1.5H), 6.89 (dd, J = 7.5, 2.0 Hz, ~0.2H), 6.86–6.82 (m, ~0.6H), 6.79–6.75 (m, ~0.8H), 6.68 (ddd, J = 8.0, 2.5, 1.0 Hz, ~0.6H), 6.61 (d, J = 7.5 Hz, ~0.3H), 6.54–6.51 (m, 0H), 6.10–5.98 (m, ~0.1H), 5.38 (d, J = 14.5 Hz, ~0.6H), 5.29–5.25 (m, ~0.6H), 5.22–5.06 (m, ~1.3H), 4.89 (s, ~0.6H), 4.83 (s, ~0.2H), 4.74 (d, J = 15.0 Hz, ~0.2H), 4.56–4.39 (m, ~0.1H), 4.01 (d, J = 14.5 Hz, ~0.6H), 3.75 (s, ~0.5H), 3.67 (s, ~0.5H), 3.58 (s, 2H) 1.80 (s, ~0.5H), 1.56–1.53 (m, 1H), 1.38–1.33 (m, 1H). **^{13}C NMR** (125 MHz, $CDCl_3$) δ 171.2, 170.4, 159.7, 159.6, 159.5, 143.8, 143.1, 140.2, 138.4, 138.3, 138.3, 138.0, 136.1, 133.1, 132.8, 130.2, 130.1, 129.5, 129.3, 129.2, 128.4, 128.0, 128.0, 127.6, 127.5, 127.4, 127.1, 126.9, 126.7, 126.2, 122.4, 120.4, 119.3, 118.7, 118.6, 116.3, 114.5, 114.0, 113.8, 113.3, 112.9, 68.5, 68.2, 55.3, 55.1 overlapping 55.1, 47.6, 47.3, 22.7, 21.3. **IR** (film, cm^{-1}) 3063, 2937, 1651, 1600, 1494, 1431, 1406, 1327, 1261, 1165, 1045. **HRMS** (ESI+) Calc'd for $C_{25}H_{25}BrNO_2$ 450.1063, found 450.1074. Enantiomeric purity was determined by HPLC analysis in comparison with racemic material t_r = 13.1 (minor) t_r = 15.7 (major) (99:1 er shown; Chiralcel AD-H column, 10% ^{i}PrOH in hexanes 1.0 mL/min, 210 nm). $[\alpha]_D^{20}$ −35.2 (c = 0.5, $CHCl_3$).

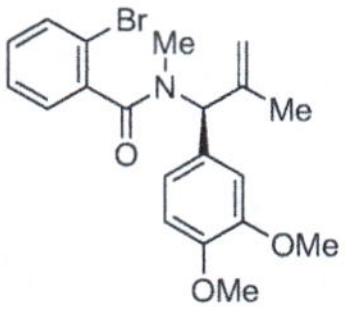

(S)-2-bromo-N-(1-(3,4-dimethoxyphenyl)-2-methylallyl)-N-methylbenzamide (2.105i)—Synthesized according to **general procedure 1** using 2-bromobenzoyl chloride (148 mg, 0.739 mmol) and **1.235d** (180 mg, 0.814 mmol). The crude

[94]This compound was prepared by Hyung Yoon.

amide was obtained in analytically pure form by silica gel flash column chromatography using hexanes:EtOAc (2:1 v:v) as the mobile phase. The title compound was obtained as a white solid (198 mg, 0.491 mmol, 66%, MP = 92–93 °C). *NMR analysis showed the clean desired compound to be present as a complex mixture of at least 4 rotamers, and only approximate integration values are displayed.* **^{1}H NMR** (500 MHz, $CDCl_3$) δ 7.63 (dt, J = 8.0, 1.0 Hz, ~0.3H), 7.61–7.52 (m, ~0.6H), 7.43–7.13 (m, 3H), 7.10 (td, J = 7.5, 1.0 Hz, ~0.1H), 6.94–6.77 (m, 3H), 6.64 (dt, J = 8.5, 2.0 Hz, ~0.1H), 6.43 (d, J = 2.0 Hz, ~0.1H), 6.36–6.28 (m, ~0.5H), 5.28–4.56 (m, ~2.4H), 3.96–3.80 (m, 6H), 3.78 (s, ~0.4H), 3.15 (s, ~0.4H), 2.83 (s, 1H), 2.56 (m, ~1.6H), 1.91–1.74 (m, 2H), 1.52–1.34 (m, 1H). **^{13}C NMR** (125 MHz, $CDCl_3$) δ 170.3, 169.4, 169.3, 149.0, 148.9, 148.7, 148.6, 148.5, 148.2, 143.2, 142.8, 141.9 (br), 141.7 (br), 138.7 (br), 138.2, 137.9, 132.9, 132.9, 132.7, 131.2, 130.2, 130.1 overlapping 130.0, 128.7, 127.8, 127.5, 127.3, 127.1, 127.0, 126.8, 121.7, 121.2, 119.5, 119.2, 118.9, 118.7, 117.8, 114.2, 113.7, 113.0, 112.5, 110.9, 110.8 overlapping 110.7, 67.5, 61.6, 55.9, 55.8, 33.1, 32.8, 30.3, 30.0, 22.3, 21.7, 21.4, 21.1. **IR** (neat, cm^{-1}) 3003, 2936, 1635, 1515, 1394, 1253, 1141, 1026. **HRMS** (DART) Calc'd for $C_{20}H_{23}BrNO_3$ 404.08613, found 404.08675. Enantiomeric purity was determined by HPLC analysis in comparison with racemic material t_r = 8.0 (minor) t_r = 12.7 (major) (99:1 er shown; Chiralcel AD-H column, 25% ^{i}PrOH in hexanes 1.0 mL/min, 210 nm). $[\alpha]_D^{20}$ −97.4 (c = 0.54, $CHCl_3$).

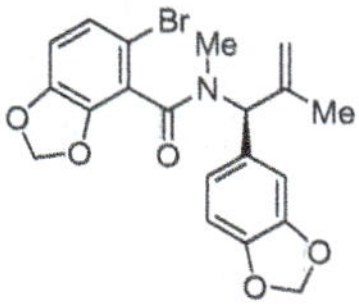

(*S*)-*N*-(1-(benzo[*d*][1,3]dioxol-5-yl)-2-methylallyl)-5-bromo-*N*-methylbenzo[*d*][1,3]dioxole-4-carboxamide (2.105j)—Synthesized according to **general procedure 1 using 5-bromobenzo[d][1,3]dioxole-4-carboxylic acid** (1.82 g, 7.50 mmol) and **1.226** (1.69 g, 8.25 mmol). The crude amide was obtained in analytically pure form by silica gel flash column chromatography using hexanes:EtOAc (10:3 v:v) as the mobile phase. The title compound was obtained as a hygroscopic white foam (2.58 g, 5.9 mmol, 79%). *NMR analysis showed the clean desired compound to be present as a complex mixture of 4 rotamers, and only approximate integration values are displayed.* **^{1}H NMR** (600 MHz, $CDCl_3$) δ 7.07–6.98 (m, 1H), 6.87–6.52 (m, 4H), 6.25 (m, ~0.7H), 6.09–5.55 (m, 4H), 5.16 (m, ~0.9H), 5.11–5.06 (m, ~0.2H), 5.05–4.92 (m, ~0.9H), 4.86–4.71 (m, ~0.3H), 3.07–2.59 (m, 3H), 1.87–1.74 (m, ~2.6H), 1.62–1.53 (m, ~0.5H). **^{13}C NMR** (125 MHz, $CDCl_3$) δ 165.9, 165.8, 165.0, 164.9, 147.7, 147.6, 147.5, 147.2 overlapping 147.2, 147.1 overlapping 147.1, 147.0, 146.9 overlapping 146.8, 146.7, 145.2, 144.7, 144.6, 142.3, 142.1, 141.6, 141.4, 131.7, 130.5, 130.0, 130.0, 125.6, 125.4, 125.3 overlapping 125.3, 123.0, 122.7, 122.4, 121.5, 120.4, 120.3, 120.1, 120.0, 117.0, 114.8, 114.2, 113.1, 110.2, 110.0, 109.9 overlapping 109.9, 109.8, 109.5 overlapping 109.5 overlapping 109.5, 109.34 overlapping 109.4, 108.5, 108.1 overlapping 108.1, 108.0, 107.8, 102.3, 102.2, 102.0,

101.9, 101.1, 100.99, 100.98, 100.96, 67.2, 67.1, 61.7, 61.6, 32.3 overlapping 32.3, 30.5, 30.4, 21.8, 21.6, 21, 20.9. **IR** (neat, cm^{-1}) 3010, 1647, 1489, 1444, 1394, 1240, 1037. **HRMS** (DART) Calc'd for $C_{20}H_{19}BrNO_5$ 432.04466, found 432.04434. Enantiomeric purity of this compound could not be determined due to severe peak broadening. However, the er of the corresponding iodo derivative was determined to be >99:1. $[\alpha]_D^{20}$ −69.2 (c = 0.5, $CHCl_3$).

(*S*)-*N*-benzyl-2-bromo-*N*-(5-((4-methoxybenzyl)oxy)-2-methylpent-1-en-3-yl) benzamide (2.105k)[95]—Synthesized according to **general procedure 1** using 2-bromobenzoyl chloride (136 mg, 0.68 mmol) and **1.235g** (244 mg, 0.752 mmol, 1.1 equiv). The crude amide was obtained in analytically pure form by silica gel flash column chromatography using hexanes:EtOAc (3:1 v:v) as the mobile phase. The title compound was obtained as a clear and colourless oil (219 mg, 0.43 mmol, 63%). *NMR analysis showed the clean desired compound to be present as a complex mixture of at least 3 rotamers, and only approximate integration values are displayed.* **^{1}H NMR** (600 MHz, $CDCl_3$) δ 7.60–7.58 (m, ~0.2H), 7.53–7.46 (m, ~1.4H), 7.42–7.39 (m, ~0.4H), 7.32–7.22 (m, ~3.4H), 7.21–7.09 (m, ~3.6H), 7.04 (d, *J* = 7.5 Hz, ~0.2H), 7.02–6.94 (m, 2H), 6.88 (d, *J* = 7.5 Hz, 1H), 6.83–6.77 (m, 1H), 5.29 (d, *J* = 15.0 Hz, ~0.3H), 5.23 (t, *J* = 7.5 Hz, ~0.3H), 5.14–5.01 (m, ~0.5H), 4.96 (s, ~0.2H), 4.91 (s, ~0.3H), 4.84 (s, ~0.5H), 4.50–4.43 (m, ~0.5H), 4.40–4.30 (m, 1H), 4.22–4.06 (m, ~2.5H), 3.79 (s, 2H), 3.78 (s, ~1H), 3.61–3.46 (m, 1H), 3.29–3.19 (m, ~0.5H), 3.10–2.98 (m, ~0.5H), 2.21–2.12 (m, ~0.4H), 2.05–1.92 (m, ~1H), 1.92–1.76 (m, ~2H), 1.68 (s, ~0.6H), 1.63 (s, ~0.9H). **^{13}C NMR** (125 MHz, $CDCl_3$) δ 170.5, 170.21, 170.16, 159.1 overlapping 159.0, 158.9, 142.8, 142.6, 141.9, 141.0, 138.9 overlapping 138.8, 138.7, 138.1, 137.5, 137.4, 137.1, 133.3, 133.0, 132.8, 132.5, 130.5 overlapping 130.4 overlapping 130.4, 130.1, 130.0, 130.0, 129.8, 129.4 overlapping 129.4, 129.0, 128.9, 128.6, 128.4, 128.2, 128.28 overlapping 128.17 overlapping 128.15 overlapping 128.11, 128.0, 127.9, 127.8, 127.7, 127.3 overlapping, 127.3, 127.1, 126.96 overlapping 126.95, 126.91, 119.7, 119.4, 119.3, 118.8, 114.8, 114.6 overlapping 114.6, 114.3, 113.7, 113.6, 113.5, 72.6, 72.3, 72.1, 67.4 overlapping 67.4, 66.9, 66.5, 61.0, 60.5, 56.3, 55.5, 55.20 overlapping 55.18, 49.8, 49.2, 45.6, 44.9, 32.4, 31.5, 31.2, 30.6, 22.5, 22.3, 21.6. **IR** (film, cm^{-1}) 3063, 3030, 3003, 2936, 2861, 1636, 1516, 1410, 1302, 1248, 1173, 1144, 1101, 1080, 1030. **HRMS** (DART) Calc'd for $C_{28}H_{31}BrNO_3$ 508.14873, found 508.15041. Enantiomeric purity was determined by HPLC analysis in comparison with racemic material t_r = 20.3 (major) t_r = 22.8 (minor) (>99:1 er shown; Chiralcel AD-H column, 10% ^{i}PrOH in hexanes 1.0 mL/min, 210 nm). $[\alpha]_D^{20}$ −86.5 (c = 0.555, $CHCl_3$).

[95]This compound was prepared by Hyung Yoon.

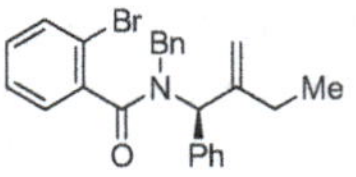

(*S*)-2-bromo-*N*-methyl-*N*-(2-methylene-1-phenylbutyl)benzamide (2.1051)—Syn thesized according to **general procedure 1** using 2-bromobenzoyl chloride (451 mg, 1.82 mmol) and **2.108** (450 mg, 2.00 mmol). The crude amide was obtained in analytically pure form by silica gel flash column chromatography using hexanes:EtOAc:DCM (24:1:16 v:v:v) as the mobile phase. The title compound was obtained a clear and colourless oil (654 mg, 1.34 mmol, 74%). *NMR analysis showed the clean desired compound to be present as a complex mixture of at least 4 rotamers, and only approximate integration values are displayed.* **^{1}H NMR** (500 MHz, $CDCl_3$) δ 7.82–7.76 (m, ~0.3H), 7.65–7.52 (m, ~1.2H), 7.44–7.14 (m, 8H), 7.08 (td, *J* = 7.5, 1.0 Hz, ~0.2H), 7.05–7.00 (m, ~0.3H), 6.79 (dd, *J* = 7.5, 1.5 Hz, 0.2H), 6.46 (s, ~0.6H), 5.28–4.91 (m, ~2.5H), 4.68 (m, ~0.2H), 3.12 (s, ~0.5H), 2.83 (s, 1H), 2.65–2.51 (m, ~1.6H), 2.21–2.06 (m, ~1.4H), 1.91–1.81 (m, ~0.3H), 1.56–1.47 (m, ~0.3H), 1.15 (t, *J* = 7.4 Hz, ~1.7H), 1.12–1.08 (m, ~0.4H), 1.00 (t, *J* = 7.4 Hz, ~0.5H), 0.71 (t, *J* = 7.4 Hz, 1H). **^{13}C NMR** (125 MHz, $CDCl_3$) δ 170.3, 169.5, 169.3 (br) overlapping 169.3 (br), 148.8, 148.5, 147.7, 138.8, 138.7 (br), 138.1, 137.7, 137.6, 136.7, 135.7, 135.5, 135.4, 135.4, 133.0, 132.7 overlapping 132.6 (br), 130.2, 130.1, 130.0, 129.8, 129.5, 128.7, 128.5, 128.3 (br) overlapping 128.3, 128.0, 127.7 (br) overlapping 127.6 (br), 127.5 (br) overlapping 127.5, 127.4 overlapping 127.4 overlapping 127.4, 126.9, 126.7, 119.3, 118.8, 118.7, 118.6, 115.6, 112.6, 111.7, 110.7, 66.6, 66.5, 61.0 overlapping 60.7, 33.1 overlapping 33.0, 30.3, 30.1, 27.6, 27.4, 27.1, 26.8, 12.2, 12.0, 11.8. **IR** (film, cm^{-1}) 3003, 2966, 2933, 1635, 1479, 1394, 1338, 1078, 1026. **HRMS** (DART) Calc'd for $C_{19}H_{21}BrNO$ 358.08065, found 358.08137. Enantiomeric purity was determined by HPLC analysis in comparison with racemic material t_r = 7.1 (minor) t_r = 12.5 (major) (99:1 er shown; Chiralcel OD-H column, 15% ^{i}PrOH in hexanes 1.0 mL/min, 210 nm). $[\alpha]_D^{20}$ −83.3 (c = 0.54, $CHCl_3$).

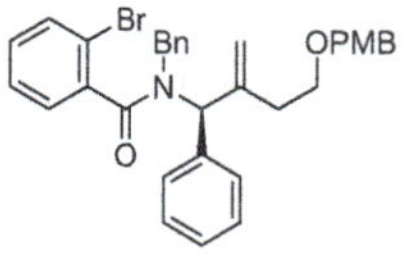

(*S*)-*N*-benzyl-2-bromo-*N*-(4-((4-methoxybenzyl)oxy)-2-methylene-1-phenylbutyl) benzamide (2.105m)—Synthesized according to **general procedure 1** using 2-bromobenzoyl chloride (203 mg, 0.82 mmol), and **2.113** (349 mg, 0.90 mmol). The crude amide was obtained in analytically pure form by silica gel flash column chromatography using hexanes:EtOAc (3:1 v:v) and 1% Et_3N as the mobile phase. The title compound was obtained as a crystalline white solid (335 g, 0.589 mmol, 72%, MP = 92–93 °C). *NMR analysis showed the clean desired compound to be present as a complex mixture of at least 3 rotamers, and only approximate integration values are displayed.* **^{1}H NMR** (500 MHz, $CDCl_3$) δ 7.59–7.52 (m, ~0.8H), 7.51–7.16 (m, ~6.5H), 7.16–7.04 (m, ~4.3H), 7.04–6.95 (m, ~2.8H), 6.95–6.79 (m, ~4H), 6.65–

6.11 (m, ~0.4H), 5.39–5.12 (m, ~3.4H), 4.93 (s, ~0.2H), 4.89 (d, J = 7.5 Hz, ~0.1H), 4.65 (d, J = 14.5 Hz, ~0.2H), 4.56–4.39 (m, ~0.6H), 4.26–4.17 (m, ~0.4H), 4.12 (s, ~1H), 4.05 (d, J = 15.0 Hz, ~0.7H), 3.83–3.78 (m, 3H), 3.61 (td, J = 7.0, 4.5 Hz, ~0.4H), 3.39 (q, J = 6.5 Hz, ~0.1H), 3.32–3.15 (m, ~0.4H), 3.14–3.05 (m, ~0.6H), 2.91 (ddd, J = 9.5, 7.5, 6.5 Hz, ~0.6H), 2.25–2.08 (m, ~1H), 1.90 (dt, J = 14.0, 6.5 Hz, ~0.2H), 1.80–1.72 (m, ~0.6H), 1.63–1.59 (m, ~0.6H). **^{13}C NMR** (125 MHz, CDCl$_3$) δ 171.2, 170.4, 163.4, 159.2, 159.1 overlapping 159.1, 159.1, 145.4, 145.3, 144.1, 138.7, 138.4, 138.3, 138.1, 137.9, 136.9, 136.7, 134.8, 133.2, 132.8, 130.8, 130.6, 130.4, 130.3, 130.2, 130.2, 129.3, 129.3, 129.1, 128.9, 128.8 overlapping 128.8, 128.7, 128.6, 128.4, 128.3 overlapping 128.3, 128.2, 128.1, 127.9, 127.9, 127.5, 127.4, 127.3, 127.1, 126.8, 126.7, 126.1, 119.4, 119.0, 118.8, 116.8, 114.1, 113.8, 113.9, 113.70 overlapping 113.68, 72.6, 72.6, 72.4, 72.2, 69.0, 68.9, 68.4, 68.3, 67.4, 67.0, 64.9, 55.3 overlapping 55.3, 47.5, 47.3, 46.9, 35.1, 34.4. **^{13}C NMR**. **IR** (film, cm^{-1}) 2931, 2854, 1639, 1512, 1406, 1247, 1095, 1031. **HRMS** (DART) Calc'd for $C_{33}H_{33}BrNO_3$ 570.16438, found 570.16392. Enantiomeric purity was determined by HPLC analysis in comparison with racemic material t_r = 27.8 (minor), t_r = 34.6 (major) (>99:1 er shown; Chiralcel AD-H column, 10% ^{i}PrOH in hexanes 1.0 mL/min, 210 nm). $[\alpha]_D^{20}$ −18.8 (c = 0.5, CHCl$_3$).

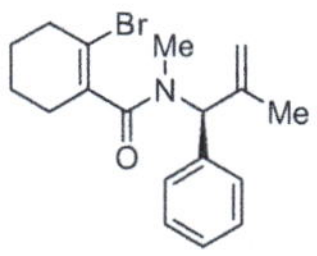

(*S*)-2-bromo-*N*-methyl-*N*-(2-methyl-1-phenylallyl)cyclohex-1-enecarboxamide (2.105n)—Synthesized according to **general procedure 1** using 2-bromocyclohex-1-enecarboxylic acid (277 mg, 1.36 mmol) and **1,235a** (241 mg, 1.5 mmol). The crude amide was obtained in analytically pure form by silica gel flash column chromatography using hexanes:EtOAc (3:1 v:v) as the mobile phase. The title compound was obtained as a clear, colourless oil (441 mg, 1.27 mmol, 94%). *NMR analysis showed the clean desired compound to be present as a complex mixture of at least 3 rotamers, and only approximate integration values are displayed.* **^{1}H NMR** (500 MHz, CDCl$_3$) δ 7.43–7.38 (m, 0.4H), 7.38–7.24 (m, ~4H), 7.24–7.18 (m, ~0.6H), 7.14–7.09 (m, ~0.3H), 6.21 (s, ~0.6H), 5.39 (d, J = 12.5 Hz, ~0.3H), 5.18–5.16 (m, ~0.2H), 5.16–5.13 (m, ~0.5H), 5.13–5.09 (m, ~0.4H), 4.99–4.95 (m, ~0.3H), 4.95–4.93 (m, ~0.1H), 4.79 (s, ~0.36H), 4.56–4.53 (m, ~0.1H), 3.04 (s, ~0.4H), 2.77 (s, ~1H), 2.71 (s, ~0.6H), 2.68 (s, 1H), 2.63–2.38 (m, ~2.5H), 2.37–2.23 (m, ~0.5H), 2.19–2.01 (m, ~1H), 1.97–1.92 (m, ~0.4H), 1.85–1.33 (m, ~7H). **^{13}C NMR** (125 MHz, CDCl$_3$) δ 172.0, 171.2, 170.7, 170.6, 143.5, 143.4, 141.9, 141.9, 139.8, 137.5, 136.5, 136.0, 135.5, 135.2, 135.0, 134.7, 129.69 overlapping 129.66, 128.9, 128.52 overlapping 128.50, 128.4, 128.3, 128.1, 127.6, 127.48 overlapping 127.46, 127.2, 119.7, 119.5, 119.4, 119.2, 118.2, 114.7, 114.1, 112.9, 66.9, 61.3, 61.2, 35.4, 35.4, 35.2, 35.1, 31.9, 31.7, 30.4, 30.1, 29.1, 28.8, 28.8, 28.6, 24.2, 24.1, 24.0, 22.7, 21.9, 21.5, 21.4, 21.3, 21.3, 21.1. **IR** (film, cm^{-1}) 2933, 2883, 1639, 1446, 1319, 1165, 1030. **HRMS** (DART) Calc'd for $C_{18}H_{23}BrNO$ 348.0951, found 348.0958. Enantiomeric purity was determined by HPLC analysis in comparison with racemic material t_r = 8.3 (minor) t_r = 11.1 (major) (>99:1 er shown; Chiralcel AD-H column, 10% ^{i}PrOH in hexanes 1 mL/min, 210 nm). $[\alpha]_D^{20}$ −76.3 (c = 0.75, $CHCl_3$).

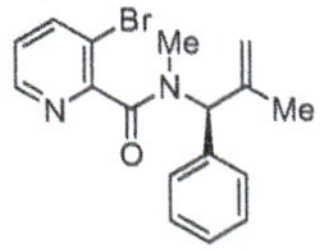

(*S*)-3-bromo-*N*-methyl-*N*-(2-methyl-1-phenylallyl)picolinamide (2.105o)[96]—Synthesized according to **general procedure 1** using 3-bromopicolinic acid (202 mg, 1.0 mmol) and **1.235a** (177 mg, 1.1 mmol). The crude amide was obtained in analytically pure form by silica gel flash column chromatography using hexanes:EtOAc (3:2 v:v) as the mobile phase. The title compound was obtained as a clear orange oil (183 mg, 0.53 mmol, 53%). *NMR analysis showed the clean desired compound to be present as a complex mixture of at least 3 rotamers, and only approximate integration values are displayed.* **^{1}H NMR** (500 MHz, $CDCl_3$) δ 8.56 (dd, *J* = 4.5, 1.5 Hz, ~0.7H), 8.52 (dd, *J* = 4.5, 1.5 Hz, ~0.3H), 7.92–7.89 (m, 1H), 7.38–7.23 (m, 5H), 7.19 (ddd, *J* = 8.0, 4.5, 3.0 Hz, 1H), 6.40–6.38 (m, ~0.7H), 5.21 (h, *J* = 1.5 Hz, ~0.7H), 5.17 (h, *J* = 1.5 Hz, ~0.3H), 5.09 (s, ~0.3H), 5.04–5.03 (m, ~0.7H), 4.88–4.87 (m, ~0.3H), 2.98 (s, ~1H), 2.57 (s, ~2H), 1.85 (dt, *J* = 1.5, 1.0 Hz, ~2H), 1.66 (dt, *J* = 1.5, 1.0 Hz, ~1H). **^{13}C NMR** (126 MHz, $CDCl_3$) δ 168.2, 167.6, 155.1, 154.6, 148.1, 147.5, 141.9, 141.5, 140.8, 140.5, 136.7, 136.6, 129.4, 128.9, 128.5, 128.4, 127.8, 127.7, 124.7, 117.9, 117.0, 115.9, 114.2, 67.0, 61.8, 32.5, 30.2, 21.7, 21.5. **IR** (film, cm^{-1}) 3086, 3028, 2974, 2918, 1645, 1456, 1335, 1128, 1088, 1053. **HRMS** (DART) Calc'd for $C_{17}H_{18}BrN_2O$ 345.06025, found 345.05981. Enantiomeric purity was determined by HPLC analysis in comparison with racemic material t_r = 28.3 (minor) t_r = 32.1 (major) (99:1 er shown; Chiralcel AD-H column, 10% ^{i}PrOH in hexanes 0.35 mL/min, 210 nm). $[\alpha]_D^{20}$ −76.2 (c = 0.64, $CHCl_3$).

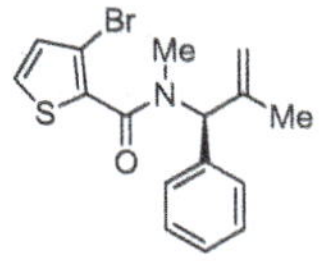

(*S*)-3-bromo-*N*-methyl-*N*-(2-methyl-1-phenylallyl)thiophene-2-carboxamide (2.105p)[97]—Synthesized according to **general procedure 1** using 3-bromothiophene-2-carboxylic acid (207 mg, 1 mmol) and **1.235a** (177 mg, 1.1 mmol). The crude amide was obtained in analytically pure form by silica gel flash column chromatography using hexanes:EtOAc:DCM (12:1:6 v:v:v) as the mobile phase. The title compound was obtained as a clear and colourless oil (223 mg,

[96]This compound was prepared by Hyung Yoon.

[97]This compound was prepared by Hyung Yoon.

0.639 mmol, 64%). *NMR analysis showed the clean desired compound to be present as a complex mixture of 2 rotamers, and only approximate integration values are displayed.* **^{1}H NMR** (500 MHz, $CDCl_3$) δ 7.49–7.13 (m, 6H), 6.96 (d, J = 5.0 Hz, 1H), 6.28 (s, ~0.4H), 5.24 (s, ~0.2H), 5.20 (d, J = 1.0 Hz, 1H), 4.93 (s, 1H), 3.08–2.64 (m, 3H), 1.95–1.60 (m, 3H). **^{13}C NMR** (125 MHz, $CDCl_3$) δ 164.6 (br), 163.8 (br), 142.5 (br), 141.5 (br), 136.6, 132.8 (br), 131.6 (br), 130.0, 129.2 (br) overlapping 128.5, 127.8, 126.7 (br), 115.5, 114.0, 110.0 (br), 109.2 (br), 68.0, 62.9, 33.5, 30.8, 21.6. **IR** (neat, cm^{-1}) 3086, 2918, 1635, 1444, 1388, 1321. **HRMS** (ESI+) Calc'd for $C_{16}H_{17}BrNOS$ 350.02142, found 350.02125. Enantiomeric purity was determined by HPLC analysis in comparison with racemic material t_r = 8.0 (minor) t_r = 9.6 (major) (99:1 er shown; Chirapak AS column, 10% ^{i}PrOH in hexanes 1.0 mL/min, 30 °C, 225 nm). $[\alpha]_D^{20}$ −84.3 (c = 0.7, $CHCl_3$).

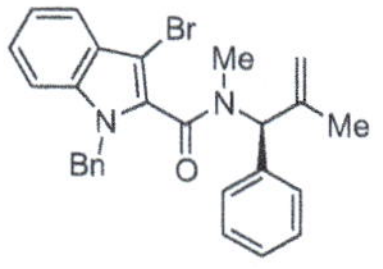

(*S*)-1-benzyl-3-bromo-*N*-methyl-*N*-(2-methyl-1-phenylallyl)-1*H*-indole-2-carbox amide (2.105q)[98]—Synthesized according to **general procedure 1** using 3-bromo-1-benzyl-1H-indole-2-carboxylic acid (330 mg, 1 mmol) and **1.235a** (177 mg, 1.1 mmol). The crude amide was obtained in analytically pure form by silica gel flash column chromatography using hexanes:EtOAc:DCM (20:1:10 v:v:v) as the mobile phase. The title compound was obtained as a hygroscopic white foam (410 mg, 0.87 mmol, 87%). *NMR analysis showed the clean desired compound to be present as a complex mixture of at least 3 rotamers, and only approximate integration values are displayed.* **^{1}H NMR** (500 MHz, $CDCl_3$) δ 7.66–7.58 (m, 1H), 7.42–7.13 (m, ~10H), 7.06–6.92 (m, 3H), 6.33 (s, ~0.5H), 6.18 (s, ~0.5H), 5.58 (d, J = 17.0 Hz, ~0.5H), 5.53–5.44 (m, 1H), 5.41 (d, J = 17.0 Hz, ~0.5H), 5.21 (q, J = 1.5 Hz, ~0.5H), 5.05 (s, ~0.5H), 4.92–4.90 (m, ~0.5H), 4.64 (t, J = 8.5 Hz, ~0.1H), 4.24–4.19 (m, ~0.5), 3.26 (s, ~0.1H), 2.96 (s, ~0.1H), 2.66 (s, ~1.5), 2.63 (s, ~1.3H), 1.96 (s, ~0.1H), 1.91 (s, ~1.5H), 1.68–1.66 (m, ~1.3H), 1.46 (s, ~0.1H). **^{13}C NMR** (125 MHz, $CDCl_3$) δ 164.7, 164.3, 163.5, 163.1, 143.9, 142.7, 141.9, 140.8, 139.4, 137.5, 137.3, 136.8, 136.8, 136.7, 136.3, 132.4, 131.3, 131.0, 129.8, 129.6, 128.9, 128.8, 128.7, 128.5, 128.4, 127.9, 127.6, 127.6, 127.5, 126.9, 126.4, 126.3, 126.3, 126.2, 124.4, 124.3, 123.8, 123.7, 121.1, 119.97, 119.95, 115.4, 114.1, 112.8, 111.3, 111.1, 110.2, 110.2, 91.6, 91.1, 68.2, 67.5, 63.0, 62.3, 49.8, 49.0, 47.6, 47.6, 33.2, 32.8, 31.2, 29.7, 22.4, 22.1, 21.3. **IR** (film, cm^{-1}) 3086, 3061, 3030, 2922, 1634, 1526, 1452, 1402, 1320, 1250, 1161, 1117, 1071. **HRMS** (DART) Calc'd for $C_{27}H_{26}BrN_2O$ 473.12285, found 473.12294. Enantiomeric purity was determined by HPLC analysis in comparison with racemic material t_r = 13.1 (minor) t_r = 19.9 (major) (99:1 er shown; Chiralcel AD-H column, 15% ^{i}PrOH in hexanes 1.0 mL/min, 210 nm). $[\alpha]_D^{20}$ −22.6 (c = 0.58, $CHCl_3$).

[98]This compound was prepared by Hyung Yoon.

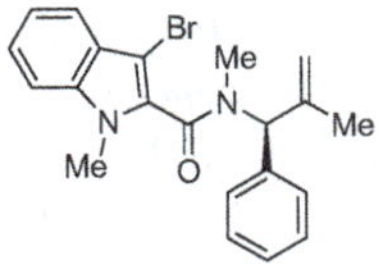

(*S*)-3-bromo-*N*,1-dimethyl-*N*-(2-methyl-1-phenylallyl)-1*H*-indole-2-carboxamide (2.105r)[99]—Synthesized according to **general procedure 1** using 3-bromo-1-methyl-1H-indole-2-carboxylic acid (254 mg, 1 mmol) and **1.235a** (181 mg, 1.1 mmol). The crude amide was obtained in analytically pure form by silica gel flash column chromatography using hexanes:EtOAc (3:1 v:v) as the mobile phase. The title compound was obtained as a pale yellow oil (181 mg, 0.456 mmol, 45%). *NMR analysis showed the clean desired compound to be present as a complex mixture of 4 rotamers, and only approximate integration values are displayed.* **^{1}H NMR** (500 MHz, $CDCl_3$) δ 7.62–7.51 (m, 1H), 7.47–7.12 (m, 7H), 6.90–6.80 (m, ~0.4H), 6.43 (s, ~0.4H), 6.31 (s, ~0.4H), 5.52–5.04 (m, 2H), 4.96–4.87 (m, ~0.5H), 4.59 (s, ~0.2H), 3.86–3.62 (m, ~2.4H), 3.26 (s, ~0.5H), 3.10–2.71 (m, 3H), 2.02–1.80 (m, ~2.4H), 1.47 (s, ~0.4H). **^{13}C NMR** (125 MHz, $CDCl_3$) δ 164.9, 164.4, 163.4, 163.1, 144.0, 142.5, 141.6, 141.5, 139.4, 137.1, 136.5, 136.4, 136.2, 136.1, 135.6, 131.9, 131.6, 129.7, 129.4, 128.8, 128.58 overlapping 128.55, 128.4, 128.1, 127.9, 127.7, 127.5, 126.29 overlapping 126.25, 123.8 overlapping 123.7 overlapping 123.6 overlapping 123.5, 120.8, 120.7, 119.70 overlapping 119.68 overlapping 119.6, 118.1, 115.0, 113.8, 112.54, 109.77 overlapping 109.74 overlapping 109.70, 89.6, 89.1, 88.5, 88.1, 68.2, 67.4, 62.8, 62.3, 33.2, 32.8, 31.8, 31.4, 31.2, 31.03 overlapping 31.00, 30.8, 22.2, 21.8, 21.4, 21.2. **IR** (neat, cm^{-1}) 3086, 3059, 3028, 2970, 2940, 2922, 2245, 1636, 1530,1464, 1447, 1399, 1319, 1065. **HRMS** (DART) Calc'd for $C_{21}H_{22}BrN_2O$ 397.09155, found 397.09117. Enantiomeric purity was determined by HPLC analysis in comparison with racemic material t_r = 9.1 (minor) t_r = 15.7 (major) (>99:1 er shown; Chiralcel AD-H column, 15% ^{i}PrOH in hexanes 1.0 mL/min, 210 nm). $[\alpha]_D^{20}$ −62.8 (c = 0.5, $CHCl_3$).

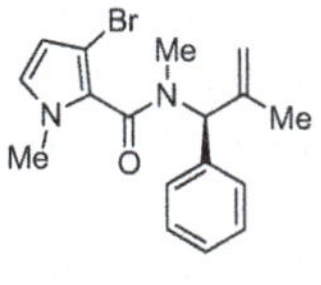

(*S*)-3-bromo-*N*,1-dimethyl-*N*-(2-methyl-1-phenylallyl)-1*H*-pyrrole-2-carboxamide (2.105s)[100]—Synthesized according to general procedure **1** using 3-bromo-1-methyl-1H-pyrrole-2-carboxylic acid (245 mg, 1.2 mmol) and **1.235a** (213 mg, 1.32 mmol). The crude amide was obtained in analytically pure form by silica gel flash column chromatography using hexanes:EtOAc (3:1 v:v) as the mobile phase. The title compound was obtained as an off-white solid (375 mg, 1.08 mmol, 90% (MP = 78–80 °C). ***NMR analysis of the pure title compound in***

[99]This compound was prepared by Hyung Yoon.

[100]This compound was prepared by Hyung Yoon.

$CDCl_3$ at 300 K revealed a broad series of peaks which precluded assignment. **IR** (film, cm^{-1}) 33109, 3061, 3028, 1628, 1526, 1485, 1441, 1304, 1067, 1017. **HRMS** (DART) Calc'd for $C_{17}H_{20}BrN_2O$ 347.07590, found 347.07593. Enantiomeric purity was determined by HPLC analysis in comparison with racemic material t_r = 5.3 (minor) t_r = 7.2 (major) (99:1 er shown; Chiralcel AD-H column, 15% ^{i}PrOH in hexanes 1.0 mL/min, 210 nm). $[\alpha]_D^{20}$ −69.8 (c = 1.0, $CHCl_3$).

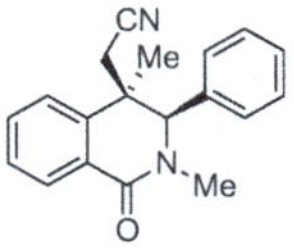

2-((3*R*,4*R*)-2,4-dimethyl-1-oxo-3-phenyl-1,2,3,4-tetrahydroisoquinolin-4-yl)acetonitrile (2.117a)—The title compound was synthesized from **2.105a** (103 mg, 0.3 mmol) using **general procedure 2a**, and was isolated via silica gel flash column chromatography using hexanes:EtOAc (3:1 v:v) as the mobile phase (78 mg, 0.27 mmol, 90%, 95:5 dr (*cis:trans*)).

Major: Isolated as a white solid (MP = 129–130 °C). **^{1}H NMR** (500 MHz, $CDCl_3$) δ 8.31–8.25 (m, 1H), 7.54–7.45 (m, 2H), 7.31–7.25 (m, 1H), 7.24–7.17 (m, 2H), 7.08–6.97 (m, 3H), 4.43 (s, 1H), 3.09 (s, 3H), 2.89 (d, *J* = 16.5 Hz, 1H), 2.23 (d, *J* = 16.5 Hz, 1H), 1.72 (s, 3H). **^{13}C NMR** (125 MHz, $CDCl_3$) δ 163.0, 140.4, 135.7, 132.6, 129.0, 128.9, 128.8, 128.4, 128.0, 127.9, 122.6, 117.6, 71.5, 40.5, 34.4, 29.7, 25.8. **IR** (cm^{-1}, film). **HRMS** (ESI+) Calc'd for $C_{18}H_{19}INO$, 392.0506; found, 392.0511. Enantiomeric purity was determined by HPLC analysis in comparison with racemic material t_r = 8.1 (minor) t_r = 8.8 (major) (>99:1 er shown; Chiralcel AD-H column, 20% ^{i}PrOH in hexanes 1.0 mL/min, 210 nm). $[\alpha]_D^{20}$ +312.6 (c = 0.5, $CHCl_3$).

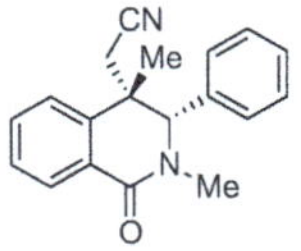

2-((3*S*,4*S*)-2,4-dimethyl-1-oxo-3-phenyl-1,2,3,4-tetrahydroisoquinolin-4-yl)acetonitrile (*ent*-2.117)—The title compound was synthesized from ***ent*-2.105a** (103 mg, 0.3 mmol) using **general procedure 2a**, and was isolated via silica gel flash column chromatography using hexanes:EtOAc (3:1 v:v) as the mobile phase (83 mg, 0.285 mmol, 95%, 95:5 dr (*cis:trans*)). **All analytic data was in accordance with that reported for the (3*R*,4*R*) analog shown above** Enantiomeric purity was determined by HPLC analysis in comparison with racemic material t_r = 8.1 (major) t_r = 9.0 (minor) (>99:1 er shown; Chiralcel AD-H column, 20% ^{i}PrOH in hexanes 1.0 mL/min, 210 nm). $[\alpha]_D^{20}$ −310.7 (c = 1.0, $CHCl_3$).

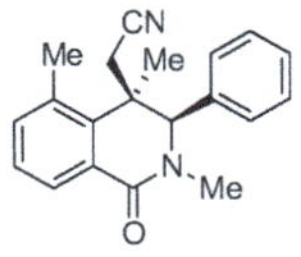

2-((3*R*,4*R*)-2,4,5-trimethyl-1-oxo-3-phenyl-1,2,3,4-tetrahydroisoquinolin-4-yl) acetonitrile (2.117b)[101]—The title compound was synthesized from **2.105b** (107 mg, 0.3 mmol) using **general procedure 2a**, and was isolated via silica gel flash column chromatography using hexanes:EtOAc (3:2 v:v) as the mobile phase (81 mg, 0.266 mmol, 89%, >95:5 dr (*cis:trans*)) An analytically pure sample of the major diastereomer was obtained by recrystallization in EtOAc/hexanes at −20 °C. **Major**: Isolated as a crystalline white solid (MP = 174–175 °C). **^{1}H NMR** (500 MHz, $CDCl_3$) δ 8.20 (dd, *J* = 7.5, 1.5 Hz, 1H), 7.37–7.32 (m, 1H), 7.31–7.26 (m, 2H), 7.24–7.18 (m, 2H), 7.12–7.06 (m, 2H), 4.33 (s, 1H), 3.37 (d, *J* = 16.5 Hz, 1H), 3.06 (s, 3H), 2.52 (d, *J* = 16.5 Hz, 1H), 2.43 (s, 3H), 1.76 (s, 3H). **^{13}C NMR** (125 MHz, $CDCl_3$) δ 163.3, 137.9, 137.4, 136.3, 133.7, 130.1, 129.0, 128.9, 128.7, 127.9, 127.8, 117.5, 72.6, 42.7, 34.3, 27.6, 27.3, 23.7. **IR** (cm^{-1}, film) 3063, 3007, 2974, 2930, 2874, 2247, 1636, 1589, 1454, 1427, 1402, 1329, 1271, 1072. **HRMS** (DART) Calc'd for $C_{19}H_{21}INO$, 406.06678; found, 406.06645. Enantiomeric purity was determined by HPLC analysis in comparison with racemic material t_r = 8.6 (minor) t_r = 7.8 (major) (>99:1 er shown; Chiralcel AD-H column, 20% ^{i}PrOH in hexanes 1.0 mL/min, 225 nm). $[\alpha]_D^{20}$ +356.5 (c = 0.58, $CHCl_3$).

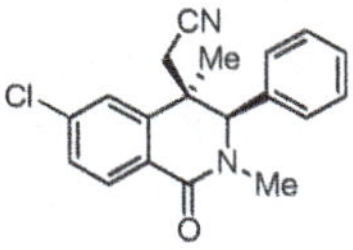

2-((3*R*,4*R*)-6-chloro-2,4-dimethyl-1-oxo-3-phenyl-1,2,3,4-tetrahydroisoquinolin-4-yl)acetonitrile (2.117c)[102]—The title compound was synthesized from **2.105** (113.6 mg, 0.3 mmol) using **general procedure 2a**, and was isolated via silica gel flash column chromatography using hexanes:EtOAc (3:2 v:v) as the mobile phase (88 mg, 0.27 mmol, 90%, 94:6 dr (*cis:trans*)). **Major**: Isolated as a white solid (MP = 178–180 °C). **^{1}H NMR** (500 MHz, $CDCl_3$) δ 8.22 (d, *J* = 8.5 Hz, 1H), 7.45 (dd, *J* = 8.5, 2.0 Hz, 1H), 7.33–7.19 (m, 3H), 7.07–6.98 (m, 3H), 4.44 (s, 1H), 3.07 (s, 3H), 2.87 (d, *J* = 16.5 Hz, 1H), 2.22 (d, *J* = 16.5 Hz, 1H), 1.72 (s, 3H). **^{13}C NMR** (125 MHz, $CDCl_3$) δ 162.1, 142.3, 138.8, 135.3, 130.6, 129.2, 129.0, 128.3, 127.8, 126.9, 123.1, 117.1, 71.4, 40.6, 34.4, 29.5, 25.7. **IR** (cm^{-1}, film) 3067, 3032, 2967, 2930, 2247, 1651, 1595, 1568, 1493, 1452, 1281, 1262, 1165. **HRMS** (DART) Calc'd for $C_{19}H_{18}ClN_2O$, 325.11077; found, 325.10950. Enantiomeric purity was determined by HPLC analysis in comparison with racemic material t_r = 14.3 (major) t_r = 15.7 (minor) (99:1 er shown; Chiralcel AD-H column, 20% ^{i}PrOH in hexanes 1.0 mL/min, 210 nm). $[\alpha]_D^{20}$ +335.2 (c = 0.59, $CHCl_3$).

---

[101]This compound was prepared by Hyung Yoon.

[102]This compound was prepared by Hyung Yoon.

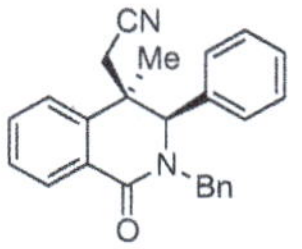

2-((3*R*,4*R*)-2-benzyl-4-methyl-1-oxo-3-phenyl-1,2,3,4-tetrahydroisoquinolin-4-yl) acetonitrile (2.117d)[103]—The title compound was synthesized from **2.105d** (126 mg, 0.3 mmol) using **general procedure 2b**, and was isolated via silica gel flash column chromatography using hexanes:EtOAc:DCM (20:1:20 v:v:v) as the mobile phase (94 mg, 0.26 mmol, 85%, 91:9 dr (*cis:trans*)). **Major**: Isolated as a white powder (MP = 185–186 °C). **^{1}H NMR** (500 MHz, $CDCl_3$) δ 8.41–8.29 (m, 1H), 7.54–7.46 (m, 2H), 7.40–7.25 (m, 6H), 7.24–7.17 (m, 2H), 7.08–7.03 (m, 1H), 7.03–6.95 (m, 2H), 5.67 (d, J = 14.5 Hz, 1H), 4.34 (s, 1H), 3.52 (d, J = 14.5 Hz, 1H), 2.75 (d, J = 16.5 Hz, 1H), 2.16 (d, J = 16.5 Hz, 1H), 1.34 (s, 3H). **^{13}C NMR** (125 MHz, $CDCl_3$) δ 162.8, 140.4, 136.2, 135.8, 132.8, 129.2, 129.2, 129.0, 128.9, 128.7, 128.4, 128.2, 128.1, 127.9, 122.7, 117.3, 67.2, 48.4, 40.0, 29.7, 25.9. **IR** (cm^{-1}, film) 3030, 2968, 2920, 2243, 1647, 1600, 1471, 1446, 1263. **HRMS** (DART) Calc'd for $C_{25}H_{23}N_2O$, 367.18104; found, 367.18043. Enantiomeric purity was determined by HPLC analysis in comparison with racemic material t_r = 17.3 (major) t_r = 20.8 (minor) (>98:2 er shown; Chiralpack AS column, 10% ^{i}PrOH in hexanes 1.0 mL/min, 210 nm). $[\alpha]_D^{20}$ +205.37 (c = 1.0, $CHCl_3$).

2-((3*R*,4*R*)-2-benzyl-6-chloro-4-methyl-1-oxo-3-phenyl-1,2,3,4-tetrahydroisoqui nolin-4-yl)acetonitrile (2.117e)[104]—The title compound was synthesized from **2.105e** (136 mg, 0.3 mmol) using **general procedure 2b**, and was isolated via silica gel flash column chromatography using hexanes:EtOAc (5:1 v:v) as the mobile phase (85 mg, 0.21 mmol, 70%, 91:9 dr (*cis:trans*)). **Major**: Isolated as a crystalline white solid (MP = 223–223 °C). **^{1}H NMR** (500 MHz, $CDCl_3$) δ 8.29 (d, J = 8.5 Hz, 1H), 7.47 (dd, J = 8.5, 2.0 Hz, 1H), 7.42–7.28 (m, 6H), 7.26–7.21 (m, 2H), 7.06–6.94 (m, 3H), 5.65 (d, J = 14.5 Hz, 1H), 4.35 (s, 1H), 3.52 (d, J = 14.5 Hz, 1H), 2.71 (d, J = 16.5 Hz, 1H), 2.15 (d, J = 16.5 Hz, 1H), 1.34 (s, 3H). **^{13}C NMR** (125 MHz, $CDCl_3$) δ 162.0, 142.3, 139.0, 136.0, 135.4, 130.9, 129.2, 129.2, 129.1, 128.8, 128.4, 128.1, 128.0, 127.0, 123.3, 116.8, 67.1, 48.4, 40.1, 29.5, 25.8. **IR** (cm^{-1}, film) 30885, 3065, 3028, 3011, 2968, 2930, 2249, 1636, 1593, 1568, 1456, 1418, 1262, 1155. **HRMS** (DART) Calc'd for $C_{25}H_{22}ClN_2O$, 401.14207; found, 401.14256. Enantiomeric purity was determined by HPLC analysis in comparison with racemic

[103]This compound was prepared by Hyung Yoon.

[104]This compound was prepared by Hyung Yoon.

material t_r = 20.4 (major) t_r = 23.6 (minor) (>99:1 er shown; Chiralcel AD-H column, 10% ^{i}PrOH in hexanes 1.0 mL/min, 210 nm). $[\alpha]_D^{20}$ +200.58 (c = 1.0, $CHCl_3$).

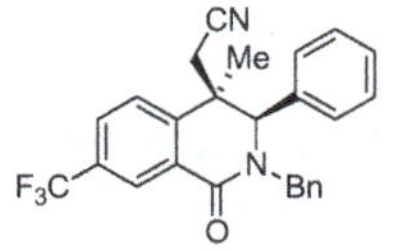

2-((3*R*,4*R*)-2-benzyl-4-methyl-1-oxo-3-phenyl-7-(trifluoromethyl)-1,2,3,4-tetrahydroisoquinolin-4-yl)acetonitrile (2.117f)[105]—The title compound was synthesized from **2.105f** (143 mg, 0.3 mmol) using **general procedure 2b**, and was isolated via silica gel flash column chromatography using hexanes:EtOAc:DCM (20:1:10 v:v:v) as the mobile phase (102 mg, 0.234 mmol, 78%, 81:11 dr (*cis:trans*)). **Major**: Isolated as a white powder (MP = 239–241 °C). **^{1}H NMR** (500 MHz, $CDCl_3$) δ 8.67–8.61 (m, 1H), 7.80–7.73 (m, 1H), 7.41–7.29 (m, 6H), 7.27–7.19 (m, 3H), 7.02–6.92 (m, 2H), 5.67 (d, *J* = 14.5 Hz, 1H), 4.37 (s, 1H), 3.55 (d, *J* = 14.5 Hz, 1H), 2.75 (d, *J* = 16.5 Hz, 1H), 2.22 (d, *J* = 16.5 Hz, 1H), 1.35 (s, 3H). **^{13}C NMR** (125 MHz, $CDCl_3$) δ 161.4, 144.2, 135.8, 135.2, 130.7 (q, *J* = *33.5* Hz), 129.4 (q, *J* = 3.5 Hz), 129.35 (overlapping quartet), 129.31, 129.2, 129.1, 128.8, 128.1, 128.1, 126.4 (q, *J* = 4.0), 124.7 (q, *J* = 272 Hz) 123.7, 116.8, 67.1, 48.5, 40.3, 29.6, 25.9. **^{19}F NMR** (565 MHz, $CDCl_3$) δ-62.81. **IR** (cm^{-1}, film) 3090, 3069, 3026, 2967, 2243, 1657, 1651, 1618, 1452, 1333, 1296, 1252, 1227, 1175, 1128. **HRMS** (DART) Calc'd for $C_{26}H_{22}F_3N_2O$, 435.16842; found, 435.16976. Enantiomeric purity was determined by HPLC analysis in comparison with racemic material t_r = 18.9 (major) t_r = 20.1 (minor) (99:1 er shown; Chiralcel AD-H column, 15% ^{i}PrOH in hexanes 0.5 mL/min, 210 nm). $[\alpha]_D^{20}$ +164.0 (c = 1.0, $CHCl_3$).

2-((3*R*,4*R*)-2-benzyl-3-(3-fluorophenyl)-4-methyl-1-oxo-1,2,3,4-tetrahydroisoquinolin-4-yl)acetonitrile (2.117g)—The title compound was synthesized from **2.105g** (109.3 mg, 0.25 mmol) using **general procedure 2b**, and was isolated via silica gel flash column chromatography using hexanes:EtOAc:DCM (10:1:10 v:v:v) as the mobile phase (90 mg, 0.233 mmol, 93%, 90:10 dr (*cis:trans*)). **Major**: Isolated as a white solid. **^{1}H NMR** (600 MHz, $CDCl_3$) δ 8.37–8.31 (m, 1H), 7.56–7.47 (m, 2H), 7.39–7.28 (m, 5H), 7.24–7.16 (m, 1H), 7.09–7.02 (m, 1H), 6.98 (tdd, *J* = 8.5, 2.5, 1.0 Hz, 1H), 6.90–6.56 (m, 2H), 5.65 (d, *J* = 14.5 Hz, 1H), 4.34 (s, 1H), 3.58 (d, *J* = 14.5 Hz, 1H), 2.78 (d, *J* = 16.5 Hz, 1H), 2.17 (d, *J* = 16.6 Hz, 1H), 1.34 (s, 3H). **^{13}C NMR** (150 MHz, $CDCl_3$) δ 162.6 (d, *J* = 248.0 Hz, 162.6, 140.1, 138.5 (d, *J* = 6.4 Hz), 135.9, 133.0, 130.5 (d, *J* = 8.2 Hz), 129.3, 129.1, 128.7, 128.3, 128.2, 128.0, 124.2, 122.7, 117.1, 116.2, 116.0, 115.0, 66.7, 48.5, 39.9, 29.6, 25.8.

[105]This compound was prepared by Hyung Yoon.

^{19}F NMR (470 MHz, $CDCl_3$) δ-110.9. **IR** (cm^{-1}, film) 3063, 3030, 2967, 2938, 2930, 2243, 1647, 1601, 1472, 1447, 1435, 1261, 1250. **HRMS** (DART): Calc'd for $C_{25}H_{22}FN_2O$, 385.17162; found, 385.17072. Enantiomeric purity was determined by HPLC analysis in comparison with racemic material t_r = 18.0 (major) t_r = 20.6 (minor) (99:1 er shown; Chiralcel AD-H column, 10% ^{i}PrOH in hexanes 1.0 mL/min, 210 nm). $[\alpha]_D^{20}$ +195.2 (c = 0.54, $CHCl_3$).

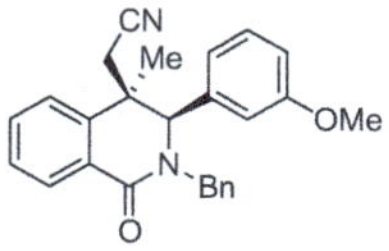

2-((3*R*,4*R*)-2-benzyl-3-(3-methoxyphenyl)-4-methyl-1-oxo-1,2,3,4-tetrahydroiso quinolin-4-yl)acetonitrile (2.117h)—The title compound was synthesized from **2.105h** (135 mg, 0.3 mmol) using **general procedure 2b**, and was isolated via silica gel flash column chromatography using hexanes:EtOAc:DCM (5:1:6 v:v:v) as the mobile phase (103 mg, 0.259 mmol, 86%, 90:10 dr (*cis:trans*)). **Major**: Isolated as a white solid. **^{1}H NMR** (600 MHz, $CDCl_3$) δ 8.39–8.28 (m, 1H), 7.55–7.43 (m, 2H), 7.42–7.28 (m, 5H), 7.12 (t, *J* = 8.0 Hz, 1H), 7.09–7.03 (m, 1H), 6.80 (dd, *J* = 8.0, 2.5 Hz, 1H), 6.69–6.38 (m, 2H), 5.66 (d, *J* = 14.5 Hz, 1H), 4.30 (s, 1H), 3.58 (s, 3H), 3.55 (d, *J* = 14.5 Hz, 1H), 2.75 (d, *J* = 16.5 Hz, 1H), 2.17 (d, *J* = 16.5 Hz, 1H), 1.33 (s, 3H). **^{13}C NMR** (150 MHz, $CDCl_3$) δ 162.7, 159.6, 140.5, 137.3, 136.2, 132.7, 129.8, 129.1, 129.0, 128.7, 128.4, 128.0, 127.9, 122.7, 120.5 (br.), 117.3, 114.6, 113.3 (br), 67.1, 54.9, 48.3, 39.9, 29.5, 25.8. **IR** (cm^{-1}, film) 2970, 2929, 1635, 1577, 1467, 1261. **HRMS** (DART): Calc'd for $C_{26}H_{25}N_2O_2$, 397.19160; found, 397.19105. Enantiomeric purity was determined by HPLC analysis in comparison with racemic material t_r = 10.7 (major) t_r = 12.6 (minor) (>99:1 er shown; Chiralcel AD-H column, 20% ^{i}PrOH in hexanes 1.0 mL/min, 210 nm). $[\alpha]_D^{20}$ +192.5 (c = 0.51, $CHCl_3$).

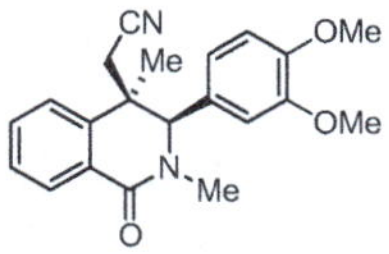

2-((3*R*,4*R*)-3-(3,4-dimethoxyphenyl)-2,4-dimethyl-1-oxo-1,2,3,4-tetrahydroisoq uinolin-4-yl)acetonitrile (2.117i)—The title compound was synthesized from **2.105i** (129 mg, 0.3 mmol) using **general procedure 2b**, and was isolated via silica gel flash column chromatography using hexanes:EtOAc (2:1 v:v) as the mobile phase (98 mg, 0.282 mmol, 94%, 94:6 dr (*cis:trans*)). **Major**: Isolated as a hygroscopic white foam. **^{1}H NMR** (600 MHz, $CDCl_3$) δ 8.26–8.23 (m, 1H), 7.50–7.46 (m, 1H), 7.46–7.43 (m, 1H), 7.03–7.00 (m, 1H), 6.70–6.66 (m, 1H), 6.64 (s (br), 1H), 6.37 (s (br), 1H), 4.36 (s, 1H), 3.79 (s, 3H), 3.50 (s (br), 3H), 3.10–3.07 (m, 3H), 2.88 (d, *J* = 16.5 Hz, 1H), 2.18 (d, *J* = 16.5 Hz, 1H), 1.69 (s, 3H). **^{13}C NMR** (150 MHz, $CDCl_3$) δ 162.9, 149.3, 148.8, 140.6, 132.6, 128.7, 128.6, 128.0, 128.0, 122.6, 117.8, 110.9, 71.1, 55.8, 55.4, 40.7, 34.4, 29.4, 25.9. **IR** (cm^{-1}, film)

3003, 2965, 2934, 2874, 2837, 2247, 1645, 1602, 1578, 1520, 1476, 1424, 1288, 1261, 1146, 1026. **HRMS** (DART) Calc'd for $C_{21}H_{23}N_2O_3$, 351.17087; found, 351.16955. Enantiomeric purity was determined by HPLC analysis in comparison with racemic material t_r = 8.0 (major) t_r = 9.3 (minor) (99:1 er shown; Chiralcel AD-H column, 25% ^{i}PrOH in hexanes 1.0 mL/min, 210 nm). $[\alpha]_D^{20}$ +291.8 (c = 0.56, $CHCl_3$).

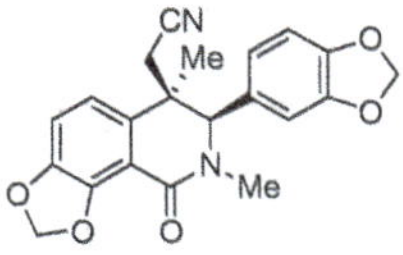

2-((6*R*,7*R*)-7-(benzo[*d*][1,3]dioxol-5-yl)-6,8-dimethyl-9-oxo-6,7,8,9-tetrahydro-[1,3]dioxolo[4,5-*h*]isoquinolin-6-yl)acetonitrile (1.222)—**0.3 mmol scale**: The title compound was synthesized from **2.105j** (113 mg, 0.3 mmol) using **general procedure 2b** using 2.5 mol% Pd(P^{t}Bu$_3$)$_2$, and was isolated via silica gel flash column chromatography using DCM:MeOH (80:1 v:v) as the mobile phase (89.8 mg, 0.237 mmol, 79%, >95:5 dr (*cis:trans*). **4.75 mmol scale**: The title compound was synthesized from **2.105j** (2.16 g, 4.75 mmol), Pd(P^{t}Bu$_3$)$_2$ (60.8 mg, 0.117 mmol, 2.5 mol% [Pd]), $Zn(CN)_2$ (307 mg, 2.61 mmol, 0.55 equiv) and DMF (23.8 mL, 0.2 M) using a modified version of **general procedure 2b**, and was isolated via silica gel flash column chromatography using DCM:MeOH (80:1 v:v) as the mobile phase (1.51 g, 3.76 mmol, 79%, >95:5 dr (*cis:trans*). **Major**: Isolated as a white powder (MP = 240–242 °C). Spectral data was in accordance with the reported literature [101]. Enantiomeric purity was determined by HPLC analysis in comparison with racemic material t_r = 13.9 (minor), t_r = 16.5 (major) (>99:1 er shown; Chiralcel AD-H column, 25% ^{i}PrOH in hexanes 1.0 mL/min, 210 nm).

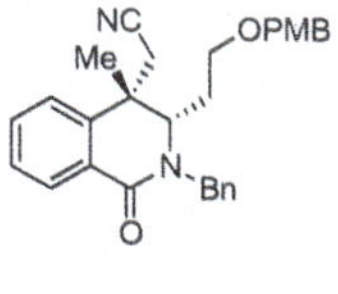

2-((3*S*,4*S*)-2-benzyl-3-(2-((4-methoxybenzyl)oxy)ethyl)-4-methyl-1-oxo-1,2,3,4-tetrahydroisoquinolin-4-yl)acetonitrile (2.117k)—The title compound was synthesized from **2.105k** (152.535 mg, 0.3 mmol) using **general procedure 2a**, and was isolated via silica gel flash column chromatography using hexanes:EtOAc (3:2 v:v) as the mobile phase (75.5 mg, 0.2 mmol, 66%, 53:47 dr (*cis:trans*). **Major**: **^{1}H NMR** (600 MHz, $CDCl_3$) δ 8.12 (dd, *J* = 8.0, 1.5 Hz, 1H), 7.50 (td, *J* = 8.0, 1.5 Hz, 1H), 7.41 (dd, *J* = 8.0, 1.0 Hz, 1H), 7.40–7.36 (m, 2H), 7.34–7.25 (m, 6H), 6.93–6.86 (m, 2H), 5.76 (d, *J* = 14.0 Hz, 1H), 4.48–4.40 (m, 2H), 3.80 (s, 3H), 3.77 (d, *J* = 14.0 Hz, 1H), 3.60 (dd, *J* = 9.5, 4.0 Hz, 1H), 3.52–3.47 (m, 1H), 3.43–3.40 (m, 1H), 2.10 (d, *J* = 17.0 Hz, 1H), 2.03–1.95 (m, 1H), 1.83 (d, *J* = 17.0 Hz, 1H), 1.55–1.45 (m, 4H). **^{13}C NMR** (150 MHz,$CDCl_3$) δ 162.7, 159.4, 140.0, 136.8, 132.5, 129.7, 129.6, 129.6, 128.8, 128.7, 128.1, 128.08, 128.06, 125.1, 116.7, 113.8, 72.9, 65.7, 59.4, 55.2, 49.1, 39.2, 31.0, 30.2, 20.7. **IR** (cm^{-1}, film)

3032, 2952, 2934, 2862, 2245, 1647, 1514, 1447, 1250, 1173, 1105, 1034. **HRMS** (DART): Calc'd for $C_{29}H_{31}N_2O_3$, 455.23347; found, 455.23473. Enantiomeric purity was determined by HPLC analysis in comparison with racemic material t_r = 12.7 (major) t_r = 17.0 (minor) (>99:1 er shown; Chiralcel OD-H column, 20% ^{i}PrOH in hexanes 1.0 mL/min, 210 nm). $[\alpha]_D^{20}$ −185.6 (c = 0.5, $CHCl_3$).

Minor: Isolated as a pale yellow semisolid. **^{1}H NMR** (600 MHz, $CDCl_3$) δ 8.14 (ddd, *J* = 8.0, 1.5, 0.5 Hz, 1H), 7.46 (td, *J* = 7.5, 1.5 Hz, 1H), 7.41–7.36 (m, 3H), 7.32–7.25 (m, 5H), 7.06 (dd, *J* = 8.0, 1.0 Hz, 1H), 6.91–6.86 (m, 2H), 5.71 (d, *J* = 14.0 Hz, 1H), 4.37 (q, *J* = 11.0 Hz, 2H), 3.85 (d, *J* = 14.0 Hz, 1H), 3.81 (s, 3H), 3.74 (dd, *J* = 9.5, 4.0 Hz, 1H), 3.51 (td, *J* = 9.5, 4.0 Hz, 1H), 3.42–3.36 (m, 1H), 2.88 (d, *J* = 16.5 Hz, 1H), 2.68 (d, *J* = 16.5 Hz, 1H), 1.98–1.90 (m, 1H), 1.60–1.50 (m, 4H). **^{13}C NMR** (150 MHz, $CDCl_3$) δ 162.8, 159.3, 141.5, 136.5, 132.5, 129.8, 129.7, 129.5, 129.1, 128.5, 128.1, 127.8, 127.6, 122.6, 117.0, 113.8, 72.6, 65.5, 59.4, 55.2, 49.5, 39.4, 30.8, 28.0, 25.5. **IR** (cm^{-1}, film) 3065, 2961, 2928, 2864, 2246, 1643, 1417, 1248, 1175, 1155, 1100, 1080, 1065, 1034. **HRMS** (DART): Calc'd for $C_{28}H_{31}N_2O_3$,455.2337; found, 455.23445. Enantiomeric ratio was not determined. $[\alpha]_D^{20}$ −145.62 (c = 0.585, $CHCl_3$).

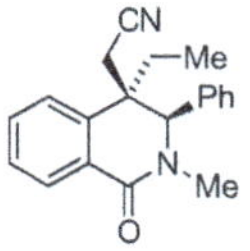

2-((3*R*,4*R*)-4-ethyl-2-methyl-1-oxo-3-phenyl-1,2,3,4-tetrahydroisoquinolin-4-yl) acetonitrile (2.117l)—The title compound was synthesized from **2.105l** (107 mg, 0.3 mmol) using **general procedure 2a**, and was isolated via silica gel flash column chromatography using hexanes:EtOAc (3:2 v:v) as the mobile phase (64 mg, 0.211 mmol, 70%, 95:5 dr (*cis:trans*)). **Major**: Isolated as a white solid (MP = 159–161 °C). **^{1}H NMR** (500 MHz, $CDCl_3$) δ 8.32–8.23 (m, 1H), 7.53–7.44 (m, 2H), 7.30–7.23 (m, 1H), 7.19 (m, 2H), 7.06–6.89 (m, 3H), 4.44 (s, 1H), 3.07 (s, 3H), 2.98 (d, *J* = 16.5 Hz, 1H), 2.29–2.16 (m, 1H), 2.15–2.03 (m, 2H), 0.75 (t, *J* = 7.5 Hz, 3H). **^{13}C NMR** (125 MHz, $CDCl_3$) δ 162.9, 138.0, 135.8, 131.8, 128.89 (2), 128.8, 128.6, 128.0, 128.0, 124.4, 117.3, 71.2, 43.9, 34.3, 32.6, 21.8, 8.3. **IR** (cm^{-1}, film) 3070, 3004, 2243, 1643, 1477, 1454, 1398, 1258. **HRMS** (ESI +) Calc'd for $C_{20}H_{21}N_2O$, 305.165139; found, 305.16608. Enantiomeric purity was determined by HPLC analysis in comparison with racemic material t_r = 6.5 (minor) t_r = 7.8 (major) (>99:1 er shown; Chiralcel AD-H column, 25% ^{i}PrOH in hexanes 1.0 mL/min, 210 nm). $[\alpha]_D^{20}$ +415.9 (c = 0.6, $CHCl_3$).

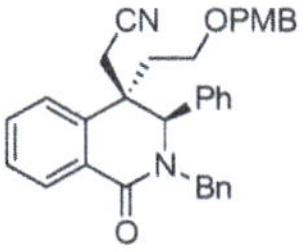

2-((3*R*,4*S*)-2-benzyl-4-(2-((4-methoxybenzyl)oxy)ethyl)-1-oxo-3-phenyl-1,2,3,4-tetrahydroisoquinolin-4-yl)acetonitrile (2.117m)—The title compound was synthesized from **2.105m** (171 mg, 0.3 mmol) using **general procedure 2b**, and was isolated via silica gel gradient flash column chromatography using hexanes: EtOAc:DCM (10:1:10 v:v:v to 10:3:10 v:v:v) as the mobile phase (119 mg, 0.231 mmol, 77%, 92:8 dr (*cis:trans*)). **Major**: Isolated as a white foam. **^{1}H NMR** (500 MHz, $CDCl_3$) δ 8.31 (dd, *J* = 8.0, 1.5 Hz, 1H), 7.48 (td, *J* = 8.0, 1.0 Hz, 1H), 7.44–7.29 (m, 6H), 7.29–7.24 (m, 1H), 7.23–7.12 (m, 4H), 7.08–6.91 (m, 2H), 6.89–6.80 (m, 3H), 5.59 (d, *J* = 14.5 Hz, 1H), 4.40 (s, 1H), 4.28 (d, *J* = 11.5 Hz, 1H), 4.12 (d, *J* = 11.5 Hz, 1H), 3.81 (s, 3H), 3.53 (d, *J* = 14.5 Hz, 1H), 3.42 (d, *J* = 16.0 Hz, 1H), 3.13 (dt, *J* = 10.5, 4.0 Hz, 1H), 2.82 (td, *J* = 10.0, 3.0 Hz, 1H), 2.01 (dd, *J* = 16.0, 2.0 Hz, 1H), 1.97–1.83 (m, 2H). **^{13}C NMR** (125 MHz, $CDCl_3$) δ 162.8, 159.2, 137.7, 136.2, 135.4, 132.1, 129.9, 129.3, 129.3, 129.1, 128.9, 128.8, 128.7, 128.6, 128.1, 127.9, 124.9, 118.0, 113.7, 72.6, 66.9, 65.7, 55.2, 48.5, 42.8, 37.5, 24.0. **IR** (cm^{-1}, film) 3726, 3628, 1647, 1467, 1247. **HRMS** (DART): Calc'd for $C_{34}H_{33}N_2O_3$, 517.24912; found, 517.24974. Enantiomeric purity was determined by HPLC analysis in comparison with racemic material t_r = 14.1 (major) t_r = 18 (minor) (>99:1 er shown; Chiralcel AD-H column, 20% ^{i}PrOH in hexanes 1.0 mL/min, 210 nm). $[\alpha]_D^{20}$ +233.2 (c = 0.5, $CHCl_3$).

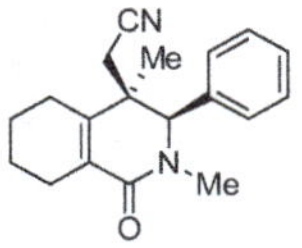

2-((3*R*,4*R*)-2,4-dimethyl-1-oxo-3-phenyl-1,2,3,4,5,6,7,8-octahydroisoquinolin-4-yl) acetonitrile (2.117n)—The title compound was synthesized from **2.105n** (142 mg, 0.3 mmol) and 1,2,2,6,6-pentamethylpiperidine (27.1 μl, 0.15 mmol) using **general procedure 2a**, and was isolated via silica gel flash column chromatography using hexanes:EtOAc (1:1 v:v) as the mobile phase (60 mg, 0.203 mmol, 68%, 93:7 dr (*cis:trans*)). **Major**: Isolated as a white foam. **^{1}H NMR** (500 MHz, $CDCl_3$) δ 7.35–7.26 (m, 5H), 4.19 (s, 1H), 2.90 (s, 3H), 2.58–2.48 (m, 1H), 2.45 (d, *J* = 16.5 Hz, 1H), 2.42–2.31 (m, 1H), 2.15–2.04 (m, 1H), 1.95 (d, *J* = 16.5, 1H), 1.92–1.81 (m, 1H), 1.81–1.71 (m, 2H), 1.61–1.48 (m, 5H). **^{13}C NMR** (125 MHz, $CDCl_3$) δ 164.0, 144.0, 136.1, 128.79, 128.76, 128.4, 128.3 (br.), 117.7, 70.6, 40.9, 33.8, 25.6, 24.9, 24.4, 23.9, 22.0, 21.6. **IR** (cm^{-1}, film) 2931, 2858, 2243, 1660, 1622, 1450. **HRMS** (DART): Calc'd for $C_{19}H_{23}N_2O$, 295.18104; found, 295.18159. Enantiomeric purity was determined by HPLC analysis in comparison with racemic material t_r = 11.7 (major) t_r = 12.8 (minor) (>99:1 er shown; Chiralcel AD-H column, 10% ^{i}PrOH in hexanes 1.0 mL/min, 210 nm). $[\alpha]_D^{20}$ +259.9 (c = 0.31, $CHCl_3$).

2-((5*R*,6*R*)-5,7-dimethyl-8-oxo-6-phenyl-5,6,7,8-tetrahydro-1,7-naphthyridin-5-yl) acetonitrile (2.117o)[106]—The title compound was synthesized from **2.105o** (104 mg, 0.3 mmol) using **general procedure 2a**, and was isolated via silica gel flash column chromatography using DCM:MeOH (20:1 v:v) as the mobile phase (61 mg, 0.21 mmol, 70%, 93:7 dr (*cis:trans*)). **Major**: Isolated as an off white crystalline solid. **^{1}H NMR**^{1}H NMR (500 MHz, Chloroform-*d*) δ 8.82 (dd, *J* = 4.5, 1.5 Hz, 1H), 7.48 (dd, *J* = 8.0, 1.5 Hz, 1H), 7.42 (dd, *J* = 8.0, 4.5 Hz, 1H), 7.30–7.26 (m, 1H), 7.23–7.19 (m, 2H), 7.02–6.97 (m, 2H), 4.46 (s, 1H), 3.12 (s, 3H), 2.82 (d, *J* = 16.5 Hz, 1H), 2.26 (dd, *J* = 16.5, 1.0 Hz, 1H), 1.76 (d, *J* = 1.0 Hz, 3H). **^{13}C NMR** (125 MHz, $CDCl_3$) δ 161.3, 149.5, 145.4, 136.5, 134.8, 131.6, 129.2, 128.9, 127.6, 126.4, 117.0, 71.0, 40.3, 34.7, 29.2, 25.5. **IR** (cm^{-1}, film) 3057, 3034, 2978, 2965, 2247, 1661, 1572, 1481, 1450, 1424, 1398, 1268, 1229. **HRMS** (DART): Calc'd for $C_{18}H_{18}N_3O$, 292.14499; found, 292.14457. Enantiomeric purity was determined by HPLC analysis in comparison with racemic material t_r = 81.9 (minor) $t_{r\ R}$ = 89.6 (major) (>99:1 er shown; Chiralcel AD-H column, 60% ^{i}PrOH in hexanes 0.125 mL/min, 210 nm). $[\alpha]_D^{20}$ +280.0 (c = 0.53, $CHCl_3$).

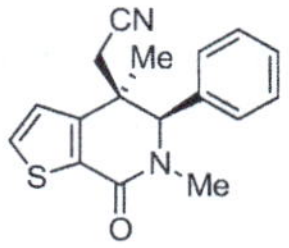

2-((4*R*,5*R*)-4,6-dimethyl-7-oxo-5-phenyl-4,5,6,7-tetrahydrothieno[2,3-*c*]pyridin-4-yl)acetonitrile (2.117p)—The title compound was synthesized from **2.105p** (104.7 mg, 0.3 mmol) using **general procedure 2a**, and was isolated via silica gel flash column chromatography using hexanes:EtOAc (3:2 v:v) as the mobile phase (77.2 mg, 0.261 mmol, 87%, 93:7 dr (*cis:trans*)). **Major**: Isolated as a white foam. **^{1}H NMR** (500 MHz, $CDCl_3$) δ 7.56 (d, *J* = 5.0 Hz, 1H), 7.34–7.23 (m, 3H), 7.12–7.06 (m, 2H), 6.92 (d, *J* = 5.0 Hz, 1H), 4.38 (s, 1H), 3.00 (s, 3H), 2.66 (d, *J* = 16.5 Hz, 1H), 2.23 (d, *J* = 16.5 Hz, 1H), 1.72 (s, 3H). **^{13}C NMR** (125 MHz, $CDCl_3$) δ 159.9, 145.8, 135.5, 132.2, 131.6, 129.1, 128.9, 127.9, 123.7, 117.2, 73.8, 40.1, 33.4, 28.5, 26.2. **IR** (cm^{-1}, film) 3090, 3066, 2960, 2245, 1635, 1537, 1452, 1398, 1321, 1305, 1047. **HRMS** (DART) Calc'd for $C_{17}H_{17}N_2OS$, 297.10616; found, 297.10741. Enantiomeric purity was determined by HPLC analysis in comparison with racemic material t_r = 7.1 (minor) t_r = 7.7 (major) (>99:1 er shown; Chiralcel AD-H column, 20% ^{i}PrOH in hexanes 1.0 mL/min, 210 nm). $[\alpha]_D^{20}$ +143.0 (c = 0.63, $CHCl_3$).

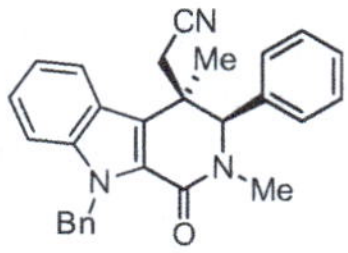

[106]This compound was prepared by Hyung Yoon.

2-((3*R*,4*R*)-9-benzyl-2,4-dimethyl-1-oxo-3-phenyl-2,3,4,9-tetrahydro-1*H*-pyrido[3,4-*b*]indol-4-yl)acetonitrile (2.117q)[107]—The title compound was synthesized from **2.105q** (142 mg, 0.3 mmol) using **general procedure 2b**, and was isolated via silica gel flash column chromatography using hexanes:EtOAc (3:2 v:v) as the mobile phase (107 mg, 0.255 mmol, 85%, 50:50 dr (*cis:trans*). **Major**: Isolated as a white foam. **^{1}H NMR** (500 MHz, $CDCl_3$) δ 7.55 (dt, *J* = 8.5, 1.0 Hz, 1H), 7.48 (dt, *J* = 8.5, 1.0 Hz, 1H), 7.34–7.20 (m, 5H), 7.19–7.08 (m, 7H), 6.28 (d, *J* = 16.0 Hz, 1H), 5.80 (d, *J* = 16.0 Hz, 1H), 4.43 (s, 1H), 3.36 (d, *J* = 16.5 Hz, 1H), 3.00 (s, 3H), 2.43 (d, *J* = 16.5 Hz, 1H), 1.82 (s, 3H). **^{13}C NMR** (125 MHz, $CDCl_3$) δ 159.7, 139.0, 138.2, 135.8, 128.9, 128.7, 128.5, 128.4, 127.3, 126.7, 125.2, 124.7, 122.7, 120.9, 120.3, 120.2, 117.6, 111.7, 73.8, 47.5, 40.0, 33.5, 28.2, 26.2. **IR** (cm^{-1}, film) 3088, 3032, 2963, 2870, 2247, 1647, 1533, 1485, 1452, 1433, 1397, 1344, 1285, 1242, 1148, 1125. **HRMS** (DART): Calc'd for $C_{28}H_{26}N_3O$, 420.20759; found, 420.20719. Enantiomeric purity was determined by HPLC analysis in comparison with racemic material t_r = 17.1 (major) t_r = 10.2 (minor) (>99:1 er shown; Chiralcel AD-H column, 20% ^{i}PrOH in hexanes 1.0 mL/min, 210 nm). $[\alpha]_D^{20}$ +175.9 (c = 0.62, $CHCl_3$).

Minor: Isolated as a white foam. **^{1}H NMR** (500 MHz, $CDCl_3$) δ 7.70 (dt, *J* = 8.5, 1.0 Hz, 1H), 7.45 (dt, *J* = 8.5, 1.0 Hz, 1H), 7.32–7.18 (m, 5H), 7.16–7.08 (m, 5H), 6.98–6.93 (m, 2H), 6.28 (d, *J* = 16.0 Hz, 1H), 5.77 (d, *J* = 16.0 Hz, 1H), 4.41 (s, 1H), 3.00 (s, 3H overlapping d, *J* = 17.0 Hz, 1H), 2.89 (d, *J* = 17.0 Hz, 1H), 1.60 (s, 3H). **^{13}C NMR** (125 MHz, $CDCl_3$) δ 159.9, 139.1, 138.1, 135.6, 128.5, 128.4, 128.4, 128.2, 127.2, 126.6, 125.0, 124.9, 123.6, 121.2, 120.9, 120.0, 117.3, 111.4, 73.2, 47.5, 39.3, 33.4, 31.4, 22.2. **IR** (cm^{-1}, film) 3063, 3030, 2990, 2976, 2247, 1647, 1609, 1553, 1485, 1452, 1433, 1395, 1383, 1344, 1300, 1285, 1244, 1159. **HRMS** (DART): Calc'd for $C_{28}H_{26}N_3O$, 420.20759; found, 420.20824. Enantiomeric purity was determined by HPLC analysis in comparison with racemic material t_r = 8.6 (minor) t_r = 12.4 (major) (>99:1 er shown; Chiralcel AD-H column, 20% ^{i}PrOH in hexanes 1.0 mL/min, 210 nm). $[\alpha]_D^{20}$ +106.5 (c = 0.58, $CHCl_3$).

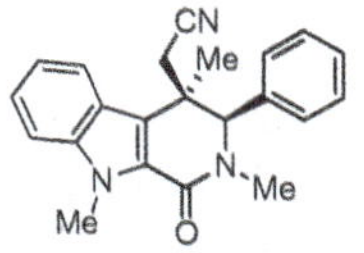

2-((3*R*,4*R*)-2,4,9-trimethyl-1-oxo-3-phenyl-2,3,4,9-tetrahydro-1*H*-pyrido[3,4-*b*]indol-4-yl)acetonitrile (2.117r)[108]—The title compound was synthesized from **2.105r** (119 mg, 0.3 mmol) using **general procedure 2a**, and was isolated via silica gel flash column chromatography using hexanes:EtOAc (2:1 v:v) as the mobile phase (79.3 mg, 0.23 mmol, 77%, 40:60 dr (*cis:trans*)). **Major**: Isolated as a white foam. **^{1}H NMR** (500 MHz, $CDCl_3$) δ 7.55 (dt, *J* = 8.5, 1.0 Hz, 1H), 7.47

[107]This compound was prepared by Hyung Yoon.

[108]This compound was prepared by Hyung Yoon.

(m, 1H), 7.39–7.35 (m, 1H), 7.33–7.27 (m, 1H), 7.26–7.22 (m, 4H), 7.14 (m, 1H), 4.45 (s, 1H), 4.25 (s, 3H), 3.36 (d, J = 16.5 Hz, 1H), 3.04 (s, 3H), 2.46 (dd, J = 16.5, 0.5 Hz, 1H), 1.80 (d, J = 1.0 Hz, 3H). **^{13}C NMR** (125 MHz, $CDCl_3$) δ 160.1, 139.3, 136.0, 129.0, 128.9, 128.5, 125.4, 124.5, 122.5, 120.7, 120.3, 119.5, 117.6, 111.0, 74.0, 39.9, 33.5, 31.5, 28.5, 26.3. **IR** (cm^{-1}, film) 2960, 2924, 2247, 1647, 1533, 1489, 1452, 1204. **HRMS** (ESI+) Calc'd for $C_{22}H_{22}N_3O$, 344.1757; found, 344.1755. Enantiomeric purity was determined by HPLC analysis in comparison with racemic material t_r = 7.4 (minor) t_r = 10.8 (major) (>99:1 er shown; Chiralcel AD-H column, 20% ^{i}PrOH in hexanes 1.0 mL/min, 210 nm). $[\alpha]_D^{20}$ +320.6 (c = 0.32, $CHCl_3$).

Minor: Isolated as a white foam. **^{1}H NMR** (500 MHz, $CDCl_3$) δ 7.71–7.67 (m, 1H), 7.47–7.43 (m, 1H), 7.39–7.33 (m, 1H), 7.29–7.18 (m, 3H), 7.16–7.10 (m, 1H), 7.10–7.05 (m, 2H), 4.43 (s, 1H), 4.25 (s, 3H), 3.04 (s, 3H), 2.98 (d, J = 17.0 Hz, 1H), 2.86 (d, J = 17.0 Hz, 1H), 1.59 (s, 3H). **^{13}C NMR** (125 MHz, $CDCl_3$) δ 160.2, 139.4, 136.0, 128.6 (2), 128.2, 125.3, 124.7, 123.4, 121.2, 120.7, 119.3, 117.4, 110.8, 73.4, 39.3, 33.5, 31.7, 31.5, 22.3. **IR** (cm^{-1}, film) 2956, 2924, 2241, 1647, 1533, 1489, 1467, 1344, 1242. **HRMS** (ESI+) Calc'd for $C_{22}H_{22}N_3O$, 344.1757; found, 344.1747. Enantiomeric purity was determined by HPLC analysis in comparison with racemic material t_r = 8.4 (major) t_r = 6.5 (minor) (>99:1 er shown; Chiralcel AD-H column, 20% ^{i}PrOH in hexanes 1.0 mL/min, 210 nm). $[\alpha]_D^{20}$ +246.0 (c = 0.33, $CHCl_3$).

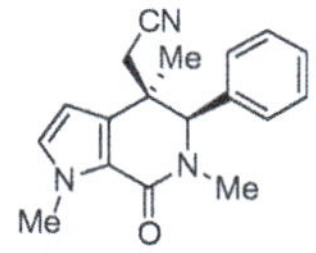

2-((4*R*,5*R*)-1,4,6-trimethyl-7-oxo-5-phenyl-4,5,6,7-tetrahydro-1H-pyrrolo[2,3-c]pyridin-4-yl)acetonitrile (2.117s)[109]—The title compound was synthesized from **2.015s** (142 mg, 0.3 mmol) using **general procedure 2a**, and was isolated via silica gel flash column chromatography using hexanes:EtOAc (1:1 v:v) as the mobile phase (81 mg, 0.276 mmol, 92%, 1:1 dr (*cis:trans*). **Major**: Isolated as an off-white semi-solid. **^{1}H NMR** (500 MHz, $CDCl_3$) δ 7.30–7.21 (m, 3H), 7.13–7.08 (m, 2H), 6.68 (d, J = 2.5 Hz, 1H), 5.90 (d, J = 2.5 Hz, 1H), 4.27 (s, 1H), 4.02 (s, 3H), 2.93 (s, 3H), 2.53 (d, J = 16.5 Hz, 1H), 2.11 (d, J = 16.5 Hz, 1H), 1.65 (s, 3H). **^{13}C NMR** (125 MHz, $CDCl_3$) δ 159.7, 136.5, 130.5, 128.76, 128.75, 128.09, 128.06, 119.8, 117.8, 102.7, 74.3, 38.2, 36.2, 33.1, 29.3, 26.2. **IR** (cm^{-1}, film) 3106, 3030, 2926, 2243, 1636, 1539, 1512, 1435, 1398, 1344, 1134, 1028. **HRMS** (DART): Calc'd for $C_{18}H_{20}N_3O$, 294.16064; found, 294.16148. Enantiomeric purity was determined by HPLC analysis in comparison with racemic material t_r = 5.9 (minor) t_r = 7.0 (major) (99:1 er shown; Chiralcel AD-H column, 20% ^{i}PrOH in hexanes 1.0 mL/min, 210 nm). $[\alpha]_D^{20}$ +193.8 (c = 0.52, $CHCl_3$).

[109]This compound was prepared by Hyung Yoon.

Minor: Isolated as a clear and colourless oil. **^{1}H NMR** (500 MHz, $CDCl_3$) δ 7.29–7.19 (m, 3H), 7.02–6.97 (m, 2H), 6.69 (d, J = 2.5 Hz, 1H), 5.93 (d, J = 2.5 Hz, 1H), 4.29 (s, 1H), 4.02 (s, 3H), 2.93 (s, 3H), 2.84 (d, J = 16.5 Hz, 1H), 2.70 (d, J = 16.5 Hz, 1H), 1.18 (s, 3H). **^{13}C NMR** (125 MHz, $CDCl_3$) δ 159.8, 136.5, 130.0, 128.5, 128.4, 128.3, 128.0, 119.7, 117.6, 104.1, 73.5, 38.0, 36.2, 33.1, 32.4, 21.5. **IR** (cm^{-1}, film) 2986, 2963, 2934, 2243, 1643, 1512, 1449, 1305, 1227, 1059. **HRMS** (DART): Calc'd for $C_{18}H_{20}N_3O$, 294.16064; found, 294.16105. Enantiomeric ratio was not determined. $[\alpha]_D^{20}$ +193.9 (c = 0.5, $CHCl_3$).

(3*R*,4*R*)-4-(cyanomethyl)-2,4-dimethyl-1-oxo-3-phenyl-1,2,3,4-tetrahydroisoquinoline-6-carbonitrile (2.119)—The title compound was synthesized from (*S*)-**2.105c** (113.6 mg, 0.3 mmol, 1 equiv) using general procedure 2 while employing 105 mol% of $Zn(CN)_2$, and was isolated via silica gel flash column chromatography using hexanes:EtOAc (1:1 v:v) as the mobile phase (63.6 mg, 0.20 mmol, 67%, 95:5 dr (*cis:trans*)) An analytically pure sample of the major diastereomer was obtained by recrystallization in EtOAc/hexanes at −20 °C. **Major**: Isolated as a crystalline white solid (MP = 153–154 °C). **^{1}H NMR** (500 MHz, $CDCl_3$) δ 8.37 (dd, J = 8.0, 0.5 Hz, 1H), 7.76 (dd, J = 8.0, 1.5 Hz, 1H), 7.37 (d, J = 1.5 Hz, 1H), 7.30–7.26 (m, 1H), 7.21 (ddd, J = 8.0, 7.5, 1.0 Hz, 2H), 6.95 (d, J = 7.0 Hz, 2H), 4.48 (s, 1H), 3.07 (s, 3H), 2.93 (d, J = 16.5 Hz, 1H), 2.24 (dd, J = 16.5, 1.0 Hz, 1H), 1.71 (d, J = 1.0 Hz, 3H). **^{13}C NMR** (125 MHz, $CDCl_3$) δ 161.2, 141.7, 134.7, 132.0, 131.8, 129.5, 129.3, 129.0, 127.6, 126.7, 117.7, 116.8, 116.0, 71.1, 40.6, 34.5, 29.4, 25.6. **IR** (cm^{-1}, film) 3067, 3034, 2969, 2932, 2247, 2232, 1651, 1609, 1456, 1427, 1399, 1331, 1289, 1261, 1082. **HRMS** (DART) Calc'd for $C_{20}H_{18}N_3O$, 316.14499; found, 316.14426. Enantiomeric purity was determined by HPLC analysis in comparison with racemic material t_r = 10.1 (minor) t_r = 9.2 (major) (99:1 er shown; Chiralcel AD-H column, 25% ^{i}PrOH in hexanes 1.0 mL/min, 210 nm). $[\alpha]_D^{20}$ +362.32 (c = 0.59, $CHCl_3$).

(*E*)-hexyl 3-((3*R*,4*R*)-4-(cyanomethyl)-2,4-dimethyl-1-oxo-3-phenyl-1,2,3,4-tetrahydroisoquinolin-6-yl)acrylate (2.220)—An oven-dried 2 dram vial was cooled under argon, charged with **2.117c** (34.7 mg, 0.107 mmol, 1 equiv), Pd(P^tBu_3)$_2$ (5.5 mg, 0.0107 mmol, 10 mol% [Pd]) and purged with argon for 10 min. The contents of the vial were taken up in distilled, dry and degassed PhMe (1.07 mL, 0.1 M), and both Cy_2NMe (68.8 μL, 0.321 mmol, 3 equiv) and hexyl acrylate (56.5 μL, 0.321 mmol, 3 equiv) were added in that order via microsyringe. The vial was fitted with a Teflon lined screw cap under a stream of argon passed through a large inverted glass funnel, sealed using Teflon tape, and placed in a pre-heated oil bath at 120 °C for 20 h. At this time TLC analysis indicated full conversion of starting material, and the reaction vial was cooled and the contents were passed through a short plug of silica eluting with 100% EtOAc. The crude material was purified via silica gel flash column chromatography using hexanes: DCM:EtOAc (6:3:1 v:v:v) as the mobile phase. The title compound was isolated as a white solid (43.2 mg, 0.0973 mmol, 91%, MP (°C) = 205–206). **^{1}H NMR** (600 MHz, $CDCl_3$) δ 8.29 (d, J = 8.0 Hz, 1H), 7.66–7.62 (m, 2H), 7.30–7.27 (m, 1H), 7.23–7.19 (m, 2H), 7.17 (d, J = 1.5 Hz, 1H), 7.01 (d, J = 7.5 Hz, 2H), 6.49

(d, J = 16.0 Hz, 1H), 4.45 (s, 1H), 4.20 (t, J = 6.5 Hz, 2H), 3.09 (s, 3H), 2.94 (d, J = 16.5 Hz, 1H), 2.26 (dd, J = 16.5, 1.0 Hz, 1H), 1.74 (s, 3H), 1.71–1.65 (m, 2H), 1.42–1.36 (m, 2H), 1.35–1.29 (m, 4H), 0.91–0.88 (m, 3H). **^{13}C NMR** (150 MHz, $CDCl_3$) δ 166.3, 162.3, 142.7, 141.2, 138.5, 135.5, 129.6, 129.6, 129.1, 129.0, 127.8, 127.4, 122.4, 120.8, 117.3, 71.5, 64.9, 40.6, 34.4, 31.4, 29.7, 28.6, 25.8, 25.6, 22.5, 14.0. **IR** (cm^{-1}, film) 2956, 2929, 2870, 2245, 1708, 1643, 1452, 1261, 1174. **HRMS** (ESI): Calc'd for $C_{28}H_{33}N_2O_3$, 445.2486; found, 445.2491. Enantiomeric purity was determined by HPLC analysis in comparison with racemic material t_r = 14.0 (minor) t_r = 24.2 (major) (99:1 er shown; Chiralcel AD-H column, 20% ^{i}PrOH in hexanes 0.75 mL/min, 210 nm). $[\alpha]_D^{20}$ +230.4 (c = 0.23, $CHCl_3$).

2-((3*R*,4*R*)-2,4-dimethyl-1-oxo-3-phenyl-1,2,3,4-tetrahydroisoquinolin-4-yl)acetamide (2.221)—A 10 mL round bottom flask was charged with **2.117a** (60.0 mg, 0.206 mmol) and a 1:1 solution of EtOH:8 M $KOH_{(aq)}$ (4 mL). The resulting solution was refluxed for 16 h at which time TLC analysis indicated full conversion of starting material. The reaction was cooled to room temperature and diluted with water, acidified to pH = using concentrated HCl, and extracted with DCM (3×). The combined organic layers were washed with brine, dried over Na_2SO_4, filtered and concentrated in vacuo. The crude primary amide was purified by silica gel flash column chromatography using EtOAc:hexanes (3:2 → 5:1 v:v) to afford the title compound as a white solid (52.7 mg, 0.171 mmol, 83% MP (°C) = 205–206). **^{1}H NMR** (500 MHz, CD_3OD) δ 8.13 (ddd, J = 7.5, 1.5, 0.5 Hz, 1H), 7.60–7.54 (m, 1H), 7.50–7.45 (m, 1H), 7.39–7.34 (m, 1H), 7.27–7.20 (m, 1H), 7.20–7.13 (m, 2H), 7.03–6.98 (m, 2H), 5.18 (s, 1H), 3.07 (dd, J = 1.0, 0.5 Hz, 3H), 2.99 (d, J = 16.0 Hz, 1H), 2.18 (dd, J = 16.0, 1.0 Hz, 1H), 1.68 (d, J = 1.0 Hz, 3H). **^{13}C NMR** (126 MHz, CD_3OD) δ 176.1, 166.0, 146.5, 139.0, 134.2, 129.8, 129.7, 129.7, 129.6, 129.3 128.6, 125.3, 71.4, 41.8, 40.2, 35.2, 29.5. **IR** (cm^{-1}, film) 3382, 3198, 2931, 1683, 1635, 1600, 1573, 1398, 1257. **HRMS** (ESI): Calc'd for $C_{19}H_{21}N_2O_2$, 309.15896; found, 309.15915. Enantiomeric purity was determined by HPLC analysis in comparison with racemic material t_r = 41.6 (minor) t_r = 45.8 (major) (97:3 er shown; Chiralcel AD-H column, 20% ^{i}PrOH in hexanes 0.15 mL/min, 210 nm). $[\alpha]_D^{20}$ +289.5 (c = 0.2, $CHCl_3$).

2-((3*R*,4*R*)-2,4-dimethyl-3-phenyl-1,2,3,4-tetrahydroisoquinolin-4-yl)ethanamine (2.222)—A flame dried 25 mL round bottom flask was cooled under argon before being charged with **2.117a** (70.0 mg, 0.241 mmol, 1 equiv), and freshly distilled THF (8 mL). The resulting solution was cooled to 0 °C before having $LiAlH_4$ added in one portion (182 mg, 4.82 mmol, 20 equiv). This suspension was stirred at this temperature for 30 min before being warmed to room temperature where it was stirred for an addition 30 min. The reaction was then fitted with an oven dried reflux condenser and the reaction was heated to reflux for 18 h at which time TLC analysis indicated clean and full conversion of starting material. The reaction was cooled to 0 °C and water (1 mL) was added drop wise. The reaction was warmed to room temperature and its volume was doubled with a saturated aqueous solution of Rochelle's salt and EtOAc (15 mL) was added. The resulting emulsion was stirred vigorously for 30 min at which time clean phase separation was evident. The layers were separated and the aqueous layer was extracted with EtOAc (2×). The combined organic layers were washed with brine, dried over Na_2SO_4, filtered, and concentrated in vacuo. The crude diamine was purified using silica gel flash column chromatography using EtOAc:MeOH:NH_4OH (25:25:1 v:v:v) as the mobile phase. The title compound was isolated as a clear and light yellow oil (56.3 mg, 0.201 mmol, 83%). **^{1}H NMR** (500 MHz, $CDCl_3$) δ 7.35–7.16 (m, 6H), 7.15–7.11 (m, 2H), 7.08–7.03 (m, 1H), 3.79 (d, J = 16.0 Hz, 1H), 3.64 (d, J = 16.0 Hz, 1H), 3.41 (s, 1H), 2.73–2.62 (m, 1H), 2.55–2.46 (m, 1H), 2.22 (s(br), 2H), 2.16 (s, 3H), 2.14–2.07 (m, 1H), 1.55–1.46 (m, 1H), 1.31 (s, 3H). **^{13}C NMR** (125 MHz, $CDCl_3$) δ 142.1, 136.0, 134.0, 130.4, 127.6, 127.2, 126.5, 126.1, 125.8, 125.7, 74.8, 56.3, 44.2, 41.7, 40.9, 37.9, 28.8. **IR** (cm^{-1}, film) 3362, 2953, 2775, 1581, 1491, 1452, 1375, 1251. **HRMS** (ESI): Calc'd for $C_{19}H_{25}N_2$, 281.20177; found, 281.20181. The enantiomeric purity of this compound was not determined by HPLC. $[\alpha]_D^{20}$ −54.0 (c = 0.2, $CHCl_3$).

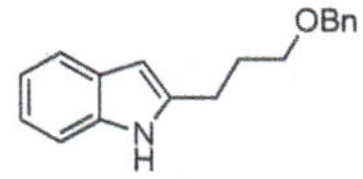

2-(3-(benzyloxy)propyl)-1*H*-indole (2.232a)—Was synthesized according to **general procedure 3** using allyl benzyl ether (688 mg, 5.46 mmol, 1.5 equiv), 9-BBN dimer (663 mg, 2.96 mmol, 0.825 equiv), **2.233** (1.00 g, 3.63 mmol, 1 equiv). The crude indole was obtained in analytically pure form by silica gel flash column chromatography using hexanes:DCM (1:1 v:v) as the mobile phase. The title compound was obtained as a clear, yellow oil (470 mg, 2.09 mmol, 58%). All spectra data were in accordance to the reported literature.[110] **^{1}H NMR** (400 MHz, $CDCl_3$) δ 8.24 (s(br), 1H), 7.54–7.48 (m, 1H), 7.43–7.31 (m, 5H), 7.18–7.14 (m, 1H), 7.13–7.00 (m, 2H), 6.22–6.19 (m, 1H), 4.55 (s, 2H), 3.60 (t, J = 6.0 Hz, 2H), 2.90 (t, J = 7.0 Hz, 2H), 2.06–1.98 (m, 2H). **^{13}C NMR** (100 MHz, $CDCl_3$) δ 139.3, 138.2, 135.9, 128.7, 128.5, 127.9, 127.8, 120.8, 119.7, 119.4, 110.4, 99.5, 73.2, 69.7, 29.2, 25.2.

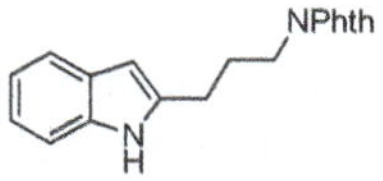

2-(3-(1*H*-indol-2-yl)propyl)isoindoline-1,3-dione (2.232b)—Was synthesized according to **general procedure 3** using 2-allylisoindoline-1,3-dione[111] (1.01 g, 5.40 mmol, 1.5 equiv), 9-BBN dimer (663 mg, 2.96 mmol, 0.825 equiv), and **2.233** (1.00 g, 3.63 mmol, 1 equiv). The crude indole was obtained in analytically pure form by silica gel flash column chromatography using hexanes:EtOAc (20:7 v:v) as the mobile phase. The title compound was obtained as a light yellow, crystalline

[110]Reference [113].

[111]Reference [114].

solid (540 mg, 1.77 mmol, 49%, MP = 150–150 °C). All spectra data were in accordance to the reported literature.[112] ^{1}**H NMR** (400 MHz, $CDCl_3$) δ 8.75 (s(br), 1H), 7.88–7.81 (m, 2H), 7.75–7.69 (m, 2H), 7.53–7.47 (m, 1H), 7.39–7.31 (m, 2H), 7.11 (ddd, *J* = 8.0, 7.0, 1.5 Hz, 1H), 7.04 (ddd, *J* = 8.0, 7.0, 1.0 Hz, 1H), 6.28–6.25 (m, 1H), 3.81 (t, *J* = 6.5 Hz 2H), 2.80 (t, *J* = 7.0 Hz, 2H), 2.14–2.06 (m, 2H). ^{13}**C NMR** (100 MHz, $CDCl_3$) δ 168.9, 138.4, 135.9, 134.1, 131.9, 128.7, 123.3, 121.0, 119.7, 119.5, 110.5, 99.8, 37.3, 28.8, 25.1.

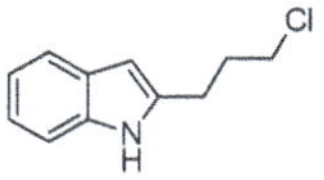

2-(3-chloropropyl)-1*H*-indole (2.232c)—Was synthesized according to **general procedure 3** using allyl chloride (442 μL, 5.40 mmol, 1.5 equiv), 9-BBN dimer (663 mg, 2.96 mmol, 0.825 equiv), and **2.233** (1.00 g, 3.63 mmol, 1 equiv). The crude indole was obtained in analytically pure form by silica gel flash column chromatography using hexanes:EtOAc (10:1 v:v) as the mobile phase. The title compound was obtained as a light brown, waxy solid (156 mg, 0.810 mmol, 22%). All spectra data were in accordance to the reported literature.[113] ^{1}**H NMR** (200 MHz, $CDCl_3$) δ 7.94 (s(br), 1H), 7.61–7.48 (m, 1H), 7.35–7.28 (m, 2H), 7.21–7.02 (m, 2H), 6.28 (dd, *J* = 2.0, 1.0 Hz, 1H), 3.60 (t, *J* = 6.5 Hz, 2H), 2.96 (t, *J* = 7. Hz, 2H), 2.25–2.10 (m, 2H).

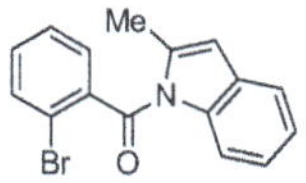

(2-bromophenyl)(2-methyl-1*H*-indol-1-yl)methanone (2.229a)—Was synthesized according to **general procedure 4** using 2-methylindole (2.44 g, 18.7 mmol, 1 equiv) and 2-bromobenzoyl chloride (7.5 g, 37.3 mmol, 2 equiv). The crude product was purified via silica gel flash column chromatography using hexanes:EtOAc (20:1 v:v) as the mobile phase, and was isolated as a clear and colorless oil (4.01 g, 12.7 mmol, 65%) which slowly solidified to a white solid upon storage at −20 °C (MP = 41–42 ° C). *Note: Prolonged storage of the title compound as an oil at room temperature lead to the product decomposing to a light yellow/pale red oil, which slowly solidifies to a pale red solid upon prolonged storage at −20 °C. This now impure compound leads to erratic results in the Pd-catalyzed title reaction, and it is therefore best to begin working with highly pure compounds in all cases.* Characterization was in accordance with the reported literature [92]. ^{1}**H NMR** (400 MHz, $CDCl_3$) δ 7.70–7.64 (m, 1H), 7.52–7.39 (m, 4H), 7.36 (d, *J* = 8.5 Hz, 1H), 7.20 (td, *J* = 7.5, 1.0 Hz, 1H), 7.11 (ddd, *J* = 8.5, 7.5, 1.5 Hz, 1H), 6.43–6.40 (m, 1H), 2.25 (d, *J* = 1.0 Hz, 3H). ^{13}**C NMR** (100 MHz, $CDCl_3$) δ 167.6, 138.4, 137.2, 136.8, 133.5, 132.0, 129.9, 129.3, 127.9, 123.7, 123.6, 120.2, 119.7, 114.9, 110.4, 16.3.

---

[112] Reference [115].

[113] Reference [116].

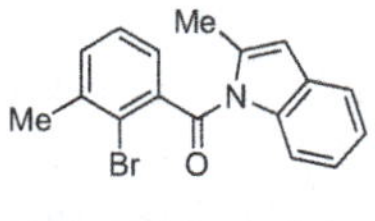

(2-bromo-3-methylphenyl)(2-methyl-1*H*-indol-1-yl)methanone (2.229b)[114]—Was synthesized according to **general procedure 3** using 2-methylindole (1.00 g, 7.63 mmol, 1 equiv) and 2-bromo-3-methylbenzoyl chloride generated from 2-bromo-3-methylbenzoic acid (3.61 g, 16.8 mmol, 2.2 equiv). The crude product was purified via silica gel flash column chromatography using hexanes:EtOAc (20:1 v:v) as the mobile phase, and was isolated as an off-white solid which was further purified by trituration using hexanes to yield a white solid (1.57 g, 4.80 mmol, 63%, MP = 106–107 °C). Characterization was in accordance with the reported literature [92]. **^{1}H NMR** (500 MHz, $CDCl_3$) δ 7.46–7.32 (m, 4H), 7.29–7.25 (m, 1H), 7.22–7.17 (m, 1H), 7.14–7.08 (m, 1H), 6.41–6.37 (m, 1H), 2.49–2.46 (m, 3H), 2.23 (s, 3H). **^{13}C NMR** (126 MHz, $CDCl_3$) δ 168.0, 139.7, 139.1, 137.3, 136.8, 132.6, 129.9, 127.7, 126.5, 123.7, 123.6, 122.2, 119.7, 115.1, 110.3, 23.2, 16.4.

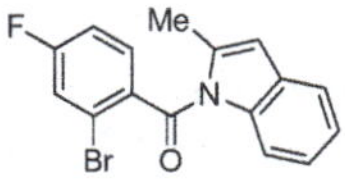

(2-bromo-4-fluorophenyl)(2-methyl-1*H*-indol-1-yl)methanone (2.229c)[115]—Was synthesized according to **general procedure 5** using 2-methylindole (1.00 g, 7.63 mmol, 1 equiv) and 2-bromo-4-fluorobenzoyl chloride generated from 2-bromo-4-fluorobenzoic acid (3.68 g, 16.8 mmol, 2.2 equiv). The crude product was purified via silica gel flash column chromatography using hexanes:EtOAc (20:1 v:v) as the mobile phase, and was isolated as a clear and light yellow oil (1.62 g, 4.88 mmol, 64%). Characterization was in accordance with the reported literature [92]. **^{1}H NMR** (500 MHz, $CDCl_3$) δ 7.49 (ddd, J = 8.5, 6.0, 0.5 Hz, 1H), 7.46 (ddd, J = 7.5, 1.5, 0.5 Hz, 1H), 7.43 (ddd, J = 8.0, 2.5, 0.5 Hz, 1H), 7.34–7.32 (m, 1H), 7.23–7.17 (m, 2H), 7.13 (ddd, J = 8.5, 7.5, 1.0 Hz, 1H), 6.42 (dt, J = 2.0, 1.0 Hz, 1H), 2.27 (d, J = 1.0 Hz, 3H). **^{13}C NMR** (125 MHz, $CDCl_3$) δ 166.8, 163.4 (d, J = 256.5 Hz), 137.1, 136.7, 134.6 (d, J = 3.5 Hz), 131.0 (d, J = 9.0 Hz), 129.9, 12367 (d, J = 10.5 Hz), 121.2 (d, J = 9.5 Hz), 121.1, 120.9, 119.8, 115.4 (d, J = 22.0 Hz), 114.7, 110.5, 16.3. **^{19}F NMR** (282 MHz, $CDCl_3$) δ-106.1 to -106.2 (m).

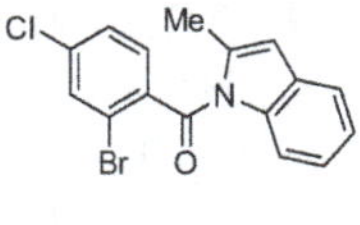

[114]This compound was synthesized by Andy Yen.

[115]This compound was synthesized by Andy Yen.

(2-bromo-4-chlorophenyl)(2-methyl-1*H*-indol-1-yl)methanone (2.229d)[116]—Was synthesized according to **general procedure 5** using 2-methylindole (1.00 g, 7.63 mmol, 1 equiv) and 2-bromo-4-chlorobenzoyl chloride generated from 2-bromo-4-chlorobenzoic acid (2.69 g, 11.4 mmol, 1.5 equiv). The crude product was purified via two rounds of silica gel flash column chromatography using hexanes:EtOAc (10:1 v:v) followed by hexanes:DCM (5:3) as the mobile phases, and was isolated as a clear and light yellow oil (946 mg, 2.92 mmol, 37%) which solidified upon storage at −20 °C to a light yellow solid (MP = 57–58 °C). **^{1}H NMR** (600 MHz, $CDCl_3$) δ 7.70 (dd, J = 2.0, 0.5 Hz, 1H), 7.47–7.45 (m, 2H), 7.37 (d, J = 8.5 Hz, 1H), 7.23–7.20 (m, 1H), 7.14 (ddd, J = 8.5, 7.0, 1.5 Hz, 1H), 6.43–6.42 (m, 1H), 2.27 (d, J = 1.0 Hz, 3H). **^{13}C NMR** (150 MHz, $CDCl_3$) δ 166.7, 137.4, 137.0, 136.8, 136.6, 133.3, 130.2, 129.9, 128.3, 123.8, 123.7, 120.9, 119.9, 114.7, 110.6, 16.3. **IR** (thin film, cm^{-1}) 3356, 3082, 3028, 2966, 2926, 1700, 1662. **HRMS** (DART, M+H) Calc'd for $C_{16}H_{12}BrClNO$ 347.97908, found 347.97943.

(2-bromo-5-(trifluoromethyl)phenyl)(2-methyl-1*H*-indol-1-yl)methanone (2.229e)[117]—Was synthesized according to general procedure **5** using 2-methylindole (550 mg, 3.81 mmol, 1 equiv) and 2-bromo-5-trifluoromethylbenzoyl chloride generated from 2-bromo-5-trifluoromethylbenzoic acid (2.25 g, 8.39 mmol, 2.2 equiv). The crude product was purified via two rounds of silica gel flash column chromatography first using hexanes:EtOAc (20:1 v:v) then 100% hexanes → hexanes:DCM (1:1 v:v) as the mobile phases, and was isolated as a clear and light yellow oil (1.02 g, 2.65 mmol, 70%). **^{1}H NMR** (500 MHz, $CDCl_3$) δ 7.82–7.77 (m, 2H), 7.68–7.65 (m, 1H), 7.50–7.46 (m, 2H), 7.25 (ddd, J = 7.5, 1.0 Hz, 1H), 7.17 (ddd, J = 8.5, 7.5, 1.5 Hz, 1H), 6.45 (qd, J = 1.0, 0.5 Hz, 1H), 2.22 (d, J = 1.0 Hz, 3H). **^{13}C NMR** (125 MHz, $CDCl_3$) δ 166.1, 139.2, 136.6, 136.6, 134.1, 130.6 (q, J = 34.0 Hz), 129.9, 128.4 (q, J = 3.5 Hz), 126.1 (q, J = 4.0 Hz), 124.2 (q, J = 1.5 Hz), 124.0, 124.0, 123.2 (q, J = 274 Hz), 119.9, 114.9, 111.1, 16.4. **^{19}F NMR** (376 MHz, $CDCl_3$) δ-62.8. **IR** (thin film, cm^{-1}) 3070, 3055, 2928, 1699, 1599. **HRMS** (DART, M+H) Calc'd for $C_{17}H_{12}BrF_3NO$ 382.00544, found 382.00525.

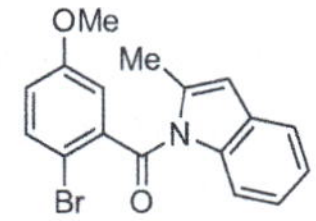

[116]This compound was synthesized by Andy Yen.

[117]This compound was synthesized by Andy Yen.

(2-bromo-5-methoxyphenyl)(2-methyl-1*H*-indol-1-yl)methanone (2.229f)[118]—Was synthesized according to general procedure 5 using 2-methylindole (500 mg, 3.82 mmol, 1 equiv) and 2-bromo-5-methoxybenzyl chloride generated from 2-bromo-5-menthoxybenzoic acid (1.93 g, 8.39 mmol, 2.2 equiv). The crude product was purified via two rounds of silica gel flash column chromatography first using hexanes:EtOAc (0 → 5%) then hexanes:PhMe:DCM (2:2:1 v:v:v) as the mobile phases, and was isolated as a translucent oil (866 mg, 2.52 mmol, 66%) which solidified upon storage to a colorless solid (MP = 75–76 °C). Characterization was in accordance with the reported literature [92]. **^{1}H NMR** (500 MHz, $CDCl_3$) δ 7.53 (d, J = 9.0 Hz, 1H), 7.48–7.45 (m, 1H), 7.43 (d, J = 8.5 Hz, 1H), 7.22 (td, J = 7.5, 1.0 Hz, 1H), 7.14 (ddd, J = 8.5, 7.5, 1.5 Hz, 1H), 7.03 (d, J = 3.0 Hz, 1H), 6.96 (dd, J = 9.0, 3.0 Hz, 1H), 6.43–6.42 (m, 1H), 3.82 (s, 3H), 2.30 (d, J = 1.0 Hz, 3H). **^{13}C NMR** (125 MHz, $CDCl_3$) δ 167.3, 159.2, 138.9, 137.2, 136.7, 134.2, 129.9, 123.7, 123.6, 119.7, 118.3, 114.9, 114.2, 110.4, 110.3, 55.7, 16.2.

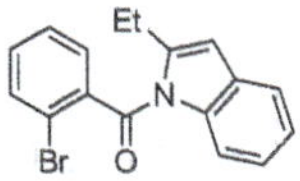

(2-bromophenyl)(2-ethyl-1*H*-indol-1-yl)methanone (2.229g)—Was synthesized according to **general procedure 4** using 2-ethylindole (333 mg, 2.32 mmol, 1 equiv) and 2-bromobenzoyl chloride (604 μL, 4.64 mmol, 2 equiv). The crude product was purified via silica gel flash column chromatography using hexanes: EtOAc (20:1 v:v) as the mobile phase, and was isolated as a clear, light yellow oil (4.01 g, 12.7 mmol, 65%) which slowly solidified to a colorless solid upon storage at −20 °C MP = 47–49 °C. **^{1}H NMR** (400 MHz, $CDCl_3$) δ 7.69–7.65 (m, 1H), 7.51–7.39 (m, 4H), 7.20–7.15 (m, 1H), 7.03 (ddd, J = 8.5, 7.0, 1.5 Hz, 1H), 6.96–6.92 (m, 1H), 6.49–6.48 (m, 1H), 2.80–2.71 (m, 2H), 1.29 (t, J = 7.5 Hz, 3H). **^{13}C NMR** (100 MHz, $CDCl_3$) δ 167.6, 144.2, 138.3, 136.8, 133.6, 132.1, 130.0, 129.5, 127.9, 123.5, 123.3, 120.3, 120.0, 114.3, 108.1, 22.9, 13.0. **IR** (thin film, cm^{-1}) 2972, 2933, 1685, 1456, 1431, 1371, 1329. **HRMS** (DART, M+H) Calc'd for $C_{17}H_{15}BrNO$ 328.03370, found 328.03311.

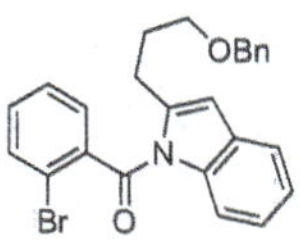

(2-(3-(benzyloxy)propyl)-1*H*-indol-1-yl)(2-bromophenyl)methanone (2.229h)—Was synthesized according to **general procedure 5** using 2-(3-(benzyloxy)propyl)-1*H*-indole (**2.232a**) (460 mg, 1.87 mmol, 1 equiv) and 2-bromobenzoyl chloride (369 μL, 3.76 mmol, 2 equiv). The crude product was purified via silica gel flash column chromatography using hexanes:EtOAc(10:1 v:v) as the mobile phase, and

[118]This compound was synthesized by Andy Yen.

was isolated a clear and light yellow oil (467 mg, 1.05 mmol, 56%). **^{1}H NMR** (500 MHz, $CDCl_3$) δ 7.66 (ddd, *J* = 7.5, 1.5, 0.5 Hz, 1H), 7.50–7.39 (m, 4H), 7.38–7.28 (m, 4H), 7.18 (ddd, *J* = 7.5, 7.0, 1.0 Hz, 1H), 7.02 (ddd, *J* = 8.5, 7.0, 1.5 Hz, 1H), 6.81–6.79 (m, 1H), 6.48 (q(ap), *J* = 1.0 Hz, 1H), 4.51–4.50 (m, 2H), 3.53 (t, *J* = 6.5 Hz, 2H), 3.01–2.88 (m(br), 3H), 2.11–1.96 (m, 2H). **^{13}C NMR** (125 MHz, $CDCl_3$) δ 167.6, 142.0, 138.5, 138.2, 136.7, 133.7, 132.2, 130.0, 129.6, 128.4, 128.0, 127.7, 127.6, 123.5, 123.3, 120.3, 120.1, 114.2, 109.3, 73.0, 69.5, 28.9, 26.4. **IR** (thin film, cm^{-1}) 3084, 3063, 2854, 1693, 1589, 1568, 1456, 1367, 1325, 1220, 1203, 1107. **HRMS** (DART, M+H) Calc'd for $C_{25}H_{22}BrNO_2$ 448.09122, found 448.09123.

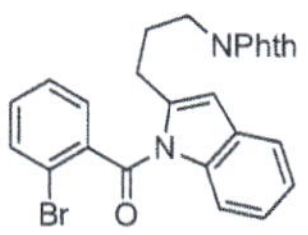

2-(3-(1-(2-bromobenzoyl)-1*H*-indol-2-yl)propyl)isoindoline-1,3-dione (2.229i)—Was synthesized according to **general procedure 5** using 2-(3-(1*H*-indol-2-yl)propyl)isoindoline-1,3-dione (**2.232b**) (490 mg, 1.61 mmol, 1 equiv) and 2-bromoben zoyl chloride (420 μL, 3.21 mmol, 2 equiv). The crude product was purified via silica gel flash column chromatography using hexanes:EtOAc:DCM (10:0.75:6 v:v:v) as the mobile phase, and was isolated as an off-white foam (494 mg, 1.01 mmol, 63%). **^{1}H NMR** (600 MHz, $CDCl_3$) δ 7.85–7.81 (m, 2H), 7.72–7.69 (m, 2H), 7.63 (ddd, *J* = 8.0, 1.0, 0.5 Hz, 1H), 7.51 (ddd, *J* = 7.5, 2.0, 0.5 Hz, 1H), 7.47–7.43 (m, 2H), 7.38 (ddd, *J* = 8.0, 7.5, 2.0 Hz, 1H), 7.15 (ddd, *J* = 7.5, 7.0, 1.0 Hz, 1H), 7.01 (ddd, *J* = 8.5, 7.0, 1.5 Hz, 1H), 6.84 (d, *J* = 8.5 Hz, 1H), 6.57 (q (ap), *J* = 1.0 Hz, 1H), 3.74 (t, *J* = 7.5 Hz, 2H), 2.84 (s(br), 2H), 2.11 (p(ap), *J* = 7.5 Hz, 1H). **^{13}C NMR** (150 MHz, $CDCl_3$) δ 168.2, 167.5, 140.8, 138.1, 136.7, 133.9, 133.6, 132.1, 132.0, 129.8, 129.5, 127.9, 123.6, 123.3, 123.1, 120.2, 120.1, 114.2, 109.5, 37.5, 27.4, 26.9. **IR** (thin film, cm^{-1}) 3063, 2939, 1705, 1685, 1485, 1396, 1327, 1165. **HRMS** (DART, M+H) Calc'd for $C_{26}H_{20}BrNO_3$ 487.06573, found 487.06521.

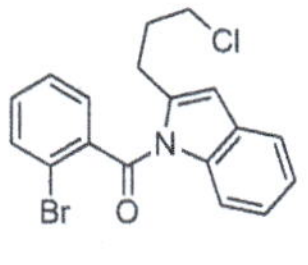

(2-bromophenyl)(2-(3-chloropropyl)-1*H*-indol-1-yl)methanone (2.229j)—Was synthesized according to **general procedure 5** using 2-(3-chloropropyl)-1*H*-indole (**2.232c**) (156 mg, 0.810 mmol, 1 equiv) and 2-bromobenzoyl chloride (212 μL, 1.62 mmol, 2 equiv). The crude product was purified via two rounds of silica gel flash column chromatography using hexanes:EtOAc (10:1 v:v) followed by hexanes:DCM (3:2 v:v) as the mobile phases, and was isolated as a clear and colorless oil (212 mg, 0.564 mmol, 70%) which solidifies slowly upon storage at −20 °C to colorless crystals (MP = 51–53 °C). **^{1}H NMR** (500 MHz, $CDCL_3$) δ 7.68 (ddd, *J* = 8.0, 1.0, 0.5 Hz, 1H), 7.53–7.43 (m, 4H), 7.17 (ddd, *J* = 7.5, 7.0, 1.5 Hz, 1H), 6.99 (ddd, *J* = 8.5, 7.0, 1.5 Hz, 1H), 6.63–6.60 (m, 1H), 6.55 (q(ap), *J* = 1.0 Hz, 1H), 3.59 (t, *J* = 6.5 Hz, 2H), 3.08 (s(br), 3H), 2.22 (p, *J* = 7.0 Hz, 2H). **^{13}C NMR** (125 MHz,

$CDCl_3$) δ 167.5, 140.7, 138.0, 136.6, 133.7, 132.3, 129.8, 129.6, 128.1, 123.6, 123.4, 120.2, 120.2, 113.9, 110.0, 44.3, 31.5, 26.9. **IR** (thin film, cm^{-1}) 3055, 2958, 2924, 1685, 1589, 1570, 1454, 1327, 1219, 1165, 1107. **HRMS** (DART, M+H) Calc'd for $C_{18}H_{16}BrClNO$ 376.01037, found 376.01019.

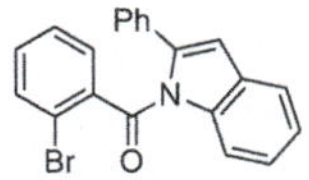

(2-bromophenyl)(2-phenyl-1*H*-indol-1-yl)methanone (2.229k)—Was synthesized according to **general procedure 4** using 2-phenylindole (300 mg, 1.55 mmol, 1 equiv) and 2-bromobenzoyl chloride (403 μL, 3.11 mmol, 2 equiv). The crude product was purified via silica gel flash column chromatography using hexanes: EtOAc (20:1 v:v) as the mobile phase, and was isolated as a light yellow, tacky oil (381 mg, 1.02 mmol, 66%). Characterization was in accordance with the reported literature [92]. **^{1}H NMR** (400 MHz, $CDCl_3$) δ 8.14–8.10 (m, 1H), 7.64–7.58 (m, 1H), 7.39–7.26 (m, 5H), 7.21–7.00 (m, 6H), 6.66 (d, J = 1.0 Hz, 1H). **^{13}C NMR** (100 MHz, $CDCl_3$) δ 168.1, 140.5, 137.8, 137.4, 133.1, 132.8, 131.7, 130.9, 129.5, 128.8, 127.7, 127.7, 127.7, 126.8, 125.0, 124.0, 121.3, 120.6, 115.3, 111.5.

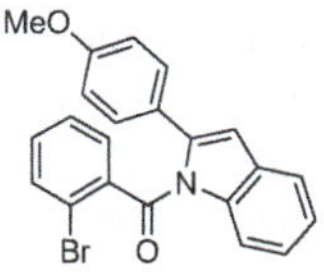

(2-bromophenyl)(2-(4-methoxyphenyl)-1*H*-indol-1-yl)methanone (2.229l)[119]—Was synthesized according to **general procedure 5** using 2-(4-methoxyphenyl) indole (400 mg, 1.79 mmol, 1 equiv) and 2-bromobenzoyl chloride (467 μL, 3.58 mmol, 2 equiv). The crude product was purified via silica gel flash column chromatography using hexanes:EtOAc (20:1) as the mobile phase, and was isolated as an off-white solid (186 mg, 0.458 mmol, 26%, MP = 134–135 °C). Characterization was in accordance with the reported literature [92]. **^{1}H NMR** (500 MHz, $CDCl_3$) δ 8.12–8.09 (m, 1H), 7.60–7.57 (m, 1H), 7.36–7.29 (m, 3H), 7.22–7.16 (m, 3H), 7.10 (td, J = 7.5, 1.5 Hz, 1H), 7.05 (ddd, J = 7.5, 2.0 Hz, 1H), 6.68–6.67 (m, 1H), 6.66–6.65 (m, 1H), 6.60 (d, J = 1.0 Hz, 1H), 3.74 (s, 3H). **^{13}C NMR** (125 MHz, $CDCl_3$) δ 168.3, 159.1, 140.3, 137.6, 137.5, 133.1, 131.7, 130.8, 130.2, 129.6, 126.8, 125.3, 124.8, 123.9, 121.2, 120.4, 115.2, 113.3, 110.9, 55.3.

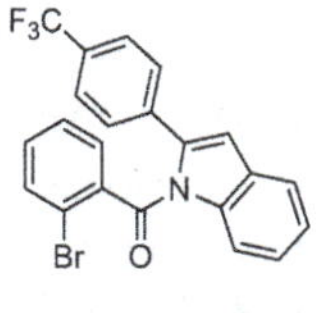

[119]This compound was synthesized by Nicolas Zeidan.

(2-bromophenyl)(2-(4-(trifluoromethyl)phenyl)-1*H*-indol-1-yl)methanone (2.229m) —Was synthesized according to **general procedure 5** using 2-(4-trifluoromethylphenyl) indole (570 mg, 2.18 mmol, 1 equiv) and 2-bromobenzoyl chloride (567 μL, 4.37 mmol, 2 equiv). The crude product was purified via silica gel flash column chromatography using hexanes:EtOAc (20:1 v:v) as the mobile phase, and was isolated as a white crystalline solid (639 mg, 1.44 mmol, 66%, MP = 71–72 °C). Characterization was in accordance with the reported literature [92]. ^{1}H NMR (500 MHz, $CDCl_3$) δ 8.03–8.00 (m, 1H), 7.64–7.62 (m, 1H), 7.41 (s, 4H), 7.40–7.31 (m, 3H), 7.24 (ddd, J = 7.5, 2.0, 0.5 Hz, 1H), 7.13 (td, J = 7.5, 1.5 Hz, 1H), 7.09 (ddd, J = 8.0, 7.5, 2.0 Hz, 1H), 6.74 (d, J = 1.0 Hz, 1H). **^{13}C NMR** (125 MHz, $CDCl_3$) δ 167.7, 138.8, 137.9, 137.1, 136.4 (q, J = 1.5 Hz), 133.2, 132.1, 130.8, 129.5 (q, J = 32.5 Hz), 129.2, 129.1, 127.1, 125.6, 124.6 (q, J = 4.0 Hz), 124.2, 123.8 (q, J = 271.5 Hz) 121.2, 120.89, 115.3, 112.8. **^{19}F NMR** (376 MHz, $CDCl_3$) δ-63.0.

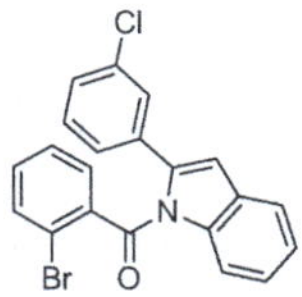

(2-bromophenyl)(2-(3-chlorophenyl)-1*H*-indol-1-yl)methanone (2.229n)[120]—Was synthesized according to **general procedure** 5 using 2-(3-chlorophenyl)indole (335 mg, 1.47 mmol, 1 equiv) and 2-bromobenzoyl chloride (405 μL, 3.10 mmol, 2 equiv). The crude product was purified via silica gel flash column chromatography using hexanes:EtOAc (40:1 v:v) as the mobile phase, and was isolated as a pale yellow oil (370 mg, 0.905 mmol, 62%). Characterization was in accordance with the reported literature [92]. **^{1}H NMR** (500 MHz, $CDCl_3$) δ 8.15–8.12 (m, 1H), 7.64–7.59 (m, 1H), 7.41–7.31 (m, 3H), 7.24–7.21 (m, 2H), 7.19–7.17 (m, 1H), 7.14 (td, J = 7.5, 1.0 Hz, 1H), 7.10–7.06 (m, 3H), 6.68 (d, J = 1.0 Hz, 1H). **^{13}C NMR** (126 MHz, $CDCl_3$) δ 167.9, 138.8, 137.8, 137.2, 134.5, 133.5, 133.1, 131.9, 130.7, 129.2, 129.0, 128.8, 127.7, 127.0, 126.9, 125.4, 124.2, 121.1, 120.7, 115.4, 112.3.

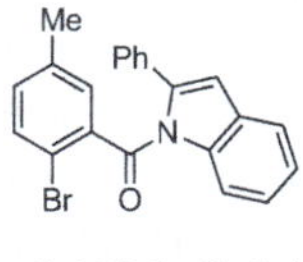

(2-bromo-5-methylphenyl)(2-phenyl-1*H*-indol-1-yl)methanone (2.229o)[121]—Was synthesized according to **general procedure** 5 using 2-phenylindole (500 mg, 2.59 mmol, 1 equiv) and 2-bromo-5-methylbenzoyl chloride prepared from 2-bromo-5-methylbenzoic acid (1.23 g, 5.69 mmol, 2.2 equiv). The crude product was purified via silica gel flash column chromatography using hexanes:EtOAc (20:1 v:v) as the mobile phase, and was isolated as a colorless solid (577 mg, 1.48 mmol, 57%, MP = 105–106 °C). **^{1}H NMR** (500 MHz, $CDCl_3$) δ 8.19–8.15

[120]This compound was synthesized by Andy Yen.

[121]This compound was synthesized by Andy Yen.

(m, 1H), 7.63 (ddd, J = 7.5, 1.5, 1.0 Hz, 1H), 7.38 (ddd, J = 8.0, 7.5, 1.5 Hz, 1H), 7.36–7.33 (m, 1H), 7.29–7.26 (m, 2H), 7.19–7.13 (m, 3H), 7.13–7.09 (m, 1H), 6.97–6.96 (m, 1H), 6.82 (ddq, J = 8.0, 2.0, 1.0 Hz, 1H), 6.67 (d, J = 1.0 Hz, 1H), 2.14–2.13 (m, 3H). **^{13}C NMR** (125 MHz, $CDCl_3$) δ 168.2, 140.5, 137.8, 136.9, 136.8, 133.0, 132.8, 132.5, 131.5, 129.4, 128.8, 127.6, 127.6, 125.0, 123.9, 120.5, 117.8, 115.3, 111.4, 20.4. **IR** (thin film, cm^{-1}) 3053, 3028, 2957, 2920, 1688, 1601, 1558, 1452. **HRMS** (ESI+, M+H) Calc'd for $C_{22}H_{17}BrNO$ 390.0488, found 390.0498.

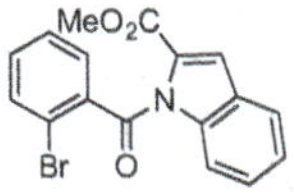

methyl 1-(2-bromobenzoyl)-1*H*-indole-2-carboxylate (2.229p)—Was synthesized according to **general procedure 4** using methyl-2-indolecarboxylate (250 mg, 1.42 mmol, 1 equiv) and 2-bromobenzoyl chloride (341 μL, 2.84, 2 equiv). The crude product was purified via two rounds of silica gel flash column chromatography first using hexanes:EtOAc:DCM (10:0.75:6) then hexanes:EtOAc (10:1 v:v) as the mobile phases, and was isolated as a light yellow solid (294 mg, 0.824 mmol, 58%, MP = 95–96 °C). Characterization was in accordance with the reported literature [92]. **^{1}H NMR** (400 MHz, CDC13) δ 8.02–7.98 (m, 1H), 7.71–7.66 (m, 2H), 7.46 (ddd, J = 8.5, 7.0, 1.5 Hz, 1H), 7.38–7.29 (m, 5H), 3.53 (s, 4H). **^{13}C NMR** (100 MHz, $CDCl_3$) δ 167.1, 161.3, 138.8, 137.1, 133.9, 132.4, 130.5, 130.4, 127.8, 127.5, 127.1, 124.2, 122.5, 121.6, 117.6, 115.0, 52.2.

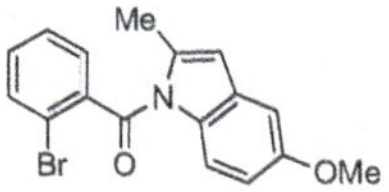

(2-bromophenyl)(5-methoxy-2-methyl-1*H*-indol-1-yl)methanone (2.229q)[122]—Was synthesized according to **general procedure 5** using 5-methoxy-2-methy lindole (500 mg, 3.10 mmol, 1 equiv) and 2-bromobenzoyl chloride (850 μL, 6.50 mmol, 2.1 equiv). The crude product was purified via two rounds of silica gel flash column chromatography first using hexanes:EtOAc (20:1 → 15:1 v:v) then DCM:hexanes (2:1 v:v) as the mobile phases, and was isolated as a clear and colourless oil (874 mg, 2.55 mmol, 82%) which solidified to white solid upon storage at −20 °C, MP = 69–70 °C). **^{1}H NMR** (500 MHz, $CDCl_3$) δ 7.67–7.64 (m, 1H), 7.47–7.45 (m, 2H), 7.39 (ddd, J = 8.0, 6.0, 3.5 Hz, 1H), 7.34 (d, J = 9.0 Hz, 1H), 6.93 (d, J = 2.5 Hz, 1H), 6.73 (dd, J = 9.0, 2.5 Hz, 1H), 6.35–6.33 (m, 1H), 3.82 (s, 3H), 2.20 (d, J = 1.0 Hz, 3H). **^{13}C NMR** (125 MHz, CDC13) δ 167.2, 156.4, 138.4, 137.8, 133.3, 131.8, 131.3, 130.9, 129.1, 127.8, 120.1, 115.8, 111.6, 110.5, 102.9, 55.5, 16.3. **IR** (thin film, cm^{-1}) 3097, 3063, 2997, 2958, 1678, 1593, 1288, 1253, 1111, 1033. **HRMS** (DART, M+H) Calc'd for $C_{17}H_{15}BrNO_2$ 344.02862, found 344.02793.

[122]This compound was synthesized by Andy Yen.

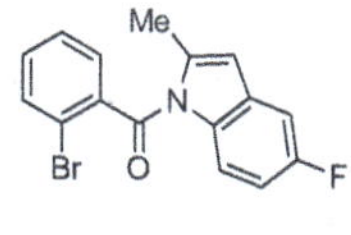

(2-bromophenyl)(5-fluoro-2-methyl-1*H*-indol-1-yl)methanone (2.229r)[123]—Was synthesized according to **general procedure 3** using 5-fluroro-2-methylindole (500 mg, 3.35 mmol, 1 equiv) and 2-bromobenzoyl chloride (870 μL, 6.71 mmol, 2.0 equiv). The crude product was purified via silica gel flash column chromatography using hexanes:EtOAc (4:1 v:v) as the mobile phase, and was isolated as a clear and colourless oil (903 mg, 2.73 mmol, 82%). **^{1}H NMR** (500 MHz, $CDCl_3$) δ 7.66–7.64 (m, 1H), 7.50–7.44 (m, 3H), 7.43–7.38 (m, 1H), 7.10 (dd, *J* = 8.5, 2.5 Hz, 1H), 6.86 (td, *J* = 9.0, 2.5 Hz, 1H), 6.36–6.35 (m, 1H), 2.17 (d, *J* = 1.0 Hz, 3H). **^{13}C NMR** (125 MHz, $CDCl_3$) δ 167.3, 159.6 (d, *J* = 240.0 Hz), 138.8, 138.1, 133.4, 133.1 (d, *J* = 1.5 Hz), 132.0, 130.9 (d, *J* = 10.0 Hz), 129.1, 127.9, 120.0, 116.0 (d, *J* = 9.0 Hz), 111.1 (d, *J* = 25.0 Hz), 110.1 (d, *J* = 4.0 Hz), 105.4 (d, *J* = 24.0 Hz), 16.2. **^{19}F NMR** (376 MHz, $CDCl_3$) δ-119.5 to -119.7 (m). **IR** (thin film, cm^{-1}) 3175, 3071, 2970, 2928, 2859, 1663, 1603, 1361, 1180. **HRMS** (DART, M+H) Calc'd for $C_{16}H_{12}BrFNO_2$ 332. 00863, found 332.00801.

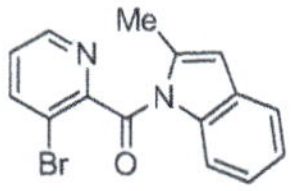

(3-bromopyridin-2-yl)(2-methyl-1*H*-indol-1-yl)methanone (2.229s)[124]—Was synthesized according to **general procedure 5** using 2-methylindole (500 mg, 3.81 mmol, 1 equiv) and 3-bromopicolinoyl chloride synthesized from 3-bromo picolinic acid (1.14 g, 5.67 mmol, 1.5 equiv). The crude product was purified via silica gel flash column chromatography using hexanes:EtOAc (10:3 v:v) as the mobile phase, followed by trituration with hexanes and was isolated as a colorless crystalline solid (310 mg, 0.986 mmol, 26%, MP = 150–151 °C). **^{1}H NMR** (500 MHz, $CDCl_3$) δ 8.64 (dd, *J* = 5.0, 1.5 Hz, 1H), 8.02 (dd, *J* = 8.0, 1.5 Hz, 1H), 7.62 (d, *J* = 8.0 Hz, 1H), 7.45 (ddd, *J* = 7.5, 1.5, 1.0 Hz, 1H), 7.35 (dd, *J* = 8.0, 4.5 Hz, 1H), 7.23 (td, *J* = 7.5, 1.0 Hz, 1H), 7.17 (ddd, *J* = 8.5, 7.5, 1.5 Hz, 1H), 6.42–6.37 (m, 1H), 2.12–2.08 (m, 3H). **^{13}C NMR** (125 MHz, $CDCl_3$) δ 165.6, 154.0, 148.1, 141.11, 136.9, 136.5, 130.0, 126.1, 124.0, 123.9, 119.8, 118.1, 115.4, 110.9, 15.9. **IR** (thin film, cm^{-1}) 2960, 2926, 1683, 1437, 1361, 1220. **HRMS** (DART, M+H) Calc'd for $C_{15}H_{12}BrN_2O$ 315.01330, found 315.01295.

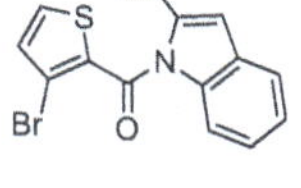

[123]This compound was synthesized by Andy Yen.

[124]This compound was synthesized by Andy Yen.

(3-bromothiophen-2-yl)(2-methyl-1*H*-indol-1-yl)methanone (2.229t)—Was synthesized according to **general procedure 4** using 2-methylindole (500 mg, 3.81 mmol, 1 equiv) and 3-bromothiophene-2-carbonyl chloride synthesized from 3-bromothiophene-2-carboxylic acid (1.56 g, 7.62 mmol, 2 equiv). The crude product was purified via two rounds of silica gel flash column chromatography first using hexanes:EtOAc (20:1 v:v) then DCM (100%) as the mobile phases, and was isolated as a yellow oil (546 mg, 1.71 mmol, 45%). **^{1}H NMR** (400 MHz, $CDCl_3$) δ 7.60 (d, J = 5.0 Hz, 1H), 7.46 (d, J = 7.5 Hz, 1H), 7.23–7.16 (m, 2H), 7.13–7.08 (m, 2H), 6.44–6.43 (m, 1H), 2.41 (d, J = 1.0 Hz, 3H). **^{13}C NMR** (125 MHz, $CDCl_3$) δ 161.7, 137.1, 136.7, 133.3, 131.9, 131.2, 129.7, 123.3, 123.1, 119.8, 115.0, 113.6, 109.4, 15.2. **IR** (thin film, cm^{-1}) 3103, 2964, 2922, 1683, 1593, 1417, 1323, 1213. **HRMS** (DART, M+H) Calc'd for $C_{14}H_{11}BrNOS$ 319.97447, found 319.97500.

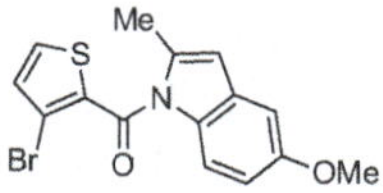

(3-bromothiophen-2-yl)(5-methoxy-2-methyl-1*H*-indol-1-yl)methanone (2.229u)—Was synthesized according to **general procedure 4** using 5-methoxy-2-methylindole (500 mg, 3.10 mmol, 1 equiv) and 3-bromo-2-thiophenecarbonyl chloride synthesized from 3-bromo-2-thiophenecarboxylic acid (1.41 g, 6.82 mmol, 2.2 equiv). The crude product was purified via silica gel flash column chromatography using hexanes:EtOAc (20:1 v:v) and was isolated as a clear yellow oil (422 mg, 1.21 mmol, 55%). **^{1}H NMR** (500 MHz, $CDCl_3$) δ 7.57 (d, J = 5.0 Hz, 1H), 7.15 (ddd, J = 9.0, 0.5 Hz, 1H), 7.07 (d, J = 5.0 Hz, 1H), 6.94 (d, J = 2.5 Hz, 1H), 6.73 (dd, J = 9.0, 2.5 Hz, 1H), 6.37–6.36 (m, 1H), 3.83 (s, 3H), 2.37 (d, J = 1.0 Hz, 3H). **^{13}C NMR** (125 MHz, $CDCl_3$) δ 161.4, 156.2, 137.8, 133.3, 131.6, 131.3, 130.8, 130.6, 114.6, 111.6, 109.6, 102.7, 55.5, 15.3. **IR** (thin film, cm^{-1}) 3103, 2926, 2833, 1683, 1361, 1321. **HRMS** (DART, M+H) Calc'd for $C_{15}H_{13}BrNO_2S$ 349.98504, found 349.98554.

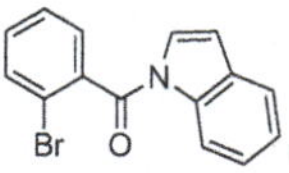

(2-bromophenyl)(1*H*-indol-1-yl)methanone (2.229v)—Was synthesized according to **general procedure 5** using indole (469 mg, 4 mmol, 1 equiv) and 2-bromobenzoyl chloride (945 μL, 8.00 mmol, 2 equiv). The crude product was purified via silica gel flash column chromatography using hexanes:EtOAc (10:1 v:v) and was isolated as a clear and colourless oil (719 mg, 2.39 mmol, 60%). **^{1}H NMR** (500 MHz, $CDCl_3$) δ 8.43 (s(br), 1H), 7.71–7.68 (m, 1H), 7.61–7.57 (m, 1H), 7.51–7.45 (m, 2H), 7.44–7.37 (m, 2H), 7.33 (td, J = 7.5, 1.0 Hz, 1H), 6.96 (s, 1H), 6.61 (d, J = 3.5 Hz, 1H). **^{13}C NMR** (125 MHz, $CDCl_3$) δ 166.6, 137.0, 135.4, 133.2, 131.7, 131.0, 128.8, 127.6, 126.6, 125.2, 124.5, 121.0, 119.6, 116.5, 109.8. **IR** (thin film, cm^{-1}) 3144, 3055, 3020, 1693, 1589, 1535, 1454, 1346, 1242. **HRMS** (DART, M+H) Calc'd for $C_{15}H_{11}BrNO$ 300.00240, found 300.00326.

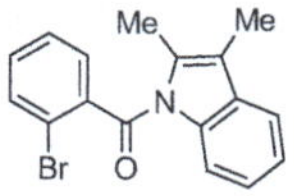

(2-bromophenyl)(2,3-dimethyl-1*H*-indol-1-yl)methanone (2.229w)—Was synthesized according to **general procedure 5** using 2,3-dimethylindole (581 mg, 4 mmol, 1 equiv) and 2-bromobenzoyl chloride (945 μL, 8.00 mmol, 2 equiv). The crude product was purified via two rounds of silica gel flash column chromatography first using hexanes:EtOAc (20:1 v:v) then using DCM:hexanes (4:1), and was isolated as a clear and colourless oil (296 mg, 0.902 mmol, 23%). **^{1}H NMR** (500 MHz, $CDCl_3$) δ 7.68–7.66 (m, 1H), 7.47–7.45 (m, 2H), 7.44–7.42 (m, 1H), 7.40 (ddd, *J* = 8.0, 5.5, 3.5 Hz, 1H), 7.31 (d, *J* = 8.4 Hz, 1H), 7.24 (ddd, *J* = 7.5, 7.0, 1.0 Hz, 1H), 7.12 (ddd, *J* = 8.5, 702, 1.0 Hz, 1H), 2.20 (s, 6H). **^{13}C NMR** (125 MHz, $CDCl_3$) δ 167.4, 138.8, 135.9, 133.4, 132.3, 131.8, 131.4, 129.3, 127.8, 123.8, 123.3, 120.2, 118.0, 116.5, 114.7, 13.3, 8.8. **IR** (thin film, cm^{-1}) 3016, 2966, 2862, 1678, 1616, 1589, 1435, 1354, 1323, 1219. **HRMS** (DART, M+H) Calc'd for $C_{17}H_{15}BrNO$ 328.03370, found 328.03317.

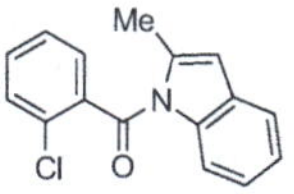

(2-chlorophenyl)(2-methyl-1H-indol-1-yl)methanone (Cl-2.229a)[125]—Was synthesized according to **general procedure 5** using 2-methylindole (500 mg, 3.81 mmol, 1 equiv) and 2-chlorobenzoyl chloride (1.06 mL, 8.38 mmol, 2.2 equiv). The crude product was purified via two rounds of silica gel flash column chromatography first using hexanes:EtOAc (20:1 v:v) then using 100% hexanes → hexanes:EtOAc (20:1 v:v) and was isolated as a turquoise oil which was further purified by adding a small amount of activated charcoal to a dilute DCM solution of this material. The resulting heterogeneous solution was filtered over Celite® eluting with 100% DCM. The solvent was removed under reduced pressure leaving a clear and colorless oil (461 mg, 1.71 mmol, 45%). **^{1}H NMR** (500 MHz, $CDCl_3$) δ 7.53–7.47 (m, 3H), 7.47–7.44 (m, 1H), 7.44–7.40 (m, 1H), 7.37 (d, *J* = 8.5 Hz, 1H), 7.21 (ddd, *J* = 7.5, 7.5, 1.0 Hz, 1H), 7.12 (ddd, *J* = 8.5, 7.5, 1.5 Hz, 1H), 6.42–6.41 (m, 1H), 2.26 (d, *J* = 1.0 Hz, 3H). **^{13}C NMR** (125 MHz, $CDCl_3$) δ 166.9, 137.2, 136.7, 136.3, 131.9, 131.6, 130.3, 129.9, 129.2, 127.3, 123.6, 123.5, 119.7, 114.8, 110.3, 16.1. **IR** (thin film, cm^{-1}) 3059, 3028, 2970, 2928, 1678, 1593, 1573, 1242, 1107. **HRMS** (DART, M+H) Calc'd for $C_{16}H_{13}ClNO$ 270.06857, found 270.06851.

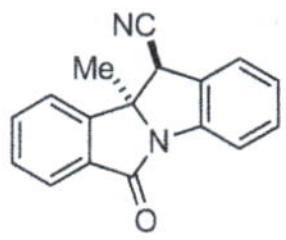

[125]This compound was synthesized by Andy Yen.

(±)-10b-methyl-6-oxo-10b,11-dihydro-6*H*-isoindolo[2,1-*a*]indole-11-carbonitrile (2.237a)—Was synthesized according to **general procedure** 6 using **2.229a** (62.5 mg, 0.2 mmol, 1 equiv). The crude product was purified via silica gel flash column chromatography using hexanes:EtOAc (5:4 v:v) and was isolated as an off-white powder which was triturated with hexanes to afford a white solid (51.0 mg, 0.196 mmol, 98%, >20:1 dr). Suitable single X-ray quality crystals of the major diastereomer were obtained by slow diffusion of hexanes into a saturated acetone solution of **2a** (MP = 180–182 °C).

Major (*cis*-**2.229a**): ^{1}**H NMR** (500 MHz, $CDCl_3$) δ 7.94–7.91 (m, 1H), 7.78–7.75 (m, 1H), 7.72 (ddd, J = 7.5, 7.5, 1.2 Hz, 1H), 7.61–7.57 (m, 2H), 7.49–7.45 (m, 1H), 7.43 (ddt, J = 7.5, 1.5, 0.5 Hz, 1H), 7.21 (td, J = 7.5, 1.0 Hz, 1H), 4.20 (s, 1H), 1.72 (s, 3H). ^{13}**C NMR** (125 MHz, $CDCl_3$) δ 167.6, 146.9, 139.2, 133.6, 132.3, 130.7, 129.9, 129.5, 126.3, 125.5, 125.3, 122.7, 118.0, 116.3, 73.4, 41.1, 27.2. **IR** (thin film, cm^{-1}) 2964, 2243, 1705, 1602, 1558, 1479, 1356, 1357. **HRMS** (ESI+, M+H) Calc'd for $C_{17}H_{13}N_2O$ 261.1022, found 261.1025.

Minor (*trans*-**2a**): ^{1}**H NMR** (500 MHz, $CDCl_3$) δ 7.90 (ddd, J = 7.5, 1.0 Hz, 1H), 7.73–7.64 (m, 3H), 7.58 (ddd, J = 7.5, 1.0 Hz, 1H), 7.47–7.43 (m, 2H), 7.27–7.20 (m, 1H), 4.33–4.31 (m, 1H), 1.87 (s, 3H). ^{13}**C NMR** (125 MHz, $CDCl_3$) δ 167.0, 148.1, 138.3, 133.6, 132.1, 130.6, 130.3, 129.9, 125.6, 125.4, 125.0, 122.1, 118.1, 116.8, 73.2, 41.9, 23.6. **IR** (thin film, cm^{-1}) 2976, 2929, 2897, 2245, 1708, 1602, 1558, 1479, 1462, 1356, 1305. **HRMS** (ESI+, M+H) Calc'd for $C_{17}H_{13}N_2O$ 261.1022, found 261.1026.

Procedure for gram-scale experiment: An oven-dried 100 mL round bottom flask containing a stir bar was charged with $Pd(OAc)_2$ (35.5 mg, 0.158 mmol, 5 mol%) and D^{t}BPF (75.1 mg, 0.158 mmol, 5 mol%) and purged with argon for 20 min. At this time, the contents were taken up in one half of the required dry and degassed MeCN (24 ml) and the solution was stirred at room temperature for 20 min. A 250 mL one-neck Schlenk flask containing a stir bar was charged with **2.229a** (1.00 g, 3.17 mmol, 1 equiv) and $Zn(CN)_2$ (204 mg, 1.74 mmol, 55 mol%) and the atmosphere was replaced with argon by four vacuum/argon purge cycles on a Schlenk line. The catalyst solution was added via cannula to the flask containing **2.229a**, and the now empty flask was rinsed with the remaining half of the MeCN (24 mL), which was also transferred via cannula to the reaction mixture. The Schlenk flask containing all components was tightly sealed using an appropriate Teflon® twist cap, and was placed into an oil bath pre-heated to 110°. After 18 h the reaction was cooled to room temperature and then contents were carefully filtered over a short pad of silica gel, eluting with 100% EtOAc (500 mL). A homogeneous aliquot was taken from the organic washings and the solvents were removed in vacuo. ^{1}H NMR analysis of this crude aliquot indicated the product to be present in >20:1 dr. All volatiles were removed from the bulk solution under reduced pressure and purification of the crude residue was carried out via silica gel flask column chromatography using hexanes:EtOAc (5:4 v:v) as the mobile phase, and the title compound was obtained as an off-white solid which was triturated with hexanes to afford **2.237a** as a white powder (715 mg, 2.77 mmol, 88%).

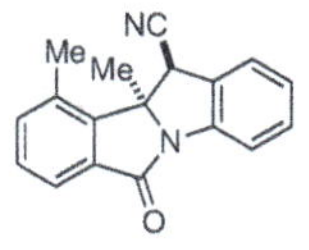

(±)-10,10b-dimethyl-6-oxo-10b,11-dihydro-6*H*-isoindolo[2,1-*a*]indole-11-carbonitrile (2.237b)[126]—Was synthesized according to **general procedure 6** using **2.229b** (65.6 mg, 0.2 mmol, 1 equiv). The crude product was purified via silica gel flash column chromatography using hexanes:EtOAc (5:4 v:v) and was isolated as a white solid (28.0 mg, 0.102 mmol, 51%, >20:1 dr, MP = 223–224 ° C). **^{1}H NMR** (500 MHz, $CDCl_3$) δ 7.79–7.75 (m, 2H), 7.50–7.43 (m, 4H), 7.21 (td, J = 7.5, 1.0 Hz, 1H), 4.27 (s, 1H), 2.59–2.56 (m, 3H), 1.75 (s, 3H). **^{13}C NMR** (125 MHz, CDC13) δ 167.0, 144.7, 138.6, 135.3, 133.0, 132.8, 130.8, 130.1, 129.3, 126.3, 125.4, 123.0, 117.7, 116.2, 73.6, 40.3, 25.3, 18.7. **IR** (thin film, cm^{-1}) 2933, 2241 m 1699, 1602, 1479, 1361. **HRMS** (DART, M+H) Calc'd for $C_{18}H_{15}N_2O$ 275.11844, found 275.11830.

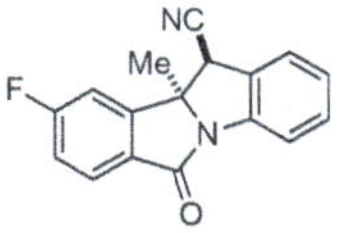

(±)-9-fluoro-10b-methyl-6-oxo-10b,11-dihydro-6*H*-isoindolo[2,1-*a*]indole-11-carbonitrile (2.237c)—Was synthesized according to **general procedure 6** using **2.229c** (66.2 mg, 0.2 mmol, 1 equiv). The crude product was purified via silica gel flash column chromatography using hexanes:EtOAc (5:4 v:v) and was isolated as a colorless solid (51.7 mg, 0.186 mmol, 93%, 20:1 dr, MP = 163–164 °C). **^{1}H NMR** (500 MHz, $CDCl_3$) δ 7.92 (dd, J = 9.0, 5.0 Hz, 1H), 7.75–7.73 (m, 1H), 7.47 (td, J = 8.0, 1.0 Hz, 1H), 7.45–7.42 (m, 1H), 7.31–7.26 (m, 2H), 7.22 (td, J = 7.5, 1.0 Hz, 1H), 4.18 (s, 1H), 1.72 (s, 3H). **^{13}C NMR** (125 MHz, $CDCl_3$) δ 167.1, 165.8 (d, J = 184.0 Hz), 149.3 (d, J = 9.5 Hz), 139.1, 130.9, 129.1, 128.4 (d, J = 2.5 Hz), 127.6 (d, J = 10.0 Hz), 126.3, 125.6, 117.9, 117.9 (d, J = 23.5 Hz), 116.0, 110.4 (d, J = 24.5 Hz), 73.0 (d, J = 2.5 Hz), 41.0, 27.1. **^{19}F NMR** (376 MHz, $CDCl_3$) δ-103.0 to -103.1 (m). **IR** (thin film, cm^{-1}) 3069, 2975, 2929, 2241, 1709, 1600, 1478, 1350, 1193. **HRMS** (DART+, M+H) Calc'd for $C_{17}H_{12}FN_2O$ 279.09337, found 279.09301.

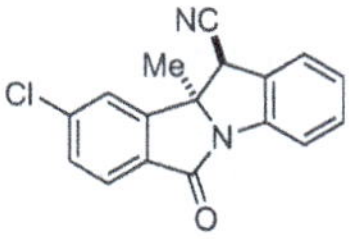

(±)-9-chloro-10b-methyl-6-oxo-10b,11-dihydro-6*H*-isoindolo[2,1-*a*]indole-11-carbonitrile (2.237d)[127]—Was synthesized according to **general procedure 6** using **2.229d** (69.4 mg, 0.2 mmol, 1 equiv). The crude product was purified via silica gel

[126]This compound was synthesized by Andy Yen.

[127]This compound was synthesized by Andy Yen.

flash column chromatography using hexanes:EtOAc (4:1 v:v) followed by trituration in hexanes, and was isolated as a colorless solid (54.2 mg, 0.186 mmol, 93%, >20:1 dr, MP = 229–231 °C). **^{1}H NMR** (500 MHz, $CDCl_3$) δ 7.85 (dd, *J* = 8.0, 0.5 Hz, 1H), 7.76–7.74 (m, 1H), 7.59–7.55 (m, 2H), 7.50–7.46 (m, 1H), 7.45 (ddt, *J* = 7.5, 1.0, 0.5 Hz, 1H), 7.23 (td, *J* = 7.5, 1.0 Hz, 1H), 4.17 (s, 1H), 1.72 (s, 3H). **^{13}C NMR** (125 MHz, $CDCl_3$) δ 166.5, 148.3, 140.1, 138.9, 130.9, 130.9, 130.7, 129.2, 126.5, 126.3, 125.8, 123.3, 118.0, 116.0, 73.1, 41.0, 27.1. **IR** (thin film, cm^{-1}) 3066, 3020, 2974, 2241, 1708, 1604, 1465, 1354, 1307. **HRMS** (DART, M+H) Calc'd for $C_{17}H_{12}ClN_2O$ 295.06382, found 295.06445.

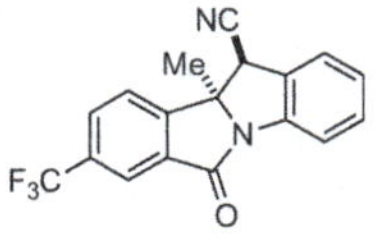

(±)-10b-methyl-6-oxo-8-(trifluoromethyl)-10b,11-dihydro-6*H*-isoindolo[2,1-*a*] indole-11-carbonitrile (2.237e)[128]—Was synthesized according to **general procedure 6** using **2.229e** (76.4 mg, 0.2 mmol, 1 equiv). The crude product was purified via silica gel flash column chromatography using hexanes:EtOAc (5:4 v:v) and was isolated as a colorless solid (63.0 mg, 0.192 mmol, 96%, 12:1 dr, MP = 224–225 °C). ^{1}H NMR (600 MHz, Acetone-d_6) δ 8.22–8.20 (m, 1H), 8.20–8.17 (m, 1H), 8.15–8.14 (m, 1H), 7.72 (ddt, *J* = 8.0, 1.0, 0.5 Hz, 1H), 7.66 (ddt, *J* = 7.65 1.5, 0.5 Hz, 1H), 7.55 (dddd, *J* = 7.65 1.5, 0.5 Hz, 1H), 7.32 (td, *J* = 7.5, 1.0 Hz, 1H), 4.84–4.83 (m, 1H), 1.85 (s, 3H). **^{13}C NMR** (125 MHz, Acetone-d_6) δ 166.1, 151.9 (q, *J* = 1.0 Hz), 139.3, 134.1, 132.4 (q, *J* = 33.0 Hz), 131.6, 131.1, 131.0 (q, *J* = 3.5 Hz), 127.5, 126.5, 125.7, 124.5 (q, *J* = 272.0 Hz), 122.1, 118.4, 117.4, 74.6, 41.2, 25.9. **^{19}F NMR** (376 MHz, Acetone-d_6) δ-63.0. **IR** (thin film, cm^{-1}) 2962, 2929, 2243, 1708, 1631, 1604, 1558, 1479, 1325, 1138. **HRMS** (ESI+, M+H) Calc'd for $C_{18}H_{12}F_3N_2O$ 329.09017, found 329.09106.

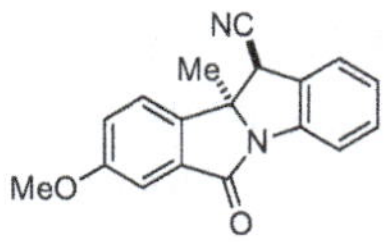

(±)-8-methoxy-10b-methyl-6-oxo-10b,11-dihydro-6*H*-isoindolo[2,1-*a*]indole-11-carbonitrile (2.237f)[129]—Was synthesized according to **general procedure 6** heating for 5 h using **2.229f** (68.8 mg, 0.2 mmol, 1 equiv). The crude product was purified via silica gel flash column chromatography using hexanes:EtOAc (5:4 v:v) and was isolated as a white solid (54.5 mg, 0.186 mmol, 94%, >20:1 dr, MP = 200–201 °C). **^{1}H NMR** (500 MHz, $CDCl_3$) δ 7.75 (ddt, *J* = 8.0, 1.0, 0.5 Hz, 1H), 7.49–7.44 (m, 2H), 7.42 (ddt, *J* = 7.5, 1.5. 0.5 Hz, 1H), 7.38–7.36

[128]This compound was synthesized by Andy Yen.

[129]This compound was synthesized by Nicolas Zeidan.

(m, 1H), 7.28–7.23 (m, 1H), 7.21 (td, J = 7.5, 1.0 Hz, 1H), 4.15 (s, 1H), 3.88 (s, 3H), 1.69 (s, 3H). **^{13}C NMR** (125 MHz, $CDCl_3$) δ 167.6, 161.2, 139.2, 139.1, 133.8, 130.7, 129.7, 126.3, 125.5, 123.5, 121.7, 118.0, 116.4, 107.9, 73.1, 55.7, 41.2, 27.2. **IR** (thin film, cm^{-1}) 2241, 1699, 1647, 1558, 1479, 1356, 1282. **HRMS** (ESI+, M+H) Calc'd for $C_{18}H_{15}N_2O_2$ 291.11335, found 291.11394.

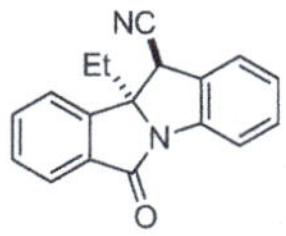

(±)-10b-ethyl-6-oxo-10b,11-dihydro-6*H*-isoindolo[2,1-*a*]indole-11-carbonitrile (2.237g)—Was synthesized according to **general procedure 6** using **2.229g** (65.4 mg, 0.2 mmol, 1 equiv). The crude product was purified via silica gel flash column chromatography using hexanes:EtOAc (5:4 v:v) and was isolated as a colorless crystalline solid (49.7 mg, 0.181 mmol, 91%, >20:1 dr, MP = 147–148 °C). **^{1}H NMR** (500 MHz, $CDCl_3$) δ 7.94–7.91 (m, 1H), 7.77–7.74 (m, 1H), 7.72 (td, J = 7.5, 1.0 Hz, 1H), 7.58 (td, J = 7.5, 1.0 Hz, 1H), 7.55 (dt, J = 7.5, 1.0 Hz, 1H), 7.47–7.43 (m, 1H), 7.43–7.41 (m, 1H), 7.20 (td, J = 7.5, 1.0 Hz, 1H), 4.20 (s, 1H), 2.16–1.98 (m, 2H), 0.66 (t, J = 7.5 Hz, 3H). **^{13}C NMR** (125 MHz, $CDCl_3$) δ 168.3, 145.2, 139.6, 133.5, 133.3, 130.6, 129.9, 129.9, 126.1, 125.4, 125.1, 122.8, 117.8, 116.4, 76.8, 40.5, 32.5, 7.7. **IR** (thin film, cm^{-1}) 3020, 2970, 2935, 2241, 1701, 1604, 1481, 1462, 1361, 1300, 1211. **HRMS** (DART, M+H) Calc'd for $C_{18}H_{15}N_2O$ 275.11844, found 275.11812.

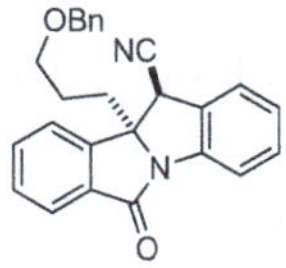

(±)-10b-(3-(benzyloxy)propyl)-6-oxo-10b,11-dihydro-6*H*-isoindolo[2,1-*a*]indole-11-carbonitrile (2.237h)—Was synthesized according to **general procedure 6** using **2.229h** (89.4 mg, 0.2 mmol, 1 equiv). The crude product was purified via silica gel flash column chromatography using hexanes:EtOAc (5:2 v:v) and was isolated as a colorless solid (65.5 mg, 0.166 mmol, 83%, >20:1 dr, MP = 109–110 °C). **^{1}H NMR** (500 MHz, $CDCl_3$) δ 7.92 (ddd, J = 7.5, 1.0, 1.0 Hz, 1H), 7.76–7.74 (m, 1H), 7.71 (td, J = 7.5, 1.0 Hz, 1H), 7.61–7.55 (m, 2H), 7.49–7.42 (m, 1H), 7.42 (ddt, J = 7.5, 1.5, 1.0 Hz, 1H), 7.33–7.24 (m, 3H), 7.23–7.18 (m, 3H), 4.35 (s, 2H), 4.21 (s, 1H), 3.29 (t, J = 6.0 Hz, 2H), 2.18–2.13 (m, 2H), 1.57–1.47 (m, 1H), 1.20–1.10 (m, 1H). **^{13}C NMR** (125 MHz, $CDCl_3$) δ 168.3, 145.4, 139.6, 138.0, 133.7, 133.0, 130.7, 130.0, 129.8, 128.4, 127.7, 127.6, 126.1, 125.5, 125.3, 122.8, 117.9, 116.2, 76.2, 72.9, 69.2, 40.8, 36.3, 23.8. **IR** (thin film, cm^{-1}) 3063, 3024, 2928, 2858, 2241, 1708, 1604, 1481, 1357, 1303, 1207. **HRMS** (DART, M+H) Calc'd for $C_{26}H_{23}N_2O_2$ 395.17595, found 395.17574.

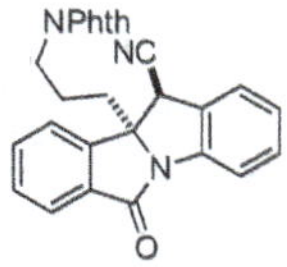

(±)-**10b-(3-(1,3-dioxoisoindolin-2-yl)propyl)-6-oxo-10b,11-dihydro-6*H*-isoindolo[2,1-*a*]indole-11-carbonitrile (2.237i)**—Was synthesized according to **general procedure 6** using **2.229i** (98.4 mg, 0.2 mmol, 1 equiv). The crude product was purified via silica gel flash column chromatography using hexanes:EtOAc (5:4 v:v) and was isolated as a colourless foam (85.0 mg, 0.196 mmol, 98%, >20:1 dr). **^{1}H NMR** (500 MHz, $CDCl_3$) δ 7.90 (dt, *J* = 7.5, 1.0 Hz, 1H), 7.79 (dd, *J* = 5.5, 3.0 Hz, 2H), 7.73–7.66 (m, 4H), 7.57 (td, *J* = 7.5, 1.0 Hz, 1H), 7.51 (dt, *J* = 7.5, 1.0 Hz, 1H), 7.44 (td, *J* = 7.5, 1.0 Hz, 1H), 7.42–7.39 (m, 1H), 7.20 (td, *J* = 7.5, 1.0 Hz, 1H), 4.20 (s, 1H), 3.53 (td, *J* = 7.0, 1.0 Hz, 2H), 2.18–2.03 (m, 2H), 1.68–1.58 (m, 1H), 1.25–1.14 (m, 1H). **^{13}C NMR** (126 MHz, CDCl) δ 168.3, 168.2, 145.0, 139.5, 134.1, 133.8, 133.0, 131.9, 130.8, 130.2, 129.5, 126.2, 125.6, 125.4, 123.3, 122.7, 118.0, 116.1, 75.9, 40.9, 37.3, 36.8, 22.8. **IR** (thin film, cm^{-1}) 3020, 2935, 2241, 1716, 1708, 1604, 1481, 1435, 1396, 1303. **HRMS** (DART, M+H) Calc'd for $C_{27}H_{20}N_3O_3$ 434.15047, found 434.15068.

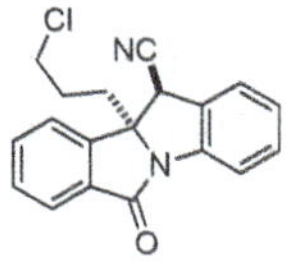

(±)-**10b-(3-chloropropyl)-6-oxo-10b,11-dihydro-6*H*-isoindolo[2,1-*a*]indole-11-carbonitrile (2.237j)**—Was synthesized according to **general procedure** 6 using **2.229j** (75.1 mg, 0.2 mmol, 1 equiv). The crude product was purified via silica gel flash column chromatography using hexanes:EtOAc (5:3 v:v) and was isolated as a colourless solid (41.0 mg, 0.127 mmol, 64%, >20:1 dr, MP = 172–173 °C). **^{1}H NMR** (500 MHz, $CDCl_3$) δ 7.93 (dt, *J* = 7.5, 1.0 Hz, 1H), 7.77–7.73 (m, 2H), 7.63–7.57 (m, 2H), 7.49–7.45 (m, 1H), 7.45–7.43 (m, 1H), 7.22 (td, *J* = 7.5, 1.1 Hz, 1H), 4.22 (s, 1H), 3.39–3.30 (m, 2H), 2.23 (dtdd, *J* = 18.5, 14.0, 11.0, 5.0 Hz, 2H), 1.75–1.66 (m, 1H), 1.32–1.21 (m, 1H). **^{13}C NMR** (125 MHz, $CDCl_3$) δ 168.3, 145.0, 139.5, 134.0, 133.0, 130.8, 130.2, 129.7, 126.2, 125.7, 125.4, 122.8, 118.0, 116.0, 75.9, 44.2, 41.1, 36.6, 26.4. **IR** (thin film, cm^{-1}) 3018, 2916, 2848, 2243, 1708, 1479, 1215. **HRMS** (DART, M+H) Calc'd for $C_{19}H_{16}ClN_2O$ 323.09512, found 323.09514.

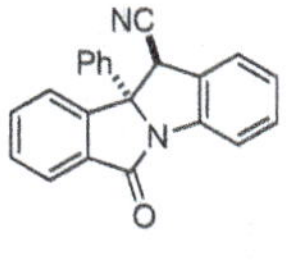

(±)-**6-oxo-10b-phenyl-10b,11-dihydro-6*H*-isoindolo[2,1-*a*]indole-11-carbonitrile (2.237k)**—Was synthesized according to **general procedure 6** using **2.229k** (74.8 mg, 0.2 mmol, 1 equiv). The crude product was purified via silica gel flash column chromatography using hexanes:EtOAc (5:2 v:v) and was isolated as a colorless foam (58.6 mg, 0.182 mmol, 91%, >20:1 dr). **^{1}H NMR** (500 MHz, $CDCl_3$) δ 7.91 (ddd, *J* = 7.5, 1.0, 1.0 Hz, 1H), 7.89–7.87 (m, 1H), 7.73 (dt, *J* = 8.0, 1.0 Hz, 1H), 7.66 (ddd, *J* = 8.0, 7.5, 1.0 Hz, 1H), 7.64–7.61 (m, 2H), 7.54 (td, *J* = 7.5, 1.0 Hz, 1H), 7.49–7.45 (m, 1H), 7.38–7.33 (m, 3H), 7.33–7.25 (m, 1H), 4.82 (s, 1H). **^{13}C NMR** (125 MHz, $CDCl_3$) δ 168.1, 146.5, 140.8, 139.7, 133.8, 131.7, 130.8, 129.9, 129.5, 129.23, 128.8, 126.0, 125.8, 125.4, 124.7, 123.6, 117.8, 116.3, 78.4, 43.1. **IR** (thin film, cm^{-1}) 3063, 3020, 3920, 2241, 1708, 1604, 1481, 1354, 1307. **HRMS** (DART, M+H) Calc'd for $C_{22}H_{15}N_2O$ 323.11844, found 323.11787.

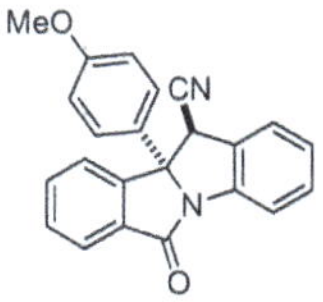

(±)-10b-(4-methoxyphenyl)-6-oxo-10b,11-dihydro-6*H*-isoindolo[2,1-*a*]indole-11-carbonitrile (2.237l)—Was synthesized according to **general procedure 6** using **2.229l** (81.1 mg, 0.2 mmol, 1 equiv). The crude product was purified via silica gel flash column chromatography using hexanes:EtOAc (5:3 v:v) and was isolated as an off-white and waxy solid (60.1 mg, 0.170 mmol, 85%, >20:1 dr). **^{1}H NMR** (500 MHz, $CDCl_3$) δ 7.91 (ddd, *J* = 7.5, 1.0, 1.0 Hz, 1H), 7.86 (ddt, *J* = 8.0, 1.0, 0.5 Hz, 1H), 7.71–7.64 (m, 2H), 7.55–7.49 (m, 3H), 7.48–7.44 (m, 1H), 7.36 (ddt, *J* = 7.5, 1.5, 1.0 Hz, 1H), 7.18 (td, *J* = 7.5, 1.0 Hz, 1H), 6.87–6.85 (m, 1H), 6.85–6.83 (m, 1H), 4.77 (s, 1H), 3.74 (s, 3H). **^{13}C NMR** (125 MHz, $CDCl_3$) δ 168.1, 159.9, 146.9, 139.7, 133.8, 132.5, 131.7, 130.8, 129.8, 129.7, 126.0, 126.0, 125.8, 125.4, 123.5, 117.9, 116.4, 114.6, 78.2, 55.3, 43.1. **IR** (thin film, cm^{-1}) 2960, 2841, 2241, 1708, 1606, 1558, 1512, 1479, 1464. **HRMS** (ESI+, M+H) Calc'd for $C_{23}H_{17}N_2O_2$ 353.1285, found 353.1289.

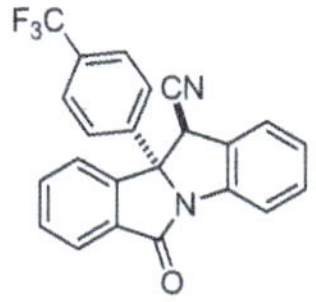

(±)-6-oxo-10b-(4-(trifluoromethyl)phenyl)-10b,11-dihydro-6*H*-isoindolo[2,1-*a*] indole-11-carbonitrile (2.237m)[130]—Was synthesized according to **general procedure 6** using **2.229m** (81.1 mg, 0.2 mmol, 1 equiv). The crude product was purified via silica gel flash column chromatography using hexanes:EtOAc (5:3 v:v) and was isolated as a colorless white foam (60.1 mg, 0.170 mmol, 85%, >20:1 dr). **^{1}H NMR** (500 MHz, $CDCl_3$) δ 7.94 (ddd, *J* = 7.5, 1.0, 1.0 Hz, 1H), 7.90 (ddt, *J* = 8.0, 1.0, 0.5 Hz, 1H), 7.80–7.76 (m, 2H), 7.73 (ddd, *J* = 8.0, 1.0, 0.5 Hz, 1H), 7.69 (ddd, *J* = 8.0, 7.0, 1.0 Hz, 1H), 7.64–7.60 (m, 2H), 7.58 (ddd, *J* = 7.5, 7.0,

[130]This compound was synthesized by Andy Yen.

1.0 Hz, 1H), 7.49 (dddd, $J = 8.0, 7.5, 1.0, 0.5$ Hz, 1H), 7.37 (ddt, $J = 7.5, 1.2, 1.0$ Hz, 1H), 7.20 (td, $J = 7.5, 1.0$ Hz, 1H), 4.80 (s, 1H). **^{13}C NMR** (125 MHz, $CDCl_3$) δ 167.9, 145.6, 144.8 (q, $J = 1.5$ Hz), 139.5, 134.0, 131.7, 131.2 (q, $J = 33.0$ Hz), 131.1, 130.3, 129.0, 126.3 (q, $J = 3.5$ Hz), 126.1, 126. 1, 125.7, 125.3, 123.5, 123.5 (q, $J = 272.5$ Hz), 117.9, 116.0, 78.1, 43.1. **^{19}F NMR** (376 MHz, $CDCl_3$) δ-62.9. **IR** (thin film, cm^{-1}) 3074, 3022, 2924, 2850, 2243, 1712, 1479, 1352, 1329, 1168, 1122, 1070. **HRMS** (DART, M+H) Calc'd for $C_{23}H_{15}F_3N_2O$ 391.10582, found 391.10642.

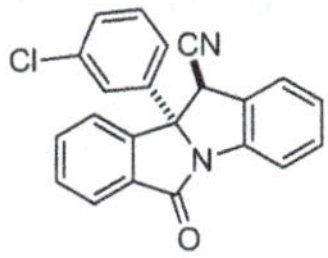

(±)-10b-(3-chlorophenyl)-6-oxo-10b,11-dihydro-6*H*-isoindolo[2,1-*a*]indole-11-carbonitrile (2.237n)[131]—Was synthesized according to **general procedure 6** using **2.229n** (81.8 mg, 0.2 mmol, 1 equiv). The crude product was purified via two rounds of silica gel flash column chromatography first using hexanes:EtOAc (5:4 v:v) followed by 100% hexanes → 100% DCM and was isolated as a colorless solid (68.5 mg, 0.192 mmol, 96%, >20:1 dr, MP = 88–89 °C). **^{1}H NMR** (500 MHz, $CDCl_3$) δ 7.93 (ddd, $J = 7.5, 1.0$ Hz, 1H), 7.89 (ddt, $J = 8.0, 1.0, 0.5$ Hz, 1H), 7.73–7.67 (m, 2H), 7.60 (td, $J = 2.0, 0.5$ Hz, 1H), 7.57 (ddd, $J = 7.5, 7.0, 1.5$ Hz, 1H), 7.55–7.52 (m, 1H), 7.49 (dddd, $J = 8.0, 7.5, 1.0, 0.5$ Hz, 1H), 7.37 (ddt, $J = 7.5, 1.5, 0.5$ Hz, 1H), 7.32–7.26 (m, 2H), 7.20 (td, $J = 7.5, 1.0$ Hz, 1H), 4.79 (s, 1H). **^{13}C NMR** (125 MHz, $CDCl_3$) δ 168.0, 145.9, 142.9, 139.6, 135.3, 134.0, 131.6, 131.0, 130.6, 130.2, 129.1, 129.1, 126.0, 126.0, 125.6, 125.1, 123.6, 123.0, 118.0, 116.0, 78.0, 43.1. **IR** (thin film, cm^{-1}) 3066, 2960, 2245, 1712, 1593, 1573, 1479, 1464. **HRMS** (DART, M+H) Calc'd for $C_{22}H_{14}ClN_2O$ 357.07947, found 357.07909.

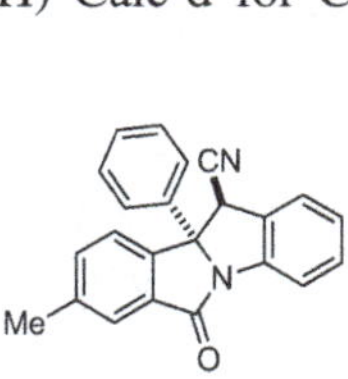

(±)-8-methyl-6-oxo-10*b*-phenyl-10*b*,11-dihydro-6*H*-isoindolo[2,1-*a*]indole-11-carbonitrile (2.237o)[132]—Was synthesized according to **general procedure 6** using **2.229o** (78.0 mg, 0.2 mmol, 1 equiv). The crude product was purified via silica gel flash column chromatography using hexanes:EtOAc (5:2 v:v) and was isolated as a colorless foam (59.4 mg, 0.177 mmol, 89%, >20:1 dr). **^{1}H NMR** (500 MHz, $CDCl_3$) δ 7.87 (ddt, $J = 8.0, 1.0, 0.5$ Hz, 1H), 7.71–7.70 (m, 1H), 7.62–7.59 (m, 3H), 7.48–7.44 (m, 2H), 7.36–7.32 (m, 3H), 7.30–7.26 (m, 1H), 7.17 (td, $J = 7.5, 1.0$ Hz, 1H), 4.79 (s, 1H), 2.44–2.42 (m, 3H). **^{13}C NMR** (125 MHz,

[131] This compound was synthesized by Andy Yen.

[132] This compound was synthesized by Andy Yen.

$CDCl_3$) δ 168.3, 143.9, 141.1, 140.3, 139.8, 134.8, 131.8, 130.8, 129.5, 129.2, 128.7, 126.0, 125.7, 125.6, 124.6, 123.3, 117.8, 116.5, 78.3, 43.2, 21.3. **IR** (thin film, cm^{-1}) 2956, 2924, 2850, 2241, 1708, 1588, 1481, 1356. **HRMS** (DART, M +H) Calc'd for $C_{27}H_{17}N_2O$ 337.13409, found 337.13410.

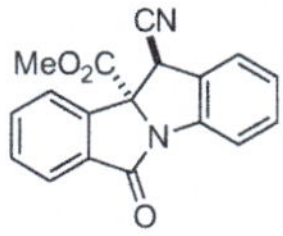

(±)-methyl 11-cyano-6-oxo-10b,11-dihydro-6*H*-isoindolo[2,1-*a*]indole-10b-carboxylate (2.237p)—Was synthesized according to **general procedure 6** using **2.229p** (71.2 mg, 0.2 mmol, 1 equiv). The crude product was purified via two rounds of silica gel flash column chromatography first using hexanes:EtOAc (5:3 v:v) to obtained the pure major/minor mixture (43.6 mg, 0.143 mmol, 72%, 11:1 dr), followed by hexanes:DCM:EtOAc (15:15:1 v:v:v) to separate the diastereomers. The major diastereomer was isolated as a colorless solid (MP = 146–147 °C). **^{1}H NMR** (500 MHz, $CDCl_3$) δ 7.95 (ddd, *J* = 7.5, 1.0, 1.0 Hz, 1H), 7.79 (ddt, *J* = 8.0, 1.0, 0.5 Hz, 1H), 7.74 (ddd, *J* = 7.5, 7.0, 1.0 Hz, 1H), 7.69 (ddd, *J* = 7.5, 1.0, 1.0 Hz, 1H), 7.65 (ddd, *J* = 7.0, 1.0 Hz, 1H), 7.51–7.45 (m, 1H), 7.44 (ddd, *J* = 7.5, 1.0, 0.5 Hz, 1H), 5.22 (dt, *J* = 1.0, 0.5 Hz, 1H), 3.70 (s, 3H). **^{13}C NMR** (125 MHz, $CDCl_3$) δ 170.0, 167.8, 140.7, 139.4, 134.0, 133.0, 131.1, 130.9, 128.4, 125.9, 125.8, 125.6, 123.5, 117.4, 115.6, 77.8, 54.3, 38.5. **IR** (thin film, cm^{-1}) 3020, 2955, 2245, 1716, 1604, 1481, 1350, 1303, 1261, 1211, 1145. **HRMS** (DART, M+H) Calc'd for $C_{18}H_{13}N_2O_3$ 305.09262, found 305.09241.

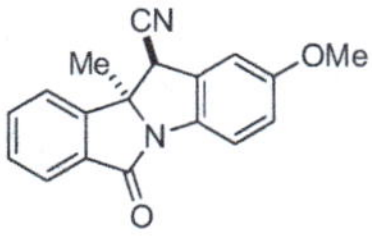

(±)-2-methoxy-10b-methyl-6-oxo-10b,11-dihydro-6*H*-isoindolo[2,1-*a*]indole-11-carbonitrile (2.237q)[133]—Was synthesized according to **general procedure 6** heating for 5 h using **2.229q** (68.6 mg, 0.2 mmol, 1 equiv). The crude product was purified via silica gel flash column chromatography using hexanes:EtOAc (5:4 v:v), and was isolated as a colorless crystalline solid (55.8 mg, 0.192 mmol, 93%, >20:1 dr, MP = 166–167 °C). **^{1}H NMR** (500 MHz, $CDCl_3$) δ 7.92–7.90 (m, 1H), 7.72–7.69 (m, 1H), 7.68–7.65 (m, 1H), 7.60–7.56 (m, 2H), 7.02–6.97 (m, 2H), 4.14–4.13 (m, 1H), 3.81 (s, 3H), 1.71 (s, 3H). **^{13}C NMR** (125 MHz, $CDCl_3$) δ 167.7, 157.7, 146.8, 133.4, 132.6, 132.4, 130.8, 129.9, 125.2, 122.6, 118.6, 116.2, 115.8, 112.2, 73.8, 55.8, 41.4, 27.0. **IR** (thin film, cm^{-1}) 2945, 2245, 1705, 1558, 1489, 1278, 1236. **HRMS** (DART, M+H) Calc'd for $C_{18}H_{15}N_2O_2$ 291.11335, found 291.11384.

[133]This compound was synthesized by Andy Yen.

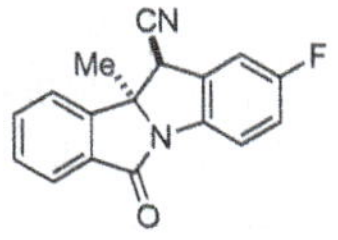

(±)-**2-fluoro-10b-methyl-6-oxo-10b,11-dihydro-6*H*-isoindolo[2,1-*a*]indole-11-carbonitrile (2.237r)**[134]—Was synthesized according to **general procedure 6** heating for 5 h using **2.229r** (66.2 mg, 0.2 mmol, 1 equiv). The crude product was purified via silica gel flash column chromatography using hexanes:EtOAc (5:4 v: v), and was isolated as an off-white solid (47.9 mg, 0.172 mg, 86%, >20:1 dr, MP = 154–155 °C). **^{1}H NMR** (500 MHz, $CDCl_3$) δ 7.94–7.90 (m, 1H), 7.75–7.69 (m, 2H), 7.59 (ddd, *J* = 9.0, 7.5, 1.0 Hz, 2H), 7.20–7.15 (m, 2H), 4.18 (s, 1H), 1.72 (s, 3H). **^{13}C NMR** (125 MHz, $CDCl_3$) δ 167.8, 160.24 (d, *J* = 246.0 Hz), 146.7, 135.5 (d, *J* = 2.5 Hz), 133.8, 132.1, 131.1 (d, *J* = 9.0 Hz), 130.1, 125.4, 122.7, 119.0 (d, *J* = 8.5 Hz), 117.6 (d, *J* = 23.5 Hz), 115.8 (d, *J* = 1.0 Hz), 113.9 (d, *J* = 25.0 Hz), 74.0, 41.3 (d, *J* = 2.0 Hz), 27.1. **^{19}F NMR** (376 MHz, $CDCl_3$) δ-115.8 (td, *J* = 8.0, 4.5 Hz). **IR** (thin film, cm^{-1}) 3076, 3061, 2974, 2929, 2243, 1699, 1614, 1485, 1356, 1134. **HRMS** (DART, M+H) Calc'd for $C_{17}H_{12}FN_2O$ 279.09337, found 279.09360.

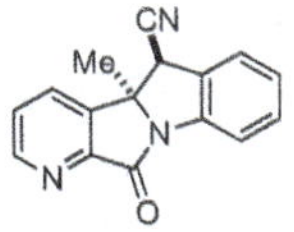

(±)-**4b-methyl-11-oxo-5,11-dihydro-4b*H*-pyrido[3′,2′:3,4]pyrrolo[1,2-*a*]indole-5-carbonitrile (2.237s)**[135]—Was synthesized according to **general procedure 6** (exception: $Pd(P^tBu_3)_2$ (10 mol%) was used) using **2.229s** (63.2 mg, 0.2 mmol, 1 equiv). The crude product was purified via silica gel flash column chromatography using hexanes:EtOAc (5:2 v:v), and was isolated as a colorless crystalline solid (21.1 mg, 0.0808 mmol 40%, >20:1 dr, MP = 199–201 °C). **^{1}H NMR** (500 MHz, CDC13) δ 8.93 (dd, *J* = 5.0, 1.5 Hz, 1H), 8.00 (dd, *J* = 8.0, 1.5 Hz, 1H), 7.86 (ddt, *J* = 8.0, 1.0, 0.5 Hz, 1H), 7.61 (dd, *J* = 8.0, 5.0 Hz, 1H), 7.52–7.48 (m, 1H), 7.47 (ddt, *J* = 7.5, 1.5, 0.5 Hz, 1H), 4.27 (s, 1H), 1.75 (s, 3H). **^{13}C NMR** (125 MHz, $CDCl_3$) δ 165.3, 152.8, 150.2, 141.5, 138.7, 131.1(2), 128.7, 126.7, 126.3, 126.0, 118.3, 115.9, 71.3, 41.0, 27.1. **IR** (thin film, cm^{-1}) 2926, 2846, 1717, 1479, 1464, 1369, 1219. **HRMS** (DART, M+H) Calc'd for $C_{16}H_{12}N_3O$ 262.098045, found 262.09796.

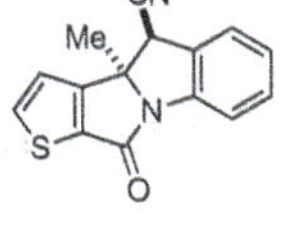

[134]This compound was synthesized by Andy Yen.

[135]This compound was synthesized by Andy Yen.

(±)-3b-methyl-10-oxo-4,10-dihydro-3bH-thieno[3′,2′:3,4]pyrrolo[1,2-a]indole-4-carbonitrile (2.237t)—Was synthesized according to **general procedure 6** using **2.229t** (63.8 mg, 0.2 mmol, 1 equiv). The crude product was purified via silica gel flash column chromatography using hexanes:EtOAc (2:1 v:v) followed by trituration with hexanes and was isolated as a colorless solid (41.1 mg, 0.155 mmol 78%, >20:1 dr, MP = 134–135 °C). **^{1}H NMR** (500 MHz, $CDCl_3$) δ 7.87 (d, J = 5.0 Hz, 1H), 7.69–7.67 (m, 1H), 7.47–7.43 (m, 1H), 7.42–7.39 (m, 1H), 7.22–7.18 (m, 2H), 4.15 (s, 1H), 1.72 (s, 3H). **^{13}C NMR** (125 MHz, $CDCl_3$) δ 163.7, 157.0, 139.7, 138.8, 135.8, 130.8, 129.0, 126.3, 125.4, 120.5, 117.7, 116.3, 72.2, 40.7, 26.6. **IR** (thin film, cm^{-1}) 3010, 2929, 2241, 1705, 1602, 1479, 1342, 1307, 1120, 1070. **HRMS** (DART, M+H) Calc'd for $C_{15}H_{12}N_2OS$ 267.05921, found 267.05887.

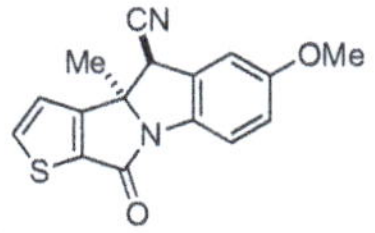

(±)-6-methoxy-3b-methyl-10-oxo-4,10-dihydro-3b*H*-thieno[3′,2′:3,4]pyrrolo[1,2-*a*] indole-4-carbonitrile (2.237u)—Was synthesized according to **general procedure 6** using **2.237u** (69.8 mg, 0.2 mmol, 1 equiv). The crude product was purified via silica gel flash column chromatography using hexanes:EtOAc (2:1 v:v) followed by trituration with hexanes and was isolated as an off-white solid (40.8 mg, 0.138 mmol 69%, >20:1 dr, MP = 178–179 °C). **^{1}H NMR** (500 MHz, $CDCl_3$) δ 7.84 (d, J = 5.0 Hz, 1H), 7.59–7.54 (m, 1H), 7.18 (d, J = 5.0 Hz, 1H), 6.98–6.95 (m, 2H), 4.10 (s, 1H), 3.80 (s, 3H), 1.70 (s, 3H). **^{13}C NMR** (125 MHz, $CDCl_3$) δ 163.9, 157.6, 157.6, 138.5, 135.9, 133.1, 130.3, 120.4, 118.4, 116.2, 115.9, 112.1, 72.5, 55.8, 40.9, 26.3. **IR** (thin film, cm^{-1}) 3099, 2968, 2943, 2837, 2243, 1695, 1595, 1489, 1280, 1238. **HRMS** (DART, M+H) Calc'd for $C_{16}H_{14}N_2O_2S$ 297.06977, found 297.06959.

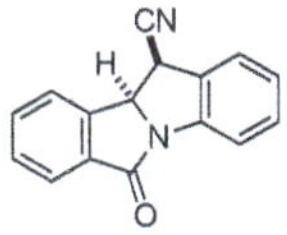

(±)-6-oxo-10b,11-dihydro-6*H*-isoindolo[2,1-*a*]indole-11-carbonitrile (2.237v)—Was synthesized according to **general procedure 6** using **2.229v** (59.8 mg, 0.2 mmol, 1 equiv). The crude product was purified via silica gel flash column chromatography using hexanes:EtOAc (5:4 v:v) followed by trituration with hexanes and was isolated as an colourless crystalline solid solid (15.6 mg, 0.0640 mmol 32%, >20:1 dr, MP = 223–235 °C). **^{1}H NMR** (500 MHz, DMSO-d_6) δ 7.89–7.80 (m, 3H), 7.69–7.59 (m, 3H), 7.50–7.47 (m, 1H), 7.25 (td, J = 7.5, 1.0 Hz, 1H), 6.01 (d, J = 8.5 Hz, 1H), 5.18 (d, J = 8.5 Hz, 1H). **^{13}C NMR** (125 MHz, DMSO-d_6) δ 167.3, 142.5, 139.6, 133.5, 133.1, 131.4, 130.2, 129.9, 126.7, 125.4, 124.8, 124.3, 117.4, 116.5, 66.0, 34.5. **IR** (thin film, cm^{-1}) 2955, 2848, 2237, 1699, 1600, 1477. **HRMS** (DART, M+H) Calc'd for $C_{16}H_{11}N_2O_2$ 247.08714, found 247.08715.

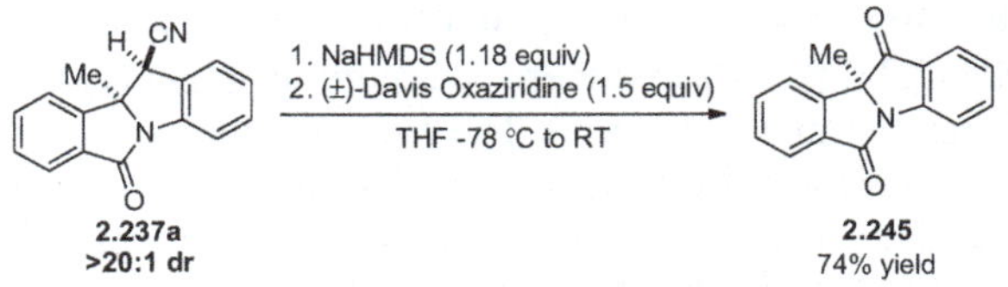

(±)-10b-methyl-6*H*-isoindolo[2,1-*a*]indole-6,11(10b*H*)-dione (2.245)—A oven-dried 2 dram vial containing a stir bar was charged with **2.237a** (120 mg, 0.464 mg, 1 equiv) and purged with argon for 10 min. THF (3.5 mL) was added and the resulting solution was cooled to −78 °C. At this time NaHMDS (557 μL, 0.557 mmol, 1.18 equiv, 1 M in THF) was added drop wise over 5 min. After the addition of the first drop an intense dark purple colour was obtained which was remained. The cooled solution of the sodium α-cyano anion was stirred at this temperature for 10 min at which time a THF solution (1 mL) of (±)-Davis oxaziridine[136] (1.77 g, 0.696 mmol, 1.5 equiv) cooled to −78 °C was added rapidly via cannula. After 5–10 s the deep purple colour dissipated and a transparent, and a light yellow/brown solution remained. The reaction was warmed to room temperature where it was stirred for 1 h. *Note: Even immediately after warming to room temperature a significant amount of* ***2.245*** *could be observed by TLC analysis.* At this time the reaction was quenched with a saturated aqueous solution of NH_4Cl (1 mL), and the aqueous layer was extracted with EtOAc (3×). The combined organic layers were washed sequentially with water and brine, dried over Na_2SO_4, filtered, and concentrated under reduced pressure. The crude residue was purified via silica gel flash column chromatography using hexanes:EtOAc (5:1 v:v) to afford the title compound as a white solid (85.1 mg, 0.344 mmol, 74%, MP = 133–134 °C). **^{1}H NMR** (500 MHz, $CDCl_3$) δ 7.97 (dt, *J* = 8.0, 1.0 Hz, 1H), 7.84 (ddd, *J* = 7.5, 1.0 Hz, 1H), 7.81 (dt, *J* = 7.5, 1.0 Hz, 1H), 7.78–7.73 (m, 2H), 7.70 (ddd, *J* = 7.5, 1.0 Hz, 1H), 7.52 (td, *J* = 7.5, 1.0 Hz, 1H), 7.32–7.25 (m, 1H), 1.80 (s, 3H). **^{13}C NMR** (125 MHz, $CDCl_3$) δ 197.4, 169.6, 151.7, 145.6, 137.4, 134.4, 131.2, 129.5, 127.5, 125.3, 125.2, 125.2, 122.8, 118.6, 73.6, 26.0. **IR** (thin film, cm^{-1}) 1716, 1683, 1602, 1464, 1319, 1296, 1209. **HRMS** (DART, M+H) Calc'd for $C_{16}H_{12}NO_2$ 250.08680, found 250.08696.

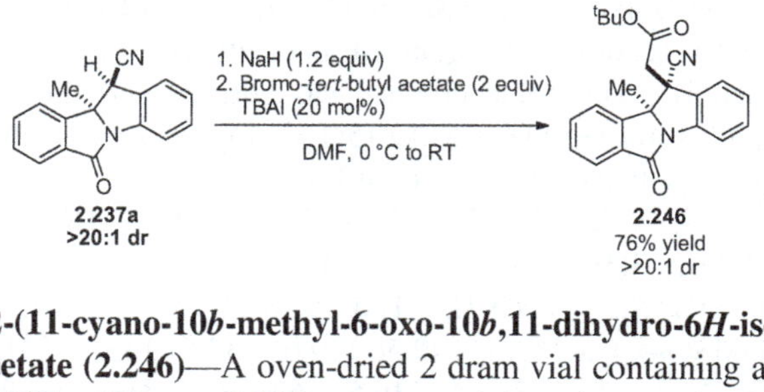

(±)-*tert*-butyl 2-(11-cyano-10*b*-methyl-6-oxo-10*b*,11-dihydro-6*H*-isoindolo[2,1-*a*]indol-11-yl)acetate (2.246)—A oven-dried 2 dram vial containing a stir bar was charged with **2.237a** (75 mg, 0.286 mg, 1 equiv) and purged with argon for

[136]Reference [117].

10 min. DMF (1.0 mL) was added and the resulting solution was cooled 0 °C. At this time, 60% dispersion of NaH in mineral oil (13.7 mg, 0.343 mmol, 1.2 equiv) was added at once and the resulting deep purple solution of the sodium α-cyano anion was stirred at this temperature for 10 min. At this time, TBAI (21.1 mg, 0.0572 mmol, 20 mol%) and bromo*tert*-butyl acetate (85.0 μL, 0.572, 2 equiv) were added in that order. The resulting solution was stirred at this temperature for 30 min before warming to room temperature where it was stirred for an additional 15 min. At this time the deep purple color had dissipated and the reaction was quenched a saturated aqueous solution of NH_4Cl (1 mL), and the aqueous layer was extracted with EtOAc (3×). The combined organic layers were washed sequentially with water (3×) and brine, dried over Na_2SO_4, filtered, and concentrated under reduced pressure. ^{1}H NMR analysis of a homogenous sample of the crude residue indicated the product to be present as a single diastereomer (>20:1). The crude residue was purified via silica gel flash column chromatography using hexanes: EtOAc (5:3 v:v) followed by trituration with hexanes to afford the title compound as a colorless solid (82.2 mg, 0.217 mmol, 76%, MP = 174–175 °C). Suitable single X-ray quality crystals of the major diastereomer were obtained by slow diffusion of hexanes into a saturated acetone solution of **2.246**. **^{1}H NMR** (500 MHz, $CDCl_3$) δ 7.96–7.91 (m, 1H), 7.73–7.68 (m, 2H), 7.67–7.57 (m, 3H), 7.46 (td, *J* = 7.5, 1.0 Hz, 1H), 7.22 (td, *J* = 7.5, 1.0 Hz, 1H), 2.20 (d, *J* = 16.0 Hz, 1H), 1.84 (s, 3H), 1.76 (d, *J* = 16.0 Hz, 1H), 1.29 (s, 9H). **^{13}C NMR** (126 MHz, CDCl) δ 166.6, 166.1, 144.9, 136.4, 134.3, 133.3, 133.2, 130.6, 130.0, 126.4, 125.3, 125.2, 123.4, 118.8, 117.8, 82.4, 76.5, 48.1, 42.1, 27.9, 24.7. **IR** (thin film, cm^{-1}) 3001, 2979, 2931, 2245, 1736, 1704, 1602, 1475, 1357, 1165. **HRMS** (DART, M+H) Calc'd for $C_{23}H_{23}N_2O_3$ 375.17087, found 375.17110.

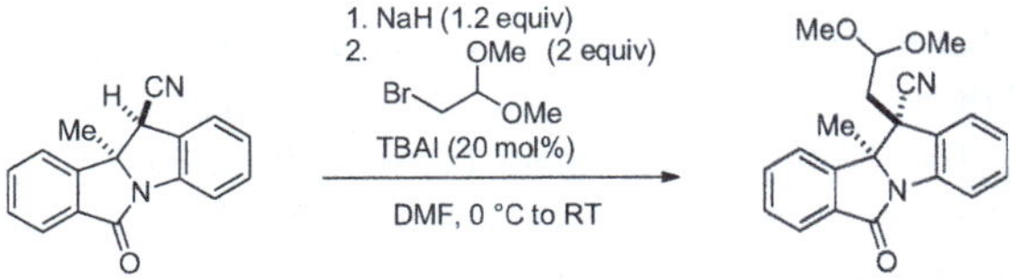

2.237a
>20:1 dr
2.247
77% yield

(±)-11-(2,2-dimethoxyethyl)-10*b*-methyl-6-oxo-10*b*,11-dihydro-6*H*-isoindolo [2,1-*a*]indole-11-carbonitrile (2.247)—A oven-dried 2 dram vial containing a stir bar was charged with **2.237a** (75 mg, 0.286 mg, 1 equiv) and purged with argon for 10 min. DMF (1.0 mL) was added and the resulting solution was cooled 0 °C. At this time, a 60% dispersion of NaH in mineral oil (13.7 mg, 0.343 mmol, 1.2 equiv) was added at once and the resulting deep purple solution of the sodium α-cyano anion was stirred at this temperature for 10 min. At this time, TBAI (21.1 mg, 0.0572 mmol, 20 mol%) and bromoacetaldehyde dimethyl acetal (67.7 μL, 0.572, 2 equiv) were added in that order. The resulting solution was warmed to room temperature and then 60 °C where it was stirred for 1.5 h. At this time the deep colour purple had dissipated and the reaction was quenched a saturated aqueous solution of NH_4Cl (1 mL), and the aqueous layer was extracted with EtOAc (3×). The combined organic layers were washed sequentially with water (3×) and brine, dried over Na_2SO_4, filtered, and concentrated under reduced pressure. ^{1}H NMR analysis of a homogenous sample of the crude residue indicated the product to be present as a single diastereomer (>20:1). The crude residue was purified via silica gel flash column chromatography using 100% DCM to afford the title compound as a colorless solid (76.5 mg, 0.220 mmol, 77%, MP = 104–105 °C). Suitable single X-ray quality crystals of the major diastereomer were obtained by slow diffusion of hexanes into a saturated DCM solution of **2.247**. **^{1}H NMR** (500 MHz, $CDCl_3$) δ 7.92 (ddd, J = 7.5, 1.0 Hz, 1H), 7.72–7.68 (m, 2H), 7.63–7.57 (m, 2H), 7.52 (ddd, J = 7.5, 1.0, 0.5 Hz, 1H), 7.46 (td, J = 7.5, 1.0 Hz, 1H), 7.25 (td, J = 7.5, 1.0 Hz, 1H), 4.12 (dd, J = 7.5, 4.0 Hz, 1H), 3.30 (s, 3H), 2.96 (s, 3H), 1.80 (s, 3H), 1.66 (dd, J = 14.5, 8.0 Hz, 1H), 0.93 (dd, J = 14.5, 4.0 Hz, 1H). **^{13}C NMR** (126 MHz, $CDCl_3$) δ 166.1, 145.2, 136.5, 133.7, 133.3, 133.3, 130.5, 130.0, 126.1, 125.3, 125.0, 122.9, 119.3, 118.1, 77.2, 53.0, 52.3, 48.2, 38.0, 24.6. **IR** (thin film, cm^{-1}) 3016, 2960, 2933, 2835, 2243, 1708, 1602, 1477, 1464, 1361, 1332, 1209, 1122, 1074, 1062. **HRMS** (DART, M+H) Calc'd for $C_{21}H_{21}N_2O_3$ 349.15522, found 349.15518.

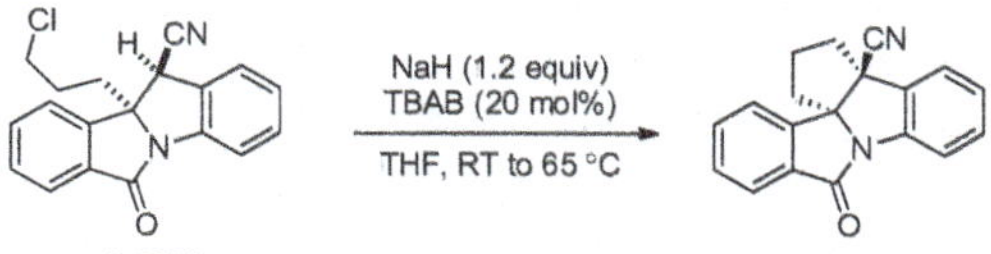

2.237j
>20:1 dr

2.248
89% yield
>20:1 dr

(±)-8-oxo-2,3,8,13*b*-tetrahydro-1*H*-cyclopenta[*b*]isoindolo[2,1-*a*]indole-13*b*-carbonitrile (2.248)—A oven-dried 2 dram vial containing a stir bar was charged with **2.237j** (15 mg, 0.0466 mg, 1 equiv) and TBAB (3.0 mg, 0.00932 mmol, 20 mol%) and was purged with argon for 10 min. THF (1.5 mL) was added and NaH as a 60% dispersion in mineral oil (2.3 mg, 0.0589 mmol, 1.2 equiv) was added at once. The vial was sealed with a Teflon® lined screw cap and placed into an oil bath pre-heated to 65 °C where it was stirred for 16 h. At this time the reaction was quenched a saturated aqueous solution of NH_4Cl (1 mL), and the aqueous layer was extracted with EtOAc (3×). The combined organic layers were washed sequentially with water and brine, dried over Na_2SO_4, filtered, and concentrated under reduced pressure. ^{1}H NMR analysis of a homogenous sample of the crude residue indicated the product to be present as a single diastereomer (>20:1). The crude residue was purified via two rounds of silica gel flash column chromatography first using hexanes:EtOAc (5:2) then 100% DCM to afford the title compound as a colorless foam (11.8, 0.0413, 89%). Suitable single X-ray quality crystals of the title compound were obtained by slow diffusion of hexanes into a dilute acetone solution of **2.248**. **^{1}H NMR** (500 MHz, $CDCl_3$) δ 7.87 (ddd, J = 7.5, 1.0 Hz, 1H), 7.74–7.70 (m, 2H), 7.59–7.54 (m, 2H), 7.41 (ddd, J = 8.0, 7.5, 1.5 Hz, 1H), 7.34 (ddd, J = 7.5, 1.5, 0.5 Hz, 1H), 7.20 (ddd, J = 7.5, 1.0 Hz, 1H), 2.69–2.66 (m, 2H),

2.46–2.32 (m, 2H), 2.06–1.99 (m, 1H), 1.72–1.61 (m, 1H). **^{13}C NMR** (125 MHz, $CDCl_3$) δ 169.8, 148.0, 141.4, 133.9, 133.2, 131.7, 130.4, 129.7, 125.6, 125.0, 124.3, 123.0, 119.3, 116.7, 83.6, 52.0, 43.4, 40.7, 24.6. **IR** (thin film, cm^{-1}) 2966, 2916, 2848, 2252, 1708, 1481, 1296. **HRMS** (DART, M+H) Calc'd for $C_{19}H_{15}N_2O$ 287.11844, found 287.11926.

References

1. Yoon, H., Petrone, D.A., Lautens, M.: Org. Lett. **16**, 6420 (2014)
2. Petrone, D.A., Yen, A., Zeidan, N., Lautens, M.: Org. Lett. **17**, 4838 (2015)
3. Knah, N.-H., Kureshy, R.I., Abdi, S.H.R., Agarwal, S., Jasra, R.V.: Coor. Chem. Rev. **252**, 593 (2008)
4. Kurono, N., Ohkuma, T.: ACS. Catal. **6**, 989 (2016)
5. Pongratz, A.: Monatsh. Chem. **48**, 585 (1927)
6. Rosenmund, K.W., Struck, E.: Chem. Ber. **52**, 1749 (1919)
7. von Braun, J., Manz, G.: Liebigs Ann. Chem. **488**, 111 (1931)
8. Lindley, J.: Tetrahedron **40**, 1433 (1984)
9. Ellis, G.P., Romney-Alexander, T.M.: Chem. Rev. **87**, 779 (1987)
10. Beletskaya, I.P., Sigeev, A.S., Peregudov, A.S., Petrovskii, P.V.: J. Organomet. Chem. **689**, 3810 (2004)
11. Anbarasan, P., Schareina, T., Beller, M.: Chem. Soc. Rev. **40**, 5049 (2011)
12. Wen, Q., Jin, J., Zhang, L., Luo, L., Lu, P., Wang, Y.: Tetrahedron Lett. **55**, 1271 (2014)
13. Takagi, K., Okamoto, T., Sakakibara, Y.: Chem. Lett, 471 (1973)
14. Takagi, K., Okamoto, T., Sakakibara, Y., Ohno, A., Oka, S., Hayama, N.: Bull. Chem. Soc. Jpn **49**, 3177 (1976)
15. Cassar, L., Foà, M., Montanari, F., Marinelli, G.P.: J. Organomet. Chem. **173**, 335 (1979)
16. Takagi, K., Okamoto, T., Sakakibara, Y., Ohno, A., Oka, S., Hayama, N.: Bull. Chem. Soc. Jpn **48**, 3298 (1975)
17. Yamamura, K., Muzahashi, S.J.: Tetrahedron Lett. **18**, 4429 (1977)
18. Dalton, J.R., Regen, S.L.: J. Org. Chem. **44**, 4443 (1979)
19. Regen, S.L., Quici, S., Liaw, S.J.: J. Org. Chem. **44**, 2029 (1979)
20. Luo, F.-H., Chu, C.-I., Cheng, C.-H.: Organometallics **1998**, 17 (1025)
21. Jiang, B., Kann, Y., Zangh, A.: Tetrahedron **57**, 1581 (2001)
22. Sundermeier, M., Zapf, A., Beller, M.: Eur. J. Inorg. Chem., 3513 (2003)
23. Jin, F., Confalone, P.N.: Tetrahedron Lett. **41**, 3271 (2000)
24. Zapf, A., Beller, M.: Chem. Eur. J. **7**, 2908 (2001)
25. Gómez Andreu, M., Zapf, A., Beller, M. Chem. Commun., 2475 (2000)
26. Ehrentraut, A., Zapf, A., Beller, M.J.: Mol. Catal. **182–183**, 515 (2002)
27. Sundermeier, M., Zapf, A., Beller, M.: Tetrahedron Lett. **42**, 670 (2001)
28. Sundermeier, M., Zapf, A., Mutyala, S., Baumann, W., Sans, J., Weiss, S., Beller, M.: Chem. Eur. J. **9**, 1828 (2003)
29. Littke, A., Soumeillant, M., Kaltenback, R.F., Cherney, R.J., Tarby, C.M., Kiau, S.: Org. Lett. **9**, 1711 (2007)
30. Shareina, T., Zapf, A., Beller, M.: Chem. Commmun., 1388 (2004)
31. Murahashi, S., Naota, T., Nakajima, N.: J. Org. Chem. **51**, 898 (1986)
32. Miller, J.A.: Tetrahedron Lett. **42**, 6991 (2001)
33. Miller, J.A., Dankwardt, J.W.: Tetrahedron Lett. **2003**, 44 (1907)
34. Miller, J.A., Dankwardt, J.W., Penney, J.M.: Synthesis **11**, 1643 (2003)
35. Penney, J.M., Miller, J.A.: Tetrahedron Lett. **45**, 4989 (2004)

36. Nakao, Y., Hiyama, T.: Pure Appl. Chem. **2008**, 80 (1097)
37. Nakao, Y., Oda, S., Hiyama, T.: J. Am. Chem. Soc. **126**, 13904 (2004)
38. Nakao, Y., Oda, S., Yada, A., Hiyama, T.: Tetrahedron Lett. **62**, 7567 (2006)
39. Nakao, Y., Yada, A., Ebata, S., Hiyama, T.: J. Am. Chem. Soc. **129**, 2428 (2007)
40. Nakao, Y., Yada, A., Satoh, J., Ebata, S., Oda, S., Hiyama, T.: Chem. Lett. **35**, 790 (2006)
41. Nakoa, Y., Yukawa, T., Hirata, Y., Oda, S., Satoh, J., Hiyama, T.: J. Am. Chem. Soc. **128**, 7116 (2006)
42. Hirata, Y., Yukawa, T., Kashihara, N., Nakao, Y., Yiyama, T.: J. Am. Chem. Soc. **131**, 10964 (2009)
43. Nakao, Y., Hirata, Y., Tanaka, M., Hiyama, T.: Angew. Chem. Int. Ed. **47**, 385 (2008)
44. Hirata, Y., Iniu, T., Nakao, Y., Hiyama, T.: J. Am. Chem. Soc. **131**, 6642 (2009)
45. Hirata, Y., Yada, A., Morita, E., Nakao, Y., Hiyama, T., Ohashi, M., Ogoshi, S.: J. Am. Chem. Soc. **132**, 10070 (2010)
46. Nakao, Y., Yada, A., Hiyama, T.: J. Am. Chem. Soc. **132**, 10024 (2010)
47. Nakao, Y., Ebata, S., Yada, A., Hiyama, T., Ikawam, M., Ogoshi, S.: J. Am. Chem. Soc. **130**, 12874 (2008)
48. Hsieh, J.-C., Ebata, S., Nakao, Y., Hiyama, Y.: Synthesis **11**, 1709 (2010)
49. Watson, M.P., Jacobsen, E.N.: J. Am. Chem. Soc. **130**, 12594 (2008)
50. Fang, X., Yu, P., Morandi, B.: Science **351**, 832 (2016)
51. Burns, B., Grigg, R., Sridharan, V., Worakun, T.: Tetrahedron Lett. **29**, 4325 (1988)
52. Arcadi, A., Bernocchi, E., Burini, A., Cacchi, S., Marinelle, F., Pietroni, B.: Tetrahedron **44**, 481 (1988)
53. Burns, B., Grigg, R., Santhakumar, V., Sridharan, V., Strevenson, P., Worakun, T.: Tetrahedron **48**, 7297 (1992)
54. Grigg, R., Sridharan, V.: J. Organomet. Chem. **576**, 65 (1999)
55. Brown, S., Clarkson, S., Grigg, R., Thomas, W.A., Sridharan, B., Wilson, D.M.: Tetrahedron **57**, 1347 (2001)
56. René, O., Lapoint, D., Fagnou, K.: Org. Lett. **11**, 4560 (2009)
57. Brown, D., Grigg, R., Sridhara, V., Tambyrajah, V.: Tetrahedron Lett. **36**, 8137 (1995)
58. Ruck, R.T., Huffman, M.A., Kim, M.M., Shelvin, M., Kandur, W.V., Davies, I.W.: Angew. Chem. Int. Ed. **47**, 4711 (2008)
59. Satyanarayana, G., Maichle-Mössmer, C., Maier, M.E.: Chem. Commun. **12**, 1571 (2009)
60. Hu, Y., Yu, C., Ren, D., Hu, Q., Zhang, L., Cheng, D.: Angew. Chem. Int. Ed. **48**, 5448 (2009)
61. Jaegli, S., Erb, W., Retailleau, P., Vors, J.-P., Neuville, L., Zhu, J.: Chem. Eur. J. **16**, 5863 (2010)
62. Liu, X., Ma, X., Huang, Y., Gu, Z.: Org. Lett. **15**, 4814 (2013)
63. Liu, Z., Xia, Y., Zhou, S., Wang, L., Zhang, Y., Wang, J.: Org. Lett. **15**, 5032 (2013)
64. Vachhani, D.D., Butani, H.H., Sharma, N., Bhoya, U.C., Shah, A.K., Van der Eycken, E.V.: Chem. Commun. **51**, 14862 (2015)
65. Yoon, H., Jang, Y.-J., Lautens, M.: Synthesis **48**, 1483 (2016)
66. Nájera, C., Sansano, J.M.: Angew. Chem. Int. Ed. **48**, 2452 (2009)
67. Torii, S., Okumoto, H., Ozaki, H., Nakayasu, S., Tadokoro, T., Kotani, T.: Tetrahedron Lett. **33**, 3499 (1992)
68. Grigg, R., Santhakumar, V., Sridharan, V.: Tetrahedron Lett. **34**, 3163 (1993)
69. Pinto, A., Jia, Y., Leuville, L., Zhu, J.: Chem. Eur. J. **13**, 961 (2007)
70. Jaegli, S., Vors, J.-P., Leuville, L., Zhu, J.: Synlett **18**, 2997 (2009)
71. Jaegli, S., Vors, J.-P., Leuville, L., Zhu, J.: Tetrahedron **66**, 8911 (2010)
72. Lee, H.S., Kim, K.H., Lim, J.W., Kim, J.N.: Bull. Chem. Soc. Jpn **2011**, 32 (1082)
73. Lu, Z., Hu, C., Guo, J., Li, J., Zui, Y., Jia, Y.: Org. Lett. **12**, 280 (2010)
74. Chatani, N., Takeyasu, T., Noriuchi, N., Hanafusa, T.: J. Org. Chem. **53**, 3539 (1988)
75. Nozaki, K., Sato, N., Takaya, H.: J. Org. Chem. **59**, 2679 (1994)

76. Nozaki, K., Sato, N., Takaya, H.: Bull. Chem. Soc. Jpn **69**, 1629 (1996)
77. Nishihara, Y., Inoue, Y., Itazaki, M., Takagi, K.: Org. Lett. **7**, 2639 (2005)
78. Nishihara, Y., Inoue, Y., Izawa, S., Miyasaka, M., Tanemure, K., Nakajima, K., Takagi, K.: Tetrahedron **62**, 9872 (2006)
79. Kobayashi, Y., Kamisaki, H., Yanada, R., Takemoto, Y.: Org. Lett. **8**, 2711 (2006)
80. Kobayashi, Y., Kamisaki, H., Takeda, H., Yasui, Y., Yanada, R., Takemoto, Y.: Tetrahedron **63**, 2978 (2007)
81. Kobayashi, Y., Kamisaki, H., Takemoto, Y.: Org. Lett. **10**, 3303 (2008)
82. Cheng, Y.-N., Duan, Z., Yu, L., Li, Z., Zhu, Y., Wu, Y.: Org. Lett. **10**, 901 (2008)
83. Hopkins, C.R., O'Neil, S.V., Laufersweiler, M.C., Wang, Y., Pokross, M., Mekel, M., Evdokimov, A., Walter, R., Kontoyianni, M., Petrey, M.E., Sabatakos, G., Roesgen, J.T., Richardson, E., Demuth, T.P.: Bioorg. Med. Chem. Lett. **16**, 5659 (2006)
84. Schareina, T., Zapf, A., Beller, M.: J. Organomet. Chem. **689**, 4576 (2004)
85. Tschaen, D.M., Desmond, R., King, A.O., Fortin, M.C., Pipik, B., King, S., Verhoeven, T.R.: Synth. Commun. **24**, 887 (1994)
86. Venkatraman, S., Li, C.-J.: Org. Lett. **1**, 1133 (1999)
87. Venkatraman, S., Li, C.-J.: Tetrahedron Lett. **41**, 4831 (2000)
88. Li, J.-H., Xie, Y.-X., Yin, D.-L.: J. Org. Chem. **68**, 9867 (2003)
89. Wang, L., Zhang, Y., Liu, L., Wang, Y.: J. Org. Chem. **71**, 1284 (2006)
90. Littke, A.F., Fu, G.C.: J. Am. Chem. Soc. **123**, 6989 (2001)
91. Zhao, L., Li, Z., Chang, L., Xu, J., Yao, H., Wu, X.: Org. Lett. **14**, 2066 (2012)
92. Shen, C., Liu, R.-R., Fan, R.-J., Li, Y.-L., Xu, T.-F., Gao, J.-R., Jia, Y.-X.: J. Am. Chem. Soc. **137**, 4936–4939 (2015)
93. Hastings, D.J., Weedon, A.C.: Can. J. Chem. **69**, 1171 (1991)
94. Fang, Y.-Q., Lautens, M.: Org. Lett. **7**, 3549 (2005)
95. Negishi, E.-I., Coperet, C., Ma, S., Liou, S.-Y., Liu, F.: Chem. Rev. **96**, 365 (1996)
96. Bipp, H., Kieczka, H.: Formamides. In: *Ullmann's Encyclopedia of Industrial Chemistry*. Wiley-VCH, Weinheim (**2010**)
97. Weinstein, A.B., Stahl, S.S.: Catal. Sci. Technol. **4**, 4301 (2014)
98. Kozikowski, A.P., Ma, D.: Tetrahedron Lett. **32**, 3317 (1991)
99. Freerksen, R.W., Selikson, S.J., Wroble, R.R., Kyler, K.S., Watt, D.S.: J. Org. Chem. **48**, 4087 (1983)
100. Davis, F.A., Nadir, U.K.: Tetrahedron Lett. **18**, 1721 (1977)
101. Petrone, D.A., Yoon, H., Weinstabl, H., Lautens, M.: Angew. Chem. Int. Ed. **53**, 7908 (2014)
102. Fang, Y.-Q., Lautens, M.: J. Org. Chem. **73**, 538 (2008)
103. Basset, J., Romero, M., Serra, T., Pujol, M.D.: Tetrahedron **68**, 356 (2012)
104. Stokes, B.J., Dong, H., Leslie, B.E., Pumphrew, A.L., Driver, T.G.: J. Am. Chem. Soc. **129**, 7500 (2007)
105. Weinstabl, H., Suhartono, M., Qureshi, Z., Lautens, M.: Angew. Chem. Int. Ed. **52**, 5305 (2013)
106. Lian, J.-J., Odedra, A., Wu, C.-J., Liu, R.S.: J. Am. Chem. Soc. **127**, 4186 (2005)
107. Karaki, F., Ohgane, K., Fukuda, H., Nakamura, M., Dodo, K., Hashimoto, Y.: Bioorg. Med. Chem. **22**, 3587 (2014)
108. Suresh, R.R., Kumara Swamy, K.C.: J. Org. Chem. **77**, 6959 (2012)
109. Ferraccioli, R., Carenzi, D., Motti, E., Catellani, M.: J. Am. Chem. Soc. **128**, 722 (2006)
110. Axford, L.C., Holden, K.E., Hasse, K., Banwell, M.G., Steglich, W., Wagler, J., Willis, A.C.: Aust. J. Chem. **61**, 80 (2008)
111. Barton, T.J., Lin, J., Ijadi-Maghsoodi, S., Power, M.D., Zhang, X., Ma, Z., Shimizu, H., Gordon, M.S.: J. Am. Chem. Soc. **117**, 11695 (1995)
112. Bunch, L., Jensen, A.A., Nielsen, B., Braeuner-Osborn, H.: J. Med. Chem. **49**, 172 (2006)
113. Yamagishi, M., Nishigai, K., Ishii, A., Hata, T., Urabe, H.: Angew. Chem. Int. Ed. **51**, 6471–6474 (2012)

114. Mosley, C.A., Myers, S.J., Murray, E.E., Santangelo, R., Tahirovic, Y.A., Kurtkaya, N., Mullasseril, P., Yuan, H., Lyuboslavsky, P., Le, P., Wilson, L.J., Yepes, M., Dingledine, R., Traynelis, S.F., Liotta, D.C.: Bioog. Med. Chem. **17**, 6463 (2009)
115. Wang, Y., Ye, L., Zhang, L.: Chem. Commun. **47**, 7815 (2011)
116. Ito, Y., Kobayashi, K., Seko, N., Saegusa, T.: Bull. Chem. Soc. Jpn **57**, 73 (1984)
117. Vishwakarma, L.C., Stringer, O.D., Davis, F.A.: Org. Syn. **66**, 203 (1988)

Chapter 3
Development of a Pd-Catalyzed Dearomative 1,2-Diarylation of Indoles Using Aryl Boron Reagents

3.1 Introduction

The Suzuki-Miyaura cross-coupling reaction is undoubtedly one of the most important carbon–carbon bond-forming processes. The extremely broad scope of this transformation has made it one of the most employed carbon–carbon bond-forming reactions in the pharmaceutical industry. This reaction traditionally involves $C(sp^2)$–$C(sp^2)$ coupling, however, virtually all permutations of nucleophile and electrophile hybridizations have been explored. Recent effort has been focused on stereospecific and stereoselective Suzuki-Miyaura cross-couplings which allows for the formation of stereodefined carbon–carbon bonds. These transformations utilize either a chiral nucleophile or electrophile in the presence of an achiral metal catalyst (Scheme 3.1a), or a racemic/achiral nucleophile or electrophile in the presence of a chiral metal catalyst (Scheme 3.1b), respectively.[1]

However, in most of these cases the synthesis of sp^3 hybridized nucleophiles and electrophiles is required. As discussed in Chap. 2, Mizoroki-Heck-type domino reactions may facilitate the generation of stereodefined Pd(II) species which can be functionalized by employing a variety of terminating transformations (i.e. reduction, cyanation, borylation). These reactions can be used to carry out alkene and alkyne diarylations by utilizing aryl or vinyl boron reagents as nucleophiles in a terminating Suzuki-Miyaura cross-coupling. One can assume that the development of these transformations using unconventional unsaturated substrates like indoles may lead to a greatly increased scope.

Portions of this chapter have appeared in print. See: Ref. [1].

[1]Reference [2].

D. A. Petrone, *Stereoselective Heterocycle Synthesis via Alkene Difunctionalization*, Springer Theses, https://doi.org/10.1007/978-3-319-77507-4_3

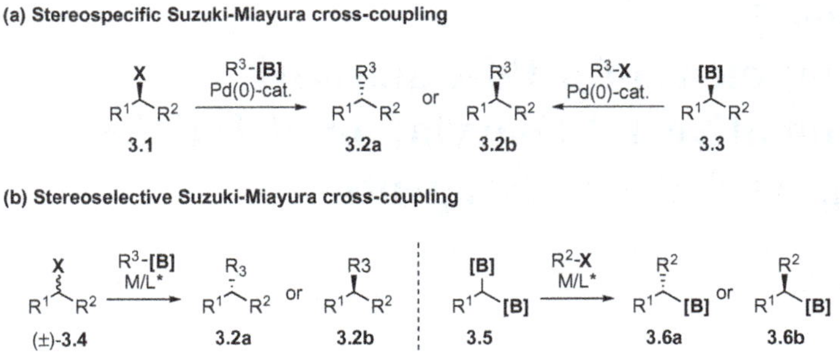

Scheme 3.1 General representations of stereospecific and stereoselective Suzuki-Miyaura cross-couplings

3.1.1 Pd-Catalyzed Domino Mizoroki-Heck/Suzuki-Miyaura Reactions of Alkenes and Alkynes

When aryl or vinyl (pseudo)halides **3.7** and aryl or vinyl boron reagents **3.8** are reacted with alkenes or alkynes in the presence of Pd salts, the resulting diarylation, arylvinylation, or divinylation products **3.9** can be obtained. This transformation proceeds via a mechanism that resembles a hybrid of the Suzuki-Miyaura cross-coupling and the Mizoroki-Heck reaction (Scheme 3.2).[2]

In 1989, Grigg reported the first examples of this reaction which involved an intramolecular cyclization.[3] Therein, they showed that alkynes **3.10** and **3.11** and alkenes **3.15** and **3.16** underwent a cyclization/anion capture sequence using alkenyl boronate ester **3.12** in the presence of a Pd(0) catalyst. These reactions proceed either via the vinylation of a vinyl (Scheme 3.3a) or an alkyl Pd(II) species (Scheme 3.3b). Since this seminal report, these authors have made many contributions which have led to a significant broadening of the reaction scope.[4] This class of cyclization has since been used to construct many types of carbo- and heterocyclic products. These reactions have facilitated the synthesis of oxindoles,[5] isoindolones,[6] dihydroindenes,[7] dibenzoxamines,[8] benzazepines,[9] and others.[10]

[2]Aryl iodides, as well as aryl and vinyl tin and zinc reagents have been used to terminate these sequences. See: Refs. [3–9].

[3]Reference [10].

[4]References [11–13].

[5]References [14, 15].

[6]References [16, 17].

[7]Reference [18].

[8]References [19, 20].

[9]Reference [21].

[10]References [22–25].

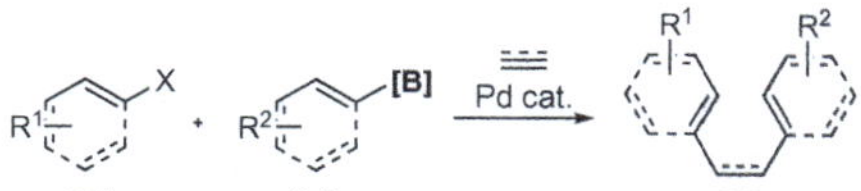

Scheme 3.2 Pd-catalyzed Mizoroki-Heck/Suzuki-Miyaura domino reaction using organoboron reagents

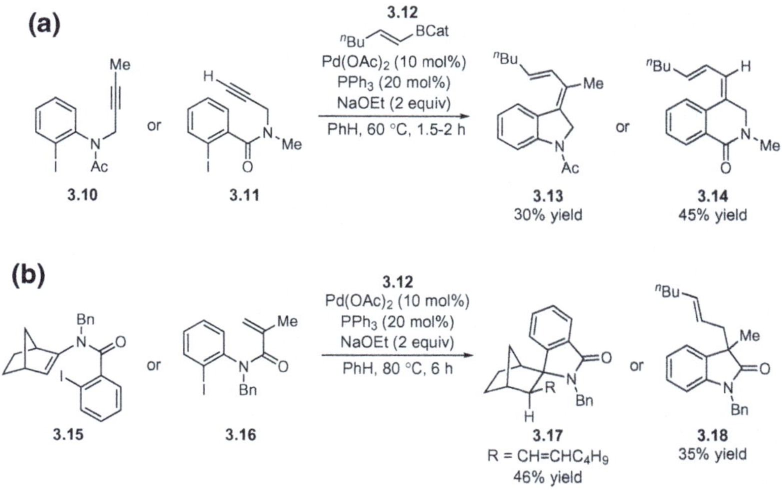

Scheme 3.3 The use of vinyl boron reagents in seminal intramolecular Pd-catalyzed Mizoroki-Heck-type domino reactions of **a** alkynes; and **b** alkenes

Also in 1989, Catellani et al. reported the intermolecular *syn* 1,2-diarylation of norbornene **3.19** or norbonadiene **3.20** using aryl bromides **3.21** and sodium tetraphenylborate **3.22** (Scheme 3.4).[11] The authors observed the formation of biaryls **3.25** which arise from a competing direct Suzuki-Miyaura cross-coupling. The corresponding *cis*-stillbenes could be obtained by a retro-Diels-Alder reaction by thermal treatment of the *syn* 1,2-diarylation products in a silica tube at 500 °C.

Based on this work, Goodson reported a Pd-catalyzed three component coupling between Diels-Alder adducts **3.26**, aryl iodides **3.27** and arylboronic acids **3.28** in 2002 (Scheme 3.5a).[12] The use of aryl boronic acids allowed them to slightly broaden the scope with respect to the previous method, which only described the

[11]Reference [26].

[12]Reference [27].

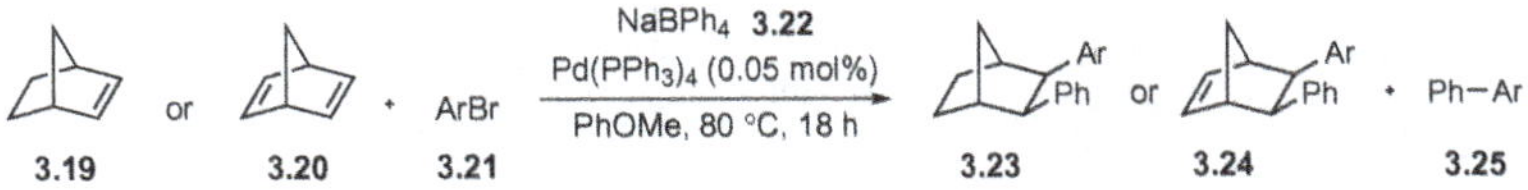

Scheme 3.4 Pd-catalyzed intramolecular *syn* 1,2-diarylation of strained alkenes

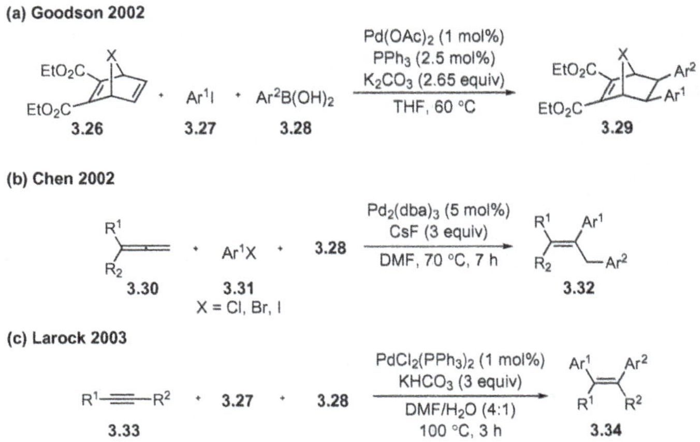

Scheme 3.5 Advances in Pd-catalyzed three component Mizoroki-Heck/Suzuki-Miyaura coupling reactions

use of $NaBPh_4$. Chen reported that allenes **3.30** could be used instead of alkenes in this class of reactions (Scheme 3.5b).[13] The organohalide **3.31** and the aryl boronic acid **3.28** components add in a 1,2 fashion across the terminal double bond of the allene **3.30**. In 2003, Larock reported that alkynes could undergo a similar three-component coupling process using alkynes **3.33** to generate tetrasubstituted alkenes **3.34** (Scheme 3.5c).[14] These authors also developed a methodology which used vinyl iodides instead of aryl iodides in the same year.[15]

Sigman and co-workers reported a Pd-catalyzed three component coupling reaction of 1,3-dienes **3.35**, vinyl triflates **3.37** and aryl boronic acids in 2011 (Scheme 3.6a).[16] This reaction relies on the formation of a key π-allyl ArPd(II) species (i.e. **3.40**) and a subsequent carbon–carbon bond-forming reductive elimination in order to generate the 1,2-difunctionalized products. Therein, they also

[13] Reference [28].
[14] Reference [29].
[15] References [30, 31].
[16] Reference [32].

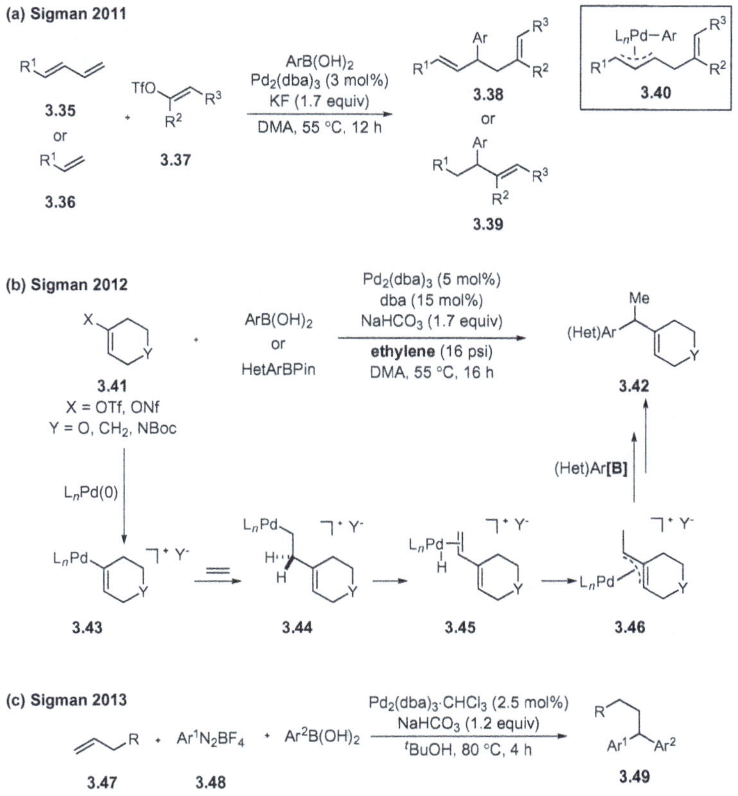

Scheme 3.6 Pd-catalyzed 1,1-alkene difunctionalization by Sigman

showed that these reaction conditions could be used to carry out the 1,1-difunctionalization of simple α-olefins **3.36**. In 2012, the same authors showed that ethylene could undergo a Pd-catalyzed 1,1-difunctionalization using vinyl sulfonates **3.41** and aryl boronic acids or heteroaryl pinacol boronates (Scheme 3.6b).[17] The authors proposed a mechanism which involves oxidative addition of **3.41** to generate cationic Pd(II) species **3.43** which can undergo migratory insertion with ethylene to generate alkyl–Pd(II) intermediate **3.44**. Due to the presence of *syn*-β-hydrogen atoms, species **3.44** can undergo β-hydride elimination to generate olefin bound Pd(II)H **3.45**, and then hydropalladate to form π-ally Pd(II) speceies **3.42**. Transmetallation with the (hetero)aryl boron reagent to **3.46** followed subsequent carbon–carbon bond-forming reductive elimination (Scheme 3.6b). The same authors extended this methodology to allylically substituted alkenes **3.47** using aryl diazonium salts **3.48** in 2013 (Scheme 3.6c).[18]

---

[17]Reference [33].

[18]Reference [34].

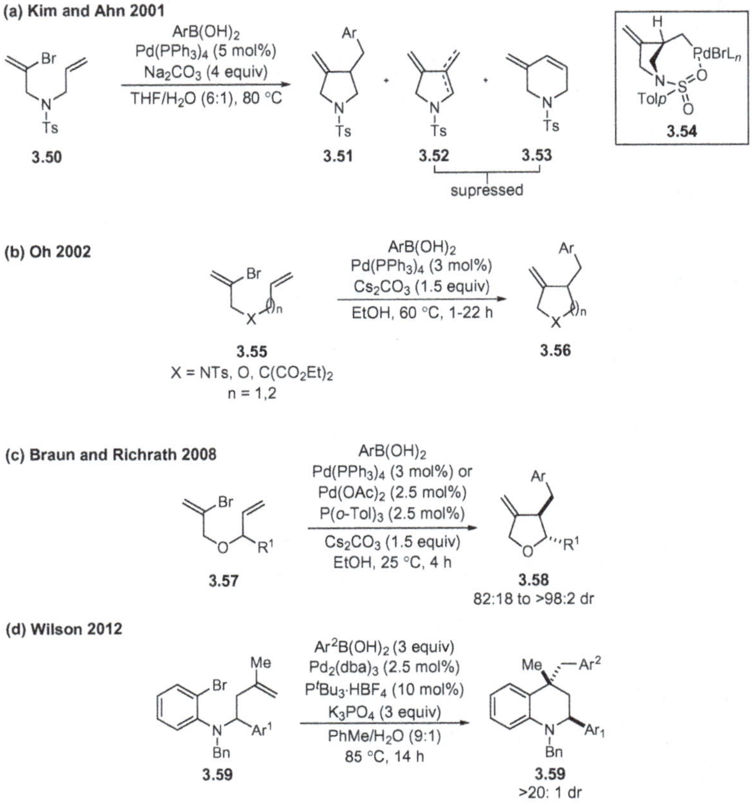

Scheme 3.7 Pd-catalyzed intramoelcular alkene diarylation reactions

Despite the large number of reports concerning the use of alkynes in intramolecular domino Miroroki-Heck/Suzuki-Miyaura diarylation reactions, the use of alkenes in analogous processes has lagged considerably. There have only been a small number of these reports which highlight synthetic utility of these reactions, and only a fraction of these are diastereoselective transformations. In 2000, Kim and Ahn reported that brominated 1,6-diene **3.50** could undergo a Pd-catalyzed intramolecular vinylarylation in the presence of arylboronic acids to generate pyrrolidines **3.51** (Scheme 3.7a).[19] Even though this reaction uses monosubstituted alkene tethers, products arising from β-hydride elimination **3.52** are suppressed due to a proposed coordination of the Ts protecting group, which stabilizes the alkyl Pd(II) species resulting from alkene insertion (i.e. **3.53**). The authors also noted that the efficiency of this reaction partly depends on a high propensity of the arylboronic acid to undergo transmetallation with **3.54**. If for any

[19]Reference [35].

reason this step is slow, larger amounts of byproducts such as **3.52** and **3.53** are observed. The latter product **3.53** is thought to arise from **3.54** undergoing an intramolecular cyclopropanation/cyclopropylcarbinyl-to-homoallyl Pd rearrangement sequence followed by β-hydride elimination.[20] In 2002, some of the same authors reported that 6- and 7-membered rings could be synthesized by using a slightly altered protocol (Scheme 3.7b).[21] They found EtOH and Cs_2CO_3 to be the optimal solvent and base, respectively, although no explanation for their positive influence was given.

Braun and Richrath extended this mode of reactivity to a diastereoselective variant which provides access to *trans*-disubstituted methylenetetrahydrofurans **3.58** (Scheme 3.7c).[22] Although the authors refer to this transformation as the first example of a diastereoselective domino Mizorok-Heck/Suzuki-Miyaura reaction, this classification is not accurate. The first example of this reaction proceeding in a diastereoselective fashion was by Grigg, wherein the *syn* nature of the carbopalladation step sets two stereocenters simultaneously in a highly diastereoselective fashion (cf. Scheme 3.3b from **3.15** to **3.18**).[23]

The report of Braun and Richrath does however represent the first example of a diastereoselective Mizorok-Heck/Suzuki-Miyaura reaction on a substrate possessing a pre-existing stereocenter using an *aryl* boron regent. It is also the first example of this transformation leading to this class of compounds. In 2012, Wilson reported the diastereoselective synthesis of substituted tetrahydroquinolinones **3.59** via a Pd-catalyzed Mizoroki-Heck/Suzuki-Miyaura cascade reaction (Scheme 3.7d).[24] Therein, a combination of $Pd_2(dba)_3$ and $P^tBu_3{\cdot}HBF_4$ provided optimal yields and diastereoselectivities for this transformation. Other ligands such as XPhos, PCy_3, PPh_3, and BINAP were also effective; however, they mostly provided the product with inferior levels of stereoselectivity. This reaction was limited to the use of aryl boronic acids, and when alkenyl and heteroaryl boronic acids were tested, only the direct Suzuki-Miyaura cross-coupling products were observed. The stereochemistry of the products was determined by NOE studies, although no hypothesis was put forward to explain the observed stereochemical outcome.

3.1.2 *Pd(II)-Catalyzed Diarylation of Alkenes and Alkynes*

There have been several reports using Pd(II) salts to catalyze various classes of diarylation reactions. In 2009, Sigman reported the oxidative Pd-catalyzed 1,2-diarylation of styrenes and 1,3-dienes **3.60** using arylstannanes **3.61** and

[20]Reference [36].

[21]Reference [37].

[22]Reference [38].

[23]Reference [10].

[24]Reference [39].

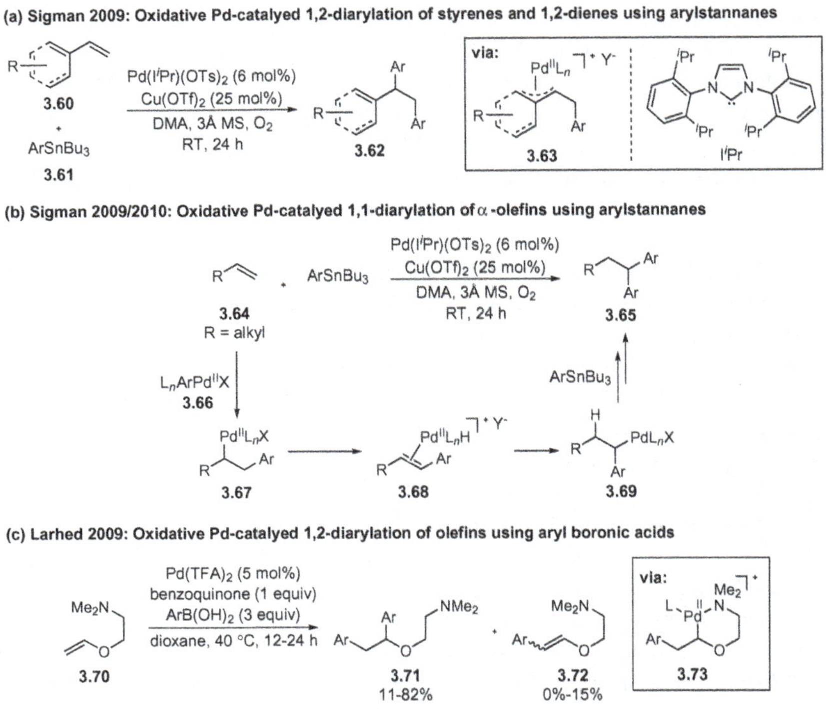

Scheme 3.8 Oxidative Pd(II)-catalyzed diarylation reactions

molecular oxygen as the terminal oxidant (Scheme 3.8a).[25] The authors proposed that the reaction mechanism involves the transmetallation of a second equivalent of the aryl stannane reagent onto π-allyl Pd(II) species **3.63** which is stabilized to such an extent that β-hydride elimination is greatly suppressed. In this same report, it was shown that simple α-olefins **3.64** such as 1-nonene provided the corresponding 1,1-diarylation products **3.65** in good yields with high levels of regioselectivities (Scheme 3.8b). The high selectivity for the 1,1-diarylation products was thought to originate from the lack of stabilization of the Pd(II) intermediate resulting from olefin insertion. The proposed mechanism involves the initial formation of the $L_nArPd(II)X$ species **3.66** by transmetallation of the $ArSnBu_3$ with the Pd(II) catalyst. Species **3.66** undergoes alkene insertion to generate non-stabilized alkyl Pd (II) species **3.67** which undergoes β-hydride elimination to generate olefin bound Pd(II) hydride **3.68**. Species **3.68** regioselectively hydropalladates to generate **3.69** which undergoes penultimate transmetallation with a second equivalent of the aryl stannane, followed by reductive elimination. The same authors reported the expansion of the scope for the 1,1-diarylation reaction in 2010.[26] Larhed and

[25]Reference [40].

[26]Reference [41].

co-workers reported an analogous reaction using aryl boronic acids and enol ether **3.70** which generated 1,2-diarylation products **3.71** in good yields with only trace amounts of the undesired oxidative Mizoroki-Heck products **3.72** (Scheme 3.8c).[27] They proposed that the tertiary amine functionality of **3.70** can coordinate to the Pd (II) species after alkene insertion to generate palladacycle **3.73** which aids in suppressing β-hydride elimination. They noted that when the reoxidation reagent was changed from benzoquinone to O_2 or $Cu(OAc)_2$, only monoarylated products **3.72** were observed. Furthermore, only **3.72** was observed when the reaction was run using 1 equivalent of $Pd(TFA)_2$ in the absence of benzoquinone. Based on these experiments, they proposed that benzoquinone must play a key role by stabilizing **3.73** through coordination (i.e. L = benzoquinone).

3.1.3 Palladium-Catalyzed Dearomatization Reactions

Aromatic compounds are ubiquitous in nature and represent important building blocks in both industrial and academic laboratories. There are countless methodologies which pertain to the functionalization of aromatic molecules. These methods, which include the Friedel-Crafts acylation/alkylation,[28] the Chichibabin amination,[29] and the Fries rearrangement,[30] proceed through an intermediate where the planar aromatic structure has been broken, to provide a product which has regained aromaticity. However, dearomatization reactions of flat aromatic molecules can yield products which possess increased topological complexity in an efficient manner (Scheme 3.9).[31]

The Birch reduction is one of the earliest dearomatization reactions (Scheme 3.10).[32] This classic reaction accomplishes the 1,4-reduction of aromatic rings **3.74** to give the corresponding unsaturated 1,4-cyclohexadienes (**3.75** or **3.76**) by using alkali metals dissolved in liquid ammonia in the presence of an alcohol.[33]

Dearomatization reactions have also been used in the total synthesis of complex natural products. In 1978, Corey and co-workers reported the total synthesis of erythronolide B **3.80** which began with a dearomative *para*-allylation of 2,4,6-trimethylphenol **3.77** using allyl bromide to provide cyclohexadienone **3.78** in moderate yield (Scheme 3.11).[34] Several other examples of dearomatization as a

---

[27]Reference [42].

[28]Reference [43].

[29]Reference [44].

[30]Reference [45].

[31]Reference [46].

[32]Reference [47].

[33]For a review on nucleophilic dearomatization of aromatic compounds, see: Ref. [48].

[34]References [49, 50].

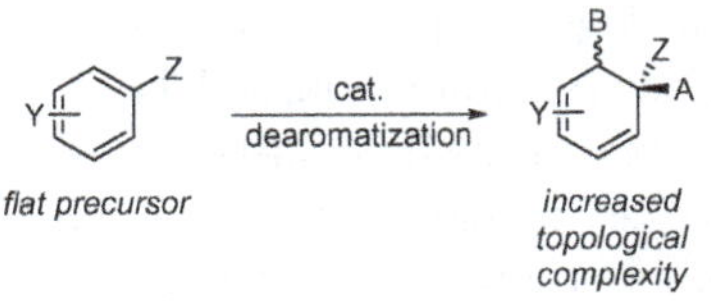

Scheme 3.9 Dearomatization of aromatic compounds. Adapted with permission from Ref. [30]. Copyright (2011) RSC Publishing

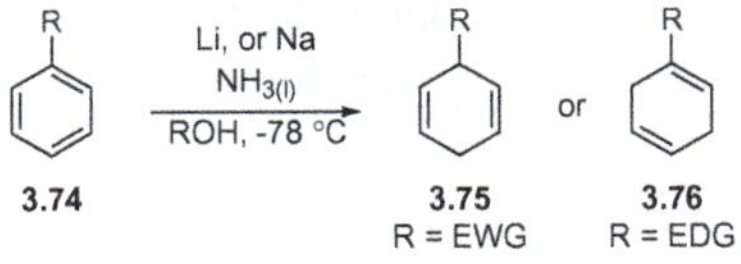

Scheme 3.10 General representation of the Birch reduction

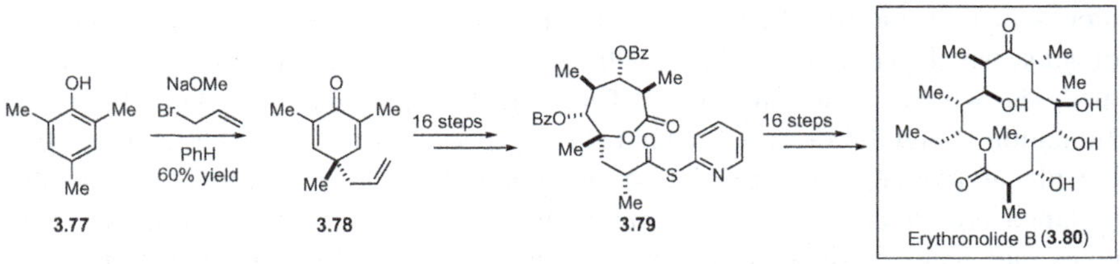

Scheme 3.11 The use of a non-catalyzed dearomatization in the total synthesis of Erythronolide B

strategy in natural product synthesis have been outlined in a number of excellent reviews.[35]

There have been many examples of organo-, transition metal-, and enzyme catalyzed dearomatization reactions reported in the literature.[36,37] These reactions can also be rendered asymmetric when a chiral catalyst is employed, leading to complex products with high levels of selectivity.[38] Aside from rhodium and copper, one of the most prolific metals used in catalytic dearomatization reactions is palladium.[39,40] The ability of Pd to react via many different manifolds, in addition to

[35] References [51, 52].

[36] For a review on transition metal-mediated dearomatization reaction, see: Ref. [53].

[37] For a review on asymmetric hydrogenations of aromatic compounds, see: Ref. [54].

[38] Reference [55].

[39] Reference [55].

[40] Reference [56].

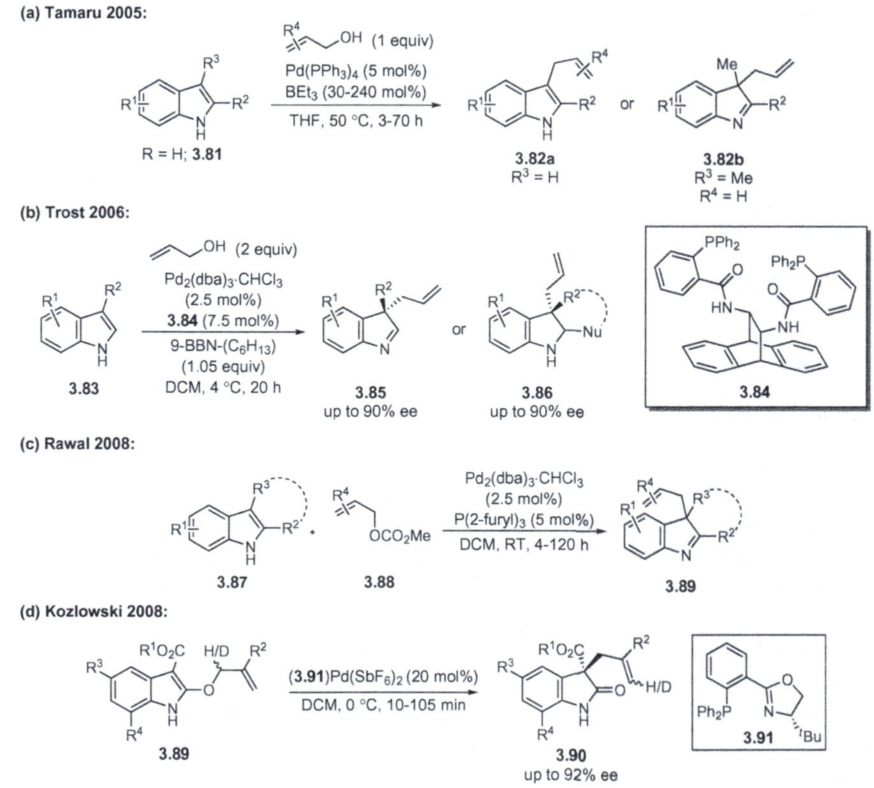

Scheme 3.12 Pd-catalyzed allylative dearomatization of indoles

its high functional group tolerance, have made it a prime candidate to explore various dearomatization reactions.[41]

Tamaru reported a BEt_3-promoted, Pd-catalyzed C3 allylation of indoles **3.81** using allylic alcohols in 2005 (Scheme 3.12a).[42] When the C3-unsubstituted indoles were used, the corresponding 3-allyl indoles **3.82a** were obtained in good yields. However, when the C3 position was methylated, the dearomatized *3H*-indole **3.82b** was observed to be the major product. Trost reported an enantioselective extension of this chemistry in 2006 where a series of C3-alkylated indoles **3.83** could undergo the dearomative allylation using allylic alcohol and obtained excellent enantioselectivites using the chiral bisphphosphine **3.84** (Scheme 3.12b).[43] When the R^2 substituent possessed a terminal nucleophile, a subsequent cyclization onto the *3H*-indole motif occured to generate polycyclic indoles **3.86**. In 2008, Rawal and co-workers reported

[41]For an example of dearomative propargylation of benzyl chlorides, see: Ref. [57].

[42]Reference [58].

[43]Reference [59].

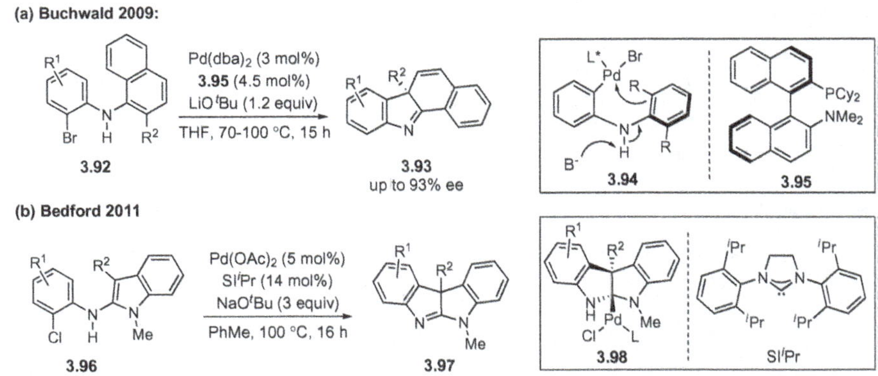

Scheme 3.13 Pd-catalyzed dearomatization of aniline derivatives

the Pd-catalyzed dearomative allylation of 2- and 2,3-substituted indoles **3.87** using allyl carbonates **3.88** (Scheme 3.12c).[44] In 2008, Kozlowski reported an enantioselective Pd-catalyzed Meerwein-Eschenmoseter Claisen rearrangement en route to allylated oxindoles (Scheme 3.12d).[45] By employing a dicationic Pd(II) salt combined with PHOX ligand **3.92**, indolylvinyl ethers **3.90** could undergo a highly efficient [3,3]-sigmatropic rearrangement at 0 °C. By carrying out deuterium labeling studies they were able to rule out the possibility of mechanism involving a π-allyl Pd (II) cation. In 2012, the same authors extended this mode of reactivity to include enantioselective dearomative Saucy-Marbet Claisen rearrangements of indolyl propargyl ethers.[46]

Aniline derivatives have been shown to be good candidates for metal-catalyzed dearomatization reactions. In 2009 Buchwald and co-workers reported a Pd-catalyzed dearomatization of naphthalene derivatives **3.92** (Scheme 3.13a).[47] This reaction provided heterocyclic products **3.93** bearing an chiral all-carbon quaternary center by utilizing chiral monophosphine ligand **3.95**. The authors proposed that the reaction proceeds via ArPd(II)Br species **3.94**, where deprotonation of the N–H bond is expected increase the electron density in the adjacent aromatic ring, and facilitates an intramolecular aromatic substitution reaction with the Pd(II) center.[48] Bedford and co-workers reported that 2-amino indole derivatives **3.96** could undergo a Pd-catalyzed dearomative cyclization under basic conditions (Scheme 3.13b).[49] They proposed a similar mechanism to that of Buchwald which involves the deprotonation of the free N–H bond to affect an intramolecular

[44]Reference [60].
[45]Reference [61].
[46]Reference [62].
[47]Reference [63].
[48]Bedford and co-workers reported a racemic variant involving diarylamines, see: Ref. [64].
[49]Reference [65].

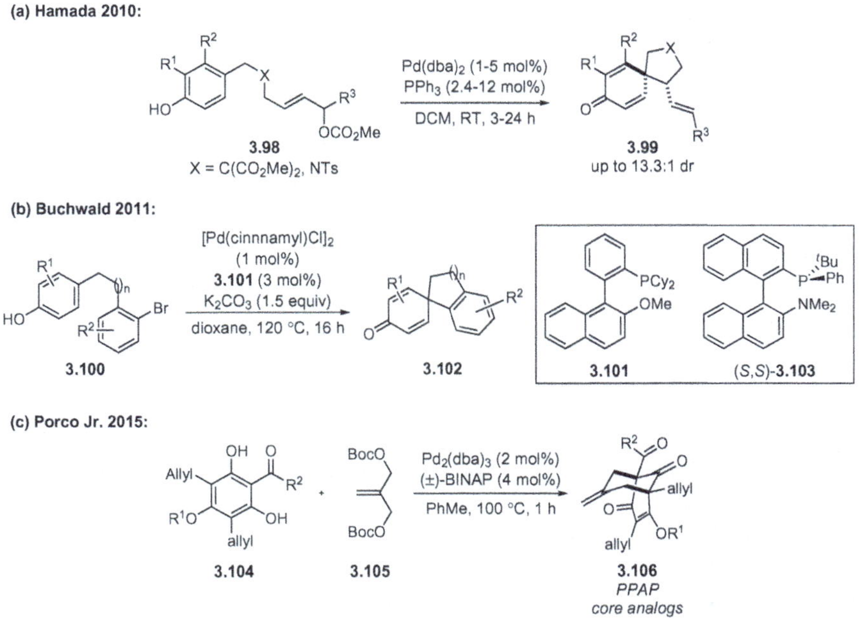

Scheme 3.14 Pd-catalyzed dearomatization of phenol derivatives

nucleophilic substitution reaction at Pd. The authors said that a mechanism involving carbopalladation of the indole carbon-carbon double bond was less likely since the *syn* nature of this insertion would form highly strained *trans*-fused intermediate **3.98**.

Phenols have also been shown to be good candidates for several Pd-catalyzed dearomatization reactions. In 2010 Hamada reported a Pd-catalyzed intramolecular dearomative allylation reaction of phenol derivatives **3.98** (Scheme 3.14a).[50] Although the majority of the scope for this *ipso*-Friedel-Crafts allylic alkylation was conducted in the presence of achiral PPh_3, a single enantioselective example was reported using **3.84** as the ligand and the allylation product could be obtained in 80% yield with 9.2:1 dr and 89% ee. The same authors later expanded the scope of the enantioselective process in 2012.[51] Buchwald reported an intramolecular Pd-catalyzed dearomative *ipso* arylation in 2011 which utilized biaryl phosphine ligand **3.101** (Scheme 3.14b).[52] By using ligand (*S,S*)-**3.103** which possesses both point and axial chirality, they were able to obtain up to 91% ee in selected examples when the reaction temperature was lowered to 80 °C. Porco Jr. reported a dearomative conjunctive allylic annulation (DCAA) reaction between elaborated phenol

[50]Reference [66].

[51]Reference [67].

[52]Reference [68].

Scheme 3.15 Pd(II)-catalyzed intramolecular oxidative dearomative Mizoroki-Heck reaction

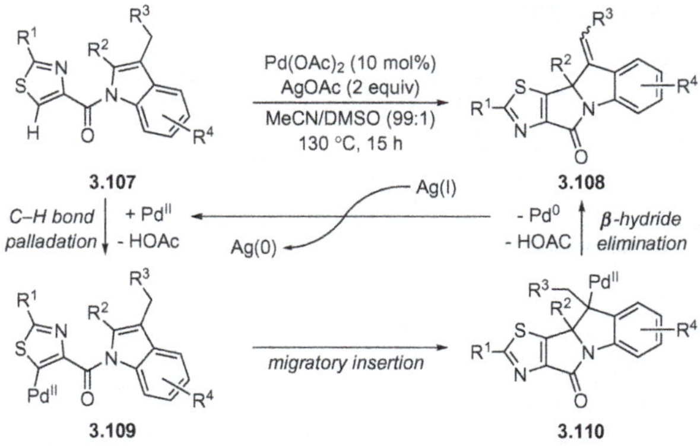

derivatives **3.104** and bis-carbonate **3.105**. This reaction provided access to racemic polyprenylated acylphloroglucinol (PPAP) analogs **3.106** (Scheme 3.13c).[53]

Chapter 2 discussed the work of Grigg,[54] Yao and Wu[55] as well as Jia[56] on Pd-catalyzed intramolecular indole dearomatization reaction which inspired our discovery of the first dearomative indole bisfunctionalization by arylcyanation.[57] Recently, Yao and Lin reported that the carbon-halogen bond in the starting material can be replaced with a C–H bond of a thiazole **3.107**. This becomes an efficient substrate for a Pd(II)-catalyzed oxidative dearomative Mizoroki-Heck reaction en route to tetracyclic indole cores **3.108** (Scheme 3.15).[58] The authors proposed that this reaction occurs via an initial C–H palladation to generate Pd(II) species **3.109**. This intermediate undergoes dearomative migratory insertion to generate 3° benzylic Pd(II) species **3.110** which can undergo β-hydride elimination to generate the product and Pd(0). Ag(I) reoxidizes Pd(0) to the active Pd(II) species with concomitant formation of Ag(0).

3.1.4 Research Goals

In the Pd-Catalyzed dearomative arylcyanation presented in Chap. 2, $Zn(CN)_2$ was used as the cyanide source. During the optimization of this reaction only trace amounts of the direct aryl halide cyanation product was observed, and it was absent under the optimized reaction conditions. One possible explanation for the high selectivities could

---

[53]Reference [69].

[54]Reference [70].

[55]Reference [71].

[56]Reference [72].

[57]Portions of this chapter have appeared in print. See: Refs. [73, 74].

[58]Reference [75].

Scheme 3.16 Proposed dearomative 1,2-diarylation of indoles

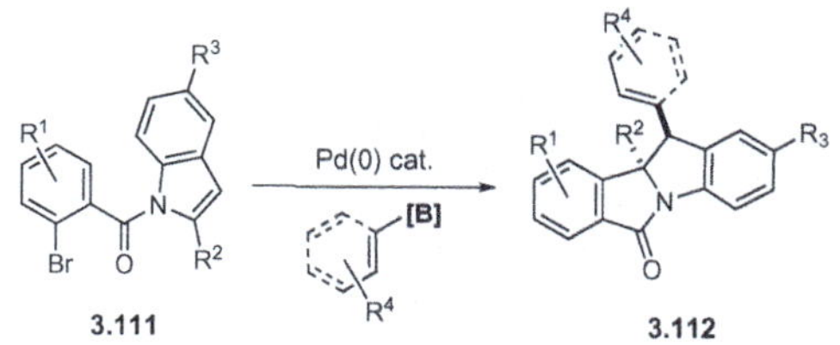

reside in the fact $Zn(CN)_2$ is a highly insoluble inorganic salt.[59] This translates to an extremely low active concentration of this cyanation agent at any given time, which may attenuate the rate of transmetallation to the ArPd(II)X species resulting from oxidative addition to the aryl halide. Instead, a dearomative migratory insertion could occur at a rate which reflects the actual product distribution. By using more soluble nucleophilic trapping reagents, the boundaries of the intramolecular indole dearomatization could be probed. Would a high active concentration of a soluble trapping reagent make the direct aryl halide coupling out-compete the dearomatization? By taking inspiration from both the diarylation and dearomatization literature, an intramolecular dearomative *syn* 1,2-diarylation of indoles **3.11** was proposed (Scheme 3.16). This would represent the first example of a domino Mizoroki-Heck reaction being terminated with a Suzuki-Miyaura coupling at a 2° benzylic Pd(II) center.

Organoboranes are highly soluble in organic solvents, especially at temperatures that this class of reaction is known to operate. Therefore, it was thought that exploring their use as nucleophilic trapping reagents in this reaction would allow us to probe the capabilities of the dearomative migratory insertion step. The structural diversity of these compounds would also allow for the possible recognition of correlations between the reactivity of the organoboranes and the selectivity of the reaction.

3.1.5 Starting Material Preparation

The majority of substrates utilized in this chemistry were identical to those used in the dearomative arylcyanation of indoles from part 2 of Chap. 2. However, some substrates were found to be successful in this *syn* 1,2-diarylation that were not compatible with the former. Only these miscellaneous compound syntheses are outlined below. These compounds were synthesized by converting a series of 2-bromobenzoic acid derivatives **3.113** to the corresponding acid chlorides **2.104** under standard conditions, followed by reaction with 2-methyl indoles **2.232** (Scheme 3.17).

Furthermore, substrates containing *N*-benzenesulfonyl and *N*-benzyl linkages were also prepared to test the effects of varying the electronic properties of the indole (Scheme 3.18). *N*-benzyl **3.117** was prepared in 40% yield by reacting 2-methylindole **3.115** with 2-bromobenzyl chloride in the presence of KOH in

[59] $Zn(CN)_2$ has a solubility in water of 0.00005 g/L at 20 °C and is insoluble in alcohol. Taken from: Ref. [76].

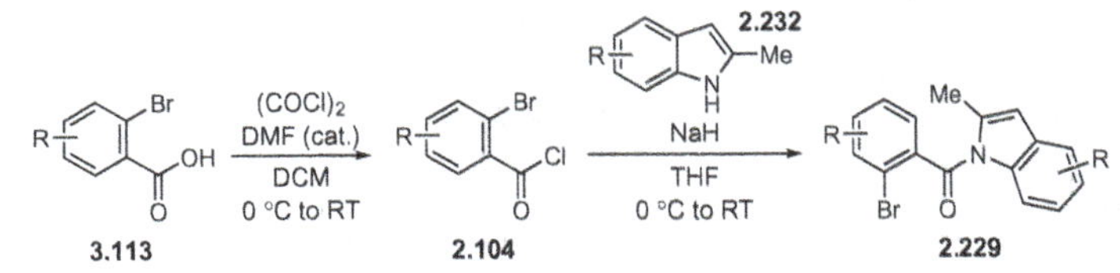

Scheme 3.17 Synthesis of miscellaneous *N*-benzoyl indole substrates

Scheme 3.18 Preparation of *N*-benzenesulfonyl and *N*-benzyl indoles

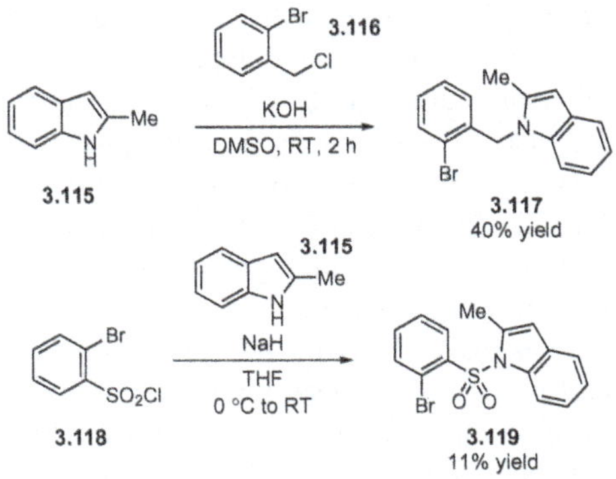

DMSO. *N*-Benzenesulfonyl **3.119** was prepared in only 11% yield by treating 2-bromobenzenesulfonyl chloride **3.118** with the sodium salt of **3.115**.

Commercially available boronic acids **3.120** may contain various quantities of the corresponding dehydro trimers known as boroxines **3.121**. Depending on the electronics or sterics of the carbon ligand on boron, the conditions of storage, age of the commercial compound and other contributing factors, the ratio of boronic acid to boroxine can change. It was initially hypothesized that maintaining an accurate equivalency of aryl boron species in this reaction would potentially allow for a higher degree of control in obtaining the desired dearomatized diarylation production over the direct Suzuki-Miyaura coupling products. For purposes of consistency and reproducibly, we opted to use the pre-formed boroxines, which could be prepared by either heating solid samples of **3.120** under high-vacuum overnight or by refluxing **3.120** in toluene overnight using a Dean-Stark trap to remove the liberated water (Scheme 3.19).[60]

[60]The majority of the boroxines used in this chemistry were prepared and donated by Thomas Johnson, a current Ph.D. student in the Lautens group and Bo Luo, who is a visiting Ph.D. student in the Jinming Gao Group at Northwest Agriculture & Forestry University in China.

Scheme 3.19 Preparation of boroxines

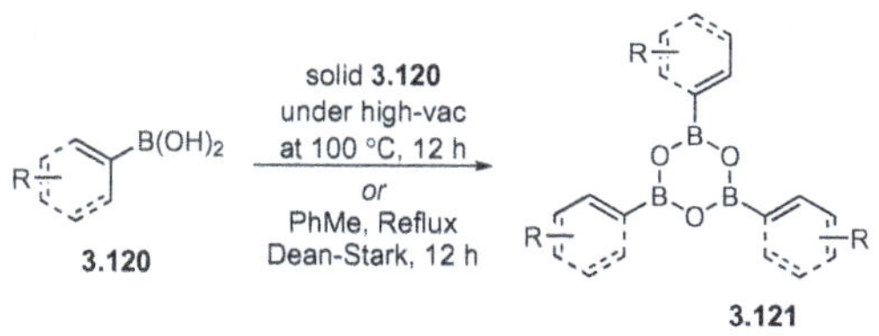

3.1.6 Optimization

N-Benzoylated indole **2.229a** was used as the model substrate to optimize the dearomative 1,2-diarylation reaction (Table 3.1). The initial conditions were chosen to be 0.2 mmol of **2.229a** with $Pd(P^tBu_3)_2$ (10 mol%), K_2CO_3 (2 equiv), $(PhBO)_3$ (0.47 equiv) in PhMe (0.1 M) at 110 °C for 18 h (entry 1). This experiment led to full conversion of starting material and the desired product **3.112a** being obtained in 20% yield as a single diastereomer in the presence of 63% of the undesired direct Suzuki-Miyaura coupling product **3.122a**. When a 9:1 PhMe:H_2O solvent mixture was employed, a poorer mass balance was observed and **3.122a** was still observed as the major product (entry 2). However, the presence of 2-methylindole in the

Table 3.1 Optimization of the Pd-Catalyzed dearomative *syn* 1,2-diaryation of **2.229a**[a]

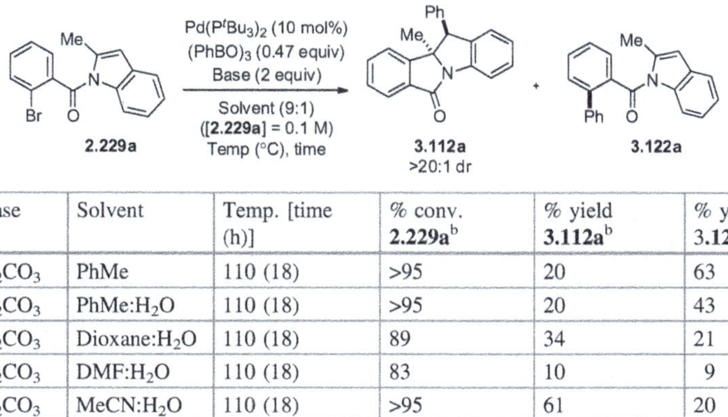

Entry	Base	Solvent	Temp. [time (h)]	% conv. **2.229a**[b]	% yield **3.112a**[b]	% yield 3.122a[b]
1[h]	K_2CO_3	PhMe	110 (18)	>95	20	63
2[h]	K_2CO_3	PhMe:H_2O	110 (18)	>95	20	43
3[h]	K_2CO_3	Dioxane:H_2O	110 (18)	89	34	21
4[h]	K_2CO_3	DMF:H_2O	110 (18)	83	10	9
5[h]	K_2CO_3	MeCN:H_2O	110 (18)	>95	61	20
6[h]	K_2CO_3	MeCN:H_2O	100 (18)	>95	64	29
7[c]	K_2CO_3	MeCN:H_2O	100 (18)	>95	64	26
8[d]	K_2CO_3	MeCN:H_2O	100 (18)	>95	47	44
9	K_2CO_3	MeCN:H_2O	100 (2.5)	>95	71	28
10[h]	KOH	MeCN:H_2O	100 (1)	>95	28	8
11[h]	KOAc	MeCN:H_2O	100 (1)	53	29	4
12	K_3PO_4	MeCN:H_2O	100 (1)	>95	58	29
13[h]	Na_2CO_3	MeCN:H_2O	100 (1)	>95	67	16

(continued)

Table 3.1 (continued)

Entry	Base	Solvent	Temp. [time (h)]	% conv. **2.229a**[b]	% yield **3.112a**[b]	% yield 3.122a[b]
14[h]	Cs_2CO_3	MeCN:H_2O	100 (1)	>95	75	25
15	CsF	MeCN:H_2O	100 (1)	>95	80	10
16[e,f,g]	**CsF**	**MeCN:H_2O**	**100 (1)**	**>95**	**(88)**	**7**

[a]All reactions were run on a 0.2 mmol scale
[b]Determined by ^{1}H NMR analysis of the crude reaction mixture using 1,3,5-trimethoxybenzene as an internal standard
[c][**2.229a**] = 0.05 M
[d][**2.229a**] = 0.2 M
[e]Average value over three runs
[f]Value in parentheses represents the average isolated yield over three runs
[g]Reaction was run using 2.5 mol% $Pd(P^tBu_3)_2$
[h]Reaction was performed by Nicolas Zeidan

crude reaction mixture suggests that decomposition of **2.229a** and/or **3.122a** by hydrolysis of the weak amide C–N bond was occurring.

The use of a 9:1 dioxane:H_2O mixture led **3.112a** product being obtained as the major product in 34% yield (vs. 21% of **3.113a**) (entry 3). Although a 9:1 DMF:H_2O mixture provided inferior results with respect to dioxane:H_2O (entry 4), employing a 9:1 mixture of MeCN:H_2O afforded full conversion of **2.229a**, **3.112a** and **3.122a** were obtained in 61 and 20% yield, respectively (entry 5). Lowering the reaction temperature to 100 °C led to better overall mass recovery and a slight increase in the yield of **3.112a** to 64% (entry 6), whereas deviating from a 0.1 M concentration of **2.229a** only led to inferior results (entries 7 and 8). Decreasing the reaction time to 2.5 h led to an increase in yield of **3.112a** to 71%, and nearly quantitative mass recovery (entry 9). The use of other potassium bases such as KOH, KOAc or K_3PO_4 all led to marked decreases in yields and/or conversions (entries 10–12). Using KOH led to severe decomposition of the starting material. Although Na_2CO_3 did not lead to better results (entry 13), Cs_2CO_3 provided **3.112a** in 75% yield (entry 14). To further explore the effect of Cs counterions on the reaction, CsF was tested. This led to **3.112a** being obtained in 80% yield with 10% of **3.122a** after 1 h (entry 15). Compound **3.112a** could be isolated in 88% yield with only 7% of **3.122a** when the catalyst loading was lowered to 2.5 mol% (entry 16). The final optimal conditions were found to be $Pd(P^tBu_3)_2$ (2.5 mol%), CsF (2 equiv), phenylboroxine (0.47 equiv) in MeCN:H_2O (9:1) at 100 °C for 1 h. This set of reaction parameters will be referred to as the *standard conditions.*

The effect of altering various reaction parameters on the efficacy of the Pd-Catalyzed *syn* 1,2-diarylation was investigated next (Table 3.2). When the dinuclear complex $[Pd(P^tBu_3)(\mu\text{-}Br)]_2$ was employed instead of $Pd(P^tBu_3)_2$, nearly no change in the reaction outcome was observed (entry 2). The similarity in these reaction outcomes may suggest that $[Pd(P^tBu_3)(\mu\text{-}Br)]_2$ may be converting to the

Table 3.2 Studying the effects of altering individual reaction parameters[a]

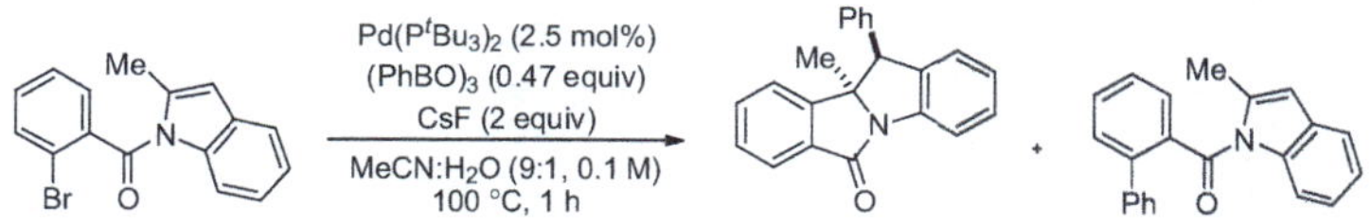

2.229a *"standard" conditions* **3.112a** >20:1 dr **3.122a**

Entry	Variation from the ***"standard" conditions***	Yield **3.112a** (%)[b,c]	Yield **3.122a** (%)[b]
1	**None**	**88 (88)**[d]	**7**
2	$[Pd(P^tBu_3)(\mu\text{-}Br)]_2$[e]	93 (88)[d]	5
3	$Pd[P(o\text{-}Tol)_3]_2$	69	3
4	$Pd(PhP^tBu_2)_4$	42	22
5	$Pd(PPh_3)_4$	5	17
6[h]	90 °C instead of 100 °C	79	10
7	ArCl instead of ArBr	63	17
8[h]	0.33 equiv of $(PhBO)_3$	67	8
9[f,g]	$PhB(OH)_2$ instead of $(PhBO)_3$	83	6
10[g]	$PhBF_3K$ instead of $(PhBO)_3$	45	0

[a]Reactions were run on a 0.2 mmol scale
[b]Determined by ^{1}H NMR analysis of the crude reaction mixture using 1,3,5-trimethoxybenzene as an internal standard
[c]Value in parentheses represents isolated yields
[d]Average value over three experiments
[e]1.25 mol% was used
[f]Freshly recrystallized phenylboronic acid was used
[g]1.4 equivalents of the aryl boron reagent was used
[h]Reaction performed by Masaru Kondo

corresponding $Pd(P^tBu_3)_2$ or $Pd(P^tBu_3)$ complex under the reaction conditions.[61] In all cases, decreasing the size of the ligand led to a decrease in yield of **3.112a** (entries 3–5).[62] In analogy to the literature findings, out hypothesis is that when extremely bulky ligands (i.e. P^tBu_3) are used, only one ligand can be accommodated in the coordination sphere of the ArPd(II)Br species resulting from oxidative addition to the carbon–halogen bond. This may result in a less coordinatively saturated metal center which can then bind the indole olefin moiety to a greater extent, thus translating to a more facile migratory insertion. Conversely, the coordination sphere of Pd can be saturated more easily when less bulky ligands (i.e. PPh_3) are used. This can decrease the propensity of the Pd(II) species to undergo olefin insertion, which manifests itself in a higher yield of the undesired direct Suzuki coupling product. When the reaction temperature was decreased below 100 °C, or when the aryl chloride analog of **2.229a** was employed, **3.122a** was obtained in higher yield (entries 6 and 7). Decreasing the amount of phenylboroxine

[61]Reference [77].
[62]Reference [78].

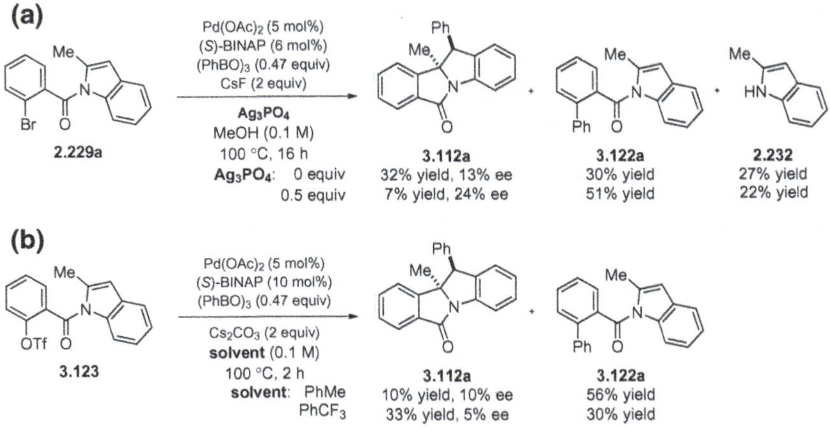

Scheme 3.20 Attempts to develop an asymmetric variant

to 0.33 equivalents (~1 equivalents of Ar[B]) led to a lower yield of **3.112a** (entry 8). The use of freshly prepared $PhB(OH)_2$ only slightly decreased the yield of **2a**, while $PhBF_3K$ markedly inhibited the formation of **3.112a** (entries 9 and 10).

3.1.6.1 Development of an Asymmetric Variant

The previous dearomative arylcyanation reaction proved to be highly ligand specific, and only P^tBu_3 and d^tbpf were found to promote the desired transformation. When chiral ligands which deviated from the highly bulky and electron-rich nature of these two ligands were employed, the reaction failed. The optimization of the 1,2-diarylation revealed a less strict tolerance towards phosphine ligands. These observations suggested that chiral phosphine ligands could still allow the reaction to proceed in an asymmetric fashion. Results obtained in these trials were generally poor under a variety of conditions using aryl bromides or aryl triflates. These reactions were plagued with hydrolysis of the starting material or direct Suzuki-Miyaura product. When **2.229a** was reacted using $Pd(OAc)_2$/(*S*)-BINAP in MeOH, **3.112a** was obtained in 32% yield with 13% ee in addition to 30% of **3.122a** and 27% of 2.232. When Ag_3PO_4 was added to facilitate the formation of a cationic Pd species, **3.112a** was obtained in 7% yield with 24% ee in addition to 51% of **3.122a** and 22% of **2.232** (Scheme 3.20a). The corresponding aryl triflate **3.123** was also tested. Using when the reaction was run in PhMe **3.112a** was obtained in 10% yield with 10% ee, whereas using $PhCF_3$ increased the yield of **3.112a** to 33% albeit with only 5% ee (Scheme 3.20b).[63]

[63]Reaction attempts using **3.123** were conducted by Masaru Kondo.

3.1.7 Examination of the Substrate Scope

In a collaborative effort with a visiting scolar Dr. Masaru Kondo, the generality of the optimized reaction conditions was evaluated using a series of substituted *N*-benzoyl indoles **2.229** (Table 3.3). Sterically hindered aryl bromide **2.229b** was found to function exceptionally well in this reaction using 3.5 mol% of Pd, and **3.112b** was obtained in 98% yield with only trace amounts prodcuts resulting from direct Suzuki-Miyaura coupling. Chloro- and fluoro-bromobenzoyl analogs **2.229c** and **2.229d** could be converted to the corresponding diarylated products in 70% (**3.112c**) and 82% yield (**3.112d**), respectively. In the case of **2.229c**, the boroxine loading was lowered to 0.4 equivalents in order to inhibit an undesired direct Suzuki-Miaura coupling of the aryl chloride in the product. Electron deficient and electron-rich aryl bromides could be converted to the corresponding products (**3.112e** and **3.112f**) in moderate to excellent yields. Various 2-alkyl (**2.229h-i**), 2-aryl (**2.229k,m,o**) and 2-carbonyl indoles (**2.229p**) were tolerated under the reaction conditions and could be converted to the corresponding products in good to excellent yields. Both electron-rich (**2.229q**) and -poor (**2.229r** and **2.229aa**)

Table 3.3 Scope of the fused indoline synthesis by *syn* 1,2-diarylation[a]

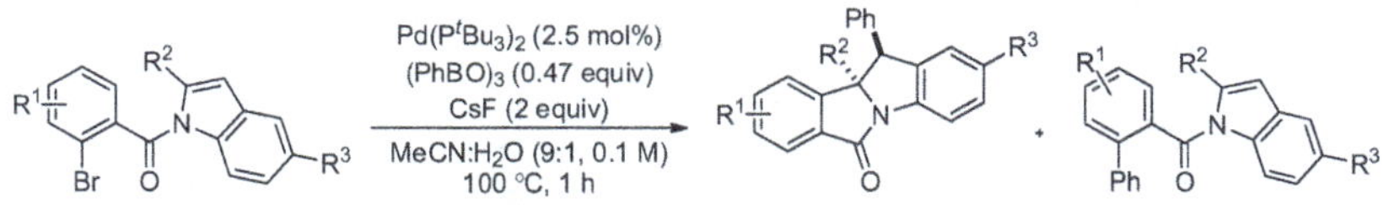

2.229a 3.112a >20:1 dr 3.122a

Entry	Substrate		Product		Yield[b] (%)
1	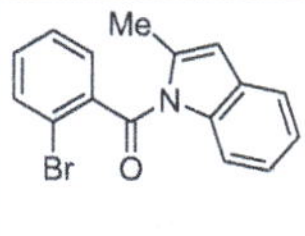	2.229a	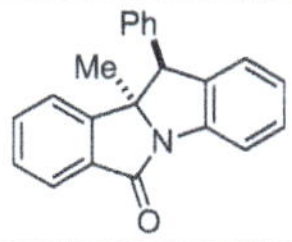	3.112a	98 (76%)[c]
2[d]	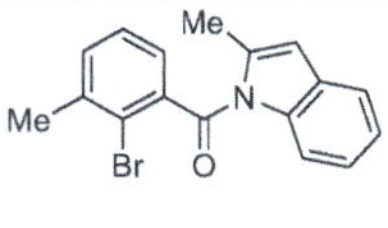	2.229b	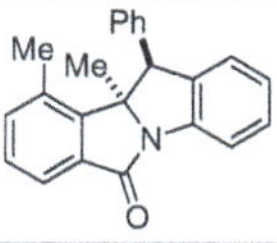	3.112b[h]	51
3	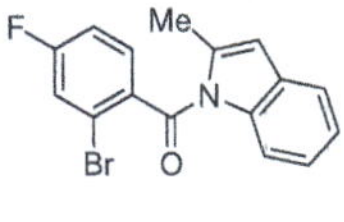	2.229c	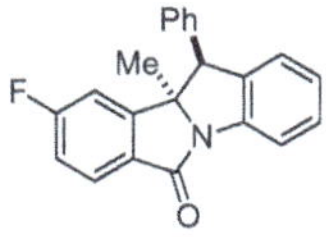	3.112c	94
4[e]		2.229d	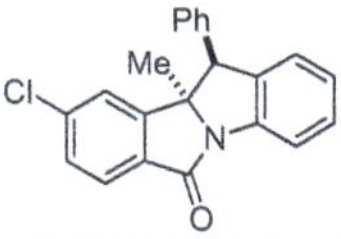	3.112d	93

(continued)

Table 3.3 (continued)

Entry	Substrate		Product		Yield[b] (%)
5	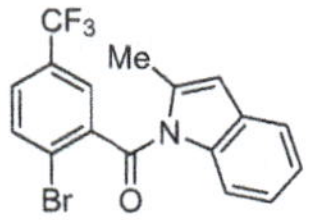	**2.229e**	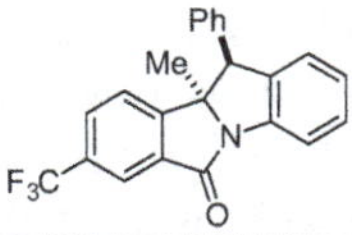	**3.112e**[h]	96
6	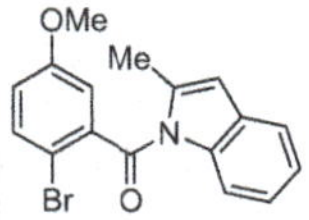	**2.229f**	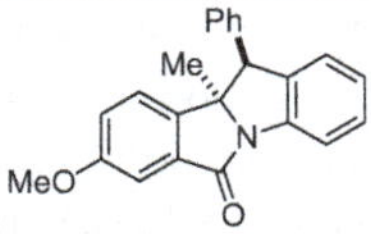	**3.112f**[h]	94
7	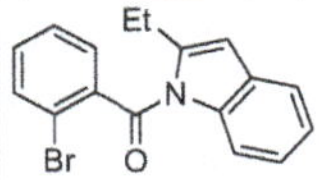	**2.229g**	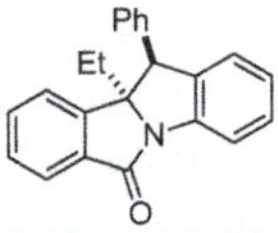	**3.112g**[h]	91
8[f]	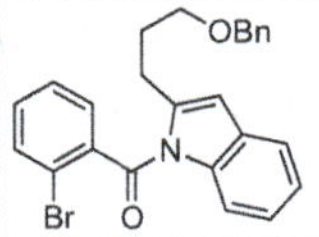	**2.229h**	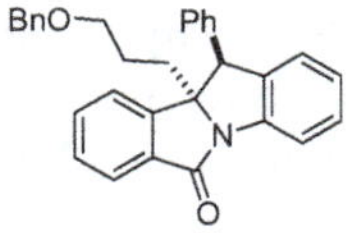	**3.112h**	83
9	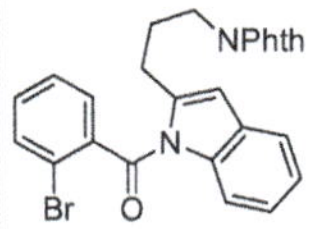	**2.229i**	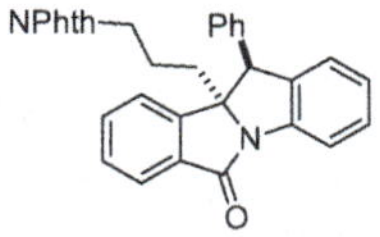	**3.112i**	98
10	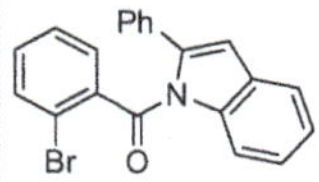	**2.229k**	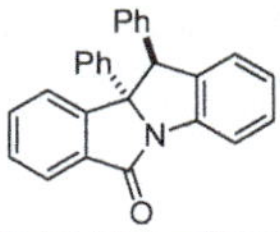	**3.112k**[h]	91
11[e]	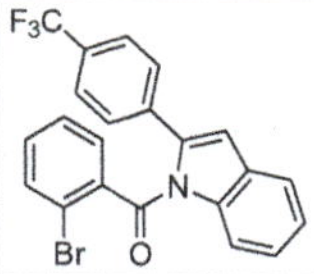	**2.229m**	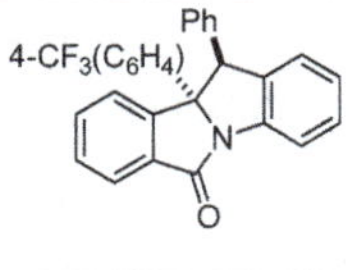	**3.112m**[h]	89
12	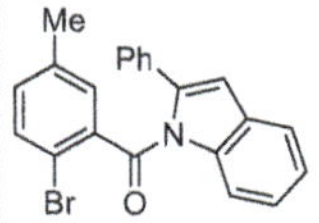	**2.229o**	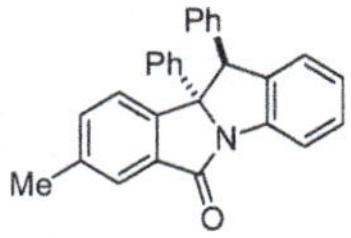	**3.112o**[h]	89
13	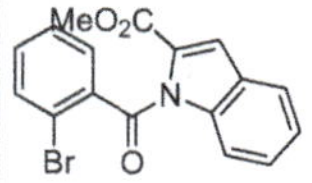	**2.229p**	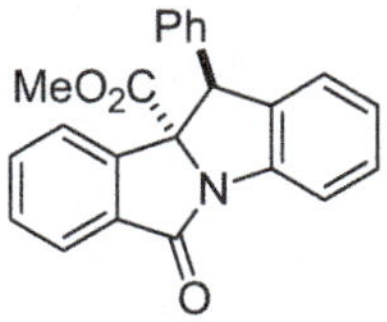	**3.112p**[h]	72
14	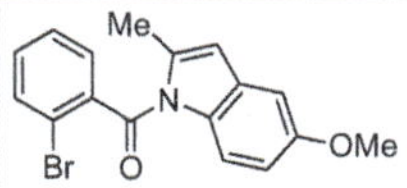	**2.229q**	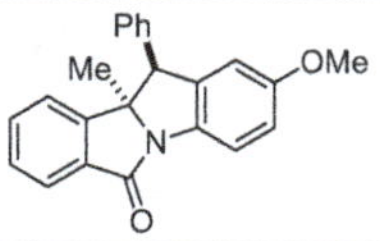	**3.112q**[h]	93

(continued)

Table 3.3 (continued)

Entry	Substrate		Product		Yield[b] (%)
15	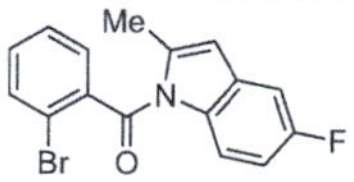	**2.229r**	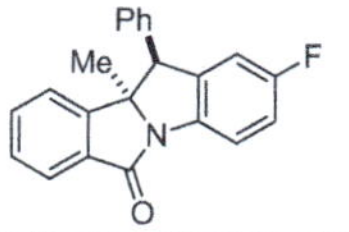	**3.112r**[h]	86
16[e]	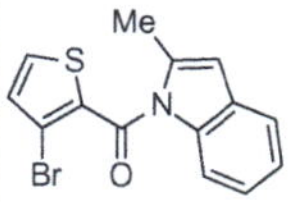	**2.229t**	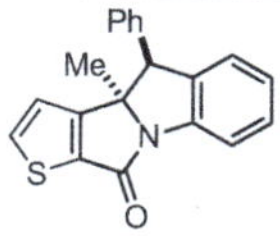	**3.112t**	45
17[d]	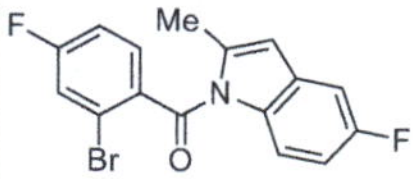	**2.229aa**	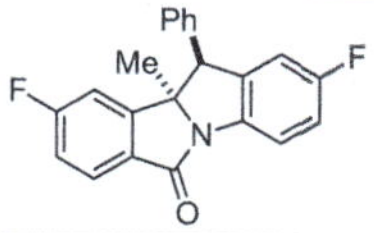	**3.112aa**[h]	60
18		**2.229ab**	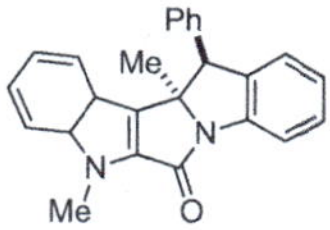	**3.112ab**[h]	68
19[g]	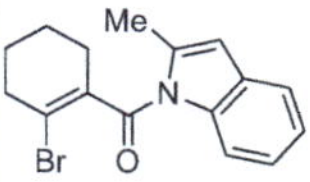	**2.229ac**	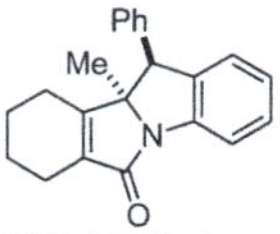	**3.112ac**[i]	74

[a]All reactions were run on a 0.2 mmol scale unless otherwise stated
[b]determined by ^{1}H NMR analysis of the crude reaction mixture using 1,3,5-trimethoxybenzene as an internal standard
[c]1 g (3.20 mmol) scale using 1.25 mol% $Pd(P^tBu_3)_2$
[d]3.5 mol% $Pd(P^tBu_3)_2$
[e]0.4 equivalents of $(PhBO)_3$
[f]Reaction was run on a 0.13 mmol scale
[g]5 mol% $Pd(P^tBu_3)_2$
[h]This compound was prepared by Masaru Kondo
[i]This compound was prepared by Nicolas Zeidan

indoles led to the desired products in moderate yields. An increased catalyst loading of 3.5 mol% was required to reach full conversion in the case of the difluorinated analog **2.229aa**. Bromothiophene derivative **2.229t** converted cleanly under the reaction conditions, yet the desired product **3.112t** could only be obtained in 45% yield with the major product being the undesired Suzuki-Miyaura coupling product (vide infra). *N*-Me indolyl (**2.229ab**) and cyclohexenyl (**2.229ac**) substrates were found to undergo the desired transformation, and the corresponding indolines could be obtained in moderate yields.

The effect of various sterically and electronically demanding boroxines was tested on the 1,2-diarylation reaction of **2.229a** (Scheme 3.21). The reaction was generally quite tolerant to many types of boroxine reagents with the 1,2-diarylation product being observed as the major product in all cases except one. Aryl groups

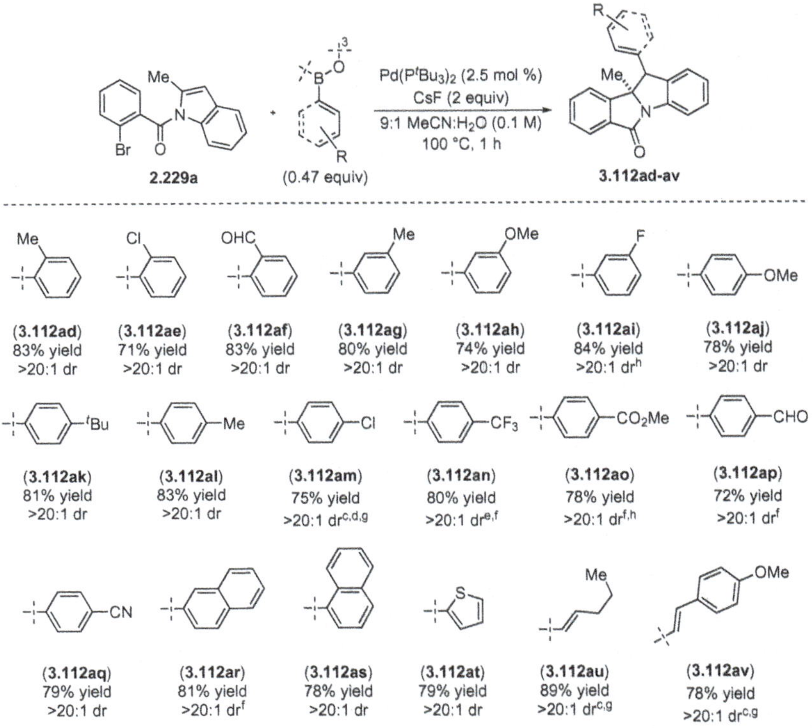

Scheme 3.21 Pd(0)-catalzyed dearomative indole diarylation: Scope of boroxines. [a]Unless otherwise stated, all reactions were run on a 0.2 mmol scale. [b]dr's were determined by ^{1}H NMR analysis of the crude reaction mixture. [c]0.4 equivalents of the corresponding boroxine. [d]Reaction time = 50 min. [e]Run using 3.5 mol% Pd(P^{t}Bu$_3$)$_2$. [f]Reaction time = 2 h. [g]This compound was prepared by David Petrone. [h]This reaction was run by Nicolas Zeidan

bearing *ortho* methyl, chloro, and aldehyde substituents could be incorporated into the indoline products in moderate to good yields (**3.112ad-af**). No Suzuki-Miyaura coupling of the arylchloride of product **3.112ae** was observed when 0.47 equivalents of the corresponding boroxine was used. *Meta* substitutions such as methyl, methoxy, and fluoro were also well tolerated (**3.112ag-ai**). The reaction functioned well when various boroxines with electronically diverse *para* substituents were employed (**3.112ak-aq**). Reaction times were increased to 2 h in order to obtain full conversion of the starting material in a few cases (**3.12an-ap**). Suzuki coupling of the aryl chloride of **3.112am** was initially observed under the standard conditions; yet decreasing the boroxine loading to 0.4 equivalents completely inhibited this undesired process. The relative stereochemistry of the diarylation was unambiguously determined to be *syn* by X-ray crystallographic analysis of **3.112an** (Fig. 3.1).

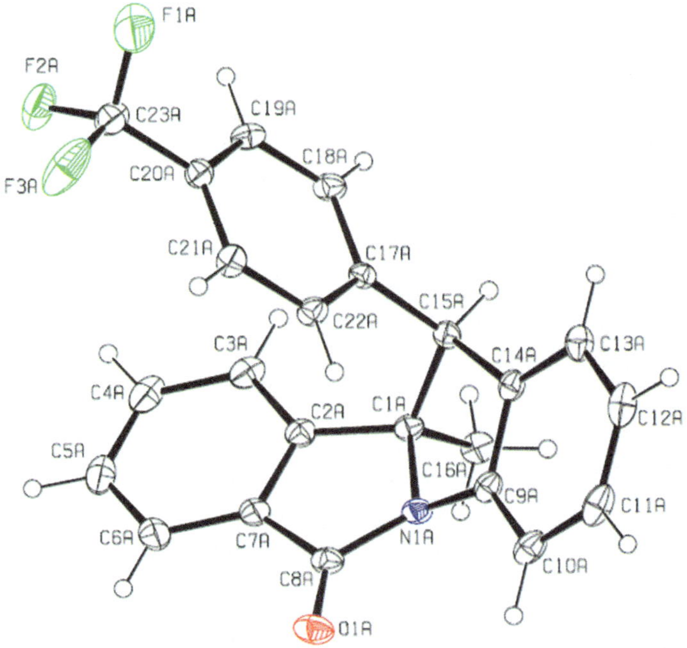

Fig. 3.1 Crystal structure of the *syn* 1.2-diarylation product **3.112an**

Naphthalene boroxines were well tolerated under the optimized reaction conditions, and the corresponding products **3.112ar** and **2.112as** were obtained in 81 and 78% yield, respectively. A thiophenyl boroxine was also tolerated, and the corresponding indole was obtained in 79% yield. Alkenyl boroxines could also be used in a *syn* 1,2-arylvinylation in 89% yield using *E*-pentenyl boroxine and in 78% yield using *para*-methoxystyrenyl boroxine. Finally, when the reaction was run on a 1 g scale **3.112a** could be obtained in 76% yield and >20:1 dr when the catalyst loading was decreased to 1.25 mol%.

3.1.8 Limitations

Despite the broad scope of indoles and boroxines which can undergo the desired transformation under the optimized conditions, this reaction does suffer a few limitations which should direct future efforts. When C2 and C4 unsubstituted indole **2.229v** was subjected to the reaction conditions, only 9% of the desired indoline product **3.112v** could be isolated in addition to 91% of the product arising from intramolecular C–H functionalization of the indole C2 position **2.243** (Scheme 3.22a).

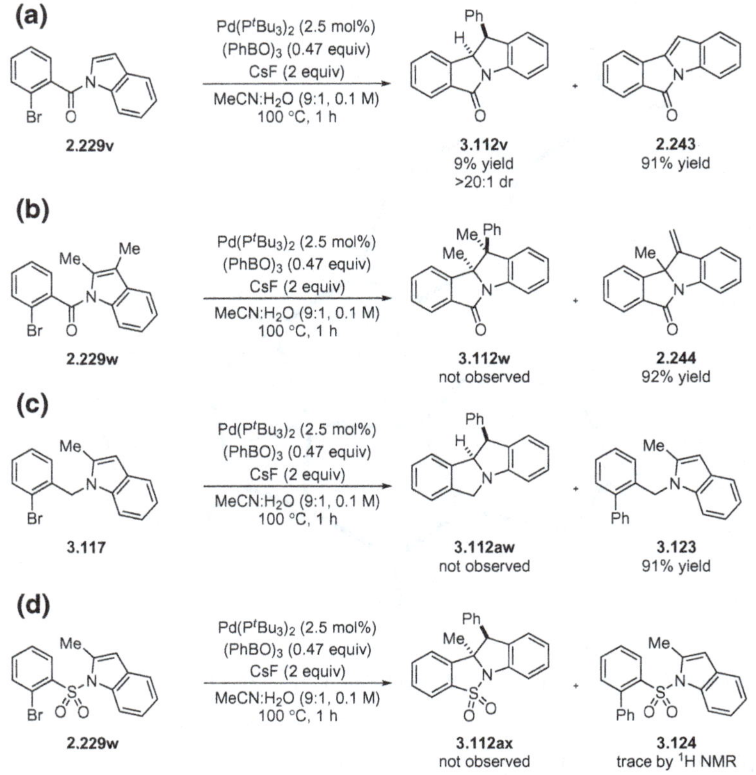

Scheme 3.22 Limitations in the *syn* 1,2-diarylation reaction

Indole substrate **2.229w** bearing methyl substituents at both the C2 and C3 positions did not lead to any desired product **3.112w**. Instead, the corresponding Mizoroki-Heck reaction product **2.244** which contains an exocyclic methylene was found to predominate and was isolated in 92% yield (Scheme 3.22b). Although only catalytic amounts of **2.243** and **2.244** were obtained in the analogous base-free arylcyanation reaction, these products can be obtained in greater quantities in the 1,2-diarylation protocol. The CsF used in the current reaction conditions can convert the Pd(II) salts formed during the reaction leading to byproduct formation into the active Pd(0) catalyst, thus enabling catalyst turnover.

The effect of changing the linking group was also examined. No desired product was observed when *N*-benzyl analog **3.117** was subjected to the reaction conditions, yet the product resulting from the direct Suzuki-Miyaura coupling **3.123** was isolated in 91% yield (Scheme 3.21c). Exchanging the amide functionality of **2.229a** for a sulfonamide (**3.119**) negatively impacted the reaction and no dearomative cyclization product was observed. Instead, only the direct Suzuki-Miyaura coupling product **3.124** was obtained along with recovered starting material (Scheme 3.21d). It is possible that the carbonyl linkage appropriately modulates the electronics of the

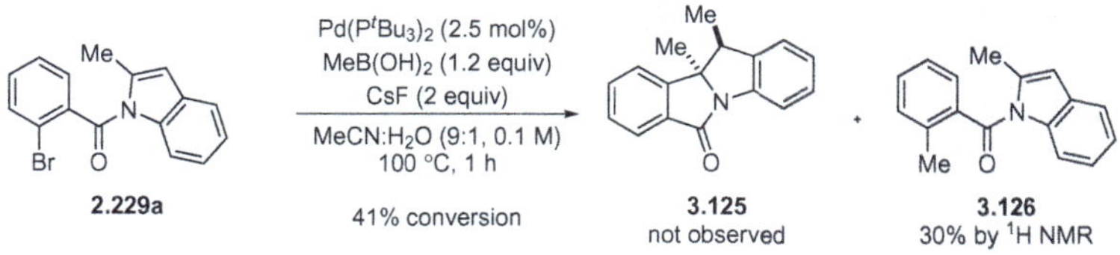

Scheme 3.23 Testing methylboronic acid

C2–C3 double bond, thus allowing the insertion involving dearomatization to proceed. It has been shown that indoles possessing carbonyl-based electron withdrawing groups at C2 have lower LUMO energies when compared with analogous indoles with a simple C2 methyl group. This LUMO is localized on the reactive site (C2–C3) which can explain their requirement to obtain the desired dearomatized products.[64] Based on literature precedence, it is likely that the dearaomtization step is proceeding in a concerted fashion via the traditional *syn* carbopalladation route. Another mechanistic scenario would invole nucleophilic attack of the indole C3 site onto the Pd center (i.e. Scheme 3.12a). However be since the proposed Ar–Pd(II)–Br species is tethered to the indoles reactive site, this would create a highly strained and distorted palladacycle. When methylboronic acid was tested in the reaction, none of the desired 1,2-diarylation product **3.125** was observed, and only traces of the direct Suzuki-Miyaura coupling product **3.126** was observed (Scheme 3.23).

3.1.9 Derivatization of Indoline Products

The tetracyclic indoline products were subjected to a series of reactions in order to probe their synthetic utility (Scheme 3.24). It was initially proposed that a stereoselective deprotonation/alkylation could forge a benzylic all-carbon quaternary center adjacent to the tetrasubstituted tertiary center. However, when **3.112a** was treated with ^{t}BuLi (1.1 equiv) at −78 °C in THF,[65] hemiaminal **3.127** was formed in 71% as a single diastereomer (Scheme 3.24a).

The relative stereochemistry of this compound was unambiguously determined by X-ray crystallographic analysis (Fig. 3.3). The unexpectedly high reactivity of the lactam leaves it vulnerable to nucleophilic attack by the organolithium reagent. This attack occurs at the face of the carbonyl that is opposite to the flanking benzylic Ph group (cf. Fig. 3.2). This nucleophilic addition appeared to be quite general across a series of organolithium reagents. When **3.112a** was treated with

[64]Reference [79].

[65]A small aliquot of the reaction mixture was quenched with D_2O and was analyzed by ^{1}H, and no deuterium enrichment at the benzylic center was observed.

Scheme 3.24 Reactivity of the indoline products towards organolithium reagents

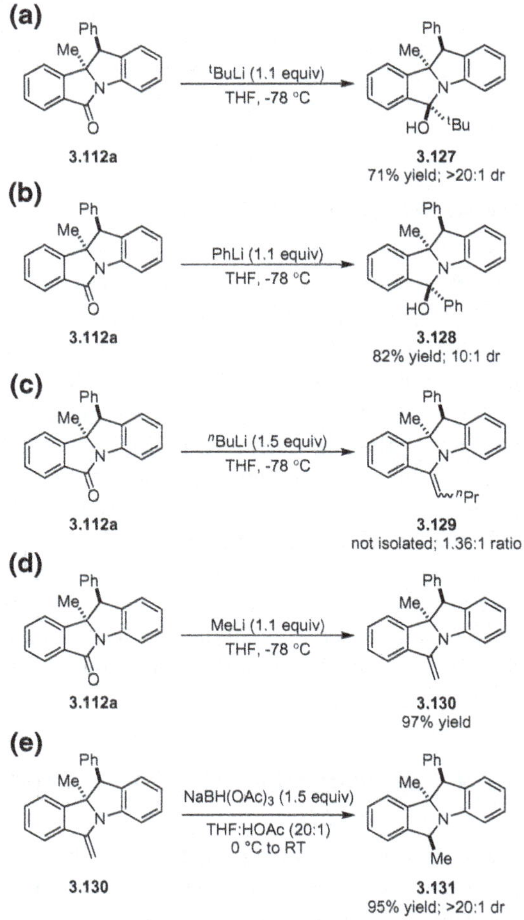

PhLi (1.2 equiv) the direct 1,2-addition product **3.128** was also observed as a 10:1 mixture of diastereomers (Scheme 3.24b). These compounds were inseparable by column chromatography despite trying numerous solvent systems, but the mixture could be isolated in 82% yield (Scheme 3.24b). When **3.112a** was treated with ^{n}BuLi (1.5 equiv), a 1.36:1 mixture of *E*/*Z* enamines was observed in the crude reaction mixture (Scheme 3.24c). These compounds appeared to be unstable, and could not be isolated using conventional methods. The enamine products are thought to arise from dehydration of the hemiaminal products under the basic reaction conditions resulting from quenching the reaction with water. Treatment of **3.112a** with MeLi (1.1 equiv), the corresponding enamine **3.130** could be isolated in 94% yield (Scheme 3.21d). We thought that enamine could undergo a

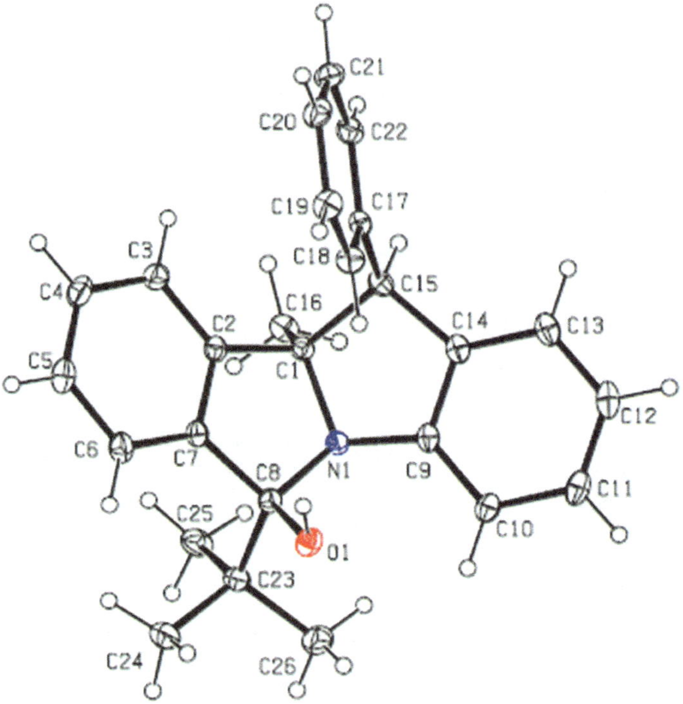

Fig. 3.2 Crystal structure of hemiaminal **3.127**

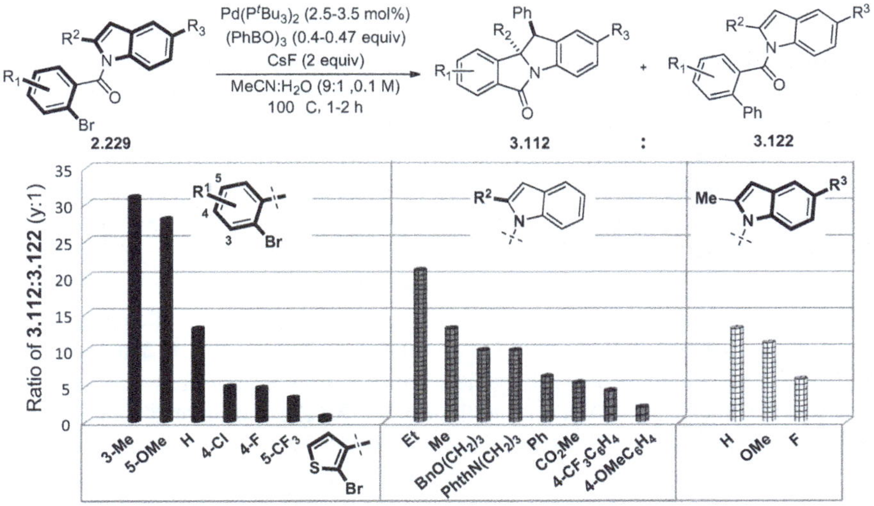

Fig. 3.3 Effect of steric and electronic perturbation of **2.229** on the ratio of **3.112**:**3.122**. Adapted with permission from Ref. [1]. Copyright (2016) John Wiley and Sons

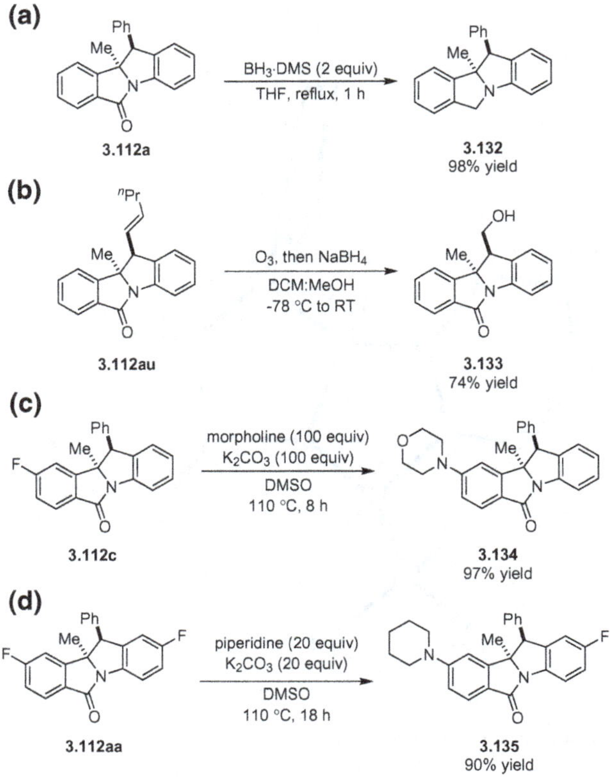

Scheme 3.25 Other derivatization of the indole products

stereoselective reduction using a hydride reducing agent under acidic conditions. When enamine **3.130** was treated with $NaBH(OAc)_3$ (1.5 equiv) in a 20:1 THF: AcOH mixture, the corresponding amine **3.131** was obtained in 95% yield as a single diastereomer (Scheme 3.24e).[66]

The amide functionality of **3.112a** could be reduced to generate amine **3.132** in 98% yield by treatment with BH_3·DMS (2 equiv) in refluxing THF (Scheme 3.25a). The use of $LiAlH_4$ as the reducing agent generally led to lower and less reproducible yields of **3.132**. The alkene moiety in **3.112au** could undergo oxonolysis in DCM:MeOH using a reductive workup to produce homobenzylic alcohol **3.133** in good yield (Scheme 3.25b). Fluorinated analogs **3.112c** and **3.112aa** underwent S_NAr reactions when treated with cyclic secondary amines under basic conditions in DMSO (Scheme 3.25c). When monofluorinated **3.112c** was treated with morpholine, aniline derivative **3.134** could be obtained in 97% yield. Due to the different electronics of the two aromatic rings in **3.111aa**, it seemed likely that a selective

[66]The stereochemistry of **3.131** was assigned by 1D NOE analysis.

S_NAr reaction could be accomplished. When this compound was reacted with excess piperidine under similar conditions, a selective S_NAr reaction of the isoindoline aromatic fluorine over the indoline aromatic fluorine could be achieved, yielding **3.135** in 90% yield (Scheme 3.25d).

3.1.10 Analysis of Selectivity for the Dearomatization Process

The undesired Suzuki-Miyaura coupling product **3.122** was observed in all instances during the examination of the substrate scope. By simply analyzing the crude reaction mixtures by ^1H NMR, trends could be identified which relate the electronic and steric perturbation of both the boroxine reagents and **2.229** to the ratio of **3.112**:**3.122**. Figure 3.3 displays the *syn* 1,2-diarylation/direct Suzuki-Miyaura coupling ratios from reactions of various *N*-benzoyl indoles **2.229** bearing substitution on the aryl bromide fragment, and both the indole C2 and C5 positions.

The introduction of local steric effects (**2.229b**) *ortho* to the carbon–bromine bond led to the lowest amount of Suzuki coupling (**3.112**:**3.122** = 31:1). The presence of an electron donating –*p*-OMe group (**1f**) bond also helped suppress the direct Suzuki-Miyaura process (28:1), whereas the presence of a *p*-CF_3 group (**2.229e**) had the opposite effect (3.4:1). The presence of activating –Cl or –F atoms (**2.229c** and **2.229d**) led to a more Suzuki-Miyaura coupling (~5:1). Heteroaryl bromides exhibited the lowest propensity to undergo the desired dearomative process. For example, the direct biaryl formation was the major pathway when the 3-bromothiophene derivative **2.229s** was employed (0.8:1). By and large, both electron rich and sterically hindered aryl bromides afforded the largest suppression of the undesired biaryl formation, while electron deficient aryl bromides led to the opposite. Varying the C2 substitution of the indole component of **2.229** had comparable effects on the outcome of the reaction. Interestingly, 2-Et **2.229** led to a decrease in the amount of biaryl formation with respect to **2.229a** (21:1), while other larger alkyl substituents (**2.229h** and **2.229i**) led to a slight increase (10:1). A C2 phenyl substituent (**2.229k**) decreased the ability to undergo dearomatization (6.4:1), and *p*-CF_3 (**2.229m**) and *p*-OMe[67] substitution of this aryl group magnified this effect (4.4:1 and 2.1:1). Perturbation of the indole backbone at C5 with either electron-donating (–OMe, **1n**) or electron-withdrawing (–CF_3, **1m**) substituents led to an increased efficiency of the biaryl formation (11:1 and 6:1). These last findings resonate with the previous discussion concerning the altered efficacy of the reaction with varying *N*-substitution, as it seems to be quite sensitive to the electronics of the enamine component of the indole.

---

[67]Under numerous conditions this substrate only reached low levels of conversion and only the ratio, which was consistent across these runs, is presented.

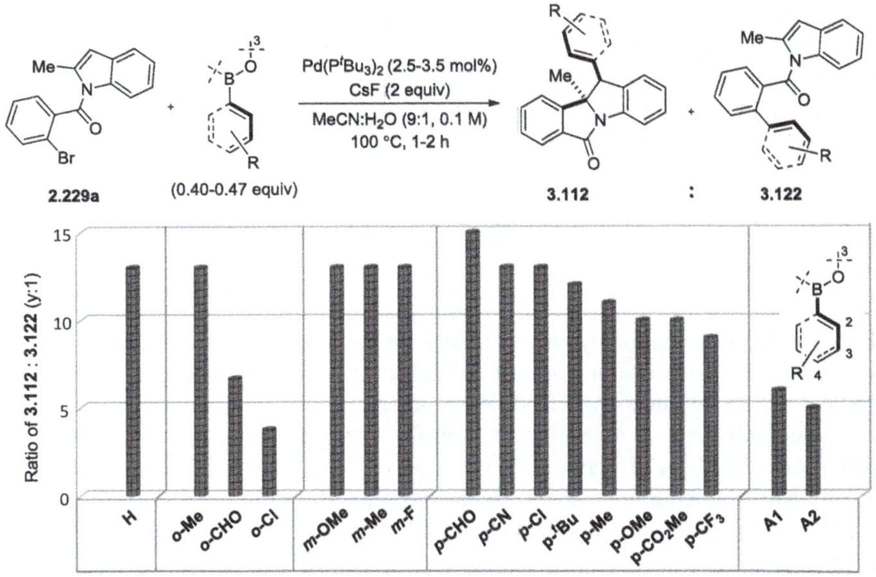

Fig. 3.4 Effect of steric and electronic perturbation of the boroxine reagent on the ratio of **3.112**:**3.112**. A1 = (*E*)-penten-1-ylboroxine, A2 = (*E*)-2-(4-methoxyphenyl)vinylboroxine. Adapted with permission from Ref. [1]. Copyright (2016) John Wiley and Sons

Figure 3.4 displays the ratios of 1,2-diarylation/direct biaryl formation from reactions of aryl and alkenyl boroxines with diverse steric and electronic properties. Even though variations can be seen within this data set, the trends within groups possessing different substituents are less obvious than those in Fig. 3.3. An *o*-Me substituted boroxine led to no change in the ratio of **3.112**:**3.122**, which suggests that some groups at this position do not have a significant effect. However, electron withdrawing *o*-CHO and *o*-Cl substituents produce significant increases in the amount of biaryl formation (6.7:1 and 3.8:1). Boroxines substituted with *meta* electron-donating and -withdrawing groups appeared to have no effect on the ratio of **3.122**:**3.122** with respect to the parent phenylboroxine. Though a *p*-CHO group impacts the ratio of **2**:**2′** positively (15:1), other electron-withdrawing groups resulted in no impact (*p*-CN and *p*-Cl) or a negative impact on the efficacy of the 1,2-diarylation process. Alkyl groups such as ^{t}Bu and Me at the *para* position led to slightly more Suzuki-Miyaura coupling (12:1 and 11:1), whereas *para* electron-withdrawing groups such as CO_2Me and CF_3 led to less diarylation product. Finally, an increase in the amount of biaryl was observed when vinyl boroxine derivatives such as (*E*)-penten-1-ylboroxine (**A1**) and (*E*)-2-(4-methoxyphenyl)vinylboroxine (**A2**) were used (6:1 and 5:1). Despite the fact that boronic acids containing electron-donating groups are known to undergo faster transmetallation with Pd(II) species than those containing electron-withdrawing groups,[68] this is not directly reflected in the observed ratio of **3.112**:**3.122**.

Although boroxines have been reported to undergo transmetallation under anhydrous conditions,[69] our standard conditions involve an aqueous biphase (MeCN:H_2O and CsF). Therefore, it is reasonable to propose that the arylboroxine reagent can be hydrolyzed to the corresponding arylboronic acid under the reaction conditions. Understanding the factors leading to activation of the boron reagent and subsequent reactivity is non-trivial, but it is likely that the ability of the dearomative process to outperform biaryl formation depends on a balance between hydrolytic susceptibility of the boroxine and the transmetallation aptitude of the corresponding boronic acid. Nonetheless, the data displayed in Figs. 3.3 and 3.4 reveal that structural variations of the indole substrate appear to have a more pronounced effect on the product distribution than do structural variations of the boroxine.

3.1.11 Summary and Future Work

Stereoselective dearomatizations catalyzed by transition metals represent a step-economic strategy to build up molecular complexity using simple aromatic precursors. By taking inspiration from both the diarylation and dearomatization literature we have been able to develop and apply a completely *syn*-selective Pd-catalyzed dearomative 1,2-diarylation and 1,2-arylvinylation of *N*-benzoyl indoles using aryl and alkenylboroxines as coupling partners, respectively. Previous efforts in the field of Pd-catalyzed domino Mizoroki-Heck/Suzuki-Miyaura couplings have focused mostly on the use of simple alkynes and alkenes, and there has been a paucity of diastereoselective variants in the latter case. The methodology described herein represents the first example of arylboroxines being used in this class of transformation, in addition to the first Pd-catalyzed diarylation proceeding via a benzylic Pd(II) species. This reaction utilized indoles as non-traditional Mizoroki-Heck acceptors, and proceeds using relatively low catalyst loadings and short reaction times (typically ≤ 2 h).

This transformation is capable of generating complex indoline-containing products which possess a unique arrangement of stereocenters consisting of a tetrasubstituted tertiary stereocenter vicinal to a tertiary benzylic stereocenter. This reaction displays a broad scope, and many substrate perturbations are well tolerated under the optimized conditions or slight variations thereof. The reaction conditions rely on several parameters which help to suppress the formation of the direct Suzuki-Miaura products, namely the use of a MeCN:H_2O mixture, CsF as base, temperatures below 110 °C and as little catalyst as possible. However, this biaryl formation occurs to some extent in all cases possibly due to the challenging nature

[68]Reference [80].

[69]Reference [81].

Scheme 3.26 Hinckdentine A as a potential synthetic target for the dearomative indole arylvinylation

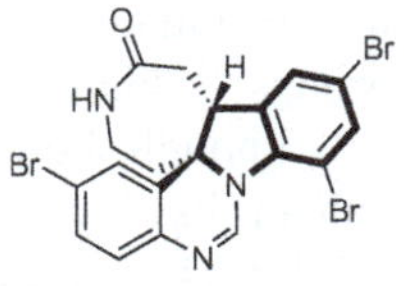

of the *syn* carbopalladation. It has also been shown that the indoline scaffolds generated by this methodology can be derivatized in several ways.

Although the development of an asymmetric variant has been unsuccessful to this point, future work could involve exploring this concept further. Furthermore, the broad scope of this reaction makes it amenable to application in the synthesis of a natural product such as Hinckdentine A (Scheme 3.26). Specifically, the use of alkenyl boroxines opens up many doors due to the scope of synthetic manipulations possible with the alkene functional group.

3.1.12 Experimental

General Considerations. Unless otherwise stated, all catalytic reactions were carried out under an inert atmosphere of dry argon or dry nitrogen utilizing glassware that was oven (120 °C) or flame dried and cooled under argon, whereas the work-up and isolation of the products from the catalytic reactions were conducted on the bench-top using standard techniques. Reactions were monitored using thin-layer chromatography (TLC) on EMD Silica Gel 60 F254 plates. Visualization of the developed plates was performed under UV light (254 nm) or using $KMnO_4$, *p*-anisaldehyde, or ceric ammonium molybdate (CAM) stain. Organic solutions were concentrated under reduced pressure on a Büchi rotary evaporator. Tetrahydrofuran was distilled over sodium, toluene was distilled over sodium and degassed via 5 freeze-pump-thaw cycles and stored over activated 4 Å molecular sieves, triethyl amine was distilled over potassium hydroxide, DCM was distilled over calcium hydride, and anhydrous *N,N*-dimethylformamide and diethyl ether were purchased from Aldrich and used as received. silica gel flash column chromatography was performed on Silicycle 230–400 mesh silica gel. All standard reagents including were purchased from Sigma Aldrich, Alfa Aesar, Combi-Blocks, or Oakwood, and were used without further purification. 2-methylindole was obtained from commercial source Alfa and purified by passing through a short plug of silica eluting with 100% hexanes then 100% DCM. Substrates **2.229a-k, 2.229i, 2.229k, 2.229m, 2.229o-r and 2.229t** were prepared by known procedures,[70] and characterization data were in accordance to the reported literature.

Instrumentation. NMR characterization data was collected at 296 K on a Varian Mercury 300, Varian Mercury 400, Varian 600, or a Bruker Avance III spectrometer operating at 300, 400, 500, or 600 MHz for ^1H NMR, and 75, 100,

[70] Reference [72].

125, for 150 MHz or ^{13}C NMR. ^{1}H NMR spectra were internally referenced to the residual solvent signal ($CDCl_3$ = 7.26 ppm, toluene-d_8 = 2.3 ppm) or TMS. ^{13}C NMR spectra were internally referenced to the residual solvent signal ($CDCl_3$ = 77.0 ppm) and are reported as observed. Data for ^{1}H NMR are reported as follows: chemical shift (δ ppm), multiplicity (s = singlet, d = doublet, t = triplet, q = quartet, m = multiplet, br = broad), coupling constant (Hz), integration. Coupling constants have been rounded to the nearest 0.5 Hz. Melting point ranges were determined on a Fisher-Johns Melting Point Apparatus and are reported uncorrected. All reported diasteromeric ratios in data section are those obtained from ^{1}H NMR analysis of the crude reaction mixtures using 5 s delay. NMR yields for the optimization section were obtained by ^{1}H NMR analysis of the crude reaction mixture using 5 s delay and 1,3,5-trimethoxybenzene as an internal standard. Enantiomertic ratios were determined using HPLC on a chiral stationary phase using a complete Agilent HPLC system. All retention times have been rounded to the nearest 0.1 min. High resolution mass spectra (HRMS) were obtained on a micromass 70S-250 spectrometer (EI) or an ABI/Sciex QStar Mass Spectrometer (ESI) or a JEOL AccuTOF medel JMS-T1000LC mass spectrometer equipped with an IONICS® Direct Analysis in real Time (DART) ion source at Advanced Instrumentation for Molecular Structure (AIMS) in the Department of Chemistry at the University of Toronto.

General Procedures

General Procedure 1: *N*-(2-bromobenzoyl)indole synthesis using NaH.

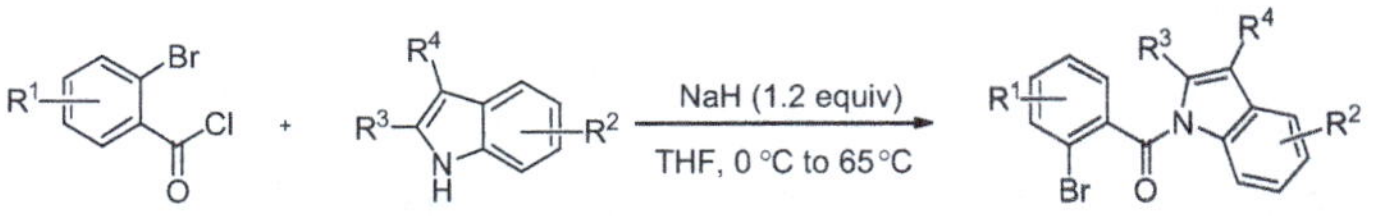

A 60% dispersion of NaH in mineral oil (1.2 equiv) was added to a stirred solution of the appropriate indole derivative **2.232** (1 equiv, ~0.5 M) in THF at 0 °C and the corresponding solution was stirred for 5 min before warming to room temperature where it was stirred for 30 min. The solution of the sodium indolate was re-cooled to 0 °C at which time a solution of appropriate 2-bromobenzoyl chloride derivative **2.104** (2 equiv or 2.2 equiv, ~1 M) in THF was added dropwise. Once the addition was complete, the reaction was allowed to warm to room temperature and then was stirred at 65 °C for 30 min. At this time the extent of completion of the reaction was determined by conversion of the indole derivate by TLC analysis. The reaction was cooled to room temperature and quenched with a saturated solution of NH_4Cl. The reaction mixture was then diluted with water and EtOAc, and after separating the layers, the aqueous layer was extracted with EtOAc (3x). The combined organic layers were washed sequentially with water and brine, dried over sodium sulfate, filtered and concentrated under reduced pressure. The crude *N*-(2-bromobenzoyl)

indole derivative **2.229** was purified by flash column silica gel chromatography using the indicated solvent system.

General Procedure 2: Pd-Catalyzed dearomative indole diarylation.

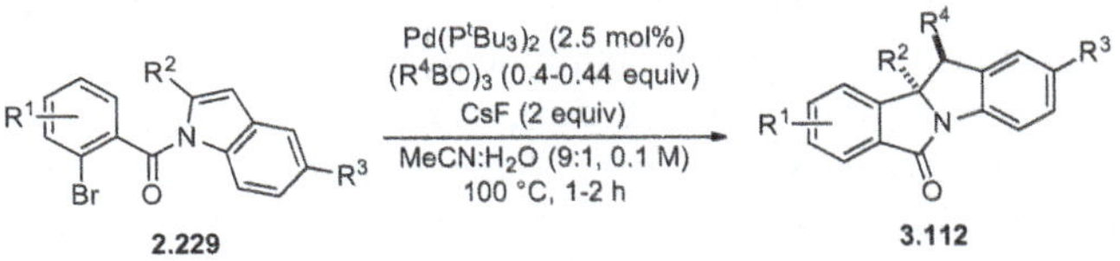

2.229 3.112

A dry 2 dram vial containing a magnetic stir bar was charged with anhydrous CsF (61.8 mg, 0.4 mmol, 2 equiv) using a flame dried spatula followed by the *N*-(2-bromobenzoyl)indole derivative **2.229** (0.2 mmol, 1 equiv), Pd(P^{t}Bu$_3$)$_2$ (2.5 mg, 0.005 mmol, 2.5 mol%), and the boroxine (0.08 mmol or 0.094 mmol, 0.4 or 0.47 equiv) and was purged with argon for 10 min. The contents of the vial were taken up in MeCN:H$_2$O (9:1, 2 mL), and the vial was sealed with a Teflon® lined screw-cap and then placed in an oil bath pre-heated to 100 °C where it was stirred for the indicated amount of time. The vial was then cooled to room temperature and anhydrous Na$_2$SO$_4$ (50 mg) was added to remove the water. The organic component was passed through a short pad of silica gel eluting with EtOAc. The diastereomeric ratio was determined by ^{1}H NMR analysis of a homogeneous aliquot of the crude reaction mixture, which was subsequently purified via silica gel flash column chromatography using hexanes/EtOAc as the mobile phase to afford the corresponding indoline **3.112**.

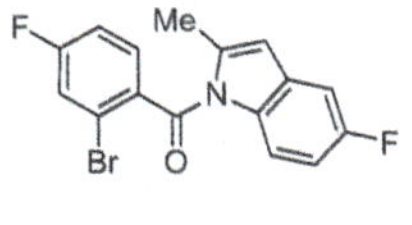

(**2.229aa**)–Was synthesized according to **general procedure 1** using 5-fluoro-2-methylindole (500 mg, 3.35 mmol, 1 equiv) and 2-bromo-4-fluorobenzoyl chloride generated from 2-bromo-4-fluorobenzoic acid (1.61 g, 7.38 mmol, 2.2 equiv). The crude product was purified via silica gel flash column chromatography using hexanes:EtOAc (10:1 v:v) as the mobile phase and was isolated as a colorless oil (810 mg, 2.31 mmol, 69%). **^{1}H NMR** (500 MHz, CDCl$_3$) δ 7.51–7.47 (m, 1H), 7.45–7.39 (m, 2H), 7.19 (ddd, *J* = 8.5, 7.8, 2.4 Hz, 1H), 7.10 (dd, *J* = 8.7, 2.4 Hz, 1H), 6.86 (td, *J* = 9.1, 2.6 Hz, 1H), 6.38–6.34 (m, 1H), 2.18 (d, *J* = 1.3 Hz, 3H). **^{13}C NMR** (125 MHz, CDCl$_3$) δ 166.6, 163.4 (d, *J* = 256.8 Hz), 159.7 (d, *J* = 240.3 Hz), 138.6, 134.5 (d, *J* = 4.0 Hz), 133.1 (d, *J* = 1.4 Hz), 130.99 (d, *J* = 9.1 Hz), 130.98 (d, *J* = 10.1 Hz), 121.2, 121.0 (d, *J* = 24.9 Hz), 115.9 (d, *J* = 9.1 Hz), 115.5 (d, *J* = 21.7 Hz), 111.3 (d, *J* = 24.8 Hz), 110.3 (d, *J* = 3.7 Hz), 105.6 (d, *J* = 24.0 Hz), 16.3. **^{19}F NMR** (376 MHz, CDCl$_3$) δ-105.77, -105.79, -105.80, -105.81, -105.82, -105.83, -119.40, -119.41, -119.42, -119.43, -119.44, -119.45. **IR** (thin film, cm^{-1}) 3095, 2978, 2929, 1699, 1657, 1593, 1361, 1311, 1257, 1164. **HRMS** (**DART**, M+H) calc'd for $C_{16}H_{11}BrF_2NO$ 350.00012, found 349.99989.

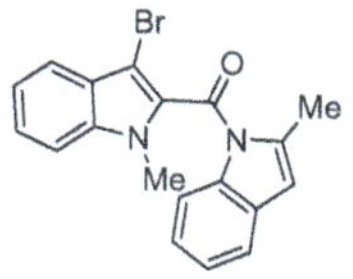

(**2.229ab**)–Was synthesized according to **general procedure 1** using 2-methylindole (108.8 mg, 0.83 mmol, 1 equiv) and 3-bromo-1-methyl-1H-indole-2-carbonyl chloride generated from 3-bromo-1-methyl-1H-indole-2-carboxylic acid (230 mg, 0.91 mmol, 1.1 equiv). The crude product was purified via silica gel flash column chromatography using hexanes:EtOAc (20:1 v:v) as the mobile phase and was isolated as an off white solid (149.7 mg, 0.409 mmol, 49%, MP = 144–145 °C). **^{1}H NMR** (500 MHz, $CDCl_3$) δ 7.63 (dt, J = 8.1, 1.0 Hz, 1H), 7.49–7.41 (m, 3H), 7.31–7.26 (m, 2H), 7.21–7.17 (m, 1H), 7.08 (ddd, J = 8.4, 7z.2, 1.3 Hz, 1H), 6.47–6.46 (m, 1H), 3.87 (s, 3H), 2.36 (d, J = 1.2 Hz, 3H). **^{13}C NMR** (125 MHz, $CDCl_3$) δ 161.0, 137.9, 136.8, 136.7, 130.8, 129.7, 126.4, 126.1, 123.6, 123.4, 121.5, 121.2, 119.8, 113.4, 110.4, 109.7, 96.8, 31.6, 15.3. **IR** (thin, film, cm^{-1}) 3059, 3028, 2922, 1678, 1668, 1614, 1545, 1574, 1506, 1394, 1316, 1334, 1211, 1202. **HRMS** (ESI+, M+H) Calc'd for $C_{19}H_{16}BrN_2O$ 367.0441, found 367.0438.

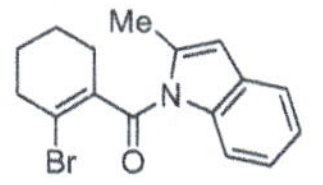

(**2.229ac**)–Was synthesized according to **general procedure 1** using 2-methylindole (200 mg, 1.53 mmol, 1 equiv) and 2-bromocyclohex-1-enecarbonyl chloride generated from 2-bromocyclohex-1-enecarboxylic acid (690 mg, 3.37 mmol, 2.2 equiv). The crude product was purified via silica gel flash column chromatography using hexanes:EtOAc (20:1 v:v) as the mobile phase, and was isolated as a colorless solid (411 mg, 1.29 mmol, 85%, MP(°C) = 84–85). **^{1}H NMR** (400 MHz, $CDCl_3$) δ 8.05–8.00 (m, 1H), 7.48–7.41 (m, 1H), 7.25–7.21 (m, 2H), 6.39 (s, 1H), 2.62 (s, 2H), 2.57 (d, J = 1.1 Hz, 3H), 2.49 (s, 2H), 1.86–1.83 (m, 4H). **^{13}C NMR** (125 MHz, $CDCl_3$) δ 169.1, 136.8, 136.5, 135.2, 129.9, 123.8, 123.6, 123.5, 119.7, 114.9, 110.2, 35.8, 29.1, 23.8, 21.4, 16.3. **IR** (thin film, cm^{-1}) 2932, 2862, 2662, 2359, 2342, 2332, 1800, 1699, 1645, 1557, 1456. **HRMS** (DART, M+H) calc'd for $C_{16}H_{17}BrNO$ 318.04880, found 318.04902.

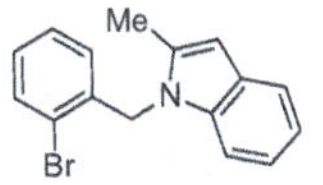

(**3.117**)–2-bromobenzyl chloride (411 mg, 2 mmol, 1 equiv) was added dropwise to a stirred solution DMSO solution (8 mL) of 2-methylindole (525 mg, 4 mmol, 2 equiv) and KOH (898 mg, 16 mmol, 8 equiv) at room temperature. The reaction was stirred at this temperature for 2 h and was then quenched with a saturated aqueous solution of NH_4Cl. The layers were separated and the aqueous layer was extracted three times with diethyl ether. The combined organic layers were sequentially washed with water and brine, dried over sodium sulfate, filtered and concentrated under reduced pressure. The crude product was purified via silica gel flash column chromatography using pentane:PhMe (10:1 v:v) to afford a white solid (240 mg, 0.8 mmol, 40%, MP = 83–84 °C). **^{1}H NMR** (500 MHz, $CDCl_3$) δ 7.64–7.60 (m, 2H), 7.17–7.05 (m, 5H), 6.40 (s, 1H), 6.28–6.22 (m, 1H), 5.34 (s, 2H), 2.37 (s, 3H). **^{13}C NMR** (125 MHz, $CDCl_3$) δ 137.0, 136.7, 136.6, 132.6, 128.7, 128.2, 127.9, 127.0, 121.6, 120.9, 119.8, 119.7, 109.1, 100.7, 46.8, 12.5. **IR** (thin film, cm^{-1}) 3055, 2980, 2916, 2859, 1612, 1555, 1462, 1398. **HRMS** (DART, M+H) Calc'd for $C_{16}H_{15}BrN$ 300.03879, found 300.03915.

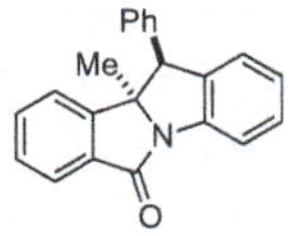

(**3.112a**)–Was synthesized according to **general procedure 2** using **2.229a** (62.8 mg, 0.2 mmol, 1 equiv). The crude product was purified via silica gel flash column chromatography using hexanes:EtOAc (10:1 to 5:1 v:v) and was isolated as a white solid (54.8 mg, 0.176 mmol, 88%, >20:1 dr, MP = 185–187 °C). **^{1}H NMR** (500 MHz, $CDCl_3$) δ 7.83 (ddt, J = 8.0, 1.0, 0.5 Hz, 1H), 7.69 (ddd, J = 7.0, 1.5, 1.0 Hz, 1H), 7.46–7.38 (m, 1H), 7.28 (td, J = 7.5, 1.5 Hz, 1H), 7.24 (td, J = 7.5, 1.5 Hz, 1H), 7.19 (ddt, J = 7.5, 1.5, 0.5 Hz, 1H), 7.13 (td, J = 7.5, 1.0 Hz, 1H), 7.06 (ddd, J = 7.5, 1.5, 1.0 Hz, 1H), 6.99–6.86 (m, 3H), 6.60–6.48 (m, 2H), 4.36 (s, 1H), 1.77 (s, 3H). **^{13}C NMR** (125 MHz, $CDCl_3$) δ 168.5, 148.7, 140.7, 139.2, 138.9, 132.9, 132.0, 128.7, 128.2, 128.1, 127.6, 126.8, 126.6, 125.0, 124.2, 123.0, 117.1, 76.0, 57.1, 28.5. **IR** (thin film, cm^{-1}) 3063, 3026, 1699, 1481, 1361, 1332. **HRMS** (DART, M+H) Calc'd for $C_{22}H_{18}NO$ 312.13884, found 312.13901.

Procedure for gram-scale experiment: An oven-dried 100 mL round bottom flask containing a stir bar was charged with anhydrous CsF (966 mg, 6.4 mmol, 2 equiv) using a flame dried spatula followed by **1a** (1.0 g, 3.2 mmol, 1 equiv), Pd($P^tBu_3)_2$ (20.5 mg, 0.04 mmol, 1.25 mol%), and phenylboroxine (399 mg, 1.28 mmol, 0.4 equiv). The flask was fitted with a rubber septum and the atmosphere was replaced with argon by four vacuum/argon purge cycles on a Schlenk line through a needle. The contents of the flask were taken up in MeCN:H_2O (9:1, 32 mL) and the flask was tightly sealed using an appropriate Teflon® twist cap, and was placed into an oil bath pre-heated to 100°. After 1 h the reaction was cooled to room temperature and anhydrous Na_2SO_4 (250 mg) was added to remove the water, and the organic component was filtered over a short pad of silica gel, eluting with 100% EtOAc (500 mL). A homogeneous aliquot was taken from the organic washings and the solvents were removed in vacuo. ^{1}H NMR analysis of this crude reaction mixture indicated the product to be present in >20:1 dr. All volatiles were removed from the bulk solution under reduced pressure and purification of the crude residue was carried out via silica gel flask column chromatography using hexanes:EtOAc (10:1–5:1 v:v) as the mobile phase, and the title compound was obtained as a white solid (751 mg, 2.42 mmol, 76%).

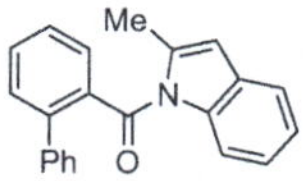

(**3.122a**)–Was obtained when applying **general procedure 2** to **2.229a**. The crude product was purified via silica gel flash column chromatography using hexanes: EtOAc (10:1 v:v) and was isolated as a colourless film. **^{1}H NMR** (500 MHz, $CDCl_3$) δ 7.68 (ddd, *J* = 7.5, 1.5, 0.5 Hz, 1H), 7.65 (td, *J* = 7.5, 1.5 Hz, 1H), 7.54 (td, *J* = 7.5, 1.0 Hz, 1H), 7.45 (ddd, *J* = 7.5, 1.5, 0.5 Hz, 1H), 7.26 (ddd, *J* = 8.0, 1.5, 0.5 Hz, 1H), 7.16–7.10 (m, 5H), 7.07 (ddd, *J* = 7.5, 7.0, 1.0 Hz, 1H), 6.97 (ddd, *J* = 8.5, 7.0, 1.5 Hz, 1H), 6.80–6.75 (m, 1H), 6.11–6.09 (m, 1H), 2.30 (d, *J* = 1.0 Hz, 3H). **^{13}C NMR** (125 MHz, $CDCl_3$) δ 170.0, 141.2, 138.9, 137.5, 136.5, 135.7, 131.4, 130.4, 129.6, 129.0, 128.0, 127.9, 127.8, 127.4, 122.7, 122.6, 119.4, 114.1, 108.9, 16.0. **IR** (thin film, cm^{-1}) 3059, 3026, 2967, 2924, 2874, 2857, 1694, 1651, 1595, 1476. **HRMS** (ESI+, M+H) Calc'd for $C_{22}H_{18}NO$ 312.1383, found 312.1387.

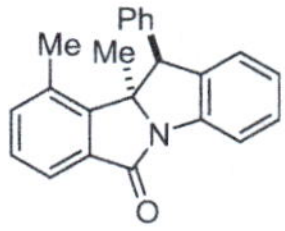

(**3.112b**)–Was synthesized according to **general procedure 2** (exception: Pd$(P^tBu_3)_2$ (3.5 mol%) was used) using **2.229b** (65.4 mg, 0.2 mmol, 1 equiv). The crude product was purified via silica gel flash column chromatography using hexanes:EtOAc (10:1 v:v) and was isolated as a white solid (62.4 mg, 0.196 mmol, 98%, >20:1 dr, MP = 186–187 °C). **^{1}H NMR** (400 MHz, $CDCl_3$) δ 7.84 (dd, *J* = 8.0, 1.0 Hz, 1H), 7.64 (dt, *J* = 7.5, 1.0 Hz, 1H), 7.37 (td, *J* = 7.5, 1.5 Hz, 1H), 7.22 (t, *J* = 7.5 Hz, 1H), 7.17–7.12 (m, 1H), 7.08 (td, *J* = 7.5, 1.0 Hz, 2H), 6.98–6.89 (m, 3H), 6.68–6.59 (m, 2H), 4.37 (s, 1H), 2.26 (s, 3H), 1.80 (s, 3H). **^{13}C NMR** (100 MHz, $CDCl_3$) δ 167.7, 146.3, 140.9, 140.0, 137.9, 134.1, 133.7, 132.5, 128.6, 128.5, 128.0, 127.3, 126.9, 126.3, 124.9, 122.0, 117.0, 76.2, 56.7, 26.5, 19.1. **IR** (thin film, cm^{-1}) 3055, 3028, 1697, 1554, 1462, 1363, 1332, 1307, 1149. **HRMS** (DART, M+H) Calc'd for $C_{23}H_{20}NO$ 326.15500, found 326.15506.

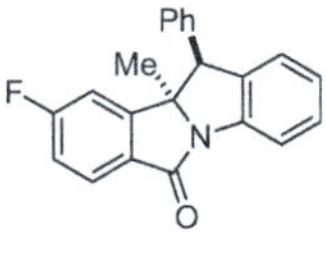

(**3.112c**)–Was synthesized according **general procedure 2** using **2.229c** (66.2 mg, 0.2 mmol, 1 equiv). The crude product was purified via silica gel flash column chromatography using hexanes:EtOAc (10:1 v:v) and was isolated as a white solid

(53.8 mg, 0.163 mmol, 82%, >20:1 dr, MP = 206–207 °C). **^{1}H NMR** (500 MHz, $CDCl_3$) δ 7.81 (dd, J = 8.0, 1.0 Hz, 1H), 7.67 (dd, J = 8.5, 5.0 Hz, 1H), 7.41 (td, J = 7.5, 1.5 Hz, 1H), 7.21–7.16 (m, 1H), 7.14 (td, J = 7.5, 1.0 Hz, 1H), 7.01–6.89 (m, 4H), 6.74 (dd, J = 8.0, 2.0 Hz, 1H), 6.60–6.52 (m, 2H), 4.35 (s, 1H), 1.76 (s, 3H). **^{13}C NMR** (125 MHz, $CDCl_3$) δ 167.5, 165.2 (d, J = 252.9 Hz), 151.3 (d, J = 9.6 Hz), 140.2, 139.2, 138.5, 128.9 (d, J = 2.1 Hz), 128.8, 128.4, 127.5, 127.0, 126.5, 126.3 (d, J = 10.0 Hz), 125.2, 117.0, 115.9 (d, J = 23.6 Hz), 110.4 (d, J = 23.9 Hz), 75.5 (d, J = 2.6 Hz), 57.1, 28.3. **^{19}F NMR** (376 MHz, $CDCl_3$) δ-105.97 to -106.03. **IR** (thin film, cm^{-1}) 3059, 3024, 2926, 1699, 1622, 1597, 1477, 1366, 1336, 1313, 1190. **HRMS** (DART, M+H) Calc'd for $C_{22}H_{17}FNO$ 318.09526, found 318.09518.

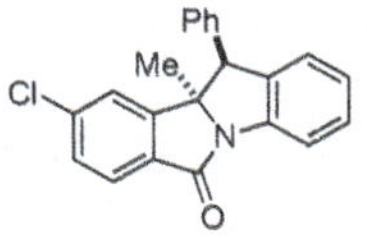

(**3.112d**)–Was synthesized according to **general procedure 2** (exception: boroxine (0.4 equiv) was used) heating for 45 min using **2.229dc** (69.2 mg, 0.2 mmol, 1 equiv). The crude product was purified via silica gel flash column chromatography using hexanes:EtOAc (10:1 v:v) and was isolated as a white solid (48.7 mg, 0.141 mmol, 70%, >20:1 dr, MP = 205–206 °C). **^{1}H NMR** (400 MHz, $CDCl_3$) δ 7.88–7.74 (m, 1H), 7.61 (dd, J = 8.0, 0.5 Hz, 1H), 7.42 (td, J = 7.5, 1.5 Hz, 1H), 7.25–7.11 (m, 3H), 7.05 (dd, J = 2.0, 0.5 Hz, 1H), 7.03–6.90 (m, 3H), 6.56 (dd, J = 8.0, 1.5 Hz, 2H), 4.35 (s, 1H), 1.76 (s, 3H). **^{13}C NMR** (100 MHz, $CDCl_3$) δ 167.3, 150.2, 140.2, 139.0, 138.5, 138.3, 131.3, 128.8, 128.7, 128.4, 127.5, 127.1, 126.6, 125.4, 125.3, 123.5, 117.1, 75.7, 57.0, 28.2. **IR** (thin film, cm^{-1}) 3061, 2966, 2924, 1697, 1477, 1363, 1323, 1143. **HRMS** (DART, M+H) Calc'd for $C_{22}H_{17}ClNO$ 342.14991, found 342.15005.

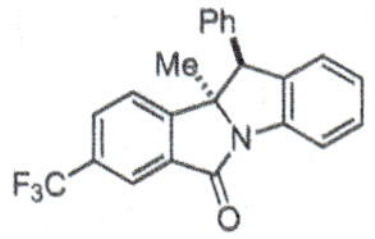

(**3.112e**)–Was synthesized according to **general procedure 2** heating for 2 h using **2.229e** (76.4 mg, 0.2 mmol, 1 equiv). The crude product was purified via silica gel flash column chromatography using hexanes:EtOAc (10:1 v:v) and was isolated as a white solid (46.7 mg, 0.124 mmol, 62%, >20:1 dr, MP = 134–135 °C). **^{1}H NMR** (500 MHz, $CDCl_3$) δ 7.98 (dt, J = 1.5, 0.5 Hz, 1H), 7.85 (ddt, J = 8.0, 1.0, 0.5 Hz, 1H), 7.53 (ddd, J = 8.0, 1.5, 0.5 Hz, 1H), 7.47–7.38 (m, 1H), 7.24–7.19 (m, 2H), 7.16 (td, J = 7.5, 1.0 Hz, 1H), 6.98–6.89 (m, 3H), 6.72–6.38 (m, 2H), 4.42 (s, 1H), 1.78 (s, 3H). **^{13}C NMR** (125 MHz, $CDCl_3$) δ 166.7, 151.9 (q, J = 1.2 Hz), 140.1, 138.7, 138.6, 133.7, 130.7 (q, J = 33.0 Hz), 128.8, 128.6 (q, J = 3.6 Hz), 128.4, 127.4, 127.1, 126.6, 125.5, 123.7, 123.5 (q, J = 272.6 Hz), 121.4 (q, J = 3.9 Hz), 117.1, 76.1, 56.9, 28.2. **^{19}F NMR** (282 MHz, $CDCl_3$) δ-62.5. **IR** (thin film, cm^{-1}) 3036,

2970, 1701, 1629, 1479, 1323, 1274, 1203, 1006. **HRMS** (DART, M+H) Calc'd for $C_{23}H_{17}F_3NO$ 380.12622, found 342.12724.

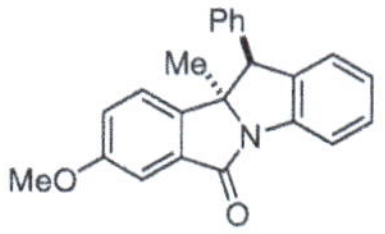

(**3.112f**)–Was synthesized according to **general procedure 2** using **2.229f** (66.0 mg, 0.2 mmol, 1 equiv). The crude product was purified via silica gel flash column chromatography using hexanes:EtOAc (10:1 v:v) and was isolated as a white solid (62.4 mg, 0.183 mmol, 91%, >20:1 dr, MP = 135–137 °C). **^{1}H NMR** (400 MHz, $CDCl_3$) δ 7.90–7.73 (m, 1H), 7.41 (td, *J* = 7.5, 1.5 Hz, 1H), 7.22–7.07 (m, 3H), 7.03–6.89 (m, 4H), 6.84 (dd, *J* = 8.0, 2.5 Hz, 1H), 6.55 (dd, *J* = 8.0, 1.5 Hz, 2H), 4.33 (s, 1H), 3.75 (s, 3H), 1.73 (s, 3H). **^{13}C NMR** (100 MHz, $CDCl_3$) δ 168.4, 159.7, 141.1, 140.8, 139.2, 139.0, 134.2, 128.6, 128.2, 127.6, 126.8, 126.6, 125.0, 123.8, 120.4, 117.0, 106.6, 75.6, 57.0, 55.5, 28.6. **IR** (thin film, cm^{-1}) 3061, 2965, 1697, 1464, 1362, 1327, 1281, 1136. **HRMS** (DART, M+H) Calc'd for $C_{23}H_{20}NO_2$ 342.14991, found 342.15005.

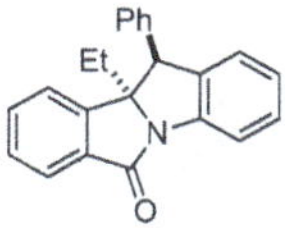

(**3.112g**)–Was synthesized according to **general procedure 2** using **2.229g** (65.4 mg, 0.2 mmol, 1 equiv). The crude product was purified via silica gel flash column chromatography using hexanes:EtOAc (10:1 v:v) and was isolated as a white solid (53.0 mg, 0.162 mmol, 81%, >20:1 dr, MP = 190–192 °C). **^{1}H NMR** (300 MHz, $CDCl_3$) δ 7.85 (d, *J* = 8.0 Hz, 1H), 7.77–7.65 (m, 1H), 7.41 (td, *J* = 7.5, 1.5 Hz, 1H), 7.34–7.21 (m, 2H), 7.21–7.07 (m, 2H), 7.06–6.99 (m, 1H), 7.00–6.88 (m, 3H), 6.58 (dd, *J* = 7.5, 2.0 Hz, 2H), 4.39 (s, 1H), 2.18 (m, 2H), 0.67 (t, *J* = 7.5 Hz, 3H). **^{13}C NMR** (100 MHz, $CDCl_3$) δ 169.1, 146.9, 140.7 139.6, 139.3, 133.8, 131.8, 128.6, 128.2, 128.0, 127.8, 126.7, 126.4, 124.9, 124.0, 123.1, 116.9, 79.5, 56.6, 33.6, 7.9. **IR** (thin film, cm^{-1}) 3028, 2966, 1701, 1602, 1477, 1462, 1344, 1300, 1141. **HRMS** (DART, M+H) Calc'd for $C_{23}H_{20}NO$ 326.15449, found 326.15421.

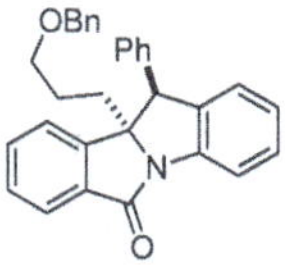

(**3.112hh**)–Was synthesized according to **general procedure 2** using **2.229h** (58.4 mg, 0.1306 mmol, 1 equiv). The crude product was purified via silica gel flash column chromatography using hexanes:EtOAc (10:1 v:v) and was isolated as a white solid (51.8 mg, 0.116 mmol, 81%, >20:1 dr, MP = 126–127 °C). **^{1}H NMR** (500 MHz, $CDCl_3$) δ 7.82 (dd, *J* = 7.5, 1.0 Hz, 1H), 7.76–7.61 (m, 1H), 7.40 (td, *J* = 7.5, 1.5 Hz, 1H), 7.35–7.29 (m, 2H), 7.29–7.20 (m, 5H), 7.20–7.08 (m, 2H), 7.05–6.98 (m, 1H), 6.98–6.87 (m, 3H), 6.60–6.51 (m, 2H), 4.39 (s, 1H), 4.37 (s, 2H), 3.32 (t, *J* = 6.5 Hz, 2H), 2.33–2.14 (m, 2H), 1.64–1.49 (m, 1H), 1.22–1.09 (m, 1H). **^{13}C NMR** (125 MHz, $CDCl_3$) δ 169.1, 147.1, 140.5, 139.6, 139.2, 138.2, 133.6, 132.0, 128.6, 128.3, 128.2, 128.1, 127.8, 127.6, 127.5, 126.7, 126.4, 125.0, 124.1, 123.2, 117.0, 78.9, 72.8, 69.8, 56.9, 37.3, 24.0. **IR** (thin film, cm^{-1}) 3063, 3028, 3009, 2854, 1699, 1604, 1494, 1479, 1359, 1306, 1301, 1139, 1101. **HRMS** (DART, M+H) Calc'd for $C_{31}H_{28}NO_2$ 446.21251, found 446.21243.

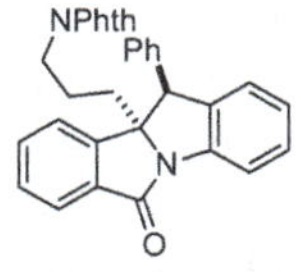

(**3.112i**)–Was synthesized according to **general procedure 2** using **2.229i** (97.5 mg, 0.2 mmol, 1 equiv). The crude product was purified via silica gel flash column chromatography using hexanes:EtOAc (10:1 v:v) and was isolated as a white solid (87.1 mg, 0.180 mmol, 90%, >20:1 dr, MP = 80–82 °C). **^{1}H NMR** (400 MHz, $CDCl_3$) δ 7.82–7.75 (m, 3H), 7.72–7.62 (m, 3H), 7.37 (td, *J* = 7.5, 1.5 Hz, 1H), 7.21 (dd, *J* = 5.5, 3.0 Hz, 2H), 7.17–7.05 (m, 2H), 7.01–6.94 (m, 1H), 6.95–6.85 (m, 3H), 6.56–6.49 (m, 2H), 4.37 (s, 1H), 3.54 (td, *J* = 7.0, 1.0 Hz, 2H), 2.20 (m, 2H), 1.66 (m, 1H), 1.21 (m, 1H). **^{13}C NMR** (100 MHz, $CDCl_3$) δ 169.1, 168.1, 146.7, 140.4, 139.5, 139.0, 133.9, 133.4, 132.1, 131.9, 128.6, 128.2, 128.1, 127.8, 126.7, 126.4, 125.0, 124.1, 123.2, 123.0, 117.1, 78.6, 56.8, 37.8, 37.6, 23.0. **IR** (thin film, cm^{-1}) 2928, 1770, 1709, 1603, 1478, 1397, 1360, 1302. **HRMS** (DART, M+H) Calc'd for $C_{32}H_{25}N_2O_3$ 485.18702, found 485.18709.

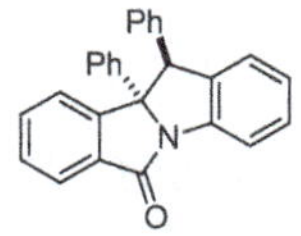

(**3.112k**)–Was synthesized according to **general procedure 2** using **2.229k** (75.3 mg, 0.2 mmol, 1 equiv). The crude product was purified via silica gel flash column chromatography using hexanes:EtOAc (10:1 v:v) and was isolated as a white solid (59.7 mg, 0.160 mmol, 80%, >20:1 dr, MP = 182–184 °C). **^{1}H NMR** (400 MHz, $CDCl_3$) δ 7.97 (dd, *J* = 8.0, 1.0 Hz, 1H), 7.82–7.72 (m, 2H), 7.72–7.61 (m, 1H), 7.42 (ddd, *J* = 8.0, 5.0, 4.0 Hz, 1H), 7.38–7.30 (m, 2H), 7.31–7.22 (m, 1H), 7.22–7.14 (m, 3H), 7.13–7.07 (m, 2H), 7.06–7.02 (m, 1H), 7.02–6.95 (m, 2H), 6.80–6.70 (m, 2H), 5.03 (s, 1H). **^{13}C NMR** (100 MHz, $CDCl_3$) δ 169.1, 147.9, 143.8, 140.3, 139.8, 138.5, 132.2, 132.1, 128.8, 128.7, 128.4, 128.0, 128.1, 127.8, 127.0, 126.2, 125.4, 125.0, 124.2, 124.1, 116.9, 81.1, 59.0. **IR** (thin film, cm^{-1})

3059, 3026, 1701, 1478, 1360, 1306, 1142. **HRMS** (DART, M+H) Calc'd for $C_{27}H_{20}NO$ 374.15500, found 374.15527.

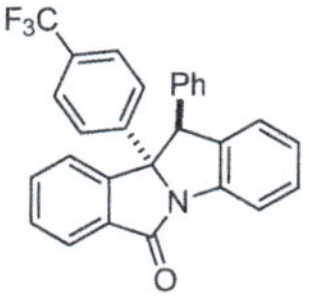

(**3.112m**)–Was synthesized according to **general procedure 2** using **2.229m** (88.8 mg, 0.2 mmol, 1 equiv). The crude product was purified via silica gel flash column chromatography using hexanes:EtOAc (10:1 v:v) and was isolated as a white solid (60.8 mg, 0.157 mmol, 79%, >20:1 dr, MP = 173–175 °C). **^{1}H NMR** (500 MHz, $CDCl_3$) δ 8.05–7.96 (m, 1H), 7.91 (d, J = 8.0 Hz, 2H), 7.75–7.66 (m, 1H), 7.62 (d, J = 8.5 Hz, 2H), 7.51–7.37 (m, 1H), 7.24–7.18 (m, 3H), 7.16–7.08 (m, 2H), 7.08–7.01 (m, 2H), 7.01–6.96 (m, 1H), 6.82–6.73 (m, 2H), 5.00 (s, 1H). **^{13}C NMR** (125 MHz, $CDCl_3$) δ 168.9, 147.8 (q, 1.2 Hz), 146.9, 139.8, 139.6, 138.0, 132.3, 132.1, 130.1 (q, J = 32.6 Hz), 128.9, 128.5, 128.4, 127.9, 127.2, 126.3, 125.9 (q, J = 3.7 Hz), 125.6, 125.5, 124.4, 123.9, 123.8 (q, J = 272.2 Hz), 116.9, 80.9, 59.0. **^{19}F NMR** (282 MHz, $CDCl_3$) δ-62.7. **IR** (thin film, cm^{-1}) 3063, 3028, 1705, 1481, 1327, 1121, 1069. **HRMS** (DART, M+H) Calc'd for $C_{23}H_{17}F_3NO$ 442.14187, found 442.14166.

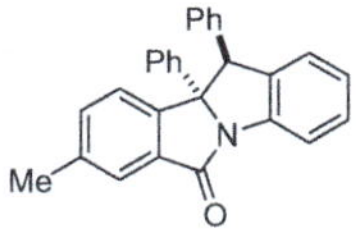

(**3.112o**)–Was synthesized according to **general procedure 2** heating for 2 h using **2.229o** (78.1 mg, 0.2 mmol, 1 equiv). The crude product was purified via silica gel flash column chromatography using hexanes:EtOAc (10:1 v:v) and was isolated as a white solid (67.1 mg, 0.173 mmol, 87%, >20:1 dr, MP = 90–92 °C). **^{1}H NMR** (400 MHz, $CDCl_3$) δ 7.97 (d, J = 8.0 Hz, 1H), 7.83–7.66 (m, 2H), 7.50–7.45 (m, 1H), 7.45–7.37 (m, 1H), 7.37–7.30 (m, 2H), 7.28–7.20 (m, 1H), 7.12–7.05 (m, 3H), 7.01 (m, 4H), 6.85–6.68 (m, 2H), 5.00 (s, 1H), 2.22 (s, 3H). **^{13}C NMR** (100 MHz, $CDCl_3$) δ 169.2, 145.2, 144.1, 140.5, 139.9, 138.5, 138.1, 133.2, 132.3, 128.8, 128.7, 128.4, 128.0, 127.7, 127.0, 126.2, 125.2, 124.9, 124.4, 123.7, 116.8, 81.0, 58.9, 21.0. **IR** (thin film, cm^{-1}) 3028, 2953, 1712, 1558, 1477, 1464, 1356, 1331, 1306, 1148. **HRMS** (DART, M+H) Calc'd for $C_{28}H_{22}NO$ 388.17014, found 388.17025.

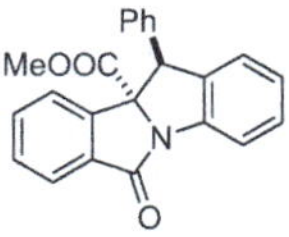

(**3.112p**)–Was synthesized according to **general procedure 2** using **2.229j** (71.6 mg, 0.2 mmol, 1 equiv). The crude product was purified via silica gel flash column chromatography using hexanes:EtOAc (10:1 v:v) and was isolated as a white solid (49.4 mg, 0.139 mmol, 70%, >20:1 dr, MP = 202–204 °C). **^{1}H NMR** (400 MHz, $CDCl_3$) δ 7.87 (d, *J* = 7.5 Hz, 1H), 7.74–7.63 (m, 1H), 7.46–7.36 (m, 1H), 7.32–7.20 (m, 3H), 7.18–7.08 (m, 2H), 7.03–6.90 (m, 3H), 6.66 (dd, *J* = 8.0, 2.0 Hz, 2H), 5.20 (s, 1H), 3.71 (s, 3H). **^{13}C NMR** (100 MHz, $CDCl_3$) δ 172.1, 168.4, 141.7, 139.8, 139.0, 137.4, 133.3, 132.3, 129.2, 128.9, 128.3, 128.1, 127.2, 126.1, 125.3, 124.3, 124.2, 116.6, 80.8, 54.0, 53.6. **IR** (thin film, cm^{-1}) 3061, 3026, 2924, 1699, 1601, 1479, 1358, 1306, 1136. **HRMS** (DART, M+H) Calc'd for $C_{23}H_{18}NO_3$ 356.12867, found 356.12863.

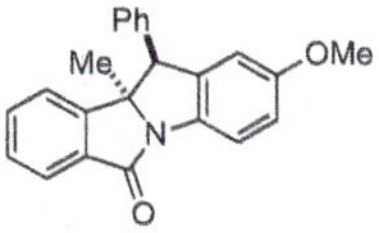

(**3.112q**)–Was synthesized according to **general procedure 2** heating for 2 h using **2.229q** (68.8 mg, 0.2 mmol, 1 equiv). The crude product was purified via silica gel flash column chromatography using hexanes:EtOAc (10:1 v:v) and was isolated as a white solid (50.2 mg, 0.148 mmol, 72%, >20:1 dr, MP = 54–55 °C). **^{1}H NMR** (400 MHz, $CDCl_3$) δ 7.74 (d, *J* = 8.5 Hz, 1H), 7.72–7.59 (m, 1H), 7.29–7.16 (m, 2H), 7.12–6.99 (m, 1H), 7.02–6.86 (m, 4H), 6.79–6.71 (m, 1H), 6.66–6.47 (m, 2H), 4.32 (s, 1H), 3.75 (s, 3H), 1.76 (s, 3H). **^{13}C NMR** (100 MHz, $CDCl_3$) δ 168.4, 157.4, 148.5, 140.4, 140.4, 133.0, 132.8, 131.7, 128.2, 128.0, 127.6, 126.8, 124.1, 122.9, 117.6, 113.8, 112.4, 76.3, 57.8, 55.6, 28.2. **IR** (thin film, cm^{-1}) 3061, 3026, 1697, 1613, 1559, 1489, 1364, 1333, 1281, 1234, 1138. **HRMS** (DART, M+H) Calc'd for $C_{23}H_{20}NO_2$ 342.14991, found 342.14980.

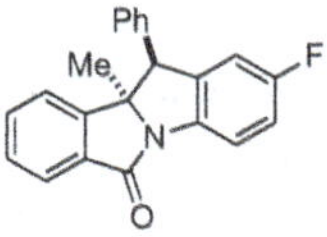

(**3.112r**)–Was synthesized according to **general procedure 2** using **2.229r** (66.4 mg, 0.2 mmol, 1 equiv). The crude product was purified via silica gel flash column chromatography using hexanes:EtOAc (10:1 v:v) and was isolated as a white solid (50.0 mg, 0.152 mmol, 76%, >20:1 dr, MP = 177–179 °C). **^{1}H NMR** (500 MHz, $CDCl_3$) δ 7.77 (dd, *J* = 8.5, 4.5 Hz, 1H), 7.73–7.64 (m, 1H), 7.31–7.19 (m, 2H), 7.10 (td, *J* = 9.0, 2.5 Hz, 1H), 7.06 (dt, *J* = 7.5, 1.0 Hz, 1H), 6.98–6.86 (m, 4H), 6.59–6.51 (m, 2H), 4.35 (s, 1H), 1.77 (s, 3H). **^{13}C NMR** (125 MHz, $CDCl_3$) δ 168.5, 160.3 (d, *J* = 243.5 Hz), 148.4, 140.7 (d, *J* = 8.6 Hz), 139.9, 135.4 (d, *J* = 2.2 Hz), 132.5, 132.0, 128.3, 128.1, 127.5, 127.0, 124.2, 123.0, 117.8 (d, *J* = 8.7 Hz), 115.3 (d, *J* = 23.6 Hz), 113.8 (d, *J* = 24.0 Hz), 76.4, 57.2 (d, *J* = 1.8 Hz), 28.2. **^{19}F NMR** (282 MHz, $CDCl_3$) δ-117.39 to -117.43. **IR** (thin film,

cm^{-1}) 3028, 2967, 1701, 1613, 1485, 1362, 1331, 1302, 1261, 1229, 1134. **HRMS** (DART, M+H) Calc'd for $C_{22}H_{17}FNO$ 330.13012, found 330.13016.

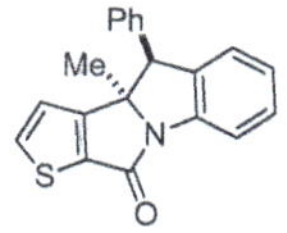

(**3.112t**)–Was synthesized according to **general procedure 2** (exception: boroxine (0.4 equiv.) was used) heating for 2 h using **2.229ts** (64.0 mg, 0.2 mmol, 1 equiv). The crude product was purified via silica gel flash column chromatography using hexanes:EtOAc (10:1 v:v) and was isolated as a white solid (28.3 mg, 0.089 mmol, 45%, >20:1 dr, MP = 174–176 °C). **^{1}H NMR** (400 MHz, $CDCl_3$) δ 7.83–7.66 (m, 1H), 7.49–7.31 (m, 2H), 7.22–7.08 (m, 2H), 7.08–6.95 (m, 3H), 6.71–6.51 (m, 3H), 4.34 (s, 1H), 1.75 (s, 3H). **^{13}C NMR** (100 MHz, $CDCl_3$) δ 164.8, 159.9, 141.1, 139.8, 138.0, 136.4, 134.8, 128.7, 128.3, 127.6, 126.9, 126.6, 124.8, 120.7, 116.7, 75.2, 56.7, 27.6. **IR** (thin film, cm^{-1}) 3028, 2969, 1697, 1601, 1479, 1350, 1306, 1269, 1121. **HRMS** (DART, M+H) Calc'd for $C_{20}H_{16}NOS$ 318.09526, found 318.09540.

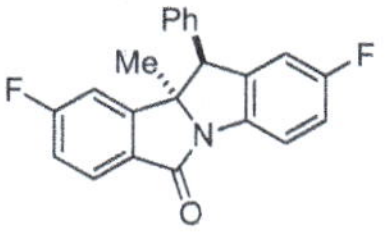

(**3.112aa**)–Was synthesized according to **general procedure 2** (exception: Pd $(P^tBu_3)_2$ (3.5 mol%) was used) heating for 2 h using **2.229aa** (70.0 mg, 0.2 mmol, 1 equiv). The crude product was purified via silica gel flash column chromatography using hexanes:EtOAc (10:1 v:v) and was isolated as a white solid (67.1 mg, 0.173 mmol, 87%, >20:1 dr, MP = 186–188 °C). **^{1}H NMR** (400 MHz, $CDCl_3$) δ 7.74 (dd, J = 8.5, 4.5 Hz, 1H), 7.66 (dd, J = 8.5, 5.0 Hz, 1H), 7.10 (td, J = 9.0, 2.5 Hz, 1H), 7.03–6.95 (m, 3H), 6.94–6.87 (m, 2H), 6.73 (dd, J = 8.2, 2.2 Hz, 1H), 6.60–6.52 (m, 2H), 4.32 (s, 1H), 1.76 (s, 3H). **^{13}C NMR** (100 MHz, $CDCl_3$) δ 167.5, 165.20 (d, J = 253.2 Hz), 160.4 (d, J = 243.8 Hz), 151.1 (d, J = 9.6 Hz), 140.3 (d, J = 8.3 Hz), 139.5, 135.4 (d, J = 2.2 Hz), 128.6 (d, J = 2.1 Hz), 128.5, 127.5, 127.3, 126.4 (d, J = 10.0 Hz), 117.9 (d, J = 8.7 Hz), 116.1 (d, J = 23.6 Hz), 115.5 (d, J = 23.6 Hz), 113.9 (d, J = 24.1 Hz), 110.4 (d, J = 24.0 Hz), 76.0 (d, J = 2.6 Hz), 57.2 (d, J = 1.8 Hz), 28.2. **^{19}F NMR** (375 MHz, $CDCl_3$) δ-105.6 to -105.7 (m), -117.08 to -117.15 (m). **IR** (thin film, cm^{-1}) 3063, 3026, 1709, 1597, 1559, 1485, 1362, 1261, 1188, 1128. **HRMS** (DART, M+H) Calc'd for $C_{22}H_{16}F_2NO$ 348.12000, found 348.11971.

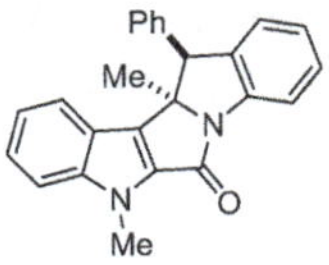

(**3.112ab**)–Was synthesized according to **general procedure 2** heating for 2 h using **3.112ab** (73.4 mg, 0.2 mmol, 1 equiv). The crude product was purified via silica gel flash column chromatography using hexanes:EtOAc (10:1 v:v) and was isolated as a white solid (49.2 mg, 0.138 mmol, 68%, >20:1 dr, MP = 166–168 °C). **^{1}H NMR** (400 MHz, $CDCl_3$) δ 7.80–7.73 (m, 1H), 7.51 (dt, *J* = 8.0, 1.0 Hz, 1H), 7.42 (td, *J* = 7.5, 1.5 Hz, 1H), 7.25–7.18 (m, 3H), 7.16–7.06 (m, 2H), 6.92–6.86 (m, 2H), 6.85–6.78 (m, 1H), 6.72–6.66 (m, 2H), 4.51 (s, 1H), 3.85 (s, 3H), 1.86 (s, 3H). **^{13}C NMR** (100 MHz, $CDCl_3$) δ 163.1, 142.9, 141.4, 139.8, 138.5, 133.4, 133.2, 128.6, 127.9, 127.8, 126.8, 126.6, 124.5, 124.4, 120.8, 120.5, 120.2, 116.3, 110.7, 73.8, 56.5, 29.9, 28.2. **IR** (thin film, cm^{-1}) 3059, 3026, 2967, 2922, 1694, 1601, 1551, 1478, 1462, 1350, 1298, 1205, 1126. **HRMS** (DART, M+H) Calc'd for $C_{25}H_{21}N_2O$ 365.16539, found 365.16546.

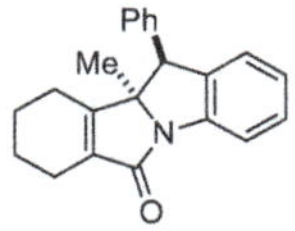

(**3.112ac**)–Was synthesized according to **general procedure 2** (exception: Pd($P^tBu_3)_2$ (3.5 mol%) was used) heating for 2 h using **2.229ac** (63.6 mg, 0.2 mmol, 1 equiv). The crude product was purified via silica gel flash column chromatography using hexanes:EtOAc (10:1 v:v) and was isolated as a white solid (46.7 mg, 0.148 mmol, 74%, >20:1 dr, MP = 176–177 °C). **^{1}H NMR** (500 MHz, $CDCl_3$) δ 7.64 (dd, *J* = 8.0, 1.0 Hz, 1H), 7.36–7.29 (m, 1H), 7.17–7.10 (m, 3H), 7.09–7.01 (m, 2H), 6.80–6.50 (m, 2H), 4.09 (s, 1H), 2.17–2.08 (m, 1H), 2.07–1.93 (m, 2H), 1.55 (s, 3H), 1.46 (m, 1H), 1.61–1.49 (m, 2H), 1.27 (m, 1H), 1.04 (m, 1H). **^{13}C NMR** (125 MHz, $CDCl_3$) δ 173.0, 160.1, 140.9, 139.7, 138.6, 132.6, 128.5, 128.1, 128.0, 127.0, 126.3, 124.5, 116.8, 77.0, 56.4, 25.5, 23.0, 21.6, 21.4, 20.1. **IR** (thin film, cm^{-1}) 3059, 3026, 2953, 2928, 2909, 1694, 1663, 1591, 1479. **HRMS** (DART, M+H) Calc'd for $C_{22}H_{22}NO$ 316.17014, found 316.16989.

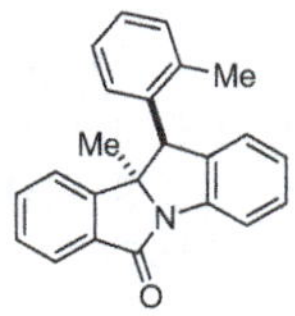

(**3.112ad**)–Was synthesized according to **general procedure 2** using **2.229a** (62.8 mg, 0.2 mmol, 1 equiv). The crude product was purified via silica gel flash column chromatography using hexanes:EtOAc (10:1 v:v) and was isolated as a white solid (53.8 mg, 0.166 mmol, 83%, >20:1 dr, MP = 204–206 °C). **^{1}H NMR**

(400 MHz, $CDCl_3$) δ 7.85 (d, *J* = 8.0 Hz, 1H), 7.74–7.64 (m, 1H), 7.47–7.35 (m, 1H), 7.33–7.20 (m, 2H), 7.18–7.10 (m, 2H), 7.11–7.04 (m, 1H), 7.05–6.97 (m, 1H), 6.79 (td, *J* = 7.5, 1.5 Hz, 1H), 6.63–6.50 (m, 1H), 5.80 (dd, *J* = 8.0, 1.5 Hz, 1H), 4.71 (s, 1H), 2.65 (s, 3H), 1.80 (s, 3H). **^{13}C NMR** (100 MHz, $CDCl_3$) δ 168.7, 148.7, 139.5, 139.4, 139.3, 133.8, 132.7, 132.1, 129.7, 128.5, 128.1, 127.6, 126.4, 126.4, 125.1, 124.2, 121.9, 117.0, 76.1, 51.9, 28.7, 20.6. **IR** (thin film, cm^{-1}) 2964, 1697, 1600, 1479, 1460, 1359, 1141. **HRMS** (DART, M+H) Calc'd for $C_{23}H_{20}NO$ 326.15500, found 326.15506.

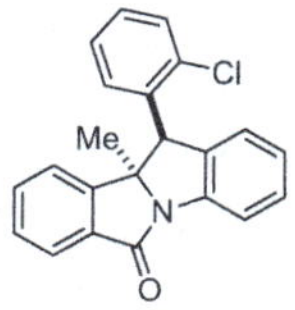

(**3.112ae**)–Was synthesized according to **general procedure 2** using **2.229a** (62.8 mg, 0.2 mmol, 1 equiv). The crude product was purified via silica gel flash column chromatography using hexanes:EtOAc (10:1 v:v) and was isolated as a white solid (49.2 mg, 0.142 mmol, 71%, >20:1 dr, MP = 140–142 °C). **^{1}H NMR** (400 MHz, $CDCl_3$) δ 7.88–7.78 (m, 1H), 7.67 (dt, *J* = 7.5, 1.0 Hz, 1H), 7.49–7.39 (m, 2H), 7.32 (td, *J* = 7.5, 1.5 Hz, 1H), 7.28–7.24 (m, 1H), 7.24–7.13 (m, 3H), 6.84 (ddd, *J* = 8.0, 7.5, 1.5 Hz, 1H), 6.67 (td, *J* = 7.5, 1.5 Hz, 1H), 5.88 (dd, *J* = 8.0, 1.5 Hz, 1H), 5.06 (s, 1H), 1.82 (s, 3H). **^{13}C NMR** (100 MHz, $CDCl_3$) δ 168.6, 148.3, 139.8, 138.5, 138.1, 132.6, 132.5, 132.3, 129.2, 128.9, 128.9, 128.4, 127.9, 127.1, 126.6, 125.2, 124.2, 122.5, 117.3, 76.3, 52.3, 28.1. **IR** (thin film, cm^{-1}) 3055, 2966, 1705, 1462, 1356, 1263, 1203, 1142. **HRMS** (DART, M+H) Calc'd for $C_{22}H_{17}ClNO$ 346.10037, found 346.10047.

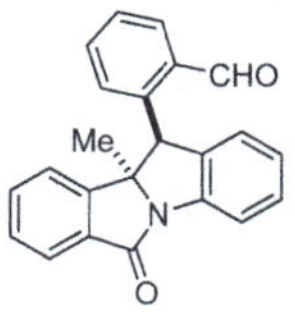

(**3.112afz**)–Was synthesized according **general procedure 2** using **2.229a** (62.8 mg, 0.2 mmol, 1 equiv). The crude product was purified via silica gel flash column chromatography using hexanes:EtOAc (10:1 v:v) and was isolated as a white solid (56.6 mg, 0.166 mmol, 83%, >20:1 dr, MP = 46–48 °C). **^{1}H NMR** (400 MHz, $CDCl_3$) δ 10.21 (s, 1H), 7.86–7.79 (m, 1H), 7.66–7.60 (m, 1H), 7.58 (dd, *J* = 7.5, 1.5 Hz, 1H), 7.47–7.39 (m, 1H), 7.25–7.16 (m, 2H), 7.17–7.13 (m, 1H), 7.13–7.08 (m, 3H), 7.01 (td, *J* = 7.5, 1.5 Hz, 1H), 6.04 (dt, *J* = 8.0, 1.0 Hz, 1H), 5.73 (s, 1H), 1.90 (s, 3H). **^{13}C NMR** (100 MHz, $CDCl_3$) δ 194.8, 168.4, 148.4, 141.9, 139.9, 139.0, 135.3, 133.9, 133.2, 132.7, 132.1, 129.6, 128.8, 128.2, 126.9, 126.5, 125.2, 124.1, 122.8, 117.3, 76.8, 50.4, 27.5. **IR** (thin film, cm^{-1}) 3066, 3016, 2862, 1699 (2), 1602, 1481, 1357, 1207, 1184. **HRMS** (DART, M+H) Calc'd for $C_{23}H_{18}NO_2$ 340.13426, found 340.13436.

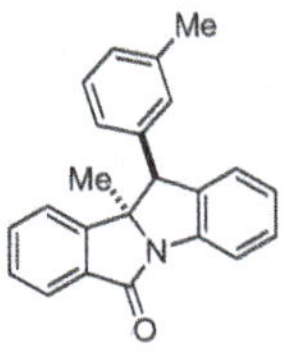

(**3.112ag**)–Was synthesized according to **general procedure 2** using **2.229a** (62.8 mg, 0.2 mmol, 1 equiv). The crude product was purified via silica gel flash column chromatography using hexanes:EtOAc (10:1 v:v) and was isolated as a white solid (53.8 mg, 0.160 mmol, 80%, >20:1 dr, MP = 145–147 °C). **^{1}H NMR** (300 MHz, $CDCl_3$) δ 7.83 (ddt, *J* = 8.0, 1.0, 0.5 Hz, 1H), 7.77–7.64 (m, 1H), 7.41 (td, *J* = 7.5, 1.5 Hz, 1H), 7.33–7.21 (m, 2H), 7.19 (ddt, *J* = 7.5, 1.5, 0.5 Hz, 1H), 7.14 (dd, *J* = 7.5, 1.0 Hz, 1H), 7.11–7.04 (m, 1H), 4.33 (s, 1H), 2.14–1.98 (m, 3H), 1.75 (s, 3H). **^{13}C NMR** (100 MHz, $CDCl_3$) δ 168.5, 148.7, 140.6, 139.2, 138.9, 137.7, 132.8, 131.9, 128.6, 128.3, 128.1, 128.0, 127.5, 126.5, 125.0, 124.7, 124.1, 123.0, 117.0, 75.9, 57.0, 28.5, 21.1. **IR** (thin film, cm^{-1}) 3016, 2970, 1697, 1602, 1477, 1462, 1351, 1141. **HRMS** (DART, M+H) Calc'd for $C_{23}H_{20}NO$ 326.15449, found 326.15468.

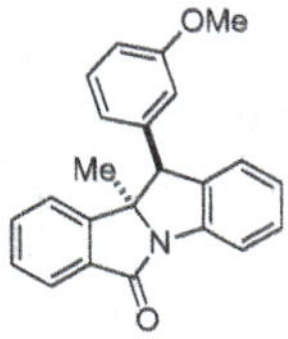

(**3.112ah**)–Was synthesized according to **general procedure 2** using **2.229a** (62.8 mg, 0.2 mmol, 1 equiv). The crude product was purified via silica gel flash column chromatography using hexanes:EtOAc (10:1 v:v) and was isolated as a white solid (50.5 mg, 0.148 mmol, 74%, >20:1 dr, MP = 171–172 °C). **^{1}H NMR** (400 MHz, $CDCl_3$) δ 7.88–7.77 (m, 1H), 7.70 (ddd, *J* = 7.0, 1.5, 1.0 Hz, 1H), 7.40 (td, *J* = 7.5, 1.5 Hz, 1H), 7.35–7.22 (m, 2H), 7.19 (ddt, *J* = 7.5, 1.5, 0.5 Hz, 1H), 7.14 (dd, *J* = 7.5, 1.0 Hz, 1H), 7.11–7.06 (m, 1H), 6.90–6.80 (m, 1H), 6.45 (ddd, *J* = 7.5, 2.5, 1.0 Hz, 1H), 6.17 (dt, *J* = 7.5, 1.0 Hz, 1H), 6.08 (dd, *J* = 2.5, 1.5 Hz, 1H), 4.33 (s, 1H), 3.50 (s, 3H), 1.75 (s, 3H). **^{13}C NMR** (100 MHz, $CDCl_3$) δ 168.5, 159.3, 148.6, 142.1, 139.2, 138.8, 132.9, 132.0, 129.2, 128.7, 128.1, 126.5, 125.0, 124.1, 123.0, 120.0, 117.1, 113.0, 112.4, 75.9, 57.0, 55.0, 28.4. **IR** (thin film, cm^{-1}) 2960, 2922, 1699, 1635, 1606, 1558, 1356, 1259. **HRMS** (DART, M+H) Calc'd for $C_{23}H_{20}NO_2$ 342.14991, found 342.15028.

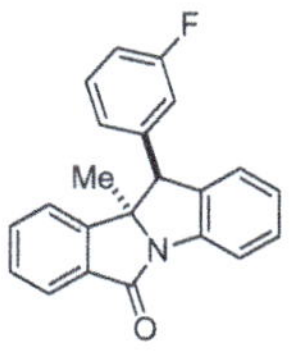

(**3.112ai**)–Was synthesized according to **general procedure 2** using **2.229a** (62.8 mg, 0.2 mmol, 1 equiv). The crude product was purified via silica gel flash column chromatography using hexanes:EtOAc (10:1 v:v) and was isolated as a white solid (55.1 mg, 0.167 mmol, 84%, >20:1 dr, MP = 185–186 °C). **^{1}H NMR** (500 MHz, $CDCl_3$) δ 7.83 (ddd, *J* = 8.0, 1.0, 0.5 Hz, 1H), 7.71 (ddd, *J* = 7.5, 1.5, 0.5 Hz, 1H), 7.43 (tdd, *J* = 7.5, 1.5, 0.5 Hz, 1H), 7.32 (td, *J* = 7.5, 1.5 Hz, 1H), 7.31–7.24 (m, 1H), 7.20–7.12 (m, 2H), 7.11–7.07 (m, 1H), 6.89 (ddd, *J* = 8.5, 7.5, 6.0 Hz, 1H), 6.61 (tdd, *J* = 8.5, 2.5, 1.0 Hz, 1H), 6.35 (m, 1H), 6.28 (dt, *J* = 7.5, 1.5 Hz, 1H), 4.35 (s, 1H), 1.76 (s, 3H). **^{13}C NMR** (125 MHz, $CDCl_3$) δ 168.3, 162.5 (d, *J* = 246.8 Hz), 148.3, 143.1 (d, *J* = 6.8 Hz), 139.2, 138.2, 132.8, 132.1, 129.8 (d, *J* = 8.1 Hz), 129.0, 128.3, 126.5, 125.1, 124.4, 123.3 (d, *J* = 2.9 Hz), 122.9, 117.2, 114.7 (d, *J* = 21.6 Hz), 113.7 (d, *J* = 21.1 Hz), 75.8, 56.7 (d, *J* = 1.6 Hz), 28.4. **^{19}F NMR** (376 MHz, $CDCl_3$) δ-113.0. **IR** (thin film, cm^{-1}) 3057, 2968, 2926, 2866, 1697, 1603, 1589, 1479. **HRMS** (DART, M+H) Calc'd for $C_{22}H_{17}FNO$ 330.12942, found 330.12898.

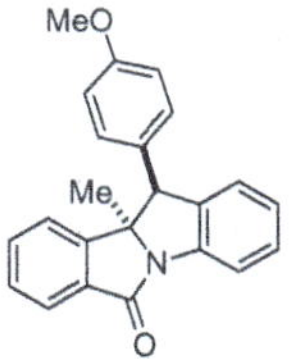

(**3.112aj**)–Was synthesized according to **general procedure 2** using **2.229a** (62.8 mg, 0.2 mmol, 1 equiv). The crude product was purified via silica gel flash column chromatography using hexanes:EtOAc (10:1 v:v) and was isolated as a white solid (63.0 mg, 0.156 mmol, 78%, >20:1 dr, MP = 192–193 °C). **^{1}H NMR** (400 MHz, $CDCl_3$) δ 7.85–7.78 (m, 1H), 7.70 (ddd, *J* = 7.5, 1.5, 1.0 Hz, 1H), 7.40 (td, *J* = 7.5, 1.5 Hz, 1H), 7.33–7.22 (m, 2H), 7.19–7.15 (m, 1H), 7.12 (td, *J* = 7.5, 1.0 Hz, 1H), 7.10–7.07 (m, 1H), 4.42–4.25 (m, 1H), 3.57 (s, 3H), 1.74 (s, 3H). **^{13}C NMR** (100 MHz, $CDCl_3$) δ 168.5, 158.1, 148.8, 139.3, 139.0, 133.1, 132.8, 132.0, 128.6, 128.5, 128.0, 126.5, 125.0, 124.2, 123.0, 117.0, 113.5, 76.0, 56.3, 55.0, 28.3. **IR** (thin film, cm^{-1}) 2964, 2924, 1697, 1585, 1512, 1477, 1462, 1357, 1176, 1033. **HRMS** (DART, M+H) Calc'd for $C_{23}H_{20}NO_2$ 342.14991, found 342.15031.

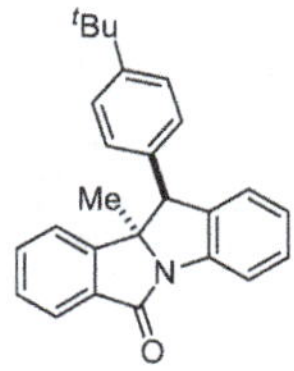

(**3.112ak**)–Was synthesized according to **general procedure 2** using **2.229a** (62.8 mg, 0.2 mmol, 1 equiv). The crude product was purified via silica gel flash column chromatography using hexanes:EtOAc (10:1 v:v) and was isolated as a white solid (59.6 mg, 0.162 mmol, 81%, >20:1 dr, MP = 178–180 °C). **^{1}H NMR** (400 MHz, $CDCl_3$) δ 7.83 (ddt, *J* = 8.0, 1.0, 0.5 Hz, 1H), 7.73–7.62 (m, 1H), 7.44–7.35 (m, 1H), 7.29–7.17 (m, 3H), 7.12 (td, *J* = 7.5, 1.0 Hz, 1H), 7.08–7.02 (m, 1H), 6.93 (d, *J* = 8.5 Hz, 1H), 6.46 (d, *J* = 8.5 Hz, 1H), 4.34 (s, 1H), 1.75 (s, 3H), 1.08 (s, 9H). **^{13}C NMR** (100 MHz, $CDCl_3$) δ 168.6, 149.6, 148.8, 139.2, 139.0, 137.6, 132.8, 131.7, 128.5, 127.8, 127.2, 126.6, 124.9, 124.9, 124.1, 123.1, 117.0, 56.7, 34.1, 31.0, 28.3. **IR** (thin film, cm^{-1}) 3047, 2962, 1701, 1464, 1361, 1203, 1132. **HRMS** (DART, M+H) Calc'd for $C_{26}H_{26}NO$ 368.20144, found 368.20150.

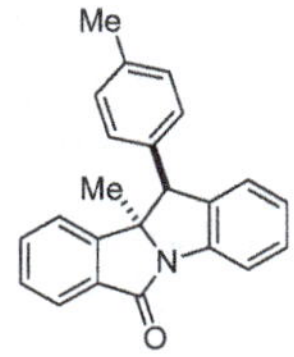

(**3.112al**)–Was synthesized according to **general procedure 2** using **2.229a** (62.8 mg, 0.2 mmol, 1 equiv). The crude product was purified via silica gel flash column chromatography using hexanes:EtOAc (10:1 v:v) and was isolated as a white solid (53.8 mg, 0.166 mmol, 83%, >20:1 dr, MP = 164–166 °C). **^{1}H NMR** (400 MHz, $CDCl_3$)δ 7.83 (dt, *J* = 8.0, 0.5 Hz, 1H), 7.71 (ddd, *J* = 7.0, 1.5, 1.0 Hz, 1H), 7.40 (td, *J* = 7.5, 1.5 Hz, 1H), 7.34–7.21 (m, 2H), 7.20–7.15 (m, 1H), 7.12 (dd, *J* = 7.5, 1.0 Hz, 1H), 7.11–7.05 (m, 1H), 6.74 (d, *J* = 8.0 Hz, 2H), 6.45 (d, *J* = 8.0 Hz, 2H), 4.33 (s, 1H), 2.08 (s, 3H), 1.75 (s, 3H). **^{13}C NMR** (100 MHz, $CDCl_3$) δ 168.4, 148.8, 139.2, 139.1, 137.7, 136.3, 132.9, 132.0, 128.9, 128.5, 28.0, 127.5, 126.5, 125.0, 124.2, 123.0, 117.0, 75.9, 56.7, 28.5, 20.8. **IR** (thin film, cm^{-1}) 3012, 1699, 1653, 1558, 1506, 1456. **HRMS** (DART, M+H) Calc'd for $C_{23}H_{20}NO$ 326.15500, found 326.15501.

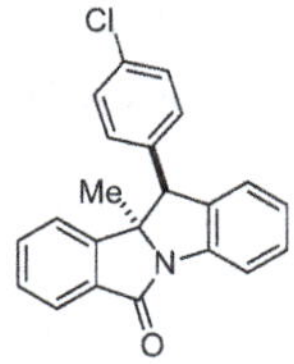

(**3.112am**)–Was synthesized according to **general procedure 2** (exception: boroxine (0.4 equiv.) was used) heating for 50 min using **2.229a** (62.8 mg, 0.2 mmol, 1 equiv). The crude product was purified via silica gel flash column chromatography using hexanes:EtOAc (10:1 v:v) and was isolated as a white solid (51.9 mg, 0.150 mmol, 75%, >20:1 dr, MP = 174–175 °C). **^{1}H NMR** (400 MHz, $CDCl_3$) δ 7.87–7.80 (m, 1H), 7.71 (ddd, J = 7.0, 1.5, 1.0 Hz, 1H), 7.45–7.37 (m, 1H), 7.35–7.23 (m, 2H), 7.18–7.10 (m, 2H), 7.10–7.04 (m, 1H), 6.91 (d, J = 8.5 Hz, 2H), 6.49 (d, J = 8.5 Hz, 2H), 4.34 (s, 1H), 1.75 (s, 3H). **^{13}C NMR** (100 MHz, $CDCl_3$) δ 168.3, 148.3, 139.2, 139.1, 138.5, 132.8, 132.5, 132.2, 128.9, 128.9, 128.4, 128.3, 126.5, 125.1, 124.3, 122.9, 117.2, 75.8, 56.3, 28.4. **IR** (thin film, cm^{-1}) 3032, 2920, 1695, 1558, 1539, 1521, 1361. **HRMS** (DART, M+H) Calc'd for $C_{22}H_{17}ClNO$ 346.10035, found 346.10027.

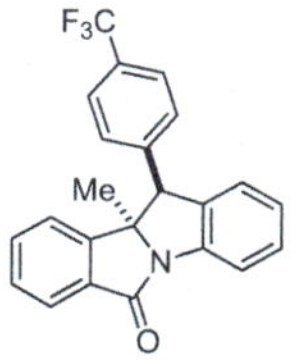

(**3.112an**)–Was synthesized according to **general procedure 2** (exception: Pd$(P^tBu_3)_2$ (3.5 mol%) was used) heating for 2 h using **2.229a** (62.8 mg, 0.2 mmol, 1 equiv). The crude product was purified via silica gel flash column chromatography using hexanes:EtOAc (10:1 v:v) and was isolated as a white solid (60.8 mg, 0.160 mmol, 80%, >20:1 dr, MP = 167–169 °C). *Note: X-ray quality crystals were obtained by slow evaporation of a $CDCl_3$ solution of the title compound.* **^{1}H NMR** (400 MHz, $CDCl_3$) δ 7.89–7.80 (m, 1H), 7.75–7.66 (m, 1H), 7.50–7.38 (m, 1H), 7.35–7.24 (m, 2H), 7.23–7.16 (m, 2H), 7.17–7.10 (m, 2H), 7.08–7.00 (m, 1H), 6.67 (d, J = 8.0 Hz, 2H), 4.42 (s, 1H), 1.78 (s, 3H). **^{13}C NMR** (125 MHz, $CDCl_3$) δ 168.2, 148.0, 144.5 (q, J = 1.4 Hz), 139.2, 138.1, 132.7, 132.3, 129.1, 128.9 (q, J = 32.5 Hz), 128.4, 128.0, 126.5, 125.2, 125.2 (q, J = 3.8 Hz), 124.4, 123.8 (q, J = 272.0 Hz), 122.8, 117.3, 75.8, 56.7, 28.4 (q, J = 21.8 Hz). **^{19}F NMR** (282 MHz, $CDCl_3$) δ-62.7. **IR** (thin film, cm^{-1}) 2964, 2924, 1701, 1512, 1357, 1325, 1305, 1203, 1124. **HRMS** (DART, M+H) Calc'd for $C_{23}H_{17}F_3NO$ 380.12622, found 380.12670.

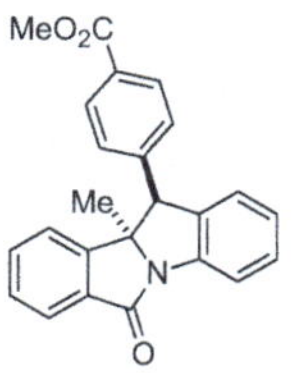

(**3.112ao**)–Was synthesized according to **general procedure 2** heating for 2 h using **2.229a** (62.8 mg, 0.2 mmol, 1 equiv). The crude product was purified via silica gel flash column chromatography using hexanes:EtOAc (10:1 v:v) and was isolated as a white solid (63.0 mg, 0.155 mmol, 78%, >20:1 dr, MP = 176–178 °C). **^{1}H NMR** (400 MHz, $CDCl_3$) δ 7.88–7.80 (m, 1H), 7.73–7.65 (m, 1H), 7.65–7.58 (m, 2H), 7.42 (ddd, *J* = 8.0, 6.5, 2.0 Hz, 1H), 7.31–7.20 (m, 2H), 7.18–7.08 (m, 2H), 7.08–7.01 (m, 1H), 6.63 (d, *J* = 8.0 Hz, 2H), 4.41 (s, 1H), 3.76 (s, 3H), 1.77 (s, 3H). **^{13}C NMR** (100 MHz, $CDCl_3$) δ 168.2, 166.5, 148.1, 145.6, 139.2, 138.2, 132.7, 132.2, 129.6, 129.0, 128.6, 128.3, 127.7, 126.5, 125.1, 124.3, 122.9, 117.2, 75.9, 56.9, 51.9, 28.5. **IR** (thin film, cm^{-1}) 3047, 3012, 1714, 1699, 1602, 1477, 1357, 1330, 1280, 1105. **HRMS** (DART, M+H) Calc'd for $C_{23}H_{20}NO_2$ 370.14483, found 370.14486.

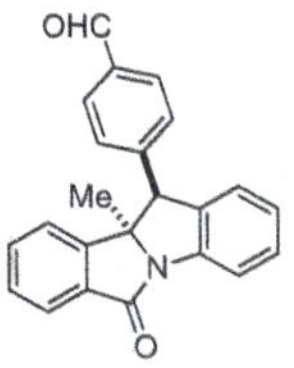

(**3.112ap**)–Was synthesized according to **general procedure 2** heating for 2 h using **2.229a** (62.8 mg, 0.2 mmol, 1 equiv). The crude product was purified via silica gel flash column chromatography using hexanes:EtOAc (10:1 v:v) and was isolated as a white solid (48.9 mg, 0.144 mmol, 72%, >20:1 dr, MP = 54–56 °C). **^{1}H NMR** (400 MHz, $CDCl_3$) δ 9.74 (s, 1H), 7.85 (d, *J* = 8.0 Hz, 1H), 7.74–7.64 (m, 1H), 7.53–7.38 (m, 3H), 7.34–7.21 (m, 2H), 7.20–7.09 (m, 2H), 7.11–7.02 (m, 1H), 6.73 (d, *J* = 8.0 Hz, 2H), 4.44 (s, 1H), 1.79 (s, 3H). **^{13}C NMR** (100 MHz, $CDCl_3$) δ 191.6, 168.2, 148.0, 147.3, 139.2, 138.0, 134.9, 132.7, 132.3, 129.7, 129.1, 128.4, 128.4, 126.5, 125.2, 124.4, 122.9, 117.3, 75.9, 57.0, 28.4. **IR** (thin film, cm^{-1}) 3016, 2970, 1697 (2), 1606, 1464, 1356, 1211. **HRMS** (DART, M+H) Calc'd for $C_{23}H_{18}NO_2$ 340.13375, found 340.13390.

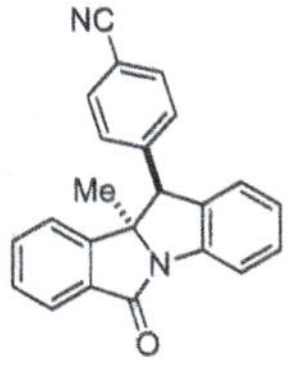

(**3.112aq**)–Was synthesized according to **general procedure 2** using **2.229a** (62.8 mg, 0.2 mmol, 1 equiv). The crude product was purified via silica gel flash column chromatography using hexanes:EtOAc (10:1 v:v) and was isolated as a white solid (53.2 mg, 0.158 mmol, 79%, >20:1 dr, MP = 150–152 °C). **^{1}H NMR** (400 MHz, $CDCl_3$) δ 7.84 (d, *J* = 8.0 Hz, 1H), 7.76–7.66 (m, 1H), 7.52–7.38 (m, 1H), 7.36–7.25 (m, 2H), 7.26–7.21 (m, 2H), 7.18–7.13 (m, 2H), 7.08–7.02 (m, 1H), 6.67 (d, *J* = 8.5 Hz, 2H), 4.41 (s, 1H), 1.78 (s, 3H). **^{13}C NMR** (100 MHz, $CDCl_3$) δ 168.1, 147.7, 145.7, 139.2, 137.6, 132.7, 132.4, 132.1, 129.2, 128.6, 128.4, 126.5, 125.3, 124.5, 122.8, 118.4, 117.4, 110.7, 75.9, 56.8, 28.3. **IR** (thin film,

cm^{-1}) 2928, 2227, 1683, 1647, 1558, 1539, 1471. **HRMS** (DART, M+H) Calc'd for $C_{23}H_{17}N_2O$ 337.13460, found 337.13492.

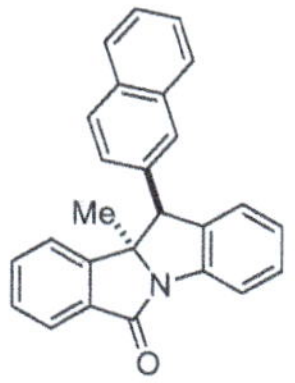

(**3.112ar**)–Was synthesized according to **general procedure 2** heating for 2 h using **2.229a** (62.8 mg, 0.2 mmol, 1 equiv). The crude product was purified via silica gel flash column chromatography using hexanes:EtOAc (10:1 v:v) and was isolated as a white solid (58.5 mg, 0.162 mmol, 81%, >20:1 dr, MP = 178–180 °C). **^{1}H NMR** (400 MHz, $CDCl_3$) δ 7.90 (dd, *J* = 7.5, 1.0 Hz, 1H), 7.72–7.66 (m, 1H), 7.65–7.60 (m, 1H), 7.59–7.54 (m, 1H), 7.44 (td, *J* = 7.5, 1.5 Hz, 1H), 7.39–7.29 (m, 3H), 7.24–7.22 (m, 1H), 7.20–7.11 (m, 4H), 7.09–7.05 (m, 1H), 4.54 (s, 1H), 1.82 (s, 3H). **^{13}C NMR** (100 MHz, $CDCl_3$) δ 168.5, 148.6, 139.2, 138.9, 138.3, 132.9, 132.8, 132.1, 132.1, 128.8, 128.1, 128.1, 127.7, 127.4, 126.7, 126.3, 125.9, 125.6 125.6, 125.0, 124.2, 122.9, 117.1, 75.9, 57.2, 28.8. **IR** (thin film, cm^{-1}) 3050, 3017, 1697, 1653, 1559, 1506, 1472, 1451, 1356 1142. **HRMS** (DART, M+H) Calc'd for $C_{26}H_{20}NO$ 362.15449, found 362.15469.

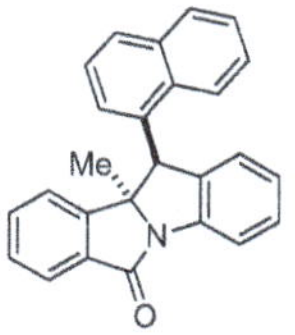

(**3.112as**)–Was synthesized according **general procedure 2** using **2.229a** (62.8 mg, 0.2 mmol, 1 equiv). The crude product was purified via silica gel flash column chromatography using hexanes:EtOAc (10:1 v:v) and was isolated as a white solid (56.4 mg, 0.156 mmol, 78%, >20:1 dr, MP = 218–220 °C). **^{1}H NMR** (400 MHz, $CDCl_3$) δ 8.46 (dt, *J* = 8.5, 1.0 Hz, 1H), 7.95–7.84 (m, 1H), 7.82–7.66 (m, 2H), 7.65–7.56 (m, 1H), 7.53 (ddd, *J* = 8.0, 7.0, 1.0 Hz, 1H), 7.47 (ddt, *J* = 6.0, 5.5, 3.0 Hz, 1H), 7.43–7.38 (m, 1H), 7.21–7.15 (m, 2H), 7.08 (td, *J* = 7.5, 1.0 Hz, 1H), 6.95–6.85 (m, 2H), 6.81–6.63 (m, 1H), 6.06 (dd, *J* = 7.5, 1.0 Hz, 1H), 5.34 (s, 1H), 1.93 (s, 3H). **^{13}C NMR** (100 MHz, $CDCl_3$) δ 168.8, 148.1, 139.9, 139.0, 136.9, 133.4, 132.5, 131.4, 131.3, 129.1, 128.7, 128.0, 127.1, 126.6, 126.3, 126.0, 125.5, 125.5, 125.2, 124.0, 122.8, 122.4, 117.2, 76.6, 51.1, 28.3. **IR** (thin film, cm^{-1}) 3059, 3005, 2964, 2920, 1701, 1641, 1481, 1307. **HRMS** (DART, M +H) Calc'd for $C_{26}H_{20}NO$ 362.15449, found 362.15437.

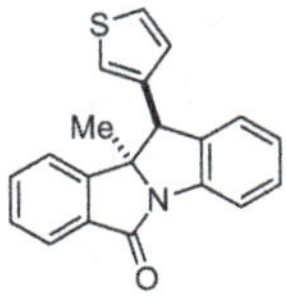

(**3.112at**)–Was synthesized according to **general procedure 2** using **2.229a** (62.8 mg, 0.2 mmol, 1 equiv). The crude product was purified via silica gel flash column chromatography using hexanes:EtOAc (10:1 v:v) and was isolated as a white solid (50.0 mg, 0.158 mmol, 79%, >20:1 dr, MP = 186–188 °C). **^{1}H NMR** (400 MHz, $CDCl_3$) δ 7.81 (ddd, *J* = 8.0, 1.0, 0.5 Hz, 1H), 7.75–7.67 (m, 1H), 7.40 (td, *J* = 7.5, 1.5 Hz, 1H), 7.35 (dd, *J* = 7.5, 1.5 Hz, 1H), 7.29 (td, *J* = 7.5, 1.0 Hz, 1H), 7.25–7.15 (m, 2H), 7.13 (td, *J* = 7.5, 1.0 Hz, 1H), 6.82 (dd, *J* = 5.0, 3.0 Hz, 1H), 6.52 (dd, *J* = 3.9, 1.5 Hz, 1H), 6.20 (dd, *J* = 5.0, 1.5 Hz, 1H), 4.49 (s, 1H), 1.73 (s, 3H). **^{13}C NMR** (100 MHz, $CDCl_3$) δ 168.4, 148.5, 141.2, 138.9, 138.7, 132.8, 132.0, 128.6, 128.2, 126.6, 126.3, 125.8, 124.9, 124.2, 122.9, 121.4, 117.3, 75.7, 52.2, 27.6. **IR** (thin film, cm^{-1}) 3032, 2969, 1697, 1653, 1602, 1559, 1479, 1356, 1308, 1142. **HRMS** (DART, M+H) Calc'd for $C_{20}H_{16}NOS$ 318.09526, found 318.09518.

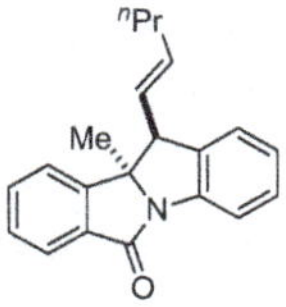

(**3.112au**)–Was synthesized according to **general procedure 2** (exception: boroxine (0.4 equiv) was used) using **2.229a** (62.8 mg, 0.2 mmol, 1 equiv). The crude product was purified via silica gel flash column chromatography using hexanes:EtOAc (10:1 v:v) and was isolated as a yellow oil (54.0 mg, 0.178 mmol, 89%, >20:1 dr). **^{1}H NMR** (500 MHz, $CDCl_3$) δ 7.96–7.78 (m, 1H), 7.70 (ddd, *J* = 8.0, 1.0, 0.5 Hz, 1H), 7.57 (td, *J* = 7.5, 1.0 Hz, 1H), 7.45 (td, *J* = 7.5, 1.0 Hz, 1H), 7.37 (dt, *J* = 7.5, 1.0 Hz, 1H), 7.36–7.31 (m, 1H), 7.22 (ddt, *J* = 7.5, 1.5, 0.5 Hz, 1H), 7.12 (td, *J* = 7.5, 1.0 Hz, 1H), 5.21 (dtd, *J* = 15.0, 7.0, 1.0 Hz, 1H), 4.71 (ddt, *J* = 15.0, 8.5, 1.5 Hz, 1H), 3.73 (d, *J* = 8.5 Hz, 1H), 1.62 (s, 3H), 1.05 (m, 2H), 0.59 (t, *J* = 7.5 Hz, 3H). **^{13}C NMR** (125 MHz, $CDCl_3$) δ 168.2, 148.9, 138.5, 138.5, 133.2, 132.2, 131.8, 131.7, 129.1, 128.3, 128.3, 126.4, 124.6, 124.5, 123.4, 117.1, 75.0, 54.0, 33.9, 27.4, 22.1, 13.2. **IR** (thin film, cm^{-1}) 2960, 2926, 1699, 1602, 1477, 1462, 1359, 1334, 1311. **HRMS** (ESI-MS, M+H) Calc'd for $C_{21}H_{22}NO$ 304.1696, found 304.1705.

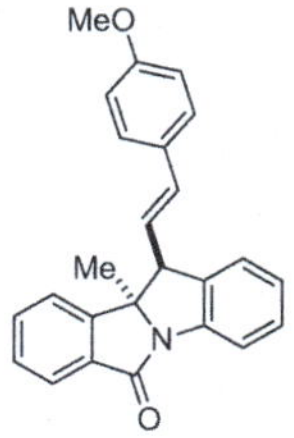

(**3.112av**)–Was synthesized according to **general procedure 2** (exception: boroxine (0.4 equiv.) was used) using **2.229a** (62.8 mg, 0.2 mmol, 1 equiv). The crude product was purified via silica gel flash column chromatography using hexanes: EtOAc (10:1 v:v) and was isolated as hygroscopic off-white solid (51.2 mg, 0.156 mmol, 78%, >20:1 dr). **^{1}H NMR** (500 MHz, $CDCl_3$) δ 7.86 (ddd, *J* = 7.5, 1.0, 1.0 Hz, 1H), 7.75 (ddd, *J* = 8.0, 1.0, 0.5 Hz, 1H), 7.53 (td, *J* = 7.5, 1.0 Hz, 1H), 7.44 (td, *J* = 7.5, 1.0 Hz, 1H), 7.41–7.34 (m, 2H), 7.29–7.26 (m, 1H), 7.13 (td, *J* = 7.5, 1.0 Hz, 1H), 6.90 (dd, *J* = 9.0, 0.5 Hz, 2H), 6.68 (d, *J* = 9.0 Hz, 2H), 6.25–6.13 (m, 1H), 5.28 (dd, *J* = 15.5, 9.0 Hz, 1H), 3.92 (dt, *J* = 9.0, 1.0 Hz, 1H). **^{13}C NMR** (125 MHz, $CDCl_3$) δ 168.2, 159.0, 148.7, 138.5, 137.9, 133.0, 132.5, 129.7, 129.7, 129.2, 128.6, 128.5, 127.4, 126.6, 126.5, 126.6, 124.7, 124.7, 123.3, 117.2, 113.7, 75.1, 55.2, 54.4, 27.6. **IR** (thin film, cm^{-1}) 3009, 2966, 2924, 2835, 1693, 1606, 1512, 1477, 1462, 1361, 1332, 1300, 1249, 1174, 1033. **HRMS** (DART, M+H) Calc'd for $C_{25}H_{22}NO_2$ 368.16422, found 368.16405.

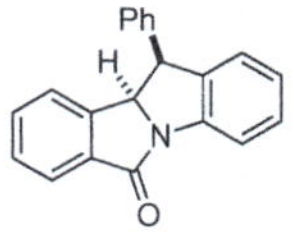

(**3.112av**)–Was synthesized according to **general procedure 2** using **2.229v** (60.0 mg, 0.2 mmol, 1 equiv). The crude product was purified via silica gel flash column chromatography using hexanes:EtOAc (10:1 v:v) and was isolated as a white solid (5.1 mg, 0.0172 mmol, 9%, >20:1 dr, MP = 188–190 °C). **^{1}H NMR** (500 MHz, $CDCl_3$) δ 7.84 (ddd, *J* = 8.0, 1.0, 0.5 Hz, 1H), 7.79–7.69 (m, 1H), 7.40 (m, 1H), 7.34–7.27 (m, 2H), 7.18 (ddt, *J* = 7.5, 1.5, 0.5 Hz, 1H), 7.14–7.07 (m, 2H), 7.02–6.91 (m, 3H), 6.64–6.60 (m, 2H), 5.79 (m, 1H), 4.77 (d, *J* = 8.5 Hz, 1H). **^{13}C NMR** (125 MHz, $CDCl_3$) δ 168.5, 143.2, 140.3, 139.7, 139.3, 134.5, 131.8, 128.7, 128.3, 128.3, 128.2, 127.9, 126.9, 126.3, 125.0, 124.2, 123.8, 116.4, 69.5, 49.8. **IR** (thin film, cm^{-1}) 3061, 2924, 1697, 1601, 1478, 1344, 1296, 1207, 1142, 1101. **HRMS** (DART, M+H) Calc'd for $C_{21}H_{16}NO$ 298.12319, found 298.12285.

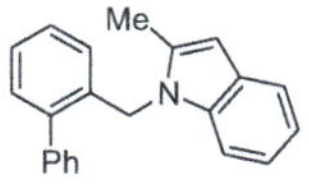

(**3.123**)–Was obtained when applying **general procedure 2** to 3.117 (60.0 mg, 0.2 mmol, 1 equiv). The crude product was purified via silica gel flash column chromatography using hexanes:EtOAc (10:1 v:v) and was isolated as a white solid (56.9 mg, 0.191 mmol, 96%, MP = 162–163 °C). **^{1}H NMR** (500 MHz, $CDCl_3$) δ 7.56–7.53 (m, 1H), 7.53–7.49 (m, 2H), 7.47–7.41 (m, 3H), 7.30–7.26 (m, 2H), 7.16–7.11 (m, 1H), 7.10–7.04 (m, 3H), 6.54–6.40 (m, 1H), 6.31 (s, 1H), 5.19 (d, J = 1.0 Hz, 2H), 2.25 (s, 3H). **^{13}C NMR** (125 MHz, $CDCl_3$) δ 140.6, 140.5, 137.1, 136.7, 135.1, 129.9, 129.0, 128.6, 128.1, 128.0, 127.5, 127.0, 125.8, 120.6, 119.6, 119.4, 109.1, 100.3, 44.7, 12.6. **IR** (thin film, cm^{-1}) 3435, 3055, 3017, 2926, 2853, 1653, 1564, 1466. **HRMS** (DART, M+H) Calc'd for $C_{22}H_{20}N$ 298.15874, found 298.15853.

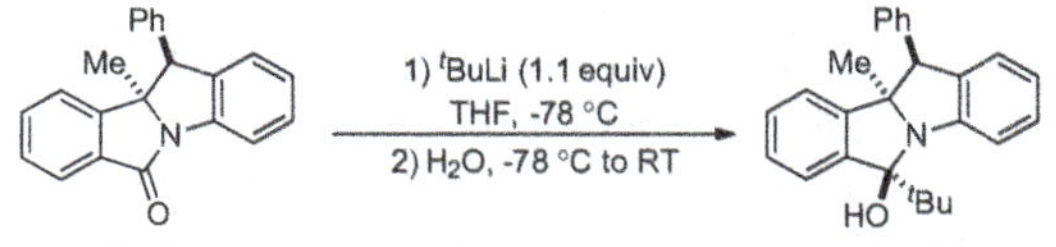

3.112a >20:1 dr

3.127 71% yield >20:1 dr

To a stirred THF (anhydrous, 2 mL) solution of **3.112a** (75 mg, 0.24 mmol, 1 equiv) at −78 °C was added ^{t}BuLi (155 µL, 0.0.264 mmol, 1.1 equiv, 1.7 M in pentane) dropwise. The solution was stirred at this time for 10 min before water (200 µL) was added dropwise to quench the reaction. The reaction vessel was removed from the cold bath and allowed to warm to room temperature, and the solution volume was doubled with water and then EtOAc was added (4 mL). The phases were separated and the aqueous layer was extracted with EtOAc (2x). The combined organic layers were washed with brine, dried over Na_2SO_4, filtered and concentrated. Crude NMR analysis indicated that the title compound was present as a single stereoisomer (>20:1). The crude residue was purified via silica gel flash column chromatography using hexanes:EtOAc (10:1 v:v) as the mobile phase and was isolated as a white solid (63.2 mg, 0.171 mmol, 71%, MP = 132–134 °C). *Note: X-ray quality crystals were obtained by slow evaporation of a DCM solution of the title compound in a vial stored in a chamber containing pentanes.* **^{1}H NMR** (500 MHz, $CDCl_3$) δ 7.30 (ddd, J = 8.0, 1.0, 0.5 Hz, 1H), 7.26–7.22 (m, 1H), 7.22–7.18 (m, 1H), 7.03–6.98 (m, 1H), 6.97–6.90 (m, 4H), 6.89–6.82 (m, 2H), 6.65–6.29 (m, 3H), 4.28 (s, 1H), 2.15 (s, 1H), 1.83 (s, 3H), 1.19 (s, 9H). **^{13}C NMR** (125 MHz, $CDCl_3$) δ 147.1, 143.3, 143.0, 142.1, 136.4, 128.7, 127.7, 127.7, 127.5, 126.4, 126.0, 125.9, 123.6, 123.4, 120.8, 116.3, 99.6, 79.2, 61.6, 40.7, 31.0, 27.1. **IR** (thin film, cm^{-1}) 3554, 3063, 3026, 2983, 2959, 2981, 2874, 1599, 1477, 1452, 1388, 1369, 1323, 1261, 1053. **HRMS** (DART, M+H) Calc'd for $C_{26}H_{28}NO$ 370.21709, found 370.21722.

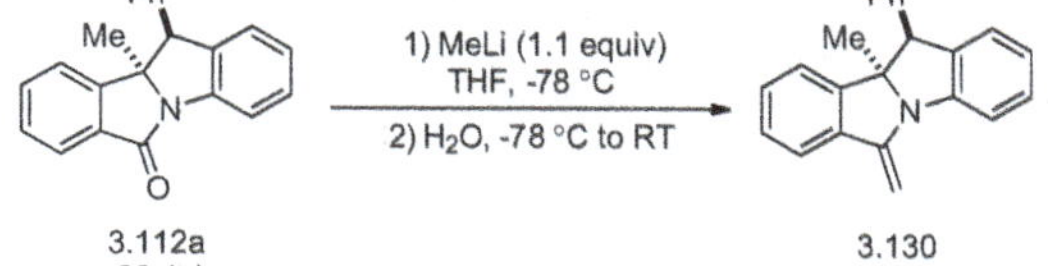

3.112a >20:1 dr

3.130 97% yield

To a stirred THF (anhydrous, 2 mL) solution of **3.112a** (75 mg, 0.24 mmol, 1 equiv) at −78 °C was added MeLi (165 μL, 0.0.264 mmol, 1.1 equiv, 1.6 M in Et_2O) dropwise. The solution was stirred at this time for 10 min before water (200 μL) was added dropwise to quench the reaction. The reaction vessel was removed from the cold bath and allowed to warm to room temperature, and the solution volume was doubled with water and EtOAc (4 mL) was added. The phases were separated and the aqueous layer was extracted with EtOAc (2x). The combined organic layers were washed with brine, dried over Na_2SO_4, filtered and concentrated. The residue was further purified via sonicating it as a suspension in cold pentanes, and **3.130** was afforded as a pure colorless crystalline solid (72.0 mg, 0.232 mmol, 97%, MP = 141–143 °C). ^{1}**H NMR** (500 MHz, CD_2Cl_2) δ 7.38 (ddd, *J* = 8.0, 1.0 Hz, 1H), 7.35–7.30 (m, 2H), 7.09 (ddd, *J* = 7.5, 7.5, 1.0 Hz, 1H), 7.07–7.05 (m, 1H), 7.02–6.95 (m, 2H), 6.95–6.87 (m, 3H), 6.82 (ddd, *J* = 7.5, 0.5 Hz, 1H), 6.59–6.54 (m, 2H), 4.78 (d, *J* = 1.5 Hz, 1H), 4.63 (d, *J* = 1.5 Hz, 1H), 4.25 (s, 1H), 1.66 (s, 3H). ^{13}**C NMR** (125 MHz, CD_2Cl_2) δ 150.69, 146.6, 144.5, 143.3, 139.2, 138.0, 129.0, 128.99, 128.97, 128.6, 128.0, 127.2, 127.0, 124.0, 123.6, 121.0, 116.3, 81.0, 79.0, 59.4, 30.7. **IR** (thin film, cm^{-1}) 3061, 3026, 2926, 2918, 1647, 1599, 1477, 1454, 1342, 1267, 1126, 1076. **HRMS** (DART, M+H) Calc'd for $C_{23}H_{20}N$ 310.15957, found 310.15993.

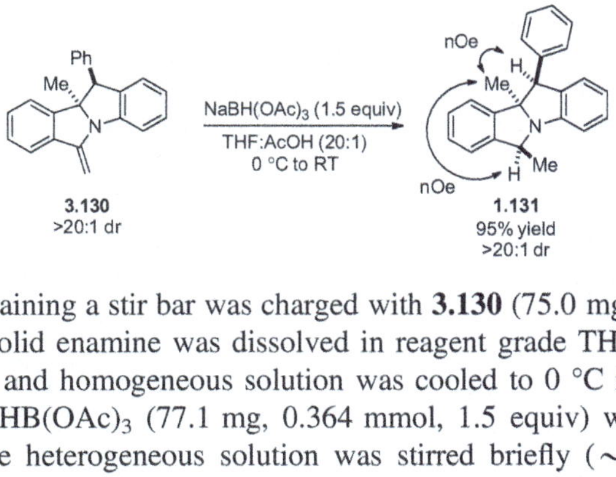

3.130
>20:1 dr

1.131
95% yield
>20:1 dr

A 2 dram vial containing a stir bar was charged with **3.130** (75.0 mg, 0.242 mmol, 1 equiv) and the solid enamine was dissolved in reagent grade THF (3 mL). The resulting colorless and homogeneous solution was cooled to 0 °C in an ice/water bath and solid $NaHB(OAc)_3$ (77.1 mg, 0.364 mmol, 1.5 equiv) was added in a single portion. The heterogeneous solution was stirred briefly (~5 min) before HOAc (150 μL, 2.38 mmol, 9.82 equiv) was added dropwise over 5 min. The reaction was warmed to room temperature where it stirred for an additional 30 min. At this time a 10% aqueous NaOH solution was added until pH = 9–10 was reached. The layers were separated and the aqueous layer was extracted with Et_2O (3x) and the combined organic layers were washed with brine, dried over Na_2SO_4, filtered, and concentrated *in vacuo*. ^{1}H NMR analysis of a homogeneous aliquot of the crude residue indicated that the desired product was present as a single diastereomer (>20:1 dr). The crude material was purified via silica gel flash column chromatography using hexanes:EtOAc (20:1 v:v) as the mobile phase. The pure amine was obtained as a free flowing, colourless solid (71.7 mg, 0.230 mmol, 95%, MP = 112–113 °C). ^{1}**H NMR** (500 MHz, CD_2Cl_2) δ 7.24–7.17 (m, 1H), 7.11–6.88 (m, 8H), 6.80 (td, *J* = 7.5, 1.0 Hz, 1H), 6.77–6.73 (m, 2H), 6.68

(dt, $J = 7.5$, 1.0 Hz, 1H), 5.03 (q, $J = 6.5$ Hz, 1H), 4.29 (s, 1H), 1.66 (s, 3H), 1.52 (d, $J = 6.5$ Hz, 3H). **^{13}C NMR** (125 MHz, CD_2Cl_2) δ 149.0, 145.8, 143.4, 143.2, 136.6, 129.9, 128.5, 128.4, 127.6, 127.3, 127.0, 126.9, 124.6, 122.9, 120.3, 115.4, 81.8, 60.2, 59.7, 33.6, 19.8. **IR** (thin film, cm^{-1}) 3063, 3026, 2968, 2926, 2854, 1600, 1479, 1454, 1371, 1300, 1265, 1109, 1024. **HRMS** (DART, M+H) Calc'd for $C_{23}H_{22}N$ 312.17522, found 312.17461.

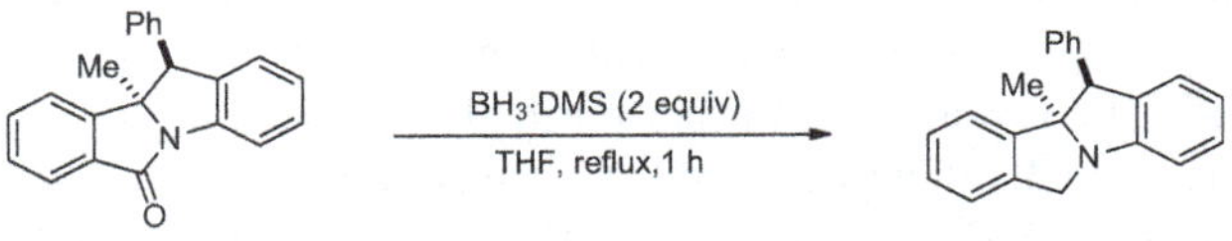

3.,112a >20:1 dr

3.132 98% yield

To a stirred THF (anhydrous, 2 mL) solution of **3.112a** (75 mg, 0.24 mmol, 1 equiv) was added neat BH_3·DMS (45.5 μL, 0.48 mol, 2 equiv) dropwise. The solution was refluxed for 1 h after which time TLC analysis indicated full conversion of the starting material. The reaction was cooled to 0 °C and MeOH (1 mL) was added. After the evolution of gas ceased, all volatiles were removed *in vacuo* and the crude reside was purified via silica gel flash column chromatography using hexanes:EtOAc (5:2 v:v) as the mobile phase and was isolated as a white solid (70.0 mg, 0.235 mmol, 98%, MP = 111–112 °C). **^{1}H NMR** (400 MHz, $CDCl_3$) δ 7.30–7.22 (m, 3H), 7.22–7.18 (m, 1H), 7.13 (dt, $J = 7.5$, 1.0 Hz, 1H), 7.06 (td, $J = 7.5$, 1.0 Hz, 1H), 7.05-6.95 (br, 2H), 6.90 (dt, $J = 7.5$, 1.0 Hz, 2H), 6.87–6.72 (m, 2H), 5.78 (d, $J = 7.5$ Hz, 1H), 4.79–4.44 (m, 3H), 1.71 (s, 3H). **^{13}C NMR** (100 MHz, $CDCl_3$) δ 154.3, 144.1, 140.3, 138.9, 132.8, 130.2, 128.1, 127.9, 127.0, 126.9, 126.0, 125.2, 124.7, 121.0 120.5, 111.9, 82.2, 59.7, 57.7, 30.6. **IR** (thin film, cm^{-1}) 3064, 3026, 2961, 2859, 1602, 1482, 1452, 1333, 1303, 1271, 1190. **HRMS** (DART, M+H) Calc'd for $C_{22}H_{20}N$ 298.16008, found 298.15998.

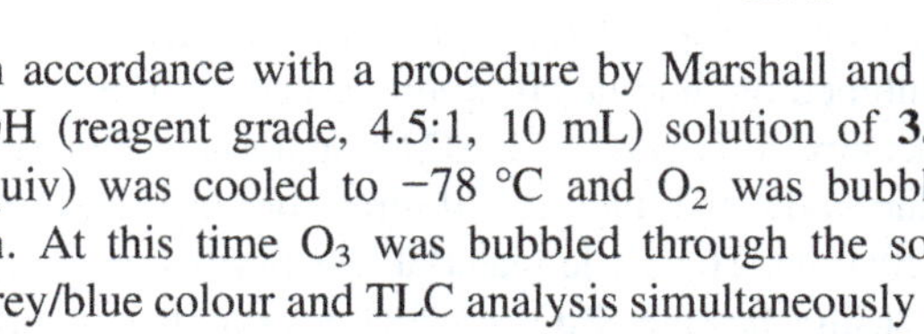

3.112au >20:1 dr

3.133 74% yield >20:1 dr

Was carried out in accordance with a procedure by Marshall and Acemoglu.[71] A stirred DCM/MeOH (reagent grade, 4.5:1, 10 mL) solution of **3.112au** (50 mg, 0.164 mmol, 1 equiv) was cooled to −78 °C and O_2 was bubbled through the solution for 5 min. At this time O_3 was bubbled through the solution until the mixture turned a grey/blue colour and TLC analysis simultaneously showed full and clean conversion of the starting alkene (approximately 30 min). O_2 was bubbled

[71]Reference [82].

through the solution until the colour dissipated and solid $NaBH_4$ (12.2 mg, 0.328 mmol, 2 equiv) was added to the mixture. The reaction was stirred at this temperature for 30 min before warming to room temperature where it was stirred for 1 h. All volatiles were removed in vacuo and the crude residue was taken up in DCM and the solution was sequentially washed with saturated aqueous NH_4Cl, water, and brine, dried over Na_2SO_4, filtered and concentrated. The crude residue was purified via silica gel flash column chromatography using hexanes:EtOAc: DCM (2:2:1 v:v:v) as the mobile phase. The pure carbinol was obtained as a free flowing, colourless solid (32.0 mg, 0.121 mmol, 74%). **^{1}H NMR** (500 MHz, $CDCl_3$) δ 7.86 (ddd, J = 7.5, 1.0, 1.0 Hz, 1H), 7.72–7.69 (m, 1H), 7.62 (ddd, J = 7.0, 1.0 Hz, 1H), 7.56–7.54 (m, 1H), 7.52–7.48 (m, 1H), 7.38–7.33 (m, 2H), 7.15–7.11 (m, 1H), 3.35–3.28 (m, 2H), 3.08 (dd, J = 10.5, 5.5 Hz, 1H), 1.60 (s, 3H). **^{13}C NMR** (125 MHz, $CDCl_3$) δ 167.8, 147.8, 138.8, 136.2, 133.5, 132.3, 128.8, 128.8, 126.1, 124.8, 124.6, 123.3, 117.3, 73.7, 63.7, 52.4, 27.8. **IR** (thin film, cm^{-1}) 3358, 2958, 2922, 2848, 1699, 1602, 1479, 1464, 1367, 1261, 1099. **HRMS** (DART, M+H) Calc'd for $C_{17}H_{15}NO_2$ 266.11810, found 266.11799.

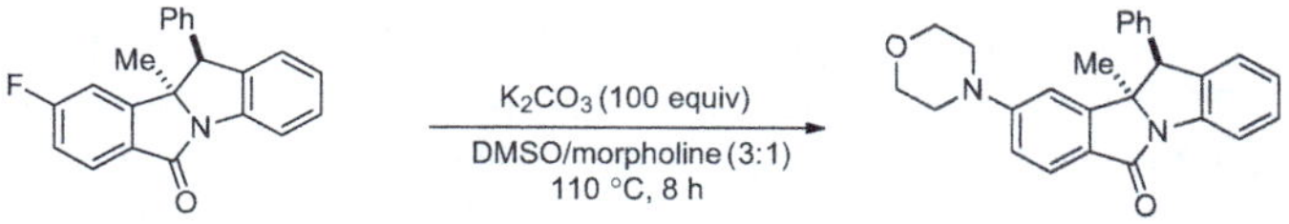

3.112c
>20:1 dr

3.134
97% yield
>20:1 dr

To a stirred suspension of **3.112c** (25 mg, 0.076 mmol, 1 equiv) and K_2CO_3 (1.04 g, 7.5 mmol, 98 equiv) in DMSO (2 mL) was added neat morpholine (647 μL, 7.5 mmol, 98 equiv) and the resulting mixture was heated at 110 °C for 8 h at which time TLC analysis indicated full and clean conversion of the starting material. The reaction mixture was poured into H_2O (15 mL) and this solution was extracted with EtOAc (3x). The combined organic layers were sequentially washed with a saturated solution of NH_4Cl, H_2O, and brine, dried over Na_2SO_4, filtered and concentrated i*n vacuo*. The crude residue was purified by passing it through a short plug of silica eluting with hexanes:EtOAc (5:4 v:v) as the mobile phase followed by trituration with hexanes. The title compound was obtained as a white solid (29.0 mg, 0.0734 mmol, 97% yield, MP = 209–210 °C). **^{1}H NMR** (400 MHz, $CDCl_3$) δ 7.79 (d, J = 8.0 Hz, 1H), 7.56 (d, J = 8.5 Hz, 1H), 7.38 (td, J = 7.5, 1.5 Hz, 1H), 7.24–7.05 (m, 2H), 7.00–6.89 (m, 3H), 6.72 (dd, J = 8.5, 2.0 Hz, 1H), 6.58 (dd, J = 7.5, 2.5 Hz, 2H), 6.38 (d, J = 2.0 Hz, 1H), 4.31 (s, 1H), 3.79 (m, 4H), 3.19–2.91 (m, 4H), 1.75 (s, 3H). **^{13}C NMR** (100 MHz, $CDCl_3$) δ 169.0, 154.2, 151.0, 141.1, 139.9, 138.2, 128.6, 128.1, 127.8, 126.7, 126.4, 125.3, 124.6, 123.7, 116.8, 115.1, 108.8, 75.5, 66.5, 57.3, 48.5, 28.8. **IR** (thin film, cm^{-1}) 3063, 2967, 2920, 2857, 1694, 1611, 1478, 1354, 1323, 1308, 1200, 1142, 1123. **HRMS** (DART, M+H) Calc'd for $C_{26}H_{25}N_2O_2$ 397.10160, found 397.19130.

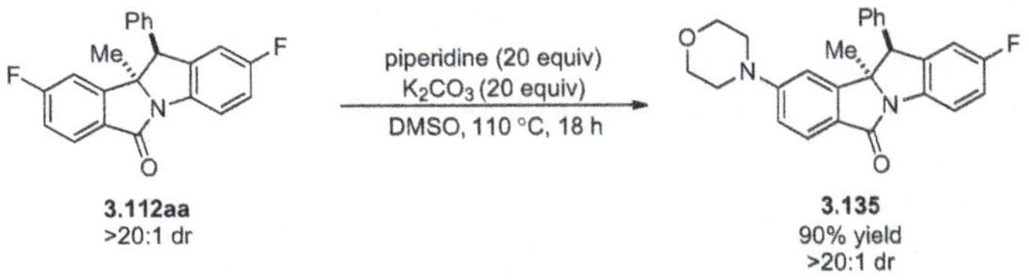

To a stirred suspension of **3.112aa** (25 mg, 0.072 mmol, 1 equiv) and K_2CO_3 (199 mg, 1.44 mmol, 20 equiv) in DMSO (2 mL) was added neat piperidine (142 μL, 1.44 mmol, 20 equiv) and the resulting mixture was heated at 110 °C for 18 h at which time TLC analysis indicated full and clean conversion of the starting material. The reaction mixture was poured into H_2O (15 mL) and this solution was extracted with EtOAc (3x). The combined organic layers were sequentially washed with a saturated solution of NH_4Cl, H_2O, and brine, dried over Na_2SO_4, filtered and concentrated i*n vacuo*. The crude residue was purified by passing it through a short plug of silica eluting with hexanes:EtOAc (5:4 v:v) as the mobile phase followed by trituration with hexanes. The title compound was obtained as a white solid (26.6 mg, 0.0645 mmol, 90% yield, MP = 131–133 °C). **^{1}H NMR** (500 MHz, $CDCl_3$) δ 7.72 (dd, *J* = 8.5, 4.5 Hz, 1H), 7.50 (d, *J* = 8.5 Hz, 1H), 7.09–7.03 (m, 1H), 7.00–6.92 (m, 3H), 6.86 (dd, *J* = 8.0, 2.5 Hz, 1H), 6.71 (d, *J* = 9.0 Hz, 1H), 6.60–6.56 (m, 2H), 6.34 (s, 1H), 4.27 (s, 1H), 3.21–3.04 (m, 4H), 1.74 (s, 3H), 1.63–1.48 (m, 6H). **^{13}C NMR** (125 MHz, $CDCl_3$) δ 169.4, 160.1 (d, *J* = 242.4 Hz), 154.5, 150.9, 140.5, 140.0 (d, *J* = 8.1 Hz), 136.3 (d, *J* = 2.0 Hz), 128.1, 127.8, 126.9, 125.2, 121.6, 117.6 (d, *J* = 8.5 Hz), 115.2, 115.2 (d, *J* = 23.5 Hz), 113.7 (d, *J* = 24.0 Hz), 108.7, 75.9, 57.5 (d, *J* = 1.8 Hz), 49.5, 28.7, 25.1, 24.3. **^{19}F NMR** (470 MHz, CDCL–) δ-118.48. **IR** (thin film, cm^{-1}) 3003, 1693, 1610, 1456, 1356, 1251, 1201. **HRMS** (DART, M+H) Calc'd for $C_{27}H_{26}FN_2O$ 413.20292, found 413.20294.

References

1. Petrone, D.A., Kondo, M., Zeidan, M., Lautens, M.: Chem. Eur. J. **22**, 5684 (2016)
2. Glasspool, B.W., Keske, E.C., Crudden, C.M.: Stereospecific and stereoselective Suzuki-Miyaura cross-coupling reactions (Chap. 11). In: Colacot, T. (ed.) New Trends in Cross-Coupling. The Royal Society of Chemistry, Cambridge (2015)
3. Burns, B., Grigg, R., Santhakumar, V., Sridharan, V., Stevenson, P., Worakun, T.: Tetrahedron **48**, 7297 (1992)
4. Kosugi, M., Kimura, T., Oda, H., Migita, T.: Bull. Chem. Soc. Jpn. **66**, 3522 (1993)
5. Oda, H., Ito, K., Kosugi, M., Migita, T.: Chem. Lett. **23**, 1443 (1994)
6. Girgg, R., Teasdalte, A., Sridharan, V.: Tetrahedron Lett. **32**, 3859 (1991)
7. Negishi, E.-I., Noda, Y., Lamaty, F., Vawter, E.J.: Tetrahedron Lett. **31**, 4393 (1990)
8. Fugami, K., Hagiwara, S., Oda, H., Kosugi, M.: Synlett 477 (1998)
9. Oda, H., Ito, K., Kosugi, M., Migita, T.: Chem. Lett. 1443 (1994)

10. Burns, B., Grigg, R., Sridharan, V., Stevenson, P., Sukirthalingam, S., Worakum, T.: Tetrahedron Lett. **30**, 1135 (1989)
11. Grigg, R., Sridharan, V.: J. Organomet. Chem. **576**, 65 (1999)
12. Brown, S., Clarkson, S., Grigg, R., Thomas, A.W., Sridharan, V., Wilson, D.M.: Tetrahedron **57**, 1347 (2001)
13. Grigg, R., Dorrity, M.J., Malone, J.F., Sridharan, V., Sukirthalingam, S.: Tetrahedron Lett. **31**, 1343 (1990)
14. Cheung, W.S., Patch, R.J., Player, M.R.: J. Org. Chem. **70**, 3741 (2005)
15. Yanada, R., Obika, S., Inokuma, T., Yanada, K., Yamashita, M., Ohta, S., Takemoto, Y.: J. Org. Chem. **70**, 6972 (2005)
16. Couty, S., Liegault, B., Meyer, C., Cossy, J.: Org. Lett. **6**, 2511 (2004)
17. Couty, S., Liegault, B., Meyer, C., Cossy, J.: Tetrahedron **62**, 3882 (2006)
18. Marchal, E., Cuprif, J.-F., Uriac, P., van de Weghe, P.: Tetrahedron Lett. **49**, 3713 (2008)
19. Yu, H., Richey, R.N., Carson, M.W., Coghlan, M.J.: Org. Lett. **8**, 1685 (2006)
20. Yu, H., Richey, R.N., Mendiola, J., Adeva, M., Somoza, C., May, S.A., Carson, M.W., Coghlan, M.J.: Tetrahedron Lett. **2008**, 49 (1915)
21. Peshkov, A.A., Peshkov, V.A., Pereshivko, O.P., Van Hecke, K., Kumar, R., van der Eycken, E.V.: J. Org. Chem. **80**, 6598 (2015)
22. Arthuis, M., Pontikis, R., Florent, J.-C.: J. Org. Chem. **74**, 2234 (2009)
23. Greenaway, R.L., Campbell, C.D., Holton, O.T., Russell, C.A., Anderson, E.A.: Chem. Eur. J. **17**, 14366 (2011)
24. Arcadi, A., Blesi, F., Cacchi, S., Fabrizi, G., Goggiamani, A., Marinelli, F.: J. Org. Chem. **78**, 4490 (2013)
25. Castanheira, T., Donnard, M., Gulea, M., Suffert, J.: Org. Lett. **16**, 3060 (2014)
26. Catellani, M., Chiusoli, G.P., Concari, S.: Tetrahedron Lett. **45**, 5263 (1989)
27. Shaulis, K.M., Hoskin, B.L., Rownsend, J.R., Goodson, F.E.: J. Org. Chem. **67**, 5860 (2002)
28. Huang, T.-S., Chang, H.-M., Wu, M.-Y., Cheng, C.-H.: J. Org. Chem. **67**, 99 (2002)
29. Zhou, C., Emrich, D.E., Larock, R.C.: Org. Lett. **5**, 1579 (2003)
30. Zhang, X., Larock, R.C.: Org. Lett. **5**, 2993 (2003)
31. Zhang, X., Larock, R.C.: Tetrahedron **66**, 4265 (2010)
32. Liao, L., Jana, R., Urkalan, K.B., Sigman, M.S.: J. Am. Chem. Soc. **133**, 5784 (2011)
33. Saini, V., Sigman, M.S.: J. Am. Chem. Soc. **134**, 11372 (2012)
34. Saini, V., Liao, L., Wang, Q., Jana, R., Sigman, M.S.: Org. Lett. **15**, 5008 (2013)
35. Lee, C.W., Oh, K.O., Kim, K.S., Ahn, K.H.: Org. Lett. **2**, 1213 (2000)
36. Owczarczyk, Z., Lamaty, F., Vawter, E.J., Negishi, E.-I.: J. Am. Chem. Soc. **114**, 10091 (1992)
37. Oh, C.H., Sung, H.R., Park, S.J., Ahn, K.H.: J. Org. Chem. **67**, 7155 (2002)
38. Braun, M., Richrath, B.: Synlett **6**, 968 (2009)
39. Wilson, J.E.: Tetrahedron Lett. **53**, 2308 (2012)
40. Urkalan, K.B., Sigman, M.S.: Angew. Chem. Int. Ed. **48**, 3146 (2009)
41. Werner, E.W., Urkalan, K.B., Sigman, M.S.: Org. Lett. **12**, 2848 (2010)
42. Trejos, A., Fardost, A., Yahiaoui, S., Larhed, M.: Chem. Commun. 7587 (2009)
43. Friedel, C., Crafts, J.M.: Compt. Rend. **84**, 1392 (1887)
44. Chichibabin, A.E., Zeide, O.A.J.: Russ. Phys. Chem. Soc. **46**, 1216 (1914)
45. Fries, K., Finck, G.: Ber. **41**, 4271 (1909)
46. Bedford, R.B., Few, N., Haddow, M.F., Sankey, R.F.: Chem. Commun. **47**, 3649 (2011)
47. Birch, A.J.: J. Chem. Soc. 430 (1944)
48. Lápez Ortiz, F., Iglesias, M.J., Fernández, I., Andújar Sánchez, C.M., Gómez, G.R.: Chem. Rev. **107**, 1580 (2007)
49. Corey, E.J., Trybulski, E.J., Melvin, L.S., Nicolaou, K.C., Secrist, J.A., Sheldrake, P.W., Palck, J.R., Brunelle, D.J.: J. Am. Chem. Soc. **100**, 4618 (1978)
50. Corey, E.J., Kim, S., Yoo, S.-E., Nicolaou, K.C., Melvin, L.S., Brunelle, D.J., Falck, J.R., Trybulski, E.J., Lett, R., Sheldrake, P.: W. J. Am. Chem. Soc. **100**, 4620 (1978)
51. Pouységu, L., Deffieus, D., Quideau, S.: Tetrahedron **107**, 1580 (2010)

52. Roche, S.P., Porco, J.A.: Angew. Chem. Int. Ed. **50**, 4068 (2011)
53. Pape, A.R., Kaliappan, K.P., Kündig, P.E.: Chem. Rev. **100**, 2917 (2000)
54. Wang, D.-S., Chen, Q.-A., Lu, S.-M., Zhou, Y.-G.: Chem. Rev. **112**, 2557 (2012)
55. Zhuo, C.-X., Zhang, W., You, S.-L.: Angew. Chem. Int. Ed. **51**, 12662 (2012)
56. Zhuo, C.-X., Zheng, C., You, S.-L.: Acc. Chem. Res. **47**, 2558 (2014)
57. Peng, B., Feng, X., Zhang, X., Bao, M.: J. Org. Chem. **75**, 2619 (2010)
58. Kimura, M., Futamata, M., Mukai, R., Tamaru, Y.: J. Am. Chem. Soc. **127**, 4592 (2005)
59. Trost, B.M., Quancard, J.: J. Am. Chem. Soc. **128**, 6314 (2006)
60. Kagawa, N., Malerich, J.P., Rawal, V.R.: Org. Lett. **10**, 1381 (2008)
61. Lindon, E.C., Kozlowski, M.C.: J. Am. Chem. Soc. **130**, 16162 (2008)
62. Cao, T., Deitch, J., Linton, E.C., Kozlowski, M.C.: Angew. Chem. Int. Ed. **51**, 2448 (2012)
63. García-Fortanet, J., Kessler, F., Bechwald, S.L.: J. Am. Chem. Soc. **131**, 6676 (2009)
64. Bedford, R.B., Butts, C.P., Haddow, M.F., Osborne, R., Sankey, R.F.: Chem. Commun. 4832 (2009)
65. Bedford, R.B., Few, N., Haddow, M.F., Sankey, R.F.: Chem. Commun. **47**, 3649 (2011)
66. Nemoto, T., Ishige, Y., Yoshida, M., Kohno, Y., Kanematsu, M., Hamada, Y.: Org. Lett. **12**, 5020 (2010)
67. Yoshida, M., Nemoto, T., Zhao, Z., Ishige, Y., Hamada, Y.: Tetrahedron Asymmetry **23**, 859 (2012)
68. Rousseaux, S., García-Fortanet, J., Del Aguila, Angel, Sanchez, M., Buchwald, S.L.: J. Am. Chem. Soc. **133**, 9282 (2011)
69. Grenning, A.J., Boyce, J.H., Porco, J.A.: J. Am. Chem. Soc. **136**, 11799 (2015)
70. Brown, S., Clarkson, S., Grigg, R., Thomas, A.W., Sridharan, V., Wilson, D.M.: Tetrahedron **57**, 1347 (2001)
71. Zhao, L., Li, Z., Chang, L., Xu, J., Yao, H., Wu, X.: Org. Lett. **14**, 2066 (2012)
72. Shen, C., Liu, R.-R., Fan, R.-J., Li, Y.-L., Xu, T.-F., Gao, J.-R., Jia, Y.-X.: J. Am. Chem. Soc. **137**, 4936 (2015)
73. Yoon, H., Petrone, D.A., Lautens, M.: Org. Lett. **16**, 6420 (2014)
74. Petrone, D.A., Yen, A., Zeidan, N., Lautens, M.: Org. Lett. **17**, 4838 (2015)
75. Gao, S., Yang, C., Huang, Y., Zhao, L., Wu, X., Yao, H., Lin, A.: Org. Biomol. Chem. **14**, 840 (2016)
76. Weast, R.C. (ed.): Handbook of Chemistry and Physics, 68th ed., p. B-144. CRC Press Inc., Boca Raton (1987–1988)
77. Proutiere, F., Aufiero, M., Schoenebeck, F.: J. Am. Chem. Soc. **134**, 606 (2012)
78. Tolman, C.A.: Chem. Rev. **77**, 313 (1977)
79. Kubota, K., Hayama, K., Iwamoto, H., Ito, H.: Angew. Chem. Int. Ed. **54**, 8809 (2015)
80. Lennox, A., Lloyd-Jones, G.: In: Colacot, T. (ed.) New Trends in Cross-Coupling: Theory and Application, vol. 1, pp. 322–354. RSC, Cambridge (2015)
81. Shintani, R., Takeda, M., Nishimura, T., Hayashi, T.: Angew. Chem. Int. Ed. **49**, 3969 (2010)
82. Acemoglu, R., Williams, J.M.J.: J. Mol. Catal. A: Chem. **196**, 3 (2003)

Chapter 4
Harnessing Reversible Oxidative Addition: Application of Diiodinated Aromatic Compounds in Aryliodination

4.1 Introduction

The previous chapters have highlighted the positive impact that Pd-catalyzed cross-coupling reactions have had on the chemical sciences. The use of substrates that are pre-functionalized with a carbon–halogen bond allows for a high degree of regiocontrol since the Pd(0) catalyst has a high propensity to undergo oxidative addition at this site. However, these reactions become less predictable once more than one carbon–halogen bond is present in the substrate. Nevertheless, extensive effort has been put forth to develop highly selective methodologies aimed at processing these useful substrates.[1]

4.1.1 Cross-Coupling Reactions of Polyhalogenated Substrates

There is a marked difference in the reactivity between the various C(sp^2)–halogen bonds of aromatic compounds (I > Br > Cl >> F). Therefore, efficient and selective cross-coupling reactions can occur by employing a substrate with two or more halogen groups possessing different reactivities (Scheme 4.1a).[2] These chemoselective cross-couplings are typically selective and high yielding, and present an opportunity to carry out a second cross-coupling (or other) reaction at the remaining carbon–halogen bond.

Portions of this chapter have appeared in print. See: Refs. [1, 2].

---

[1]For reviews on this topic, see Refs. [3–5].
[2]The use of pseudohalides (i.e. OTf or OTs) can also be used to impart selectivity.

D. A. Petrone, *Stereoselective Heterocycle Synthesis via Alkene Difunctionalization*,
Springer Theses, https://doi.org/10.1007/978-3-319-77507-4_4

Scheme 4.1 Chemoselective versus site-selective cross-coupling of substrates containing multiple halogen substituents

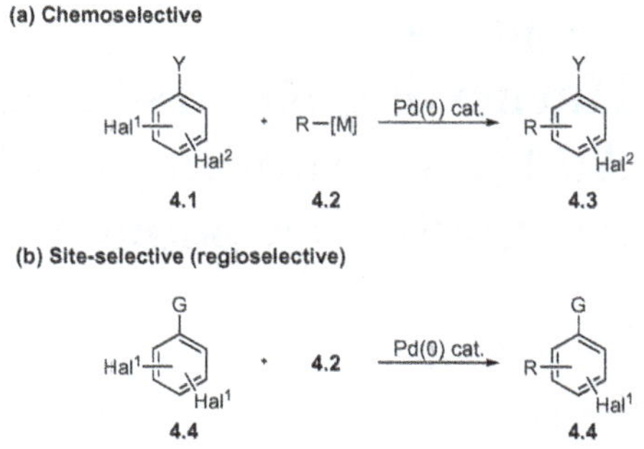

Substrates possessing the same halogen at more than one position may present challenges since the palladium catalyst may promiscuously interact at these sites, thus leading to decreased regiocontrol. However, in many cases, judicious substrate design helps to override this lack of selectivity. These reactions have been referred to as *regioselective cross-couplings*; although consensus suggests that a more accurate term is *site-selective cross-coupling* (SSCC, Scheme 4.1b) [3–5]. Site-selective cross-coupling methodologies typically take advantage of the different strengths of the carbon–halogen bonds as well as their respective steric and electronic environments [1]. Although a complete summary of this field is not possible due to the large number of advances, a few instructive examples will be shown here which best highlight this approach.

4.1.1.1 Site-Selective Cross-coupling (SSCC) Reactions of Di- and Trihalogenated Alkenes

SSCC reactions have been carried out on *gem*-dihalo alkenes **4.5** for the preparation of stereodefined polysubstituted olefins. These reactions typically yield the product arising from cross-coupling at the least sterically hindered *trans* halogen atom (i.e. **4.6**), although both the *cis* coupling product **4.7** or the decoupled product **4.8** can also be obtained (Scheme 4.2a). The first SSCC was reported by Minato in 1987 which involved the monoalkylation and arylation of 1,1-dichloroalkenes **4.9** with Grignard reagents.[3] When Pd(dppb)Cl_2 was used as the catalyst, the *trans* coupled products were obtained, whereas the use of $PdCl_2(PPh_3)_2$ led to the dicoupled products as the major species (Scheme 4.2b).[4]

[3]Reference [6].

[4]Reference [7].

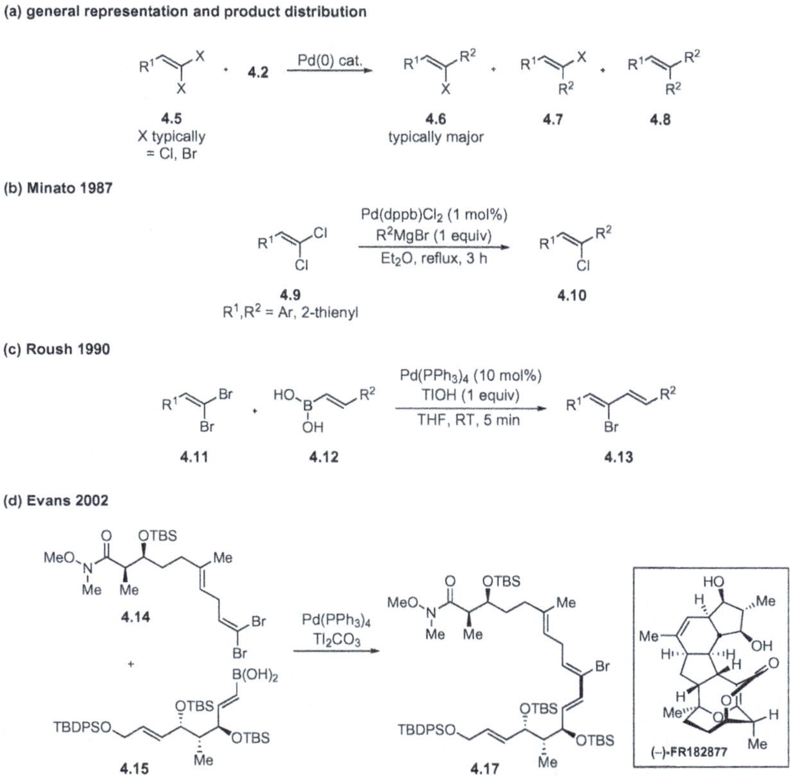

Scheme 4.2 Site-selective cross-coupling of *gem*-dihaloolefins

In 1990, Roush reported the Suzuki-Miyaura SSCC of *gem*-dibromoalkenes **4.11** and alkenyl boronic acids **4.12** which led to (*Z*,*E*)-2-bromo-1,3-dienes with high levels of selectivity (Scheme 4.2c).[5] As a result of the relative ease of starting material preparation, in addition to the highly predictable and selective nature of the transformation, the SSCC reaction of *gem*-dihaloalkenes has also been employed in the synthesis of numerous natural products. In 2002, the total synthesis of (-)-FR182877 by Evans and Starr featured an *E*-selective site-selective Suzuki-Miyaura cross-coupling of *gem*-dibromoolefin **4.14** and alkenyl boronic acid **4.15** (Scheme 4.2d).

The corresponding 1,2-dihaloalkenes have also been used in SSCC reactions, although to a much lesser extent. In 1994, Rossi and co-workers reported the

[5]Reference [8].

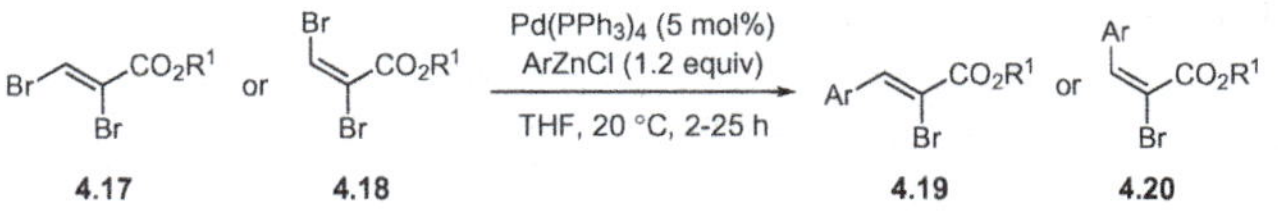

Scheme 4.3 Site-selective cross-coupling of 2,3-dibromopropenoates by Rossi

Scheme 4.4 Site-selective cross-couplings of trichloroethylene

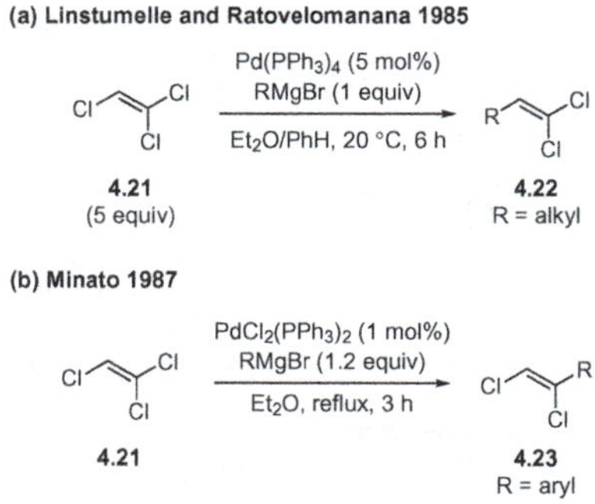

Pd-catalyzed SSCC reactions of (*Z*)- **4.17** and (*E*)-alkyl-2,3-dibromopropenoates **4.18** and arylzinc reagents (Scheme 4.3).[6] These reactions were found to occur selectively at the 3-position to afford the isomerically pure (*Z*)- **4.19** and (*E*)-alkyl-2-bromo-3-(hetero)arylpropenoates **4.20** in moderate to good yields.

Trihalogenated alkenes have also been explored in SSCC processes. In 1985, Linstrummelle and Ratocelomanana showed the trichloroethylene **4.21** could be selectively coupled with alkyl Grignard reagents to generate 1,1-dichloroalkenes **4.22** (Scheme 4.4a),[7] whereas in 1987 Minato found that the same reaction in the presence of aryl Grignard reagents led to the corresponding 1,2-dichloroalkenes (Scheme 4.4b) [6].

4.1.1.2 SSCC Reactions of Heteroaromatic Compounds with Multiple Carbon–Halogen Bonds

The use of heteroaromatic compounds such as thiophenes, furans and pyrazines containing multiple carbon–halogen bonds have been frequently explored in SSCC reactions. If the substrate contains two chemically identical carbon–halogen bonds, a statistical 1:2:1 ratio of unreacted, monosubstituted and disubstituted products

[6] Reference [9].

[7] Reference [10].

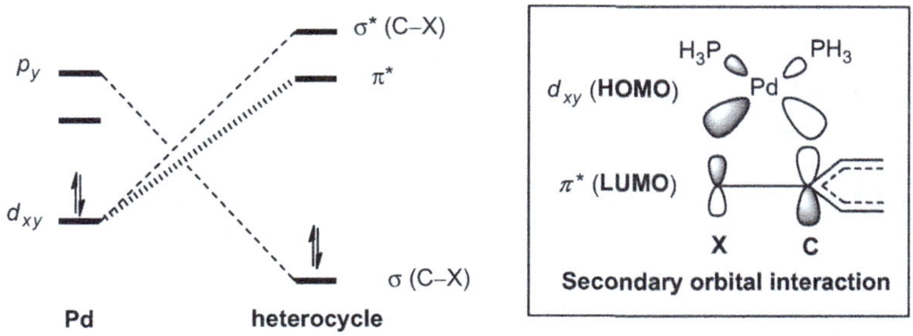

Fig. 4.1 Key frontier molecular orbitals governing the interaction energy. Adapted with permission from Ref. [12]. Copyright (2013) American Chemical Society

should be obtained if 1 equivalent of the nucleophilic coupling reagent is used, respectively. This leads to a maximum yield of 50% of the desired monosubstituted product [3]. If greater than 50% yield is obtained, there must be a difference in reactivity. In these instances, these reactions typically take place at the most electron-deficient carbon–halogen bond in the substrate.

Handy developed a prognostic guide for SSCC reaction which is based on the ^1H NMR chemical shifts of corresponding parent, non-halogenated substrates in question.[8] These authors showed that the order of reactivity in these substrates parallel the chemical shifts, which are intrinsically related to the electron-deficiency or -richness of each carbon–halogen bond. Specifically, in a substrate possessing more than one of the same carbon–halogen bond, the most electron deficient one will be the first to react. Merlic and Houk utilized DFT calculations to study the factors governing regioselectivity in oxidative additions by Pd(0) species into the carbon–halogen bonds in polyhalogenated aromatics.[9] The found that the observed site selectivities in cross-couplings of polyhalogenated heterocycles are governed by both the interaction energies (which involve the d_{xy} (Pd) → π * (C–X) FMO interactions) and the distortion energies (which are associated with the dissociation energies of carbon–halogen bond) (Fig. 4.1). These studies provide a deeper fundamental understanding of how site-selective oxidative additions of polyhalogenated aromatics can be dictated by electronic effects.

Generally, the selectivity pattern also follows the pattern for nucleophilic aromatic substitution. Kada et al. showed that both the S_NAr reaction 2,3-dibromofuran **4.24** with NaN_3 occurred at the 2-position with high levels of selectivity,[10] whereas Bach showed that the same substrate also underwent a Sonogashira cross-coupling at the C2 position (Scheme 4.5).[11]

[8]Reference [11].

[9]References [12, 13].

[10]Reference [14].

[11]Reference [15].

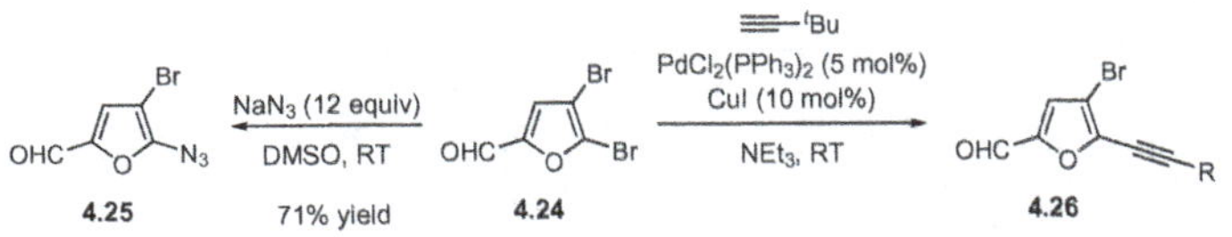

Scheme 4.5 Site-selective S_NAr and Sonogashira cross-coupling of a 2,3-dibromofuran derivative

Scheme 4.6 C5 selective cross-coupling of 3,5-diiodopyridinones

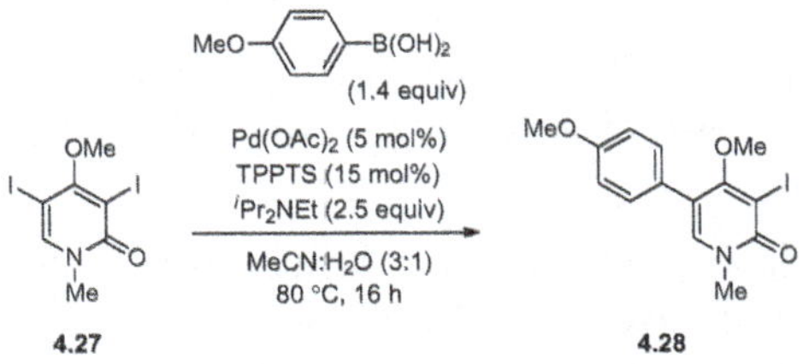

Scheme 4.7 SSCC reaction using a pyridinium directing group

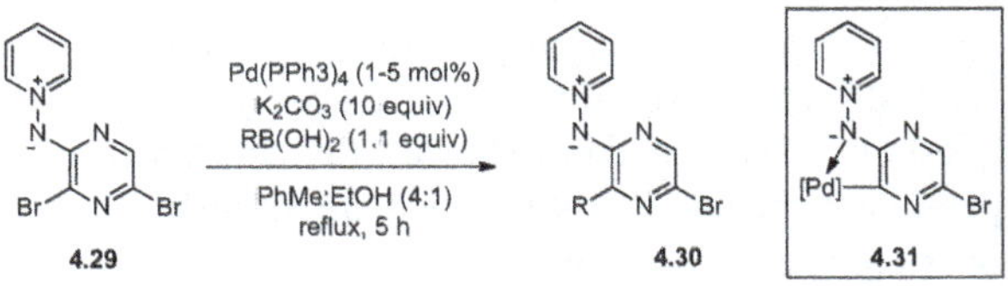

Sterics can also been known to play a role in SSCC reactions of aromatic compounds. Monteira, Vors and Balme showed that 3,5-diiodopyridinones **4.27** could undergo selective cross-couplings at the C5 position (Scheme 4.6). The authors hypothesized that steric factors intervened in the relative rates of cross-coupling and led to the observed C5 selectivity.[12]

The use of directing groups has also been shown to facilitate highly SSCC reactions. Alvarez-Builla showed that pyridinium *N*-(3,5-dibromoheteroar-2-yl) aminides **4.29** could undergo a highly site-selective Suzuki-Miyaura reaction at the carbon–halogen bond which is adjacent to the pyridinium directing group (Scheme 4.7).[13] The authors attributed the observed regiochemical outcome to the increased stability of the palladium intermediate (i.e. **4.31**).

[12]Reference [16].

[13]Reference [17].

4.1.1.3 SSCC Reactions of Benzene Derivatives with Multiple Carbon–Halogen Bonds

Due to the absence of a heteroatom in the aromatic ring, SSCC reactions of benzene derivatives are less predictable. The selectivity of these cross-couplings is affected by both the electron and steric properties of substituents on the aromatic ring [4]. Yields greater than 50% can be obtained in these reactions if the two halogens in question differ significantly. In 1970, Fahey conducted stoichiometric experiments on 1,2,4-trichlorobenzene **4.32** to determine the preferential site of oxidative addition (Scheme 4.8).[14] When **4.32** was treated with $Ni(PEt_3)_2(C_2H_4)$ and the crude reaction mixture was dissolved in Et_2O saturated with HCl, 1,4-dichlorobenzene **4.33** was obtained as the major product in 87% yield along with 1,3-dichlorobenzene **4.34** (7% yield) and 1,2-dichlorobenzene **4.35** (6% yield). These results show that carbon–chlorine bond oxidative addition occurs preferentially at the 2-position in this system.

The use of electron withdrawing groups can be used to perturb the reactivity of one halogen with respect to its counterpart(s). For example, Just reported that dibromonitrobenzenes could be used as substrates for site-selective Sonogashira cross-coupling reactions (Scheme 4.9).[15] Therein, they showed that when **4.36** was treated with heptyne under standard conditions, 92% of the *para* coupling product **4.37** could be obtained. Conversely, the *ortho* coupled product **4.39** was obtained in 82% yield when **4.38** was used.

Electron-donating groups have also been shown to control the site-selectivity of cross-coupling reactions of polyhalogenated benzene derivatives. The same authors showed that when **4.40** and **4.42** were reacted under identical conditions to those in Scheme 4.9, the *meta* coupled products **4.41** and **4.32** were formed in 70 and 72% yield, respectively (Scheme 4.10) [19].

When there are two identical, but enantiotopic (pseudo)halides in a substrate, a chiral metal catalyst can be used to selectively interact with one over the other by way of an enantioselective oxidative addition. For example, Hayashi reported an enantioselective Pd-catalyzed Kumada coupling of biaryl ditriflates **4.44** in 1995 (Scheme 4.11).[16]

4.1.1.4 Intramolecular SSCC Reactions

Issues have been reported when planning intramolecular cyclization reactions like the Mizoroki-Heck reaction using substrates containing chemically identical carbon–halogen bonds (i.e. **4.36**) which are neither enantiotopic nor amenable to

[14]Reference [18].

[15]Reference [19].

[16]Reference [20].

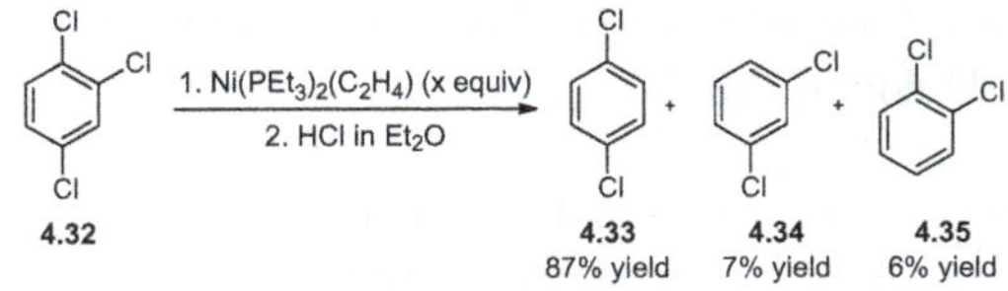

Scheme 4.8 Stoichiometric oxidative addition experiment by Fahey

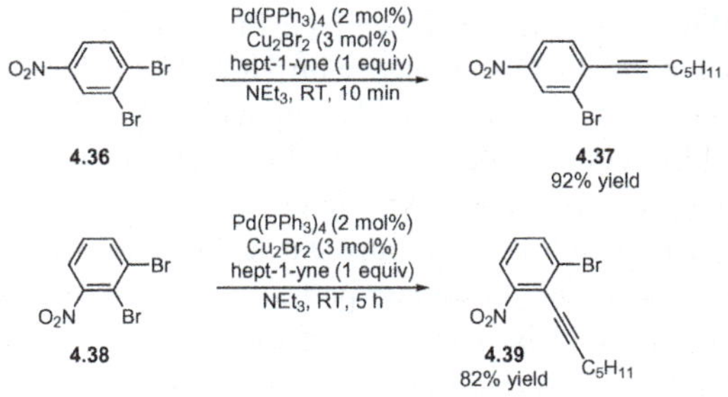

Scheme 4.9 Effect of electron-withdrawing groups on SSCC reactions

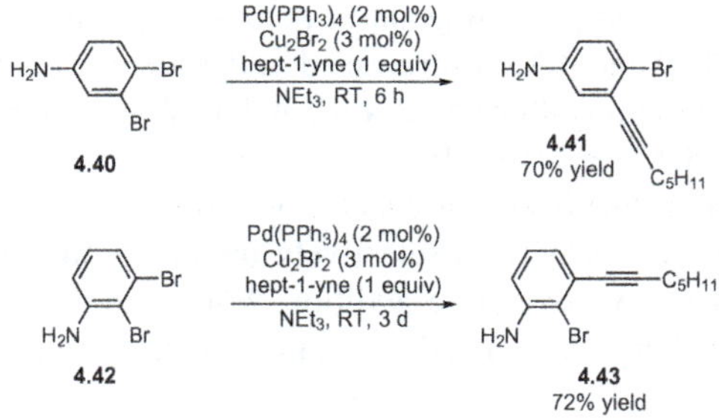

Scheme 4.10 Effect of electron-donating groups on SSCC reactions

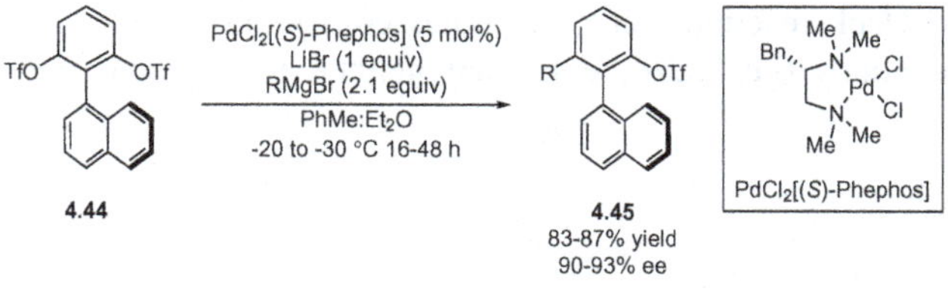

Scheme 4.11 Enantioselective Kumada coupling of symmetrical biaryl ditriflates

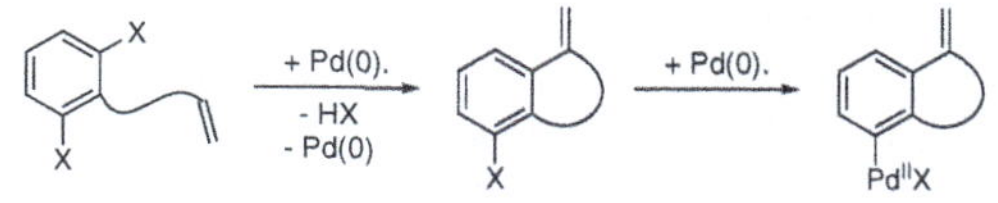

Scheme 4.12 General representation of an unproductive intramolecular SSCC reaction

discrimination by asymmetric catalysis. These reactions frequently result in catalyst deactivation since the metal catalyst can oxidatively add to the carbon–halogen bond that remains in **4.47** after the initial intramolecular cyclization. Generally, this resulting Ar–Pd(II)–X species **4.48** has no productive reaction pathways and cannot regenerate the active catalyst species (Scheme 4.12).

One strategy to avoid catalyst deactivation via species like **4.48** is to add in an external coupling partner such as an alkene to the reaction. This facilitates the conversion of otherwise unproductive ArPd(II)X species through an intermolecular cross-coupling. Bräse reported that vinyl bisnonaflate **4.49** could undergo an tandem intermolecular/intramolecular Mizoroki-Heck reaction using *tert*-butyl acrylate **4.50** to generate bicycle **4.51** in 43% yield (Scheme 4.13a). Lautens reported an intramolecular Mizoroki-Heck reaction of asymmetric ring opening products proceeding via an *anti* β-hydride elimination mechanism. When *o,o*-dibromo substrate **4.46** was subjected to the reaction conditions, none of the desired cyclized product **4.47** was observed (Scheme 4.13b).

They hypothesized that Ar–Pd(II)–Br species **4.48** was forming which cannot react further. When various Mizoroki-Heck acceptors were added to the reaction, the intramolecular/intermolecular SSCC products **4.49** could be obtained in high yields (Scheme 4.13c).[17] In 2013, Cramer reported that *o,o*-dibromoaniline derivative was not a competent substrate for an intramolecular Pd-catalyzed C–H functionalization of cyclopropanes presumably due to a similar reactions as explained by Fang and Lautens. This problem was circumvented by employing a tandem process involving the initial intramolecular cyclization by C–H activation followed by either a second intermolecular C–H activation of a heteroaromatic or an intermolecular Suzuki-Miyaura cross-coupling with a boronic acid (Scheme 4.13d).[18]

4.1.1.5 Intramolecular SSCC Reactions Using Reversible Oxidative Addition to Carbon–Halogen Bonds

As discussed in Chap. 1 of this thesis, Hartwig and co-workers showed that both ArPd(II)X dimers [23, 24] and monomers [25] could undergo carbon–halogen bond-forming reductive elimination in a stoichiometric fashion in the presence of a

[17]Reference [21].

[18]Reference [22].

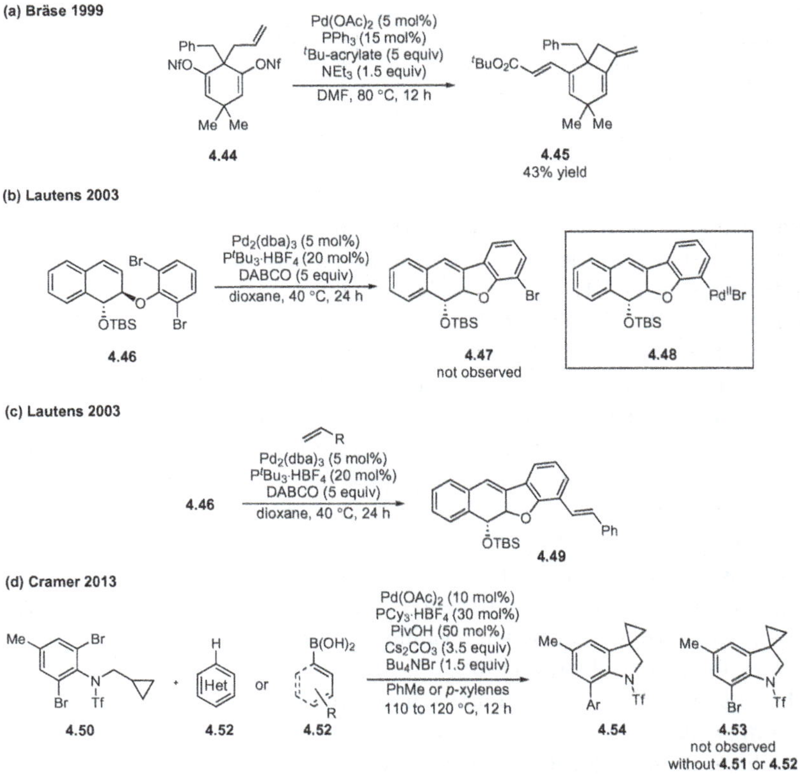

Scheme 4.13 Intermolecular/intramolecular SSCC reactions

large excess of P^tBu_3. This was the first result to suggest that the oxidative addition of specific Pd(0) complexes to C(sp^2)–halogen bonds could occur in a reversible fashion. Therefore, it can be hypothesized that a catalyst capable of both oxidative addition to carbon–halogen bonds and reductive elimination to form carbon–halogen bonds could efficiently promote SSCC reactions of substrates containing either chemically (nearly) identical halides or reactions which produce a highly reactive carbon–halogen bond in the mechanism.

In 2010, Newman and Lautens developed a Pd-catalyzed intramolecular C–N coupling of *ortho-gem*-dibromoanilines **4.54** which afforded access to various 2-bromoindoles (Scheme 4.14).[19] Screening various monodentate phosphine ligands led to the discovery that P^tBu_3 could efficiently promote this reaction. It was suggested that the steric bulk of the ligand is key for promoting this novel reactivity. Since the Pd(0) catalyst can react with the C(sp^2)–bromine bond of the product **4.55**, this could be a catalytic dead end in the absence of a suitable coupling

[19] Reference [26].

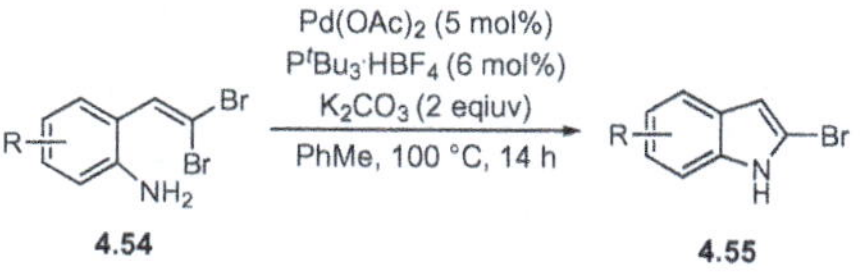

Scheme 4.14 Intramolecular C–N coupling en route to 2-bromoindoles

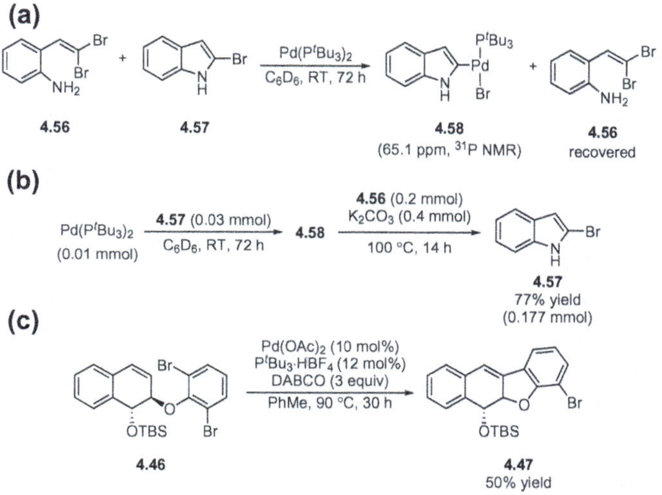

Scheme 4.15 Control experiments

partner. However, the use of P^tBu_3 renders this oxidative addition reversible such that the active Pd(0) catalyst can be easily regenerated.

It was thought that the 2-bromoindole product would be more susceptible to oxidative addition than the starting material. Furthermore, the corresponding oxidative addition complex **4.55** can form the active catalyst through carbon–bromine bond-forming reductive elimination. The authors carried out mechanistic studies in order to test these hypotheses. First, a competition experiment between **4.56** and **4.57** was conducted to determine if the product was more susceptible to oxidative addition than the starting material. Therein, **4.56** and **4.57** were treated with a stoichiometric amount of $Pd(P^tBu_3)_2$ at room temperature in deuterated benzene and for 72 h (Scheme 4.15a). The authors observed the appearance of complex **4.58** (65.1 ppm, ^{31}P NMR), while **4.56** was fully recovered. This result suggests that under the reaction conditions the formation of **4.58** is thermodynamically favored.

Complex **4.58** (5 mol%) was subjected to the standard reaction conditions in the presence of **4.56** in order to determine if it could be converted into the catalytically active Pd(0) complex (Scheme 4.15b). The experiment led to **4.57** being obtained in 77% yield which suggests that **4.58** is a competent pre-catalyst which is activated

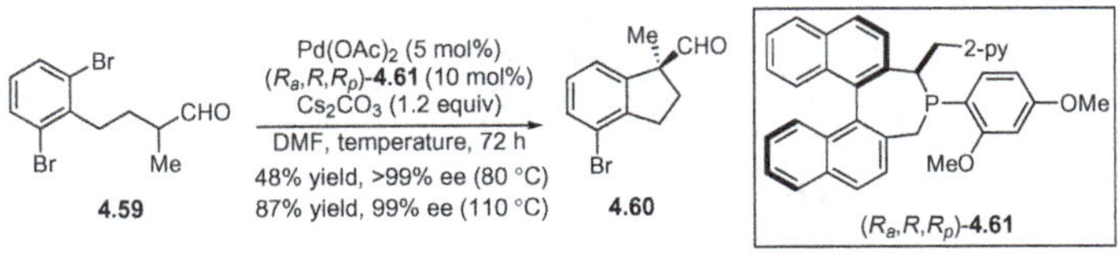

Scheme 4.16 Asymmetric intramolecular SSCC reaction via α-arylation

by carbon–bromine bond-forming reductive elimination. As a means to render the reaction from Scheme 4.14b catalytic, the authors subjected **4.46** to these catalytic conditions and were able to obtain **4.47** in 50% yield. The data suggest that the catalyst can is capable of undergoing a reversible oxidative addition with the remaining carbon–bromine bond leading to net catalyst turnover (Scheme 4.16c).

A 2012 report from Mazet and co-workers concerning the asymmetric intramolecular α-arylation of aldehydes using atropisomeric (P,N) ligands showed that *o,o*-dibromo substrate **4.59** could undergoing this transformation in high yield (Scheme 4.16).[20] At 80 °C the product was obtained in 48% yield in >99% ee, and at 110 °C the yield could be increased to 87% with no marked change in the observed enantioselectivity. The authors stated that the efficiency of this process at either temperature suggests that ligand (R_a,R,R_p)-**4.61** enables a reversible oxidative addition into the aryl bromide bonds. To ensure that the carbon–bromine bond in **4.60** was indeed reactive towards Pd catalysis, the authors carried out successful Suzuki-Miyaura couplings under both standard and catalytically relevant conditions using (R_a,R,R_p)-**4.61**.

In 2013, Arndtsen and co-workers reported a carbonylative acid chloride synthesis using aryl iodides **4.62** which involved a reversible C(sp^2)–Cl reductive elimination (Scheme 4.17).[21] In this case, benzyltriphenylphosphonium chloride was found to be the best chloride source which replaces the iodide from the substrate in situ. The authors noted that the specific combination of CO, P^tBu_3 and chloride anion is key for the high efficiency of this reaction. They hypothesized that tetracoordinate acyl Pd(II)Cl complex **4.64** is produced under the reaction conditions, where the π-acidic nature of the CO ligand in combination with the severe steric crowing of this complex prime it towards an unexpectedly efficient carbon–chlorine bond-forming reductive elimination.

This proposal was supported by stoichiometric studies (Scheme 4.18a). When $Pd(P^tBu_3)_2$ was treated with iodobenzene **4.66** in MeCN under an atmosphere of CO, complex **4.67** was generated in 87% yield followed by conversion to its chloride analog **4.68** by treatment with Bu_4NCl.[22] However, neither **4.67** nor **4.68** were reactive when treated with **4.66**. This result suggests either that these

[20]Reference [27].

[21]Reference [28].

[22]Complex **4.68** could also be generated in 85% by treatment of **4.65** with BzCl in pentane at RT for 1 h.

Scheme 4.17 Asymmetric intramolecular SSCC reaction via α-arylation

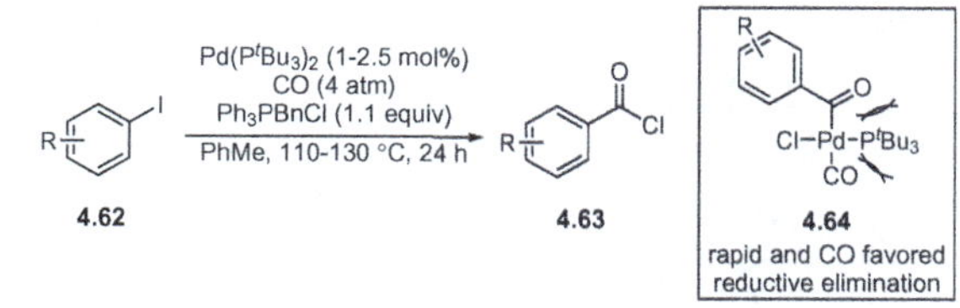

Scheme 4.18 Stoichiometric reactions of $(P^tBu_3)Pd(COPh)X$

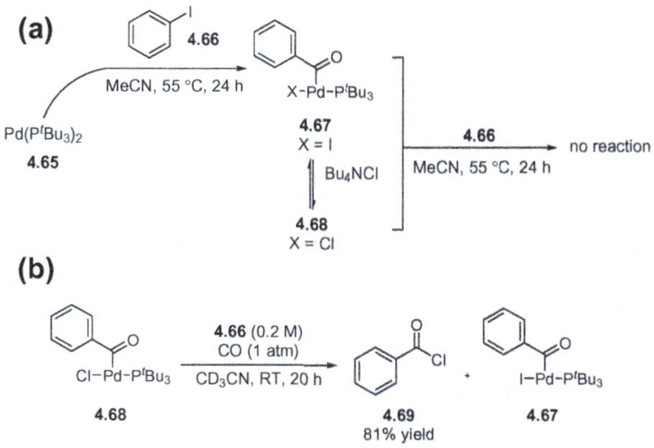

complexes cannot undergo carbon–chlorine bond-forming reductive elimination or that the equilibrium of this process lies strongly towards the Pd(II) complexes (i.e. reductive elimination is reversible). Remarkably, when complex **4.68** was treated with PhI under an atmosphere of CO, benzoyl chloride **4.69** was generated with concomitant formation of **4.67** (Scheme 4.18b). The same authors expanded the scope of this transformation to include aryl bromides in 2015.[23]

4.1.2 Research Goals

The application of polyhalogenated aromatic compounds in SSCC transformations still remains a challenge. Although many methodologies which exploit the inherent steric and electronic properties between the different carbon-halogen bonds in the substrates have been reported, most of these strategies remain limited and require pre-functionalization of the substrate in order to enforce the desired reactivity. It has been shown that oxidative addition of traditional Pd(0) catalysts to a carbon-halogen bond like that of **4.70** occurs irreversibly to provide complexes like **4.71** that lack a productive reaction pathway (Scheme 4.19a). As previously mentioned, the use of exogenous coupling partners such as alkenes, [21] as well as

[23]Reference [29].

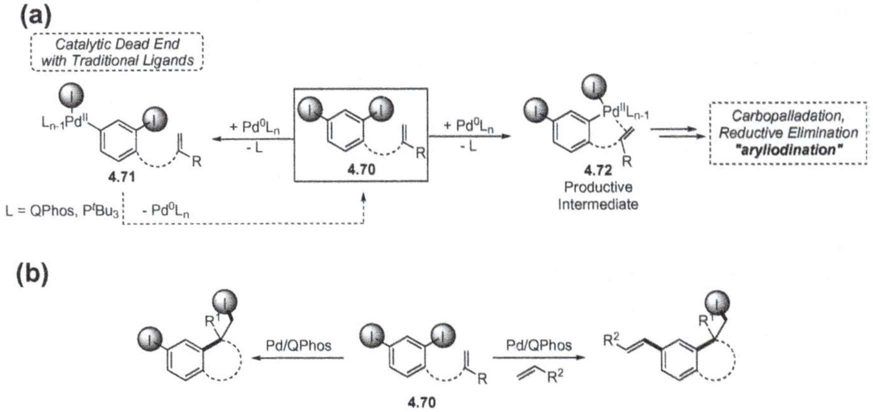

Scheme 4.19 **a** Catalyst deactivation by irreversible oxidative addition, **b** strategy to overcome catalyst deactivation (when R $\neq$ H). Used with permission from Ref. [2].

both heteroaromatic C–H bonds and boronic acids [22] represents as important strategy which fascilitates catalyst turnover.

Another solution has been to use catalysts which can reversibly oxidatively add to carbon–halogen bonds. In this regard, the scarcity of diiodinated aromatic compounds undergoing efficient intramolecular SSCC reactions prompted us to explore application of the aryliodination methodology of diiodinated aromatic compounds in a type of SSCC reaction. This would allow us to increase the scope of this class of transformation, and it would represent an opportunity to highlight the unique ability of the Pd/QPhos or Pd/P^tBu_3 combination to both undergo reversible oxidative addition to carbon–halogen bonds and promote C(sp^3)–iodine reductive elimination. This project concerned the development of improved reaction conditions for the carboiodination of diiodinated aromatic substrates as well as their use in a sequential intramolecular carboiodination/intermolecular Mizoroki-Heck reaction (Scheme 4.19b).

4.1.3 Starting Material Preparation

The diiodinated aromatic compounds used throughout this project were synthesized in a straightforward fashion from the corresponding diiodophenols (Scheme 4.20) or anilines (Scheme 4.22). Allylaryl ether **4.75** was prepared in two steps from 2-iodophenol **4.73** utilizing an oxidative aromatic iodination, followed by an *O*-allylation under standard conditions (Scheme 4.20a). Isomer **4.77** was prepared by a *meta*-selective aromatic iodination of **4.73** using Ag(I) to generate 2,5-diiodophenol **4.76** followed by a similar methallylation (Scheme 4.20b). Precursors **4.81** and **4.82** could be synthesized via a Mitsunobo reaction of **4.74** using isoprenol **4.78** and

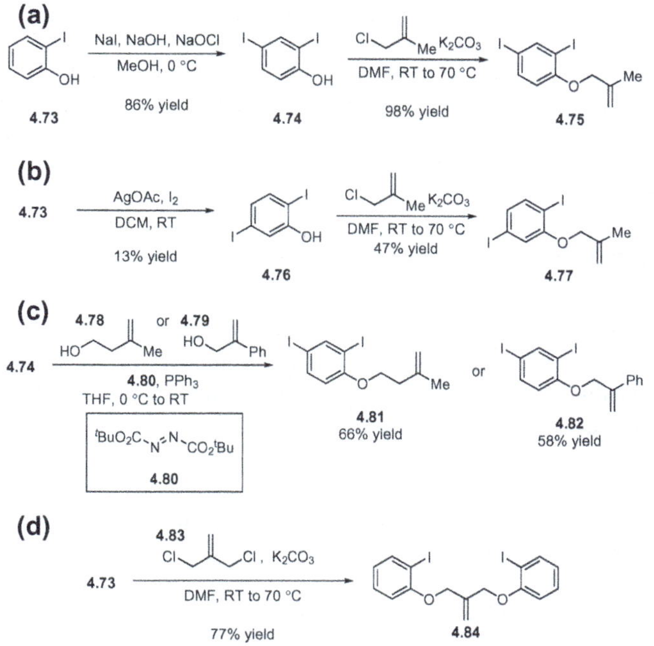

Scheme 4.20 Synthetic routes used to prepare ether starting materials

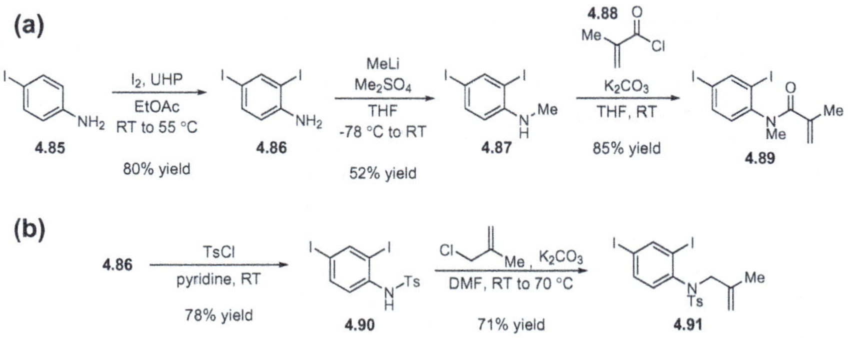

Scheme 4.21 Synthetic routes used to prepare amine starting materials

2-phenylprop-2-en-1-ol **4.79**, respectively (Scheme 4.20c, d). Symmetrical alkene **4.84** was prepared using **4.73** and 3-chloro-2-(chloromethyl)prop-1-ene **4.83** in a double alkylation under standard conditions (Scheme 4.20d).

Acrylamide substrate **4.89** was prepared in a three step protocol starting from 2-iodoaniline **4.85** (Scheme 4.21a). Aniline **4.85** was treated with iodine in the presence of urea hydrogen peroxide adduct (UHP) in EtOAc to afford 2,4-diiodoaniline **4.86** in 80% yield. This compound was methylated using MeLi as base and Me_2SO_4 as the alkylating agent at cryogenic temperatures to afford **4.87** in 52% yield. This secondary amine could be acylated in 85% yield using methacryloyl chloride **4.88** under basic conditions. Allylamine **4.91** could be prepared in 62% overall yield via a Ts protection of **4.86** in neat pyridine, followed by a standard allylation in DMF using methallyl chloride (Scheme 4.21b)

4.1.4 Reaction Optimization and Substrate Scope

This project consisted of two separate rounds of optimization. The first concerned the intramolecular aryliodination of diiodinated aromatics and the second concerned the simultaneous intramolecular aryliodination/intermolecular Mizoroki-Heck reaction of diiodoaromatics and *tert*-butyl acrylate.

4.1.4.1 Optimization for the Intramolecular Aryliodination of Diiodoaromatics

The optimization was initiated by reacting **4.75** with $Pd(P^tBu_3)_2$ (5 mol%) in PhMe (0.1 M) at 100 °C for 18 h. Full conversion of the starting material was observed, and the desired product **4.93a** was obtained in 30% yield (Table 4.1, entry 1). When the concentration was decreased to 0.05 M, the conversion was attenuated; however, the overall yield of **4.93a** was increased to 42% (entry 2). The addition of NEt_3 (1 equiv) afforded a slight increase in the yield of **4.93a** to 45% (entry 3), while the addition of extra QPhos ligand (10 mol%) had a similar effect (entry 4). The combined addition of both NEt_3 (1 equiv) and QPhos (10 mol%) led to an increase in conversion, albeit with no change in the yield of **4.93a** (entry 5). Using PMP in place of NEt_3 afforded 89% conversion of **4.75** and 72% yield of **4.93a** (entry 6). Increasing the concentration of **4.75** back to 0.1 M in the presence of both QPhos and PMP led to full conversion of the starting material and a 74% isolated yield of **4.93a** (entry 7), while a further increase in concentration to 0.2 M led to worse results (entry 8). When the reaction was run at 110 °C it was complete after 5 h and **4.93a** could be isolated in 74% yield (entry 10). The final optimized conditions were $Pd(QPhos)_2$ (5 mol%) with QPhos (10 mol%) and PMP (2 equiv) at 110 °C in PhMe at 110 °C for 5 h.

The efficiency of this reaction is quite noteworthy due to the fact that the carbon–iodine bond which is seemingly more sterically crowded (proximal) is required to interact with the Pd(0) catalyst during the productive intramolecular aryliodination. It is likely that the Pd(0) catalyst can still oxidatively add to the distal carbon–iodine bond. However, due to the known ability of this type of catalysts to undergo $C(sp^2)$–iodine bond-forming reductive elimination, this complex avoids becoming a catalytic dead end [20, 21]. The exact origin of the positive effect that extra QPhos and the addition of amine base has on the reaction is unknown. We suspect a higher

Table 4.1 Optimization of the intramolecular Pd-Catalyzed arylaiodination of **4.75**[a]

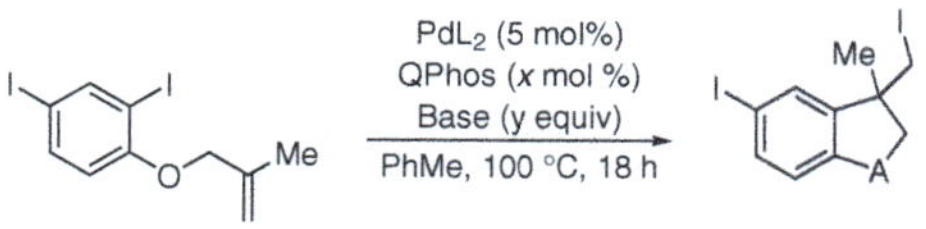

Entry	Pd source	x (mol%)	Base (y equiv)	[**4.75**] (M)	% conv.	% Yield **4.93**[b]
1	$Pd(P^tBu_3)_2$	–	–	0.1	100	30[c]
2	$Pd(QPhos)_2$	–	–	0.05	51	42
3	$Pd(QPhos)_2$	–	NEt_3 (1)	0.05	75	45
4	$Pd(QPhos)_2$	10	–	0.05	73	60
5	$Pd(QPhos)_2$	10	NEt_3 (1)	0.05	88	60
6	$Pd(QPhos)_2$	10	PMP (2)	0.05	89	72
7	$Pd(QPhos)_2$	10	PMP (2)	0.1	100	74[c]
8	$Pd(QPhos)_2$	10	PMP (2)	0.2	96	65
9[d]	$Pd(QPhos)_2$	10	PMP (2)	0.1	60	39
10[e,f]	$Pd(QPhos)_2$	10	PMP (2)	0.1	100	74[c]

[a]Reactions were run on a 0.2 mmol scale
[b]Yields determined by ^{1}H NMR analysis of the crude reaction mixture using 1,3,5-timethoxybenzene as an internal standard
[c]Isolated yield
[d]Reaction was run at 70 °C
[e]Reaction was run at 110 °C
[f]Reaction was run for 5 h

loading of bulky phosphine ligand helps direct the equilibrium of the key carbon–iodine bond-forming towards the aryl iodide in an analogous fashion to the studies by Hartwig [23, 30, 31].

4.1.4.2 Scope of the Intramolecular Aryliodination of Diiodoaromatics

The intramolecular aryliodination was tested on a series of diiodinated compounds using the optimized conditions (Table 4.2). Varying the position of the distal iodine had no effect on the reaction, as **4.77** could be efficiently cyclized to **4.93b** in 76% yield (entry 2). Chromans could be accessed under the optimized conditions, and **4.81** efficiently underwent the cyclization to **4.83c** in 74% yield (entry 3).

Dihydrobenzofuran **4.93d** and oxindole **4.93f** could be obtained in 60 and 68% yields by increasing the reaction times to 20 and 30 h, respectively (entries 4 and 6). Interestingly, symmetrical diether **4.84** which contains aryl iodides on two separate rings could undergo cyclization onto the central olefin to yield benzofuran **4.93e** in 58% isolated yield after 11 h (entry 5).

Table 4.2 Scope of the intramolecular Pd-catalyzed aryliodination of diiodoaromatics[a]

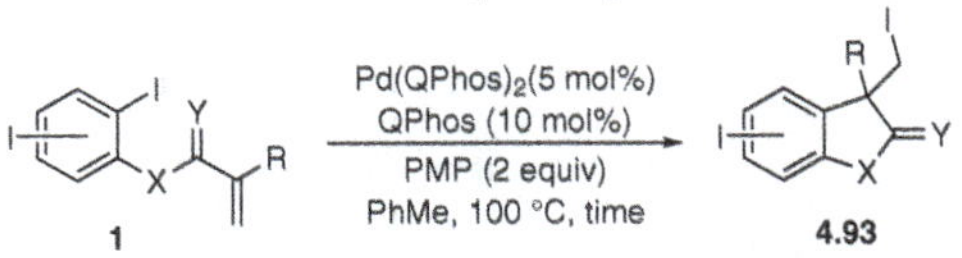

1 → 4.93

Entry	Substrate		Product		% Yield[b]
1	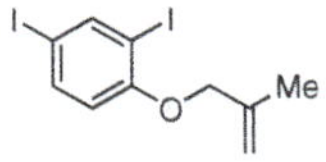	**4.75**	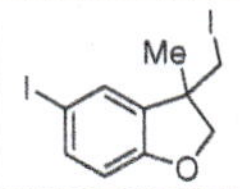	**4.93a** (5 h)	74% (67%)[c]
2	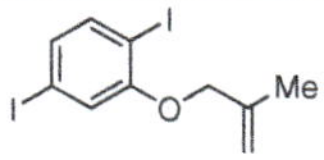	**4.77**	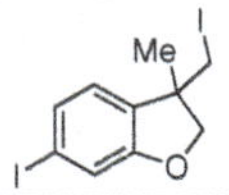	**4.93b** (7 h)	76%
3	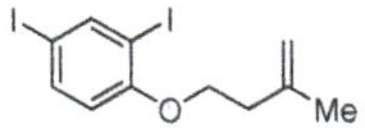	**4.81**	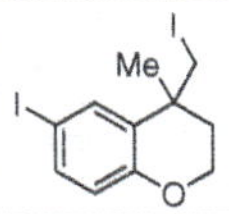	**4.93c** (8 h)	74%
4	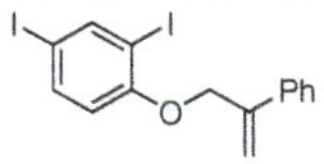	**4.82**	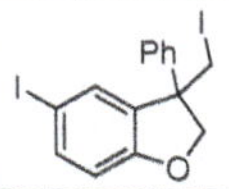	**4.93d** (20 h)	60%
5	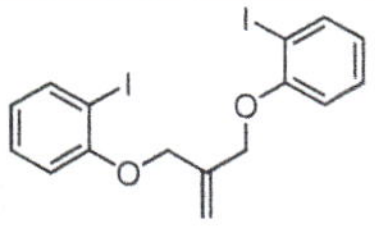	**4.84**	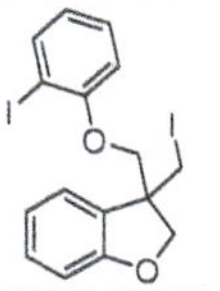	**4.93e** (11 h)	58%
6	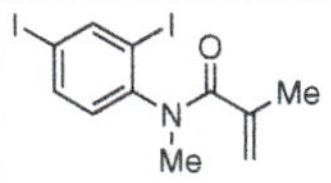	**4.89**	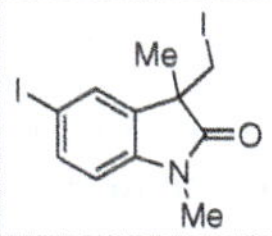	**4.93f** (30 h)	68%

[a]Reactions were run on a 0.2 mmol scale
[b]Isolated yields
[c]Reaction run on 2.5 mmol scale

4.1.4.3 Limitations

There were several substrates which either failed to produce any desired product, or led to low yields under the optimized reaction conditions (Scheme 4.22). Diene **4.94**, quinoline **4.95**, and sulfonamide **4.96** were all nearly completely recovered after subjection to the standard reaction conditions. Surprisingly, triiodoarylallyl ether **4.97** converted to an appreciable extent, and the desired intramolecular aryliodination product could be obtained in 28% yield (Fig. 4.1a). When the vinylic CH_3 in the parent substrate was replaced with an Et group **4.98**, the desired product

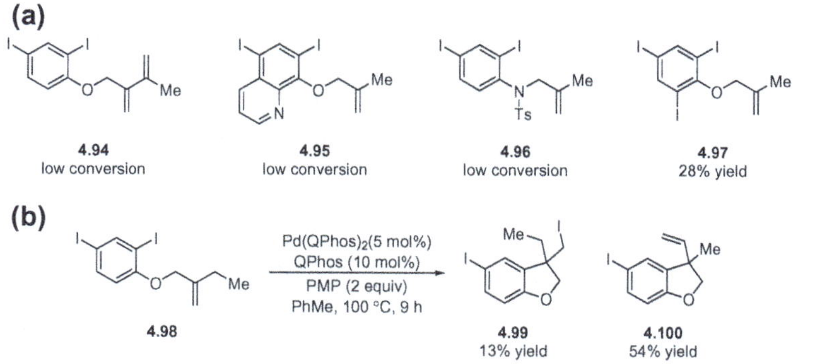

Scheme 4.22 Unsuccessful substrates in the intramolecular aryliodination

4.99 could only be obtained in 13% yield. Although **4.98** was not completely consumed under the reaction conditions, alkene **4.100** could be isolated in 54% yield. This byproduct is hypothesized to be formed via either initial isomerization of the starting material followed by 5-exo-trig cyclization and finally β-hydride elimination, or the desired carbopalladation of nonisomerized **4.98** occurs and is followed by a 1,4-Pd shift and finally β-hydride elimination (Fig. 4.1b).

4.1.4.4 Optimization for the Orthogonal Intramolecularyliodination/ Intramolecular Mizoroki-Heck Reaction

Parts of this study were conducted by Matthias Lischka under my supervision. These diiodinated compounds were then applied in an orthogonal intramolecular carboiodination/intermolecular Mizoroki-Heck sequence. The Mizoroki-Heck reaction was chosen since the reaction conditions for these reactions are typically quite similar for those utilized in the aryliodination reaction. Furthermore, the competitiveness of the intermolecular process versus the intramolecular process could be gauged more accurately by employing other alkene substrates over other coupling partners such as boronic acids or Stille reagents.

However, this reaction has some obvious potential drawbacks with respect to selectivity (Scheme 4.23). If the reaction were to undergo the desired Mizorki-Heck reaction at the distal carbon to generate **4.101**, a second and undesired intermolecular vinylation could occur to generate triene **4.102** (Scheme 4.23, path a). The desired process would involve the conversion of **4.101** into neopentyl Pd(II)I species **4.103** via an intramolecular carbopalladation. At this point **4.103** can either undergo the desired carbon–iodine bond-forming reductive elimination to generate the desired product **4.104** (path b), or it can undergo an intermolecular Mizoroki-Heck reaction to generate undesired diene **4.105** (path c).[24]

[24]Reference [32].

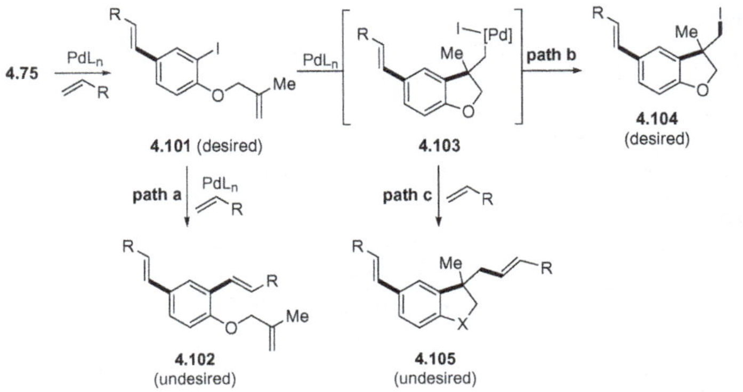

Scheme 4.23 Challenges of the simultaneous intramolecular aryliodination/intermolecular Mizoroki-Heck reaction. Used with permission from Ref. [2]. Copyright (2013) John Wiley and Sons

The optimization of this sequential reaction was conducted using **4.75** and *tert*-butyl acrylate **4.92** under various reaction conditions (Table 4.3). By treating the substrates with $Pd(QPhos)_2$ (5 mol%) and NEt_3 (2 equiv) in PhMe at 100 °C for 24 h, **4.104a** was obtained as the species in 40% yield. However, both the carboiodination only product **4.93a** and the distal intermolecular Mizoroki-Heck product **4.101a** were also obtained in 16% each (entry 1). It should be noted that this transformation occurred in a completely site-selective fashion, as none of the undesired Mizoroki-Heck products shown in Scheme 4.23 were observed (path a and path c). When the amount of NEt_3 was increased to 4 equiv, the yield of **4.104a** decreased to 21% (entry 2). Increasing the amount of **4.92** to 1.5 equiv, or adding additional QPhos (10 mol%) led to an increase in the yield of **4.104a** to 55 and 57%, respectively (entries 3–4). In both cases, the amounts of **4.93a** and **4.101a** were decreased. When NEt_3 was changed to PMP, **4.104a** was obtained in 86% yield, with only 4% of **4.93a** and none of **4.101a** (entry 6). The Pd(II) pre-catalyst Pd(crotyl)QPhosCl proved to be equally effective,[25] and **4.104a** could be obtained in 86% when 5 mol% was used (entry 8). This reaction was run in the presence of KOtBu (20 mol%), which converts the inactive Pd(II) species to active Pd(0)QPhos as a result. Furthermore, the use this Pd(II) pre-catalyst is beneficial, since it is quite stable to air and moisture and allows the overall loading of QPhos to be decreased. The final optimized conditions were found to be **4.92** (1.5 equiv), Pd(crotyl) QPhosCl (4 mol%), QPhos (8 mol%), and PMP (2 equiv) in PhMe at 100 °C for 24 h. Under these conditions, **4.104a** could be isolated in 88% yield with no trace of of **4.93a** and **4.101a** (entry 9). Due to the fact that these two intermediates have a similar R_f value to that of **4.104a**, the full conversion of both of these intermediates during the reaction is key for an efficient purification.

[25]References [33–35].

Table 4.3 Optimization of the orthogonal intramolecular aryliodination/intramolecular Mizoroki-Heck reaction of **4.75**[a,b]

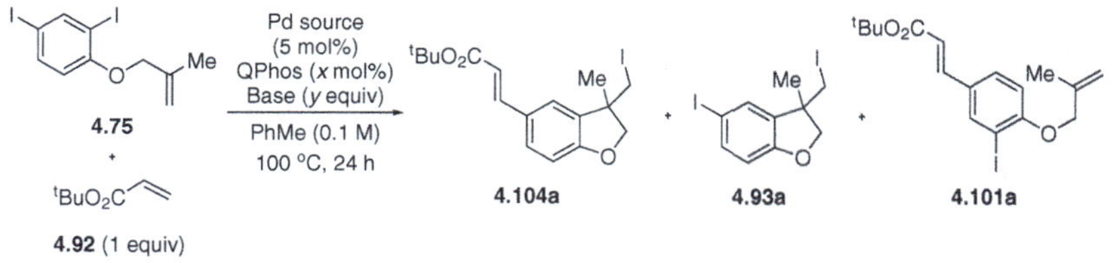

Entry	Pd source	x (mol%)	Base (y equiv)	% Yield[c]		
				4.104a	**4.93a**	**4.101a**
1	$Pd(QPhos)_2$	–	NEt_3 (2)	40	16	16
2	$Pd(QPhos)_2$	–	NEt_3 (4)	21	18	18
3[d]	$Pd(QPhos)_2$	–	NEt_3 (4)	55	5	8
4	$Pd(QPhos)_2$	10	NEt_3 (2)	57	7	10
5[d]	$Pd(QPhos)_2$	10	NEt_3 (2)	57	8	14
6[d]	$Pd(QPhos)_2$	10	PMP (2)	86	–	4
7[d]	$Pd(QPhos)_2$	5	PMP (2)	62	16	5
8[d,e]	Pd(crotyl) QPhosCl	10	PMP (2)	86	–	–
9[d–g]	Pd(crotyl) QPhosCl	8	PMP (2)	88[h] (80)[i]	–	–

[a]Reactions were run on a 0.2 mmol scale
[b]Preparative TLC was used to purify 4.93a, 4.101a and 4.101a due to the nearly identical R_f value
[c]Yields by ^1H NMR analysis of the crude reaction mixture using 1,3,5-trimethoxy benzene as an internal standard
[d]**4.92** (1.5 equiv)
[e]KO^{t}Bu used (20 mol%)
[f]4 mol% Pd(crotyl)QPhosCl
[g][**4.75**] = 0.125 M
[h]Isolated yield
[i]Reaction run on a 2.5 mmol scale

Two simple control experiments were carried out to confirm the efficiency of the reactions which converts intermediates **4.93a** and **4.101a** into **4.104a** (Scheme 4.24). Under catalytically relevant reaction conditions these intermediates could be converted to **4.104a** in 83 and 91% yields, respectively. Since both of these pathways are operational, the order in which the two steps are taking place does not matter. Furthermore, these two reactions which lead to the final product may be occurring simultaneously.

In order to test the hypothesis of a simultaneous reaction, in situ ^1H NMR spectroscopy was utilized to monitor the reaction between **4.75** and **4.92** (Fig. 4.2). To conduct this experiment, the reaction mixture was prepared in a sealed J-Young

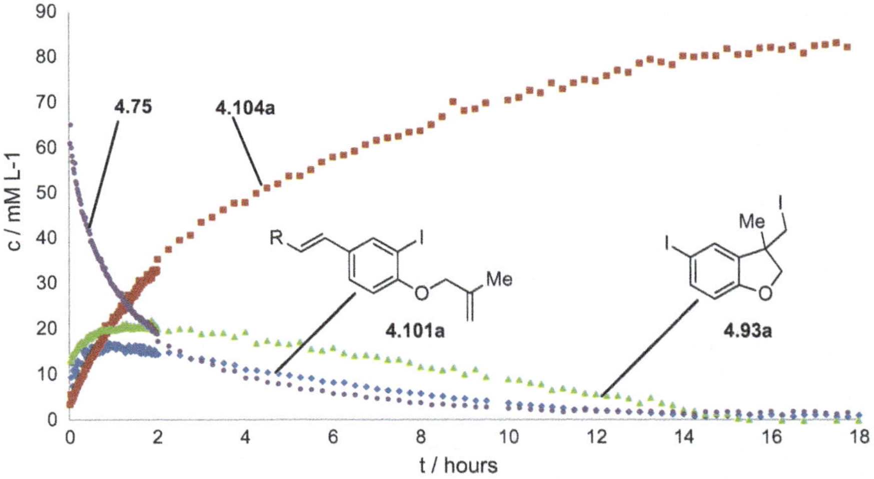

Fig. 4.2 *In situ* reaction monitoring using ^{1}H NMR spectroscopy for the reaction between **4.75** and **4.92**. Experimental conditions: Pd(crotyl)QPhosCl (0.04 equiv), QPhos (0.08 equiv), KO^{t}Bu (0.2 equiv), [**4.75**] = 0.087 M; [**4.92**] = 0.122 M, [PMP] = 0.216 M, toluene-d_8, 373 K, 700 MHz. Used with permission from Ref. [2]. Copyright (2013) John Wiley and Sons

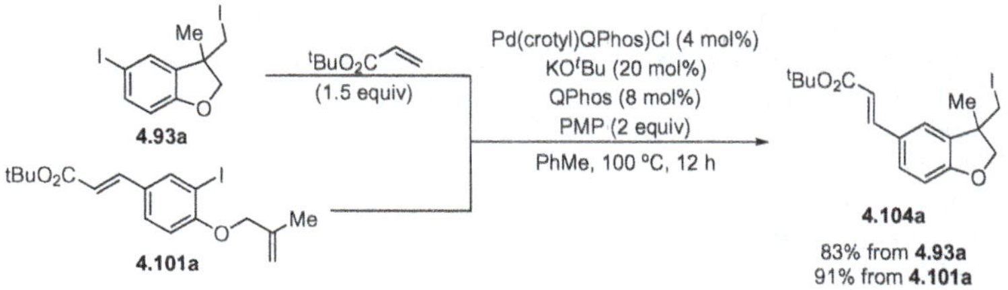

Scheme 4.24 Control experiments concerning reaction intermediates

NMR tube. There were a few experimental factors which may have resulted in a slightly lower overall yield than what were obtained under the optimized reaction conditions. First, the concentration of the reaction mixture in the NMR experiment had to be decreased to 0.087 M with respect to **4.75** since the optimal concentration (0.1 M) led to severe peak broadening in the NMR which precluded accurate calculations. Second, this reaction was heated in the spectrometer for 18 h with only sample spinning. Although this reaction is initially homogenous at 100 °C, PMP·HI is generated as the reaction proceeds which makes the reaction become heterogeneous. Without any physical stirring (i.e. magnetic stir-bar) the catalysis can become less efficient with accumulating precipitate. Nevertheless, the final observed NMR yield of **4.104a** was approximately 80% which agrees well with the isolated yield obtained using the standard protocol (88%).

During this study, similar increases in the concentration of products from both the intramolecular carboiodination (**4.93a**) and intermolecular Mizoroki-Heck reaction (**4.101a**) were observed. In contrast to Tables 4.1 and 4.2 where only selective *ortho* coupling is occurring, this experiment shows that the Pd(0) catalyst can undergo oxidative addition into both carbon–iodine bonds of **4.75** which allows both reactions to proceed simultaneously. The fact that the Pd(0) catalyst can interact with both carbon–iodine bonds in this fashion, also suggests that the intramolecular carboiodination can be occurring with reversible oxidative addition. Furthermore, similar rates of disappearance for both **4.93a** and **4.101a** echoes the results obtained in the above control experiments (Scheme 4.24).

4.1.4.5 Scope of the Orthogonal Reaction

The scope of the simultaneous reaction with respect to diiodinated aromatic substrates and Mizoroki-Heck acceptors was then examined using the optimized conditions (Table 4.4). Methyl acrylate and *N,N*-dimethyl acrylamide were well tolerated as acceptors, and their adducts **4.104b** and **4.104c** could be isolated in 83 and 73% yields, respectively (entries 2–3). Subjecting styrene and acrylonitrile to

Table 4.4 Scope of the simultaneous process[a]

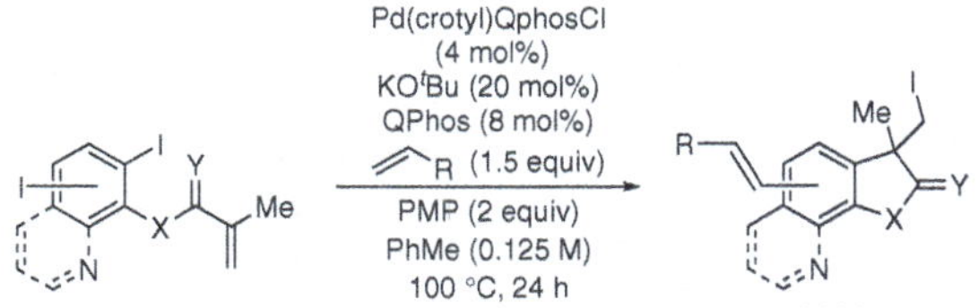

4.104

Entry	Substrate		Product		Yield[b]
1	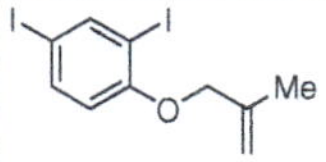	**4.75**	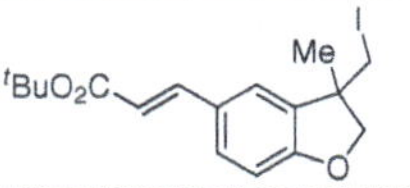	**4.104a**	88% (79%)[c]
2		**4.75**	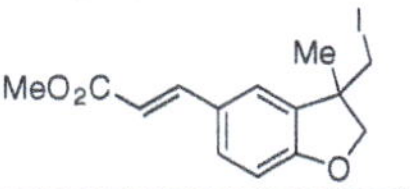	**4.104b**	83
3		**4.75**	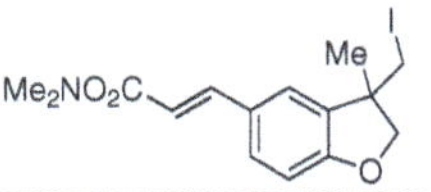	**4.104c**	73
4		**4.75**	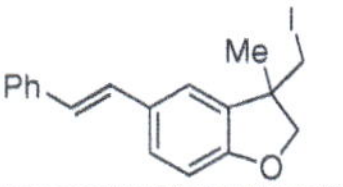	**4.104d**	82

(continued)

Table 4.4 (continued)

Entry	Substrate		Product		Yield[b]
5		**4.75**	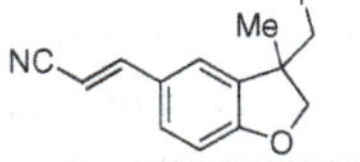	**4.104e** (4:1 *E*:*Z*)	90[d]
6		**4.75**	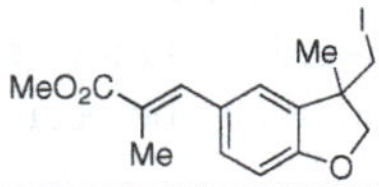	**4.104f**	77
7	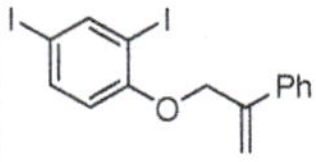	**4.82**	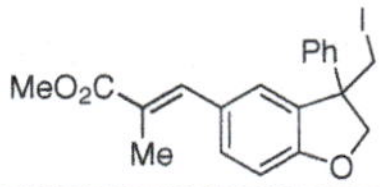	**4.104g**	25
8	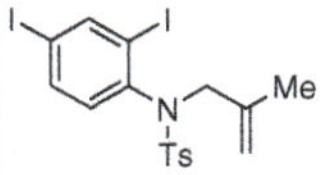	**4.105**	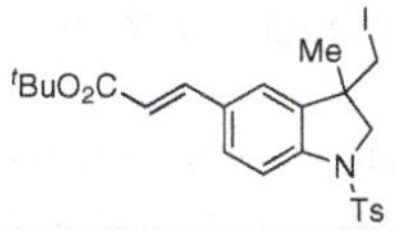	**4.104h**	86[d,f]
9	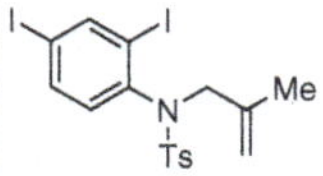	**4.91**	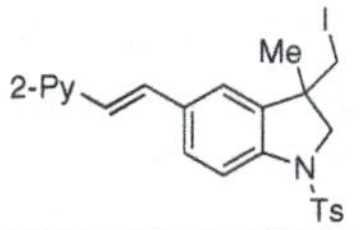	**4.104i**	84
10[e]	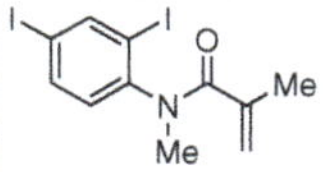	**4.89**	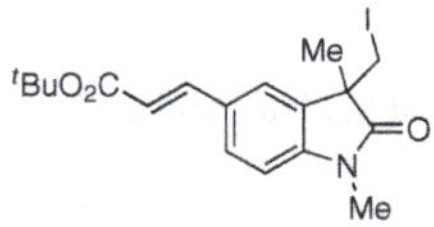	**4.104j**	87
11	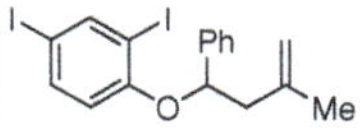	**1.169f**	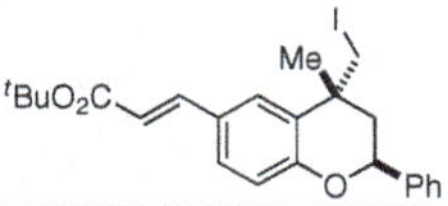	**4.104k** (87:13 dr)	86
12	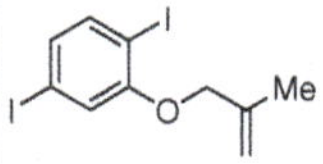	**4.77**	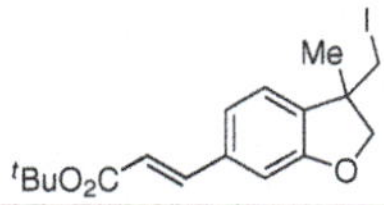	**4.104l**	91
13		**4.77**	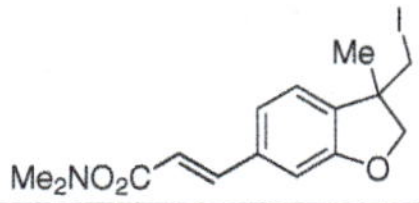	**4.104m**	88
14		**4.77**	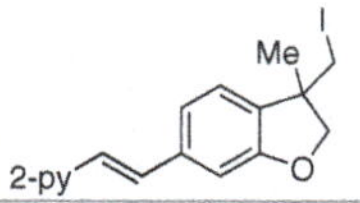	**4.104n**	74
15[e]	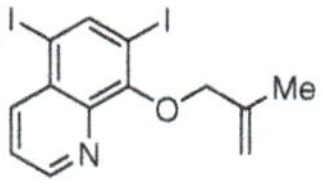	**4.99**	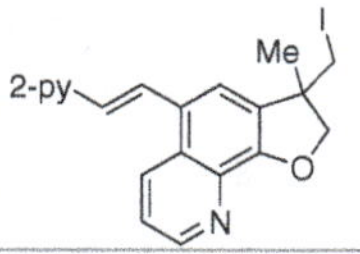	**4.104o**	70

[a]0.2 mmol scale
[b]Isolated yields
[c]**4.75** (2.5 mmol, 0.25 M), Pd(crotyl)QPhosCl (2 mol%), QPhos (4 mol%), KO*t*Bu (10 mol%), PhMe, 48 h
[d]Combined yield of both olefin isomers
[e]Reaction heated for 48 h

the standard conditions with **4.75** delivered the desired products **4.104d** and **4.104e** in 82 and 90% yield, respectively (entries 4–5). In the latter case, the product was obtained as a 4:1 mixture of *E*/*Z* isomers. Mizoroki-Heck acceptors with 1,1-disubstitution were shown to be are tolerated under the reaction conditions, and methyl methacrylate (5 equiv) led to the formation of **4.104f** in 77% yield as a single olefin isomer (entry 6). *N*-tosyl indolines such as **4.104g** and **4.104h** could be obtained in 86% and 84% yield using *tert*-butyl acrylate and 2-vinyl pyridine, respectively (entries 8–9).

Diiodinated aromatic substrates bearing electron deficient olefins were tolerated, as *N*-methyl oxindole **4.104i** could be synthesized in 87% yield (entry 10). Similarly to the previous intramolecular carboiodination, switching to a *para*-diiodo substitution pattern had no marked effect on the efficiency of the reaction (entries 12–14). Tricyclic quinolines derivative **4.104n** could also be obtained in 70% yield from the corresponding linear substrate **4.99** (entry 15).

4.1.5 Limitations

Although the vast majority of substrate combinations efficiently led to product under the optimized conditions, there were a few that were problematic.

When the reaction of **4.75** was carried out in the presence of either acrylamide or methyl vinyl ketone, poor conversion of the starting material was observed and only slight traces of products **4.104p** and **4.104q** were obtained (Scheme 4.25a). Some success was found by using allyl alcohol as the olefin partner in this reaction, and the corresponding aldehyde product **4.104r** could be isolated in 54% yield although it appeared to be a rather unstable compound (Scheme 4.25b). Unfortunately, secondary allylic alcohols were not compatible under the reaction conditions, and the corresponding ketone **4.104s** could not be isolated. Furthermore when Et-containing substrate **4.98** was treated under the standard conditions in the presence of *tert*-butyl acrylate none or the desired product **4.104t** was obtained. Instead, diene **4.104t′** was isolated in 94% yield (Scheme 4.25c).

4.1.6 Summary and Future Work

The use of di- and polyhalogenated aromatic compounds in cross-coupling reactions has been a longstanding interest in organic synthesis. Traditional methods which allow for the selective cross-coupling of a single carbon–halogen bond when using substrates with more than one type of halogen atom (i.e. Cl and Br or Br and I). This allows the Pd catalyst to undergo a highly selective oxidative addition that is dictated by carbon–halogen bond strengths. This method is known as chemoselective cross-coupling, and although it has been utilized numerous times, this strategy may require a tedious substrate synthesis. A second strategy is to use either

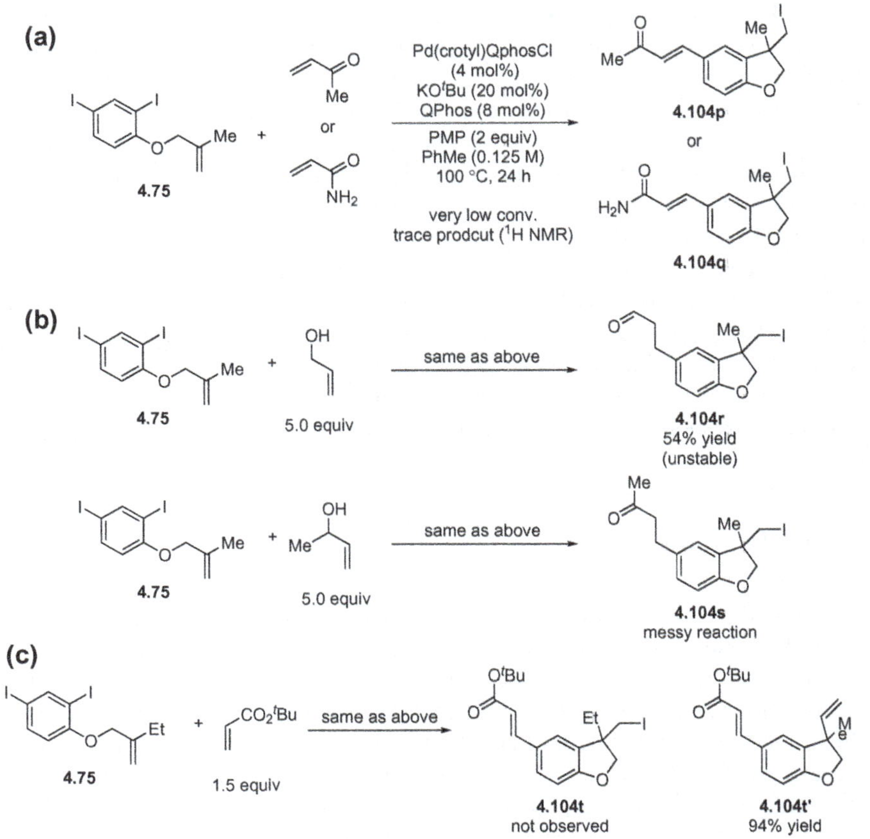

Scheme 4.25 Limitations in the simultaneous intramolecular aryliodination/intermolecular Mizoroki-Heck reaction

a directing group or a sterically encumbering group which either coordinates to or sterically interacts with the Pd(0) catalyst, thus leading to an increase or decrease in the reactivity of the halogen atom in question, respectively. This method is known as site-selective cross-coupling and has been highlighted in many successful examples.

However, if a molecule has two chemically identical carbon–halogen bonds, over coupling can occur, which is a problem that is exacerbated in intramolecular cases. The catalyst can also become deactivated by inserting into a carbon–halogen bond which cannot undergo the intramolecular reaction. Hartwig's discovery that the oxidative addition to aromatic carbon–halogen bonds can be reversible in the presence of specific Pd complexes has opened the door to a new strategy which involves site-selective cross coupling. By using catalysts which can oxidatively add to carbon–halogen bonds in a reversible manner, a substrate can possess two equally reactive carbon–halogen bonds. One can undergo the desired intramolecular reaction, while the other can interact with the catalyst via an oxidative addition mechanism without resuting in catalyst deactivation.

With this strategy in mind we have designed and developed a Pd-catalyzed intramolecular alkene aryliodination reaction of various diiodinated aromatic compounds based on reversible oxidative addition. This reaction uses a 1:4 Pd: QPhos ratio in addition to the bulky tertiary amine base PMP to impart appreciable levels of reactivity. Commonly encountered problems associated with this class of aromatic substrates are effectively avoided by judiciously choosing a ligand which allows reductive elimination from unproductive ArPd(II)I intermediates. Furthermore, conditions for a simultaneous intramolecular aryliodination/ intermolecular Mizoroki-Heck reaction have also been developed. This reaction uses a relatively air- and moisture-stable Pd(II) pre-catalyst which is converted in situ to Pd(0)QPhos by KO^{t}Bu which is present in catalytic quantities. By monitoring this latter reaction by ^{1}H NMR, we have been able to determine that the two reactions in the latter process are not time resolved. In other words, the intermolecular Mizoroki-Heck reaction and the intramolecular aryliodination process are happening simultaneously to generate two separate intermediate products. Both of these intermediates can be converted to the desired product in an efficient manner.

Future studies in this area should include the development of other inter- and intramolecular reactions capable of occurring in the presence of one or more other carbon–halogen bonds. In addition to carbon–iodine bonds, the development of conditions which can include aryl bromides and chlorides would be important from an economical and generality standpoint.

4.1.7 Experimental

General Considerations. Unless otherwise stated, all catalytic reactions were carried out under an inert atmosphere of dry argon or dry nitrogen utilizing glassware that was oven (120 °C) or flame dried and cooled under argon, whereas the work-up and isolation of the products from the catalytic reactions were conducted on the bench-top using standard techniques. Reactions were monitored using thin-layer chromatography (TLC) on EMD Silica Gel 60 F254 plates. Visualization of the developed plates was performed under UV light (254 nm) or using $KMnO_4$, *p*–anisaldehyde, or ceric ammonium molybdate (CAM) stain. Organic solutions were concentrated under reduced pressure on a Büchi rotary evaporator. Tetrahydrofuran was distilled over sodium, toluene was distilled over sodium and degassed via 5 freeze-pump-thaw cycles and stored over activated 4Å molecular sieves, triethyl amine was distilled over potassium hydroxide, DCM was distilled over calcium hydride, and anhydrous *N,N*-dimethylformamide and diethyl ether were purchased from Aldrich and used as recieved. silica gel flash column chromatography was performed on Silicycle 230–400 mesh silica gel. All standard reagents including were purchased from Sigma Aldrich, Alfa Aesar, Combi-Blocks, or Oakwood, and were used without further purification.

Instrumentation. NMR characterization data was collected at 296 K on a Varian Mercury 300, Varian Mercury 400, Varian 600, or a Bruker Avance III spectrometer operating at 300, 400, 500, or 600 MHz for ^{1}H NMR, and 75, 100, 125, for 150 MHz or ^{13}C NMR. ^{1}H NMR spectra were internally referenced to the residual solvent signal ($CDCl_3$ = 7.26 ppm, Toluene-d_8 = 2.3 ppm) or TMS. ^{13}C NMR spectra were internally referenced to the residual solvent signal ($CDCl_3$ = 77.0 ppm) and are reported as observed. Data for ^{1}H NMR are reported as follows: chemical shift (δ ppm), multiplicity (s = singlet, d = doublet, t = triplet, q = quartet, m = multiplet, br = broad), coupling constant (Hz), integration. Coupling constants have been rounded to the nearest 0.5 Hz. Melting point ranges were determined on a Fisher-Johns Melting Point Apparatus and are reported uncorrected. All reported diasteromeric ratios in data section are those obtained from ^{1}H NMR analysis of the crude reaction mixtures using 5 s delay. NMR yields for the optimization section were obtained by ^{1}H NMR analysis of the crude reaction mixture using 5 s delay and 1,3,5-trimethoxybenzene as an internal standard. Enantiomertic ratios were determined using HPLC on a chiral stationary phase using a complete Agilent HPLC system. All retention times have been rounded to the nearest 0.1 min. High resolution mass spectra (HRMS) were obtained on a micromass 70S-250 spectrometer (EI) or an ABI/Sciex QStar Mass Spectrometer (ESI) or a JEOL AccuTOF medel JMS-T1000LC mass spectrometer equipped with an IONICS® Direct Analysis in real Time (DART) ion source at Advanced Instrumentation for Molecular Structure (AIMS) in the Department of Chemistry at the University of Toronto.

General Procedures

General Procedure 1: Carboiodination of Diiodinated Compounds using Pd $(QPhos)_2$

Pd(QPhos) (5 mol%), and QPhos (10 mol%) were weighed into a 2 dram vial, and purged with argon for 10 min. To this, a toluene solution (2.0 mL) of aryliodide (0.2 mmol) and PMP (2 equiv) was added. The vial was sealed and immediately placed into a pre-heated oil bath at 110 °C. After the indicated period of time, the reaction was cooled to room temperature, filtered through a pad of silica eluting with Et_2O or EtOAc and concentrated in vacuo. The crude products were purified via silica gel flash column chromatography using the indicated mobile phase.

General Procedure 2: Carboiodination-Heck-Reaction using Pd(crotyl)QPhosCl

KO*t*-Bu (20 mol%), Pd(crotyl)QPhosCl (4 mol%), and QPhos (8 mol%) were weighed into a 2 dram vial, and purged with argon for 10 min. The contents were dissolved in toluene (0.8 mL) and stirred at room temperature for 15 min. To this, a toluene solution (0.8 mL) of aryliodide (0.2 mmol), appropriate Mizoroki-Heck acceptor (1.5–5.0 equiv), and PMP (2 equiv) was added. The vial was sealed and immediately placed into a pre-heated oil bath at 100 °C. After the indicated period of time, the reaction was cooled to room temperature, filtered through a pad of silica eluting with Et_2O or EtOAc, and concentrated in vacuo. The crude products were purified via silica gel flash column chromatography using the indicated mobile phase.

General Procedure 3: Scale-up of Carboiodination-Heck-Reaction of **4.75** using Pd(crotyl)QPhosCl

A dry pressure vessel was charged with KO*t*-Bu (56 mg, 20 mol%), Pd(crotyl) QPhosCl (91 mg, 2 mol%), QPhos (142 mg, 4 mol%), and a magnetic stirbar, and was purged with argon for 10 min. The contents were dissolved in toluene (10 mL) and stirred at room temperature for 15 min. To this, a toluene solution (10.6 mL) of **1a** (1.0 g, 2.5 mmol 1 equiv), *tert*-butyl acrylate (550 μL, 1.5 equiv), and PMP (905 μL, 2 equiv) was added. The vessel was sealed and immediately placed into a pre-heated oil bath at 100 °C. After 24 h, the reaction was cooled to room temperature, filtered through a pad of silica eluting with EtOAc, and concentrated in vacuo. The crude product were purified via repeat silica gel flash column chromatography using hexanes:EtOAc (20:3 v:v) and then hexanes:DCM (5:1 v:v) as the mobile phase to afford **4.104a** as a clear oil (805 mg, 2.01 mmol, 80%). Spectral data for **4.014a** was in accordance to that reported below. Note: reaction run with 2 mol% Pd(crotyl)QPhosCl and 4 mol% QPhos, used 10 mol% KO*t*Bu, and was run for 48 h.

General Procedure 4: Scale-up of Carboiodination of Diiodinated Compounds using $Pd(QPhos)_2$

A dry pressure vessel was charged with $Pd(QPhos)_2$ (190 mg, 5 mol%), QPhos (177 mg, 10 mol%), and a magnetic stirbar, and was purged with argon for 10 min. The contents were dissolved in toluene (12 mL). To this, a toluene solution (13 mL) of **4.75** (1.0 g, 2.5 mmol 1 equiv) and PMP (905 μL, 2 equiv) was added. The vessel was sealed and immediately placed into a pre-heated oil bath at 110 °C. After 6 h, the reaction was cooled to room temperature, filtered through a pad of silica eluting with hexanes:DCM (5:1 v:v), and the collected fraction was concentrated in vacuo. The crude product was purified via silica gel flash column chromatography using hexanes:DCM (5:1 v:v) as the mobile phase to afford **4.101a** as a clear and colourless oil (667 mg, 1.67 mmol, 67%).

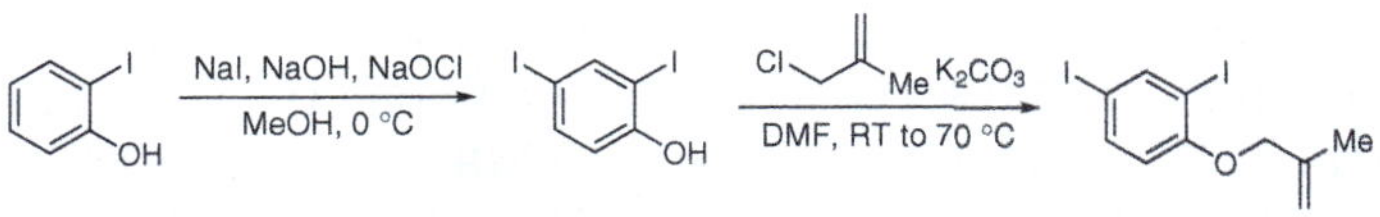

4.73 4.74 4.75

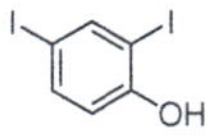

2,4-diiodophenol (4.74)—Synthesized according to the procedure of K. J. Edgar and S. N. Falling.[26] 2-iodophenol **4.73** (4.39 g, 20.0 mmol, 1 equiv), NaI (2.99 g,

[26]Reference [36].

20.0 mmol, 1 equiv), and solid NaOH (0.80 g, 20 mmol, 1 equiv) were dissolved in reagent grade MeOH (50 mL) and the solution was cooled to 0 °C. NaOCl (23 g, 6% aqueous solution) was added drop-wise over 1 h and 20 min using a syringe pump. After stirring for 1 h, the reaction was quenched by the addition of 10% (w:w) aqueous sodium thiosulfate (20 mL) and was then acidified to ~pH 7 using 5% aqueous HCl. The white precipitate was filtered and dried to constant weight in vacuo (5.96 g, 17.2 mmol, 86%). A small amount was recrystallized from hexanes to afford colourless needles (MP = 72–73 °C, lit. 73.5–74.0 °C [35]). Characterization data was in accordance with that reported by Edgar and Falling. ^{1}**H NMR** (400 MHz, $CDCl_3$) δ 7.93 (d, *J* = 2.0 Hz, 1H), 7.50 (dd, *J* = 8.5, 2.0 Hz, 1H), 6.76 (d, *J* = 8.5 Hz, 1H), 5.33 (s, 1H). ^{13}**C NMR** (100 MHz, $CDCl_3$) δ 154.9, 145.4, 139.0, 117.1, 87.0, 82.8.

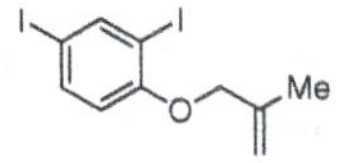

2,4-diiodo-1-((2-methylallyl)oxy)benzene (4.75)—A dry flask was charged with **4.74** (2.0 g, 5.7 mmol, 1 equiv) and anhydrous K_2CO_3 (1.57 g, 11.4 mmol, 2 equiv) and purged with nitrogen for 5 min before being dissolved in anhydrous DMF (30 mL). The resulting yellow solution had 3-chloro-2-methylprop-1-ene (660 μL, 6.8 mmol, 2 equiv) added dropwise over 2 min. The reaction was stirred at room temperature for 5 h at which time TLC analysis indicated full conversion of the starting material. The reaction was cooled and quenched with water (60 mL) and transferred to a separatory funnel using EtOAc. The aqueous layer was extracted with EtOAc (3 × 40 mL), and the combined organic layers were washed with H_2O (4 × 20 mL) and brine (20 mL), dried over $MgSO_4$, filtered, and concentrated in vacuo to afford a viscous brown oil. The crude alkene was purified using silica gel flash column chromatography using hexanes:DCM (5:1 v:v) as the mobile phase. The pure title compound was afforded as a clear and colourless oil (2.28 g, 5.59 mmol, 98%). ^{1}**H NMR** (400 MHz, $CDCl_3$) δ 8.05 (d, *J* = 2.1 Hz, 1H), 7.54 (dd, *J* = 8.6, 2.1 Hz, 1H), 6.55 (d, *J* = 8.6 Hz, 1H), 5.19–5.13 (m, 1H), 5.02 (p, *J* = 1.4 Hz, 1H), 4.45 (s, 2H), 1.85 (d, *J* = 0.7 Hz, 2H). ^{1}**H NMR** (600 MHz, Toluene-d_8, 373 K) δ 7.96 (d, *J* = 2.1 Hz, 1H), 7.21 (dd, *J* = 8.6, 2.1 Hz, 1H), 6.03 (d, *J* = 8.5 Hz, 1H), 5.01–4.96 (m, 1H), 4.82 (dd, *J* = 2.8, 1.4 Hz, 1H), 3.97 (s, 2H), 1.63 (s, 2H). ^{13}**C NMR** (100 MHz, $CDCl_3$) δ 157.2, 146.6, 139.7, 138.0, 114.0, 113.2, 87.9, 83.4, 72.7, 19.4. **IR** (cm^{-1}, thin film) 3076, 3059, 1653, 1566, 1446, 1413, 1373, 1300, 1278, 1242, 1220, 1151. **HRMS** (DART+) Calc'd for $C_{10}H_{10}I_2O$ 399.88210, found, 399.88119.

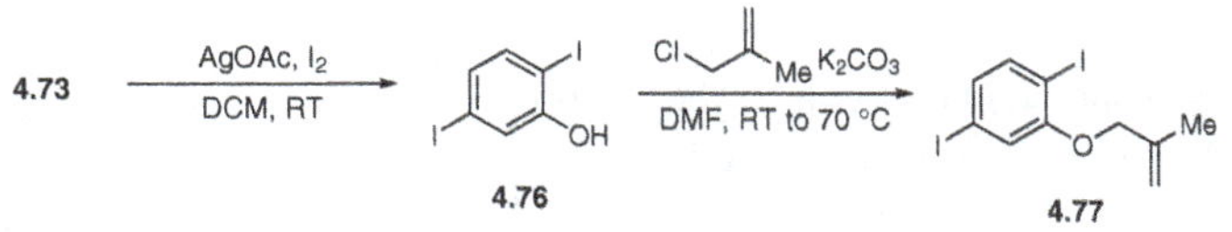

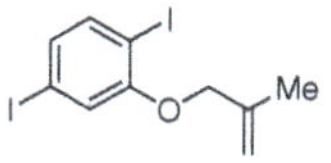

2,5-diiodo-1-((2-methylallyl)oxy)benzene (**4.77**)—A dry flask was charged with **4.76**[27] (2.0 g, 5.17 mmol, 1 equiv) and anhydrous K_2CO_3 (1.57 g, 11.4 mmol, 2 equiv) and purged with nitrogen for 5 min before being suspended in anhydrous DMF (30 mL). The yellow suspension had 3-chloro-2-methylprop-1-ene (0.66 mL, 6.84 mmol, 1.2 equiv) added dropwise over 2 min. The reaction was stirred at room temperature for 5 h at which time TLC analysis indicated full conversion of starting material. The reaction was quenched with water (50 mL) and transferred to a separatory funnel using EtOAc. The aqueous layer was extracted with EtOAc (3 × 50 mL), and the combined organic layers were washed with H_2O (4 × 30 mL) and brine (30 mL), dried over Na_2SO_4, filtered, and concentrated in vacuo to afford a clear and light yellow oil. The crude alkene was purified using silica gel flash column chromatography using 100% hexanes as the mobile phase. The pure title compound was afforded as a clear and colourless oil (603 mg, 1.51 mmol, 47%). **^{1}H NMR** (400 MHz, $CDCl_3$) δ 7.46 (d, *J* = 8.1 Hz, 1H), 7.07 (d, *J* = 1.9 Hz, 1H), 7.03 (dd, *J* = 8.1, 1.9 Hz, 1H), 5.19 (dd, *J* = 1.6, 0.9 Hz, 1H), 5.04 (p, *J* = 1.3 Hz, 1H), 4.45 (s, 2H), 1.87 (s, 3H). **^{13}C NMR** (100 MHz, $CDCl_3$) δ 157.7, 140.5, 139.6, 131.7, 121.5, 113.3, 93.6, 86.2, 72.8, 19.4. **IR** (cm^{-1}, thin film) 3078, 2884, 1577, 1468, 1436, 1383, 1245. **HRMS** (DART+) Calc'd for $C_{10}H_{11}I_2O$ 400.88993, found, 400.88947.

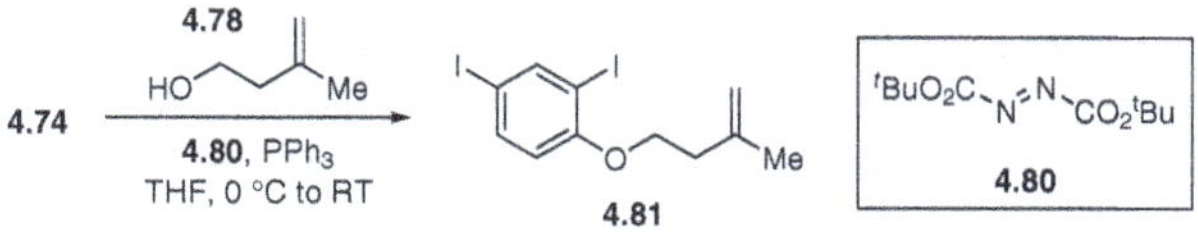

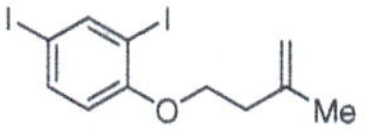

2,4-diiodo-1-((3-methylbut-3-en-1-yl)oxy)benzene (**4.81**)—A dry flask was charged with **4.74** (1.0 g, 2.8 mmol, 1.2 equiv) and isoprenol **4.78** (236 μL, 2.34 mmol, 1 equiv) and purged with nitrogen for 5 min before the mixture was dissolved in THF (20 mL). The resulting clear solution was cooled to 0 °C and had triphenylphosphine (796 mg, 3.04 mmol, 1.3 equiv) and di-*tert*-butyl azodicarboxylate **4.80** (700 mg, 3.04 mmol, 1.3 equiv) added sequentially as solids. The reaction was stirred at this temperature for 30 min before being warmed to room temperature. The reaction was allowed to stir at this temperature for 12 h at which time TLC analysis showed no further qualitative reaction progress. The reaction mixture was concentrated in vacuo, and then absorbed onto silica gel using DCM as a fascilitating solvent. The crude product was purified using silica gel flash column chromatography using 100% hexanes as the mobile phase. The pure title compound

[27]Reference [37].

was afforded as a clear and colourless oil (764 mg, 1.8 mmol, 66%). **^{1}H NMR** (400 MHz, $CDCl_3$) δ 8.04 (d, J = 2.1 Hz, 1H), 7.55 (dd, J = 8.6, 2.1 Hz, 1H), 6.56 (d, J = 8.6 Hz, 1H), 4.89–4.83 (m, 1H), 4.85–4.80 (m, 1H), 4.08 (t, J = 6.8 Hz, 2H), 2.57–2.52 (m, 2H), 1.85–1.81 (m, 3H). **^{13}C NMR** (100 MHz, $CDCl_3$) δ 157.5, 146.6, 141.8, 138.0, 113.8, 112.6, 88.1, 83.2, 68.1, 36.9, 23.0. **IR** (cm^{-1}, thin film) 3074, 2967, 2930, 2876, 1649, 1566, 1475, 1372, 1277, 1245, 1150. **HRMS** (DART+) Calc'd for $C_{11}H_{13}I_2O$ 414.90558, found, 414.90528.

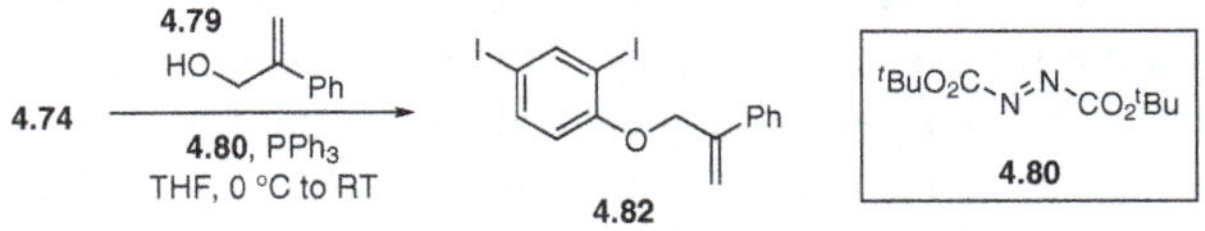

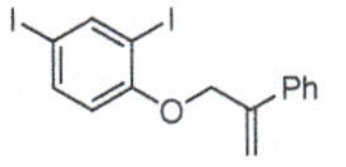

2,4-diiodo-1-((2-methylallyl)oxy)benzene (4.82)—A dry flask was charged with 2-phenylprop-2-en-1-ol **4.79** (156 mg, 1.17 mmol, 1 equiv) then **4.74** (500 mg, 1.4 mmol, 1.2 equiv) and was purged with nitrogen for 5 min before the mixture was dissolved in THF (15 mL). The resulting clear solution was cooled to 0 °C and had triphenylphosphine (398 mg, 1.52 mmol, 1.2 equiv) and **4.80** (350 mg, 1.52 mmol, 1.3 equiv) added sequentially as solids. The reaction was stirred at this temperature for 30 min before being warmed to room temperature and then to reflux at 70 °C for 12 h. At this time TLC analysis showed no further qualitative reaction progress. The reaction mixture was concentrated *in* vacuo, and then absorbed onto silica gel using DCM as a fascilitating solvent. The crude product was purified using silica gel flash column chromatography using hexanes:DCM (10:1 v:v) as the mobile phase. The pure title compound was afforded as a clear and colourless oil (313 mg, 0.68 mmol, 58%). **^{1}H NMR** (400 MHz, $CDCl_3$) δ 8.06 (d, J = 2.2 Hz, 1H), 7.56 (dd, J = 8.6, 2.2 Hz, 1H), 7.50–7.28 (m, 5H), 6.62 (d, J = 8.6 Hz, 1H), 5.62 (s (br), 1H), 5.58 (s (br), 1H), 4.92 (s (br), 2H). **^{13}C NMR** (100 MHz, $CDCl_3$) δ 157.1, 146.7, 142.0, 138.0, 128.5, 128.1, 126.1, 114.9, 114.3, 88.0, 83.7, 70.6. **IR** (cm^{-1}, thin film) 3076, 3059, 1653, 1566, 1446, 1413, 1373, 1300, 1278, 1242, 1220, 1151. **HRMS** (DART+) Calc'd for $C_{10}H_{10}I_2O$ 399.88210, found, 399.88119.

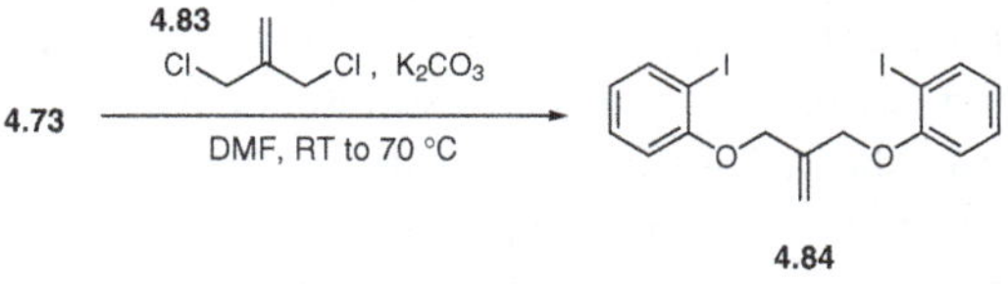

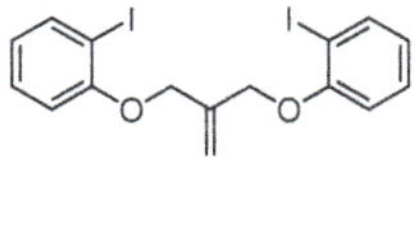

2,2′-((2-methylenepropane-1,3-diyl)bis(oxy))bis(iodobenzene) (4.84)—A dry flask was charged with **4.73** (2.5 g, 11.4 mmol, 2 equiv) and anhydrous K_2CO_3 (3.83 g, 22.8 mmol, 4 equiv) and purged with nitrogen for 5 min before being suspended in anhydrous DMF (30 mL). The resulting suspension had 3-chloro-2-(chloromethyl)prop-1-ene **4.83** (0.71 g, 5.7 mmol, 1 equiv) added dropwise over 1 min. The reaction was heated at 70 °C in an oil bath for 3 h before being cooled to room temperature. The reaction was quenched with water (50 mL) and transferred to a separatory funnel using EtOAc. The aqueous layer was extracted with EtOAc (3 × 50 mL), and the combined organic layers were washed with H_2O (4 × 30 mL) and brine (30 mL), dried over Na_2SO_4, filtered, and concentrated in vacuo to afford a clear and light yellow oil. The crude alkene was purified using silica gel flash column chromatography using hexanesEtOAc (20:1 v:v) as the mobile phase. The pure title compound was afforded as a clear and colourless oil which solidified upon standing to a free-flowing white solid (2.16 g, 4.39 mmol, 77%, MP = 49–50 °C). **^{1}H NMR** (400 MHz, $CDCl_3$) δ 7.78 (dd, J = 7.7, 1.4 Hz, 2H), 7.30 (ddd, J = 8.5, 7.4, 1.4 Hz, 2H), 6.90 (dd, J = 8.5, 1.3 Hz, 2H), 6.73 (td, J = 7.7, 1.4 Hz, 2H), 5.54–5.52 (m, 2H), 4.77 (s, 4H). **^{13}C NMR** (100 MHz, $CDCl_3$) δ 156.9, 139.4, 139.0, 129.5, 122.8, 115.9, 112.4, 86.5, 69.7. **IR** (cm^{-1}, thin film) 1581, 1471, 1456, 1438, 1278, 1244, 1230, 1060, 1049, 1016. **HRMS** (DART+NH4$^+$) Calc'd for $C_{16}H_{18}I_2NO_2$ 509.94269, found, 509.94123

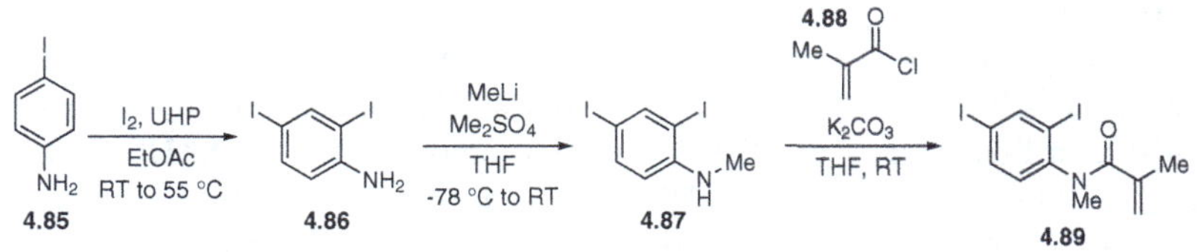

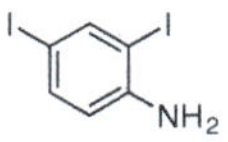

2,4-diiodoaniline (4.86)—Synthesized according to the procedure of L. Skulski and co-workers.[28] Iodine (0.508 g, 4.0 mmol, 1 equiv) and urea hydrogen peroxide adduct (UHP adduct, 0.250 g, 5.0 mmol, 1.25 equiv) were suspended in reagent grade EtOAc at room temperature. Solid 4-iodoaniline (0.963 g, 4.4 mmol, 1.1 equiv) was then added and the resulting mixture was stirred at this temperature for 30 min before heating to 55 °C in an oil bath. After 3 h, the reaction mixture was poured into an aqueous solution of Na_2SO_3 (0.5 g/70 mL) and the aqueous layer was extracted with $CHCl_3$ (3 × 20 mL). The combined organic layers were sequentially washed with water (20 mL) and brine (20 mL), dried over $MgSO_4$, filtered, and concentrated in vacuo to afford the crude aniline. This was recrystallized from

[28]Reference [38].

hexanes to afford the pure desired compound as light brown needles (1.24 g, 3.59 mmol, 80%, MP = 94–95 °C). Spectra data was in good accordance with that of the Skulski and co-workers. **^{1}H NMR** (400 MHz, $CDCl_{-3}$) δ 7.89 (d, *J* = 2.0 Hz, 1H), 7.38 (dd, *J* = 8.4, 2.0 Hz, 1H), 6.52 (d, *J* = 8.4 Hz, 1H), 4.12 (s, 2H). **^{13}C NMR** (100 MHz, $CDCl_3$) δ 146.5, 145.8, 137.9, 116.2, 84.8, 78.9.

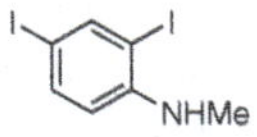

2,4-diiodo-*N*-methylaniline (4.87)—Synthesized according to the procedure of R. C. Larock and co-workers.[29] A dry flask containing a magnetic stir bar was charged with 2,4-diiodoaniline **4.86** (0.60 g, 1.74 mmol, 1 equiv), fitted with a septa, and purged with argon for 10 min. The aniline was dissolved in freshly distilled THF (4 mL) and the resulting solution was cooled to −78 °C. MeLi (1.0 mL, 1.69 mmol, 0.97 equiv, 1.70 M in diethyl ether) was then added dropwise over 5 min, and the solution was stirred at this temperature for 30 min. Dimethylsulfate (250 μL, 2.62 mmol, 1.5 equiv) was then added dropwise over 5 min, and the reaction was stirred for 15 min before being warmed to room temperature. After stirring for 2 h at this temperature, the reaction was carefully acidified to pH 6 using 5% aqueous HCl and then diluted with diethyl ether (10 mL). The collected ether layer was stirred with aqueous ammonia (2 M, 10 mL) for 30 min, and the aqeous layer was removed. The organic layer was sequentially washed with H_2O (10 mL) and brine (10 mL), dried over $MgSO_4$, filtered, and concentrated in vacuo. The crude *N*-methyl aniline was purified using silica gel flash column chromatography using hexanes:EtOAc (20:1) as mobile phase. The pure title compound was afforded as a light yellow oil (0.321 g, 0.895 mmol, 52%). **^{1}H NMR** (300 MHz, $CDCl_3$) δ 7.89 (d, *J* = 2.0 Hz, 1H), 7.47 (dd, *J* = 8.6, 2.0 Hz, 1H), 6.30 (d, *J* = 8.6 Hz, 1H), 4.25 (s, 1H), 2.86 (d, *J* = 4.9 Hz, 3H). **^{13}C NMR** (75 MHz, $CDCl_3$) δ 147.8, 145.5, 137.9, 111.7, 85.7, 77.2, 30.9. **IR** (cm^{-1}, thin film) 3402, 2983, 2929, 2904, 2881, 2823, 1575, 1546, 1496, 1471, 1446, 1425. **HRMS** (DART+) Calc'd for $C_7H_8I_2N$ 359.87461, found, 359.87359.

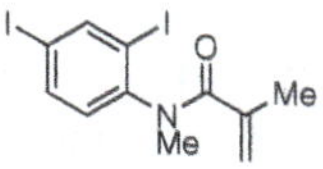

***N*-(2,4-diiodophenyl)-*N*-methylmethacrylamide (4.89)**—A THF solution (2 mL) of **4.87** (0.233 g, 0.65 mL, 1 equiv) was added via syringe to a dry flask containing anhydrous K_2CO_3 (0.269 g, 1.95 mmol, 3 equiv) at room temperature. Methacryloyl chloride (190 μL, 1.95 mmol, 3 equiv) was added dropwise to this suspension over 5 min. After 6 h, the reaction was quenched with water (3 mL) and the aqueous layer was extracted with DCM (3 × 15 mL). The combined organic layers were washed

[29]Reference [39].

with brine (15 mL), dried over $MgSO_4$, filtered, and concentrated in vacuo. The crude acrylamide was purified using silica gel flash column chromatography using hexanes:EtOAc (4:1 v:v) as the mobile phase. The pure title compound was afforded as a free-flowing white solid (0.235 g, 0.55 mmol, 85%, MP = 131–133 °C) ^{1}**H NMR** (400 MHz, $CDCl_3$) δ 8.21 (d, J = 1.9 Hz, 1H), 7.66 (dd, J = 8.4, 2.0 Hz, 1H), 6.90 (d, J = 8.2 Hz, 1H), 5.04 (s(br), 2H), 3.21 (s, 3H), 1.84 (s, 3H). ^{13}**C NMR** (100 MHz, $CDCl_3$) δ 171.4, 147.7, 146.8, 139.8, 138.5, 130.5, 119.4, 100.3, 93.3, 36.7, 20.5. **IR** (cm^{-1}, thin film) 2918, 1654, 1626, 1464, 1423, 1359, 1122. **HRMS** (DART+) Calc'd for $C_{11}H_{12}I_2NO$ 427.90082, found, 427.89922.

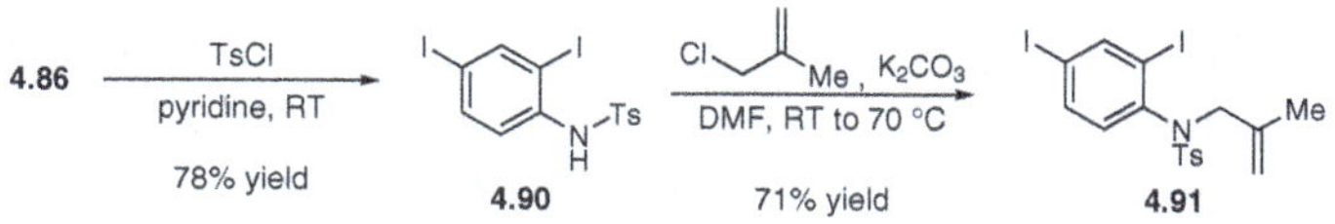

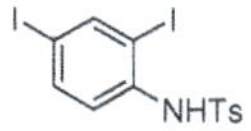

N-**(2,4-diiodophenyl)-4-methylbenzenesulfonamide (4.90)**—To a solution of 2,4-diiodoaniline (0.6 g, 1.74 mmol, 1 equiv) in pyridine (4 mL) at 0 °C was added solid tosyl chloride (0.348 g, 1.83 mmol, 1.05 equiv) in one portion. After 2 h, more solid tosyl chloride (0.348 g, 1.83 mmol, 1.05 equiv) was added. After one additional hour, the reaction was quenched with H_2O (5 mL) and the aqueous layer was extracted with DCM (3 × 10 mL). The combined organic layers were washed with a saturated aqueous solution of $CuSO_4$ (4 × 10 mL), and brine (10 mL), dried over $MgSO_4$, filtered and concentrated in vacuo. The crude sulfonamide was purified using silica gel flash column chromatography using hexanes: EtOAc (4:1 v:v) as the mobile phase. The pure title compound was afforded as a light brown solid (0.679 g, 1.36 mmol, 78%, MP = 121–123 °C). ^{1}**H NMR** (400 MHz, $CDCl_3$) δ 7.95 (d, J = 2.0 Hz, 1H), 7.65–7.61 (m, 2H), 7.58 (dd, J = 8.6, 2.0 Hz, 1H), 7.39 (d, J = 8.6 Hz, 1H), 7.25–7.19 (m, 2H), 6.77 (s, 1H), 2.39 (s, 3H). ^{13}**C NMR** (100 MHz, $CDCl_3$) δ 146.3, 144.5, 138.4, 137.4, 135.6, 129.8, 127.4, 123.5, 93.0, 89.6, 21.6. **IR** (cm^{-1}, thin film) 3417, 3063, 1633, 1612, 1537, 1487, 1464, 1330, 1224, 1197, 1161. **HRMS** (DART+) Calc'd for $C_{13}H_{12}INO_2S$ 499.86781, found, 499.86794.

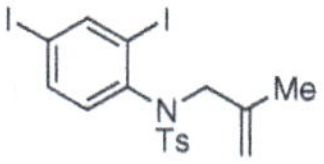

N-**(2,4-diiodophenyl)-4-methyl-*N*-(2-methylallyl)benzenesulfonamide (4.91)**—A dry flask was charged with **4.90** (0.55 g, 1.1 mmol, 1 equiv) and anhydrous K_2CO_3 (0.457 g, 3.3 mmol, 3 equiv), and then purged with dry nitrogen for 10 min. The solid mixture was then suspended in anhydrous DMF (4 mL), and then neat 3-chloro-2-methylprop-1-ene (214 μL, 2.2 mmol, 2 equiv) was added at once, and the resulting reaction mixture was placed in an oil bath preheated to 70 °C. After 1 h at this temperature the reaction was quenched with H_2O (10 mL). The aqueous layer was extracted with EtOAc (3 × 15 mL), and the combined organic layers were washed with H_2O (4 × 5 mL) and brine (15 mL), dried over $MgSO_4$, filtered, and concentrated in vacuo. The crude alkene was purified using silica gel flash column chromatrography using hexanes:EtOAc (20:3 v:v) as the mobile phase and the pure title compound was afforded as a light brown viscous oil which solidified upon standing to an off white solid. (0.429 g, 0.775 mmol, 71%, MP = 101–102 °C). **^{1}H NMR** (300 MHz, $CDCl_3$) δ 8.20 (d, J = 2.0 Hz, 1H), 7.65–7.48 (m, 3H), 7.31–7.27 (m, 2H), 6.70 (d, J = 8.4 Hz, 1H), 4.77 (t, J = 1.5 Hz, 1H), 4.61 (dd, J = 1.7, 0.9 Hz, 1H), 4.16 (d, J = 14.0 Hz, 1H), 3.98 (d, J = 13.7 Hz, 1H), 2.44 (s, 3H), 1.89–1.74 (m, 3H). **^{13}C NMR** (75 MHz, $CDCl_3$) δ 148.3, 143.9, 141.3, 139.4, 137.7, 135.9, 132.2, 129.6, 128.2, 116.7, 103.0, 94.5, 57.7, 21.6, 21.0. **IR** (cm^{-1}, thin film) 2972, 2945, 2918, 1485, 1437, 1352, 1305, 1161, 1091. **HRMS** (DART+) Calc'd for $C_{17}H_{18}I_2NO_2S$ 553.91476, found, 553.91453.

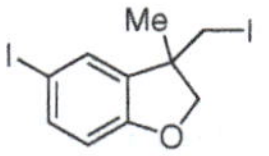

(±)-5-iodo-3-(iodomethyl)-3-methyl-2,3-dihydrobenzofuran (4.93a)—Synthesized according to **general procedure 1** using **4.75** and the reaction was run for 5 h on a 0.2 mmol scale. The crude product was purified via silica gel chromatography using hexanes: DCM (5:1 v:v) and was isolated as a white solid (58.7 mg, 0.146 mmol, 74%, MP = 40–41 °C) **^{1}H NMR** (400 MHz, $CDCl_3$) δ 7.45 (dd, J = 8.4, 1.9 Hz, 1H), 7.38 (d, J = 1.9 Hz, 1H), 6.59 (d, J = 8.4 Hz, 1H), 4.48 (d, J = 9.2 Hz, 1H), 4.16 (dd, J = 9.2, 0.7 Hz, 1H), 3.35 (d, J = 0.7 Hz, 2H), 1.50 (s, 3H). **^{1}H NMR** (600 MHz, Toluene-d_8) δ 7.20 (dd, J = 8.4, 1.9 Hz, 1H), 7.10 (d, J = 1.9 Hz, 1H), 6.29 (d, J = 8.4 Hz, 1H), 4.08 (d, J = 9.1 Hz, 1H), 3.71 (dd, J = 9.1, 1.0 Hz, 1H), 2.73 (d, J = 10.1 Hz, 1H), 2.70 (dd, J = 10.1, 1.0 Hz, 1H), 0.99 (s, 3H). **^{13}C NMR** (100 MHz, $CDCl_3$) δ 159.7, 138.1, 134.8, 131.9, 112.9, 83.5, 82.1, 77.5, 77.2, 76.8, 46.1, 25.5, 17.5. **IR** (cm^{-1}, thin film) 2964, 2926, 2883, 1635, 1600, 1581, 1471, 1456, 1404, 1377, 1207, 1246, 1209, 1178, 1159, 1118, 976, 877, 840, 810. **HRMS** (DART+) Calc'd for $C_{10}H_{11}I_2O$ 400.88993, found, 400.89038.

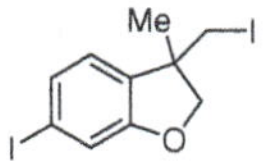

(±)-6-iodo-3-(iodomethyl)-3-methyl-2,3-dihydrobenzofuran (4.93b)—Synthesized according to **general procedure 1** using **4.77** and the reaction was run for 7 h on a 0.157 mmol scale. The crude product was purified via silica gel chromatography using hexanes:DCM (5:1 v:v) as the mobile phase and was isolated as a colourless oil (47.5 mg, 0.119 mmol, 76%) **^{1}H NMR** (400 MHz, $CDCl_3$) δ 7.24 (dd, J = 7.8, 1.5 Hz, 1H), 7.16 (d, J = 1.5 Hz, 1H), 6.85 (d, J = 7.8 Hz, 1H), 4.47 (d, J = 9.2 Hz,

1H), 4.17 (dd, $J = 9.2, 0.7$ Hz, 1H), 3.36–3.30 (m, 2H), 1.49 (s, 3H). **^{13}C NMR** (100 MHz, $CDCl_3$) δ 160.4, 131.9, 129.9, 124.2, 119.7, 93.3, 83.3, 45.6, 25.3, 17.3. **IR** (cm^{-1}, thin film) 2962, 2923, 2883, 2849, 1593, 1464, 1208, 975. **HRMS** (DART+) Calc'd for $C_{10}H_{11}I_2O$ 400.88993, found, 400.88943.

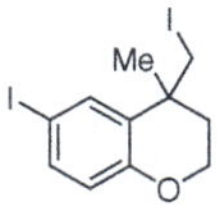

(±)-6-iodo-4-(iodomethyl)-4-methylchroman (**4.93c**)—Synthesized according to **general procedure 1** using **4.81** and the reaction was run for 8 h on a 0.2 mmol scale. The crude product was purified via silica gel chromatography using hexanes: DCM (5:1 v:v) as the mobile phase and was isolated as a clear and colourless oil (61.4 mg, 0.148 mmol, 74%) **^{1}H NMR** (400 MHz, $CDCl_3$) δ 7.48 (d, $J = 2.2$ Hz, 1H), 7.38 (dd, $J = 8.6, 2.2$ Hz, 1H), 6.59 (d, $J = 8.6$ Hz, 1H), 4.16 (ddd, $J = 11.5$, 6.1, 3.7 Hz, 1H), 4.07 (ddd, $J = 11.5, 9.1, 3.1$ Hz, 1H), 3.48 (dd, $J = 10.4, 0.9$ Hz, 1H), 3.45–3.41 (m, 1H), 2.12 (ddd, $J = 14.0, 6.1, 3.1$ Hz, 1H), 1.80 (dddd, $J = 14.0, 9.1, 3.7, 0.9$ Hz, 1H), 1.43 (s, 3H). **^{13}C NMR** (100 MHz, $CDCl_3$) δ 153.9, 137.0, 135.5, 129.0, 119.7, 82.5, 62.6, 34.7, 33.7, 28.1, 21.2. **IR** (cm^{-1}, thin film) 2959, 2917, 2879, 2848, 1480, 1472, 1277, 1222, 1177. **HRMS** (DART+) Calc'd for $C_{11}H_{13}I_2O$ 414.90558, found, 414.90625.

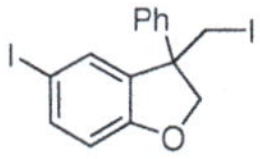

(±)-5-iodo-3-(iodomethyl)-3-phenyl-2,3-dihydrobenzofuran (**4.93d**)—Synthesized according to **general procedure 1** using **4.82** and the reaction was run for 20 h on a 0.2 mmol scale. The crude product was purified via silica gel chromatography using hexanes:DCM (5:1 v:v) as the mobile phase and was isolated as a white solid (55.4 mg, 0.12 mmol, 60%, MP = 94–96 °C). **^{1}H NMR** (400 MHz, $CDCl_3$) δ 7.51 (dd, $J = 8.4$, 1.9 Hz, 1H), 7.42 (d, $J = 1.9$ Hz, 1H), 7.40–7.27 (m, 6H), 6.68 (d, $J = 8.4$ Hz, 1H), 4.77 (d, $J = 9.5$ Hz, 1H), 4.71 (d, $J = 9.5$ Hz, 1H), 3.85 (d, $J = 10.5$ Hz, 1H), 3.74 (d, $J = 10.5$ Hz, 1H). **^{13}C NMR** (100 MHz, $CDCl_3$) δ 159.9, 141.7, 138.2, 134.4, 133.5, 128.9, 127.5, 126.4, 112.9, 84.4, 82.21, 54.0, 14.8. **IR** (cm^{-1}, thin film) 1472. 1456, 1262, 1235, 1218, 1189. **HRMS** (DART+) Calc'd for $C_{15}H_{13}I_2O$ 462.90558, found, 462.90571.

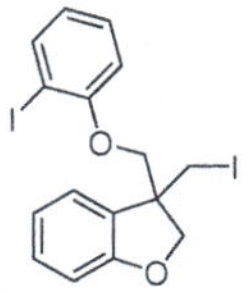

3-(iodomethyl)-3-((2-iodophenoxy)methyl)-2,3-dihydrobenzofuran (**4.93e**)—Synthesized according to **general procedure 1** using **4.84** and the reaction was run for 11 h on a 0.1 mmol scale. The crude product was purified via silica gel chromatography using hexanes:DCM (5:1 v:v) as the mobile phase and was isolated as a light yellow film (28.5 mg, 0.058 mmol, 58%). **^{1}H NMR** (400 MHz, $CDCl_3$) δ 7.79 (dd, *J* = 7.8, 1.5 Hz, 1H), 7.63 (dd, *J* = 7.5, 1.5 Hz, 1H), 7.34–7.16 (m, 2H), 6.94 (td, *J* = 7.5, 1.0 Hz, 1H), 6.86–6.69 (m, 3H), 4.54 (d, *J* = 9.7 Hz, 1H), 4.45 (d, *J* = 9.7 Hz, 1H), 4.26 (d, *J* = 8.7 Hz, 1H), 4.23 (d, *J* = 8.7 Hz, 1H), 4.06 (d, *J* = 10.0 Hz, 1H), 3.64 (d, *J* = 10.0 Hz, 1H). **^{13}C NMR** (100 MHz, $CDCl_3$) δ 160.1, 156.7, 139.4, 129.9, 129.5, 128.2, 125.4, 123.1, 120.9, 111.9, 110.3, 86.4, 79.2, 72.4, 50.0, 13.3. **IR** (cm^{-1}, thin film) 3058, 2917, 2871, 1606, 1456, 1436, 1289, 1246, 1198, 1053, 1017. **HRMS** (DART+) Calc'd for $C_{16}H_{18}$ I_2NO 509.94269, found, 509.94130.

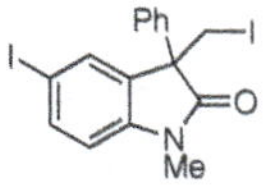

(±)-5-iodo-3-(iodomethyl)-1-methyl-3-phenylindolin-2-one (4.93f)—Synthesized according to **general procedure 1** using **4.89** and the reaction was run for 30 h on a 0.2 mmol scale. The crude product was purified via silica gel chromatography using hexanes:DCM (1:3 v:v) as the mobile phase and was isolated as a free-flowing off white solid (57.7 mg, 0.135 mmol, 68%, MP = 126–127 °C). **^{1}H NMR** (400 MHz, $CDCl_3$) δ 7.65 (dd, *J* = 8.2, 1.8 Hz, 1H), 7.54 (d, *J* = 1.8 Hz, 1H), 6.66 (d, *J* = 8.2 Hz, 1H), 3.49 (d, *J* = 9.9 Hz, 1H), 3.37 (d, *J* = 9.9 Hz, 1H), 3.21 (s, 3H), 1.50 (s, 3H). **^{13}C NMR** (100 MHz, $CDCl_3$) δ 177.1, 142.9, 137.5, 134.9, 131.5, 110.3, 85.1, 48.7, 26.3, 22.9, 9.9. **IR** (cm^{-1}, thin film) 2967, 2951, 1713, 1602, 1490, 1484, 1370, 1360, 1342, 1272, 1243, 1090. **HRMS** (DART+) Calc'd for $C_{11}H_{12}I_2NO$ 427.90082, found, 427.90173.

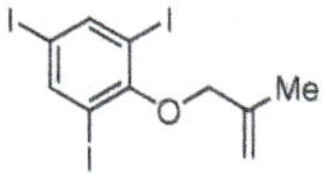

1,3,5-triiodo-2-((2-methylallyl)oxy)benzene (4.97)—A dry flask was charged with 2,4,6-triiodophenol (1.00 g, 2.12 mmol, 1 equiv) and anhydrous K_2CO_3 (586 mg, 4.13 mmol, 2 equiv) and purged with nitrogen for 5 min before being dissolved in anhydrous DMF (5 mL). The resulting suspension had 3-chloro-2-methylprop-1-ene (627 µL, 6.36 mmol, 3 equiv) added dropwise. The reaction was stirred at room temperature for 1 h before being heated at 70 °C for 6 h at which time TLC analysis indicated full conversion of the starting material. The reaction was cooled and quenched with water (60 mL) and transferred to a separatory funnel using EtOAc. The aqueous layer was extracted with EtOAc (3 × 40 mL), and the combined organic layers were washed with H_2O (4 × 20 mL) and brine (20 mL), dried over $MgSO_4$, filtered, and concentrated in vacuo to afford a viscous brown oil. The crude alkene was purified using silica gel flash column chromatography using hexanes:DCM (5:1 v:v) as the mobile phase. The pure title compound was afforded as a crystalline white solid (973 mg,

1.85 mmol, 87%). **^{1}H NMR** (400 MHz, $CDCl_3$) δ 8.06 (s, 2H), 5.25–5.21 (m, 1H), 5.07–5.03 (m, 1H), 4.37–4.34 (m, 2H), 1.98–1.95 (m, 3H). **^{13}C NMR** (100 MHz, $CDCl_3$) δ 157.7, 147.3, 140.1, 114.1, 92.1, 89.2, 76.3, 20.0. **IR** (cm^{-1}, thin film) 3076, 2977, 2744, 1652, 1535, 1516, 1426, 1358, 1257. **HRMS** (DART+NH_4) Calc'd for $C_{10}H_{13}I_3NO$ 543.81312, found, 543.81204.

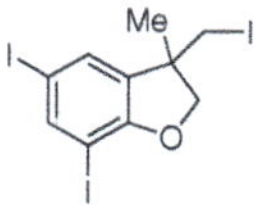

(±)-5,7-diiodo-3-(iodomethyl)-3-methyl-2,3-dihydrobenzofuran (4.93 g)—Synthesized according to **general procedure 1** using **4.97** and the reaction was run for 12 h on a 0.2 mmol scale. The crude product was purified via silica gel chromatography using hexanes:DCM (5:1 v:v) as the mobile phase and was isolated as a clear and colourless oil (29.1 mg, 0.0554 mmol, 28%). **^{1}H NMR** (500 MHz, CD_2CL_2) δ 7.84 (d, J = 1.5 Hz, 1H), 7.37 (d, J = 1.5 Hz, 1H), 4.55 (d, J = 9.5 Hz, 1H), 4.26 (dd, J = 9.5, 1.0 Hz, 1H), 3.41–3.38 (m, 1H), 3.36 (dd, J = 10.0, 1.0 Hz, 1H), 1.49 (s, 3H). **^{13}C NMR** (125 MHz, CD_2Cl_2) δ 160.2, 145.0, 134.4, 131.9, 83.0, 82.2, 75.6, 47.4, 25.4, 17.0. **IR** (cm^{-1}, thin film) 3061, 2962, 2920, 1583, 1563, 1468, 1434, 1375, 1285, 1237, 1198. **HRMS** (EI+) Calc'd for $C_{10}H_9I_3O$ 525.7788, found, 525.7789.

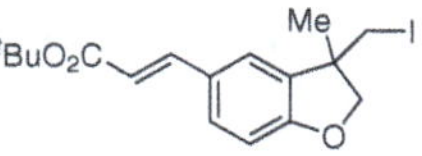

(±)-(*E*)-*tert*-butyl 3-(3-(iodomethyl)-3-methyl-2,3-dihydrobenzofuran-5-yl)acrylate (4.104a)—1) From (±)-5-iodo-3-(iodomethyl)-3-methyl-2,3-dihydrobenzofuran **4.93a**: KO*t*-Bu (20 mol%), Pd(crotyl)QPhosCl (4 mol%), and QPhos (8 mol %) were weighed into a 2 dram vial, and purged with argon for 10 min. The contents were dissolved in toluene (0.8 mL) and stirred at room temperature for 15 min. To this, a toluene solution (0.8 mL) of **4.93** (0.2 mmol), *tert*-butyl acrylate (1.46 equiv), and PMP (2 equiv) was added. The vial was sealed and immediately placed into a pre-heated oil bath at 100 °C. After 12 h, the reaction was cooled to room temperature, filtered through a pad of silica eluting with Et_2O, and concentrated in vacuo. The crude products were purified via silica gel flash column chromatography using hexanes EtOAc (100:3 v:v) as the mobile phase and was afforded as a faint pink oil, which was further purified via silica gel chromatography using DCM:hexanes (2:1 v:v) as the mobile phase to afford the title compound was afforded as a clear oil (66.4 mg, 0.166 mmol, 83%).

2) Synthesized according to **general procedure 2** using 2 equivalents of *tert*-butyl acrylate, and the reaction was run for 24 h. The crude product was purified via silica gel flash column chromatography using hexanes:EtOAc (100:0 to 100:3 v:v) as the mobile phase and was isolated as a clear and colourless oil (70.4 mg, 0.176 mmol, 88%). **^{1}H NMR** (400 MHz, $CDCl_3$) δ 7.53 (d, J = 15.9 Hz, 1H), 7.34 (dd, J = 8.3, 1.6 Hz, 1H), 7.29 (d, J = 1.6 Hz, 1H), 6.78 (d, J = 8.3 Hz, 1H), 6.23 (d, J = 15.9 Hz,

1H), 4.53 (d, J = 9.3 Hz, 1H), 4.22 (d, J = 9.3 Hz, 1H), 3.37 (s, 2H), 1.52 (s, 12H). **^{1}H NMR** (700 MHz, Toluene-d_8, 373 K) δ 7.80 (d, J = 15.8 Hz, 1H), 7.20 (d, J = 8.4 Hz, 1H), 6.74 (d, J = 8.4 Hz, 1H), 6.45 (dd, J = 15.8, 1.5 Hz, 1H), 4.38 (d, J = 9.1 Hz, 1H), 4.01 (d, J = 9.1 Hz, 1H), 3.11–2.95 (m, 2H), 1.69 (s, 9H), 1.31 (3, 3H). **^{13}C NMR** (100 MHz, $CDCl_3$) δ 166.7, 161.5, 143.3, 132.9, 130.4, 128.1, 122.3, 117.8, 110.8, 83.7, 80.4, 45.7, 28.4, 25.5, 17.8. **IR** (cm^{-1}, thin film) 2976, 2929, 1699, 1635, 1604, 1591, 1485, 1471, 1456, 1390, 1367, 1329, 1300, 1274, 1257, 1213, 1143. **HRMS** (DART+) Calc'd for $C_{17}H_{22}IO_3$ 401.06136, found, 401.06278.

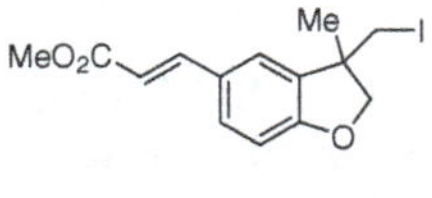

(±)-(*E*)-methyl 3-(3-(iodomethyl)-3-methyl-2,3-dihydrobenzofuran-5-yl)acrylate (4.104b)—Synthesized according **to general procedure 2** using **4.75** and 2 equivalents of methyl acrylate, and the reaction was run for 24 h. The crude product was purified via silica gel flash column chromatography using hexanes:Et_2O (100:0 to 10:1 v:v) and was isolated as a clear and light yellow oil (59.7 mg, 0.167 mmol, 83%). **^{1}H NMR** (400 MHz, $CDCl_3$) δ 7.64 (d, J = 16.0 Hz, 1H), 7.37 (dd, J = 8.3, 1.9 Hz, 1H), 7.31 (d, J = 1.9 Hz, 1H), 6.80 (d, J = 8.3 Hz, 1H), 6.30 (d, J = 16.0 Hz, 1H), 4.54 (d, J = 9.3 Hz, 1H), 4.23 (d, J = 9.3 Hz, 1H), 3.79 (s, 3H), 3.38 (s, 2H), 1.53 (s, 3H). **^{13}C NMR** (100 MHz, $CDCl_3$) δ 167.6, 161.6, 144.5, 132.9, 130.4, 127.6, 122.4, 115.2, 110.7, 83.6, 51.6, 45.6, 25.4, 17.5. **IR** (cm^{-1}, thin film) 2962, 2949, 1697, 1635, 1604, 1591, 1489, 1452, 1435, 1329, 1301, 1263, 1213, 1195. **HRMS** (DART+) Calc'd for $C_{14}H_{16}IO_3$ 359.01441, found, 359.01439.

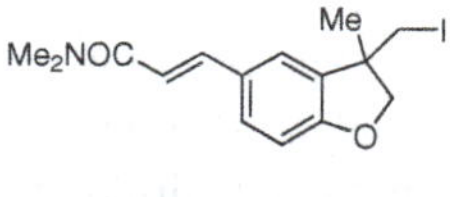

(±)-(*E*)-3-(3-(iodomethyl)-3-methyl-2,3-dihydrobenzofuran-5-yl)-*N*,*N*-dimethylacrylamide (4.104c)—Synthesized according to **general procedure 2** using **4.75** and 3.5 equivalents of *N*,*N*-dimethyl acrylamide, and the reaction was run for 48 h. The crude product was purified via silica gel flash column chromatography using hexanes:EtOAc (5:6 v:v) as the mobile phase, and was isolated as a clear and colourless oil (54.0 mg, 0.146 mmol, 73%). **^{1}H NMR** (400 MHz, $CDCl_3$) δ 7.62 (d, J = 15.4 Hz, 1H), 7.37 (ddd, J = 8.3, 1.9, 0.5 Hz, 1H), 7.27 (d, J = 1.9 Hz, 1H), 6.78 (d, J = 8.3 Hz, 1H), 6.73 (d, J = 15.4 Hz, 1H), 4.52 (d, J = 9.2 Hz, 1H), 4.21 (d, J = 9.2 Hz, 1H), 3.38 (s, 2H), 3.16 (s, 3H), 3.05 (s, 3H), 1.53 (s, 3H). **^{13}C NMR** (100 MHz, $CDCl_3$) δ 166.8, 161.0, 142.1, 132.6, 129.8, 128.6, 122.1, 114.8, 110.6, 83.6, 45.6, 37.4, 35.9, 25.4, 17.8. **IR** (cm^{-1}, thin film) 2962, 2928, 2885, 1647, 1600, 1489, 1471, 1394, 1274, 1261, 1215, 1136. **HRMS** (DART+) Calc'd for $C_{15}H_{19}INO_2$ 372.04605, found, 372.04575.

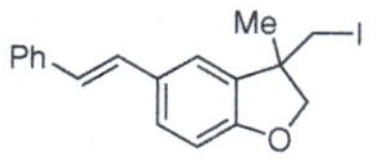

(±)-(*E*)-3-(iodomethyl)-3-methyl-5-styryl-2,3-dihydrobenzofuran (4.104d)—Synthesized according to **general procedure 2** using **4.75** 2 equivalents of styrene, and the reaction was heated for 24 h. The crude product was purified via silica gel flash column chromatography using hexanes:Et_2O (100:0 to 100:2.5 v:v) and was isolated as a white foam (62.2 mg, 0.165 mmol, 82%). **^{1}H NMR** (400 MHz, $CDCl_3$) δ 7.48 (d, *J* = 7.2 Hz, 1H), 7.39–7.18 (m, 5H), 7.06 (d, *J* = 16.2 Hz, 1H), 6.96 (d, *J* = 16.2 Hz, 1H), 6.79 (d, *J* = 8.3 Hz, 1H), 4.52 (d, *J* = 9.2 Hz, 1H), 4.23–4.16 (m, 1H), 3.42 (s, 2H), 1.55 (s, 3H). **^{13}C NMR** (100 MHz, $CDCl_3$) δ 159.4, 137.5, 132.4, 130.7, 128.6, 128.3, 128.2, 127.3, 126.6, 126.2, 120.5, 110.5, 83.4, 45.8, 25.4, 18.1. **IR** (cm^{-1}, thin film)2968, 2883, 1591, 1485, 1448, 1209, 962. **HRMS** (DART+) Calc'd for $C_{18}H_{18}IO$ 377.04023, found, 377.04096.

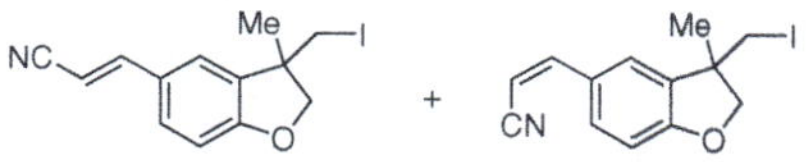

(±)-(*E*)-3-(3-(iodomethyl)-3-methyl-2,3-dihydrobenzofuran-5-yl)acrylonitrile (4.104e)—Synthesized according to **general procedure 2** using **4.75** and 1.46 equivalents of acrylonitrile, and the reaction was run for 24 h. ^{1}H NMR analysis of the crude reaction mixture showed a 4:1 ratio of the *E* and *Z* isomers, respectively. The crude product was purified via silica gel flash column chromatography using hexanes:Et_2O (20:1 to 4:1 v:v). Major isomer **4.104e-*trans*** was afforded as a colourless oil (42.0 mg, 0.129 mmol, 65%) and minor isomer **4.104e-*cis*** was afforded as a colourless oil which solidified upon storage to a white solid (16.0 mg, 0.0492 mmol, 25%, MP = 100–102 °C) for a combined yield of 58.0 mg (0.178 mmol, 90%). ***Major isomer***: **^{1}H NMR** (300 MHz, $CDCl_3$) δ 7.39–7.19 (m, 2H), 7.22 (d, *J* = 2.0 Hz, 1H), 6.81 (d, *J* = 8.4 Hz, 1H), 5.71 (dd, *J* = 16.5, 0.6 Hz, 1H), 4.55 (d, *J* = 9.4 Hz, 1H), 4.25 (d, *J* = 9.4 Hz, 1H), 3.38 (s, 2H), 1.54 (s, 3H). **^{13}C NMR** (75 MHz, $CDCl_3$) δ 162.3, 150.0, 150.0, 133.3, 129.8, 126.9, 121.8, 118.6, 110.9, 93.3, 93.3, 83.7, 45.5, 25.4, 17.3. **IR** (cm^{-1}, thin film) 3057, 2964, 2928, 2887, 2212, 1602, 1589, 1489, 1452, 1276, 1215. **HRMS** (DART+) Calc'd for $C_{13}H_{13}INO^+$ 326.00418, found, 326.00339. ***Minor isomer***: **^{1}H NMR** (300 MHz, $CDCl_3$) δ 7.74 (d, *J* = 2.0 Hz, 1H), 7.59 (ddd, *J* = 8.4, 2.0, 0.5 Hz, 1H), 7.03 (d, *J* = 12.1 Hz, 1H), 6.84 (d, *J* = 8.4 Hz, 1H), 5.29 (d, *J* = 12.1 Hz, 1H), 4.57 (d, *J* = 9.3 Hz, 1H), 4.26 (dd, *J* = 9.3, 0.7 Hz, 1H), 3.41 (s, 2H), 1.56 (s, 3H). **^{13}C NMR** (75 MHz, $CDCl_3$) δ 161.9, 148.1, 148.1, 133.0, 132.0, 127.1, 123.4, 118.0, 110.7, 91.8, 83.9, 45.7, 25.5, 17.4. **IR** (cm^{-1}, thin film) 2960, 2916, 2848, 2210, 1683, 1489, 1455. **HRMS** (DART+) Calc'd for $C_{13}H_{13}INO$ 326.00418. 326.00374.

(±)-(*E*)-methyl 3-(3-(iodomethyl)-3-methyl-2,3-dihydrobenzofuran-5-yl)-2-methylacrylate (4.104f)—Synthesized according to **general procedure 2** using **4.75** and 5 equivalents of methyl methacrylate, and the reaction was run for 48 h at 100 °C. The crude product was purified via silica gel flash column chromatography using hexanes:Et_2O (100:3 to 10:1 v:v) as the mobile phase, and was isolated as a clear and colourless oil (57.1 mg, 0.153 mmol, 77%). **^{1}H NMR** (400 MHz, $CDCl_3$) δ 7.64 (dd, *J* = 1.5, 0.8 Hz, 1H), 7.28 (ddd, *J* = 8.4, 1.9, 0.8 Hz, 1H), 7.19 (d, *J* = 1.9 Hz, 1H), 6.82 (d, *J* = 8.4 Hz, 1H), 4.53 (d, *J* = 9.2 Hz, 1H), 4.22 (d, *J* = 9.2 Hz, 1H), 3.80 (s, 3H), 3.39 (s, 2H), 2.13 (d, *J* = 1.5 Hz, 3H), 1.53 (s, 3H). **^{13}C NMR** (100 MHz, $CDCl_3$) δ 169.3, 159.8, 138.7, 132.2, 131.5, 128.9, 126.1, 124.5, 110.3, 83.4, 52.0, 45.7, 25.4, 17.8, 14.2. **IR** (cm^{-1}, thin film) 2951, 2926, 2883, 1699, 1627, 1606, 1593, 1489, 1435, 1379, 1356, 1317, 1282, 1244, 1213, 1197. **HRMS** (DART+) Calc'd for $C_{15}H_{18}IO_3$ 373.03006, found, 373.03090.

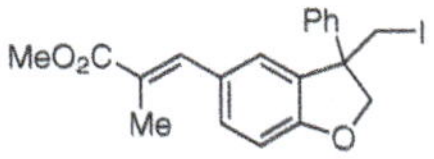

(±)-(*E*)-methyl 3-(3-(iodomethyl)-3-phenyl-2,3-dihydrobenzofuran-5-yl)-2-methylacrylate (4.104 g)—Synthesized according to **general procedure 2** using **4.82** and 3 equivalents of methyl methacrylate, and the reaction was run for 24 h at 100 °C. The crude product was purified via silica gel flash column chromatography using PhMe:EtOAc (100:1 v:v) as the mobile phase, and was isolated as a clear and colourless oil (22.0 mg, 0.05 mmol, 25%). **^{1}H NMR** (400 MHz, $CDCl_3$) δ 7.65 (d, *J* = 1.5 Hz, 1H), 7.41–7.20 (m, 9H), 6.91 (d, *J* = 8.4 Hz, 1H), 4.79 (d, *J* = 9.5 Hz, 1H), 4.76 (d, *J* = 9.5 Hz, 1H), 3.87 (d, *J* = 10.5 Hz, 1H), 3.81 (d, *J* = 10.5 Hz, 1H), 3.80 (s, 2H), 2.12 (d, *J* = 1.5 Hz, 3H). **^{13}C NMR** (100 MHz, $CDCl_3$) δ 169.3, 160.3, 142.0, 138.8, 131.8, 131.7, 128.9, 128.8, 127.4, 126.8, 126.4, 126.2, 110.4, 84.4, 54.0, 52.0, 15.3, 14.2. **IR** (cm^{-1}, thin film) 2949, 2922, 1705, 1604, 1488, 1434, 1238, 1112. **HRMS** (DART+) Calc'd for $C_{20}H_{20}IO_3$ 435.04571, found, 435.04512.

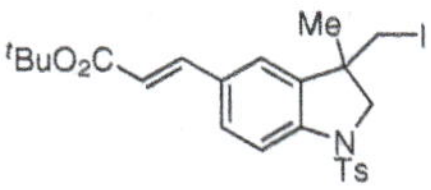

(±)-(*E*)-*tert*-butyl 3-(3-(iodomethyl)-3-methyl-1-tosylindolin-5-yl)acrylate (4.104 h)—Synthesized according to **general procedure 2** using **4.105** and 2 equivalents of *tert*-butyl acrylate, and the reaction was run for 48 h at 100 °C. The crude product was purified via silica gel flash column chromatography using hexanes:EtOAc (10:1 v:v) as the mobile phase, and was isolated as a colourless white foam (95.5 mg, 0.172 mmol, 86%). **^{1}H NMR** (400 MHz, $CDCl_3$) δ 7.73 (d, *J* = 8.2 Hz, 2H), 7.64 (d, *J* = 8.5 Hz, 1H), 7.50 (d, *J* = 15.9 Hz, 1H), 7.40 (dd, *J* = 8.5, 1.8 Hz, 1H), 7.27 (d, *J* = 8.2 Hz, 2H), 7.17 (d, *J* = 1.8 Hz, 1H), 6.26 (d, *J* = 15.9 Hz, 1H), 3.98 (d, *J* = 11.0 Hz, 1H), 3.60 (d, *J* = 11.0 Hz, 1H), 3.16 (d, *J* = 10.2 Hz, 1H), 2.88 (d, *J* = 10.2 Hz, 1H), 2.38 (s, 3H), 1.52 (s, 9H), 1.34 (s, 3H). **^{13}C NMR** (100 MHz, $CDCl_3$) δ 166.2, 144.6,

142.7, 142.5, 136.3, 133.8, 130.6, 129.9, 129.7, 127.2, 122.1, 119.1, 115.0, 80.4, 62.4, 43.9, 28.2, 26.0, 21.5, 17.6. **IR** (cm^{-1}, thin film) 2976, 2929, 2827, 1695, 1635, 1604, 1481, 1361, 1269, 1257, 1165, 1151. **HRMS** (DART+) Calc'd for $C_{24}H_{29}INO_4S$ 554.08620, found, 554.08535.

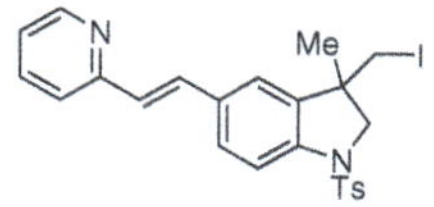

(±)-(*E*)-3-(iodomethyl)-3-methyl-5-(2-(pyridin-2-yl)vinyl)-1-tosylindoline (4.104i)—Synthesized according to **general procedure 2** using **4.91** and 2 equivalents of 2-vinyl pyridine, and the reaction was run for 24 h at 100 °C. The crude product was purified via silica gel chromatography using hexanes:EtOAc (10:3 v:v + 1% NEt_3) as the mobile phase and was isolated as an off-white foam which was triturated with hexanes:EtOAc which afforded the title compound as a white foam (88.8 mg, 0.167 mmol, 84%). **^{1}H NMR** (400 MHz, CDCl3) δ 8.67 (ddd, J = 4.9, 1.8, 0.8 Hz, 1H), 7.83 (d, J = 8.5 Hz, 2H), 7.78–7.68 (m, 2H), 7.65 (d, J = 16.0 Hz, 1H), 7.56 (dd, J = 8.5, 1.8 Hz, 1H), 7.44 (dt, J = 8.0, 1.0 Hz, 1H), 7.39–7.31 (m, 3H), 7.22 (ddd, J = 7.5, 4.9, 1.0 Hz, 1H), 7.15 (d, J = 16.0 Hz, 1H), 4.08 (d, J = 11.2 Hz, 1H), 3.69 (dd, J = 11.2, 1.0 Hz, 1H), 3.27 (dd, J = 10.2, 1.0 Hz, 1H), 2.96 (d, J = 10.2 Hz, 1H), 2.46 (s, 3H), 1.44 (s, 3H). **^{13}C NMR** (100 MHz, $CDCl_3$) δ 155.4, 149.6, 144.4, 141.2, 136.5, 136.1, 133.8, 132.7, 131.6, 129.8, 128.6, 127.2, 127.1, 122.0, 121.9, 121.2, 115.2, 62.4, 44.0, 26.0, 21.5, 18.0. **IR** (cm^{-1}, thin film) 2963, 2927, 1584, 1482, 1447, 1356, 1165, 1091. **HRMS** (ESI+) Calc'd for $C_{24}H_{24}IN_2O_2S$ 531.0598, found, 531.0594.

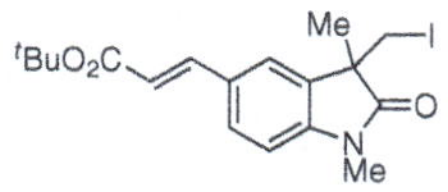

(±)-(*E*)-*tert*-butyl 3-(3-(iodomethyl)-1,3-dimethyl-2-oxoindolin-5-yl)acrylate (4.104j)—Synthesized according to **general procedure 2** using **4.89** and 2 equivalents of *tert*-butyl acrylate, and the reaction was run for 48 h. The crude product was purified via silica gel flash column chromatography using hexanes: EtOAc (10:2 v:v) as the mobile phase, and was isolated as a clear and light yellow oil (74.2 mg, 0.174 mmol, 87%). **^{1}H NMR** (500 MHz, $CDCl_3$) δ 7.57 (d, J = 15.9 Hz, 1H), 7.48–7.40 (m, 2H), 6.86 (d, J = 7.9 Hz, 1H), 6.30 (d, J = 15.9 Hz, 1H), 3.50 (d, J = 9.9 Hz, 1H), 3.40 (d, J = 9.9 Hz, 1H), 3.24 (s, 3H), 1.52 (s, 9H), 1.51 (s, 3H). **^{13}C NMR** (125 MHz, $CDCl_3$) δ 177.8, 166.3, 144.7, 143.1, 133.2, 129.7, 129.5, 121.5, 118.4, 108.4, 80.3, 48.5, 28.2, 26.4, 22.9, 10.2. **IR** (cm^{-1}, thin film) 2975, 2930, 1645, 1634, 1611, 1495, 1367, 1351, 1324, 1248. **HRMS** (DART+) Calc'd for $C_{18}H_{23}INO_3$ 428.07226, found, 428.07176.

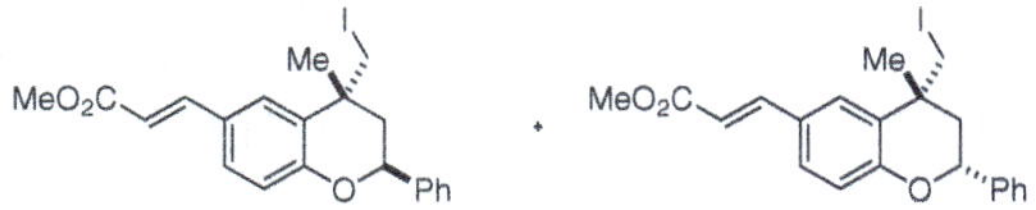

(±)-(*E*)-methyl 3-(4-(iodomethyl)-4-methyl-2-phenylchroman-6-yl)acrylate (4.104 k)—Synthesized according to **general procedure 2** using **1.169f** and 2 equivalents of methyl acrylate, and the reaction was heated for 24 h. The crude product was purified via silica gel chromatography using hexanes:Et_2O (100:1 to 10:1 v:v) as the mobile phase. Major diastereomer *anti*-**4.104k** was afforded as a clear light yellow oil (67.5 mg, 0.150 mmol, 75%) and minor isomer *syn*-**4.104k** was afforded as a clear colourless oil which solidified to a colourless solid upon storage (10.0 mg, 0.022 mmol, 11%, MP = 157–159 °C) for a combined yield of 77.5 mg (0.172 mmol, 86%). In a previously report from our lab1[a] we assigned the stereochemistry of the major product of substituted chromans synthesized by diastereoselective carboiodination to be the *trans* isomer, confirmed by X-ray analysis. Herein, the major isomer *anti*-**4.101k** is assigned based on analogy to this report, and is directly confirmed by the coupling constants of the protons on the unsaturated ring. ***Major isomer:*** **^{1}H NMR** (400 MHz, $CDCl_3$) δ 7.56 (d, *J* = 16.0 Hz, 1H), 7.45–7.20 (m, 7H), 6.87 (d, *J* = 8.5 Hz, 1H), 6.24 (d, *J* = 15.9 Hz, 1H), 4.92 (dd, *J* = 12.3, 2.2 Hz, 1H), 3.72 (s, 3H), 3.53 (d, *J* = 10.7 Hz, 1H), 3.45 (dd, *J* = 10.7, 1.6 Hz, 1H), 2.31 (dd, *J* = 14.4, 2.2 Hz, 1H), 1.83–1.73 (m, 1H), 1.37 (s, 3H). **^{13}C NMR** (100 MHz, $CDCl_3$) δ 167.6, 156.1, 144.3, 140.4, 128.6, 128.2, 128.1, 127.5, 127.2, 125.8, 125.6, 118.3, 115.5, 74.2, 51.6 42.3, 34.5, 27.3, 21.3. **IR** (cm^{-1}, thin film) 2964, 2949, 2916, 1708, 1635, 1604, 1573, 1492, 1454, 1435, 1413, 1240, 1193. **HRMS** (DART+) Calc'd for $C_{21}H_{22}IO_3^+$ 449.06163 found, 449.06078. ***Minor isomer***: **^{1}H NMR** (400 MHz, $CDCl_3$) δ 7.65 (d, *J* = 16.0 Hz, 1H), 7.52–7.27 (m, 7H), 6.94 (d, *J* = 8.5 Hz, 1H), 6.32 (d, *J* = 16.0 Hz, 1H), 5.09 (dd, *J* = 11.8, 2.0 Hz, 1H), 3.80 (s, 3H), 3.62 (d, *J* = 10.2 Hz, 1H), 3.47 (d, *J* = 10.2 Hz, 1H), 2.28–2.13 (m, 1H), 1.99 (dd, *J* = 13.9, 2.0 Hz, 1H), 1.63 (s, 3H). **^{13}C NMR** (100 MHz, $CDCl_3$) δ 167.7, 156.6, 144.5, 140.3, 128.7, 128.3, 127.6, 127.6, 127.5, 127.3, 126.2, 118.4, 115.5, 75.4, 51.6, 44.5, 35.6, 29.4, 21.4. **IR** (cm^{-1}, thin film) 2949, 2924, 1708, 1635, 1604, 1573, 1492, 1435, 1240. **HRMS** (DART+) Calc'd for $C_{21}H_{22}IO_3$ 449.06136 found, 449.06096.

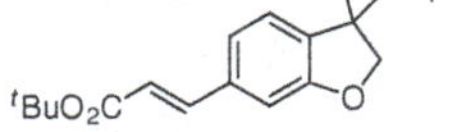

(±)-(*E*)-tert-butyl-3-(3-(iodomethyl)-3-methyl-2,3-dihydrobenzofuran-6-yl)acrylate (4.104l)—Synthesized according to **general procedure 2** using **4.77** and 1.46 equivalents of *tert*-butyl acrylate, and the reaction was run for 24 h. The crude product was purified via silica gel flash column chromatography using hexanes: EtOAc (100:0 to 100:3 v:v) as the mobile phase, and was isolated as a clear and colourless oil (72.4 mg, 0.181 mmol, 91%). **^{1}H NMR** (400 MHz, $CDCl_3$) δ 7.52 (d, *J* = 16.0 Hz, 1H), 7.10 (d, *J* = 7.7 Hz, 1H), 7.04 (dd, *J* = 7.7, 1.5 Hz, 1H), 6.94 (d, *J* = 1.5 Hz, 1H), 6.30 (d, *J* = 16.0 Hz, 1H), 4.50 (d, *J* = 9.2 Hz, 1H), 4.19 (d, *J* = 9.2 Hz, 1H), 3.37 (s, 2H), 1.52 (s, 9H), 1.51 (s, 3H). **^{13}C NMR** (100 MHz, $CDCl_3$) δ 166.2, 160.1, 143.1, 136.1, 134.0, 122.9, 121.7, 120.4, 108.7, 83.2, 80.5,

45.8, 28.2, 25.3, 17.5. **IR** (cm^{-1}, thin film) 2975, 2929, 1706, 1635, 1432, 1367, 1336, 1152. **HRMS** (DART+) Calc'd for $C_{17}H_{22}IO_3$ 401.06136, found, 401.06195.

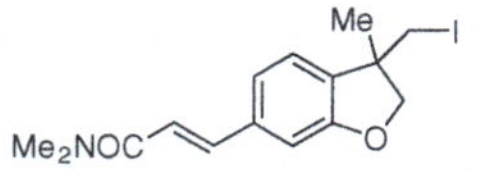

(±)-(*E*)-3-(3-iodomethyl)-3-methyl-2,3-dihydrobenzofuran-6-yl)-*N*,*N*-dimethylacrylamide (4.104m)—Synthesized according to **general procedure 2** using **4.77** and 3.5 equivalents of *N*,*N*-dimethylacrylamide, and the reaction was run for 24 h. The crude product was purified via silica gel chromatography using hexanes:EtOAc (1:2 v:v + 1% NEt_3) as the mobile phase, and was isolated as a light pink solid. This was recrystalized from hot hexanes to afford a light yellow solid (65.1 mg, 0.175 mmol, 88% MP = 117–119 °C). **^{1}H NMR** (400 MHz, $CDCl_3$) δ 7.59 (d, *J* = 15.4 Hz, 1H), 7.09 (d, *J* = 7.7 Hz, 1H), 7.04 (dd, *J* = 7.7, 1.4 Hz, 1H), 6.96 (d, *J* = 1.4 Hz, 1H), 6.82 (d, *J* = 15.4 Hz, 1H), 4.49 (d, *J* = 9.1 Hz, 1H), 4.17 (d, *J* = 9.1 Hz, 1H), 3.36 (s, 2H), 3.14 (s, 3H), 3.04 (s, 3H), 1.50 (s, 3H). **^{13}C NMR** (100 MHz, $CDCl_3$) δ 166.4, 160.0, 141.9, 136.8, 133.5, 122.9, 121.6, 117.6, 108.3, 83.2, 45.7, 37.3, 35.9, 25.2, 17.6. **IR** (cm^{-1}, thin film) 2927, 2645, 1606, 1497, 1433, 1409, 1262, 1218, 1154, 1133, 913. **HRMS** (DART+) Calc'd for $C_{15}H_{19}INO_2^+$ 372.04605, found, 372.04512.

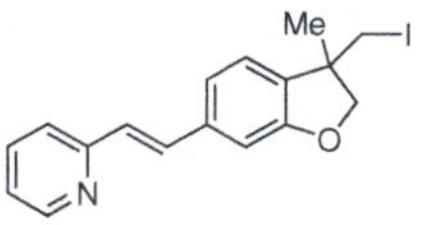

(±)-(*E*)-2-(2-(3-(iodomethyl)-3-methyl-2,3-dihydrobenzofuran-6- yl)vinyl)pyridine (4.104n)—Synthesized according to **general procedure 2** using **4.77** and 2 equivalents of 2-vinylpyridine, and the reaction was run for 24 h. The crude product was purified via silica gel chromatography using hexanes:EtOAc (10:2 v:v + 1% NEt_3) and was isolated as light pink oil. This was further purified by loading oil onto a short silica plug and first eluting with hexanes:DCM (1:1 v:v) and then 100% EtOAc, which afforded the title compound as a clear and light yellow oil (55.9 mg, 0.148 mmol, 74%). **^{1}H NMR** (400 MHz, $CDCl_3$) δ 8.60 (ddd, *J* = 4.9, 1.8, 1.0 Hz, 1H), 7.65 (td, *J* = 7.9, 1.8 Hz, 1H), 7.59 (d, *J* = 16.0 Hz, 1H), 7.36 (dt, *J* = 7.9, 1.0 Hz, 1H), 7.16–7.08 (m, 4H), 7.06–7.01 (m, 1H), 4.51 (d, *J* = 9.1 Hz, 1H), 4.19 (d, *J* = 9.1, 1H), 3.39 (s, 2H), 1.52 (s, 3H). **^{13}C NMR** (100 MHz, $CDCl_3$) δ 160.1, 155.4, 149.67, 138.2, 136.5, 132.3, 132.1, 128.2, 122.8, 122.2, 122.1, 120.7, 108.0, 83.2, 45.8, 25.3, 18.0. **IR** (cm^{-1}, thin film) 2963, 1611, 1583, 1564, 1487, 1463, 1431, 1262, 1216, 1117. **HRMS** (DART+) Calc'd for $C_{17}H_{17}INO^+$ 378.03548, found, 378.03398.

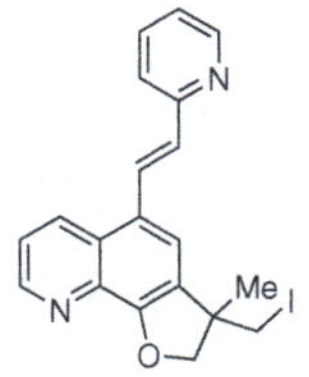

(±)-(*E*)-3-(iodomethyl)-3-methyl-5-(2-(pyridin-2-yl)vinyl)-2,3-dihydrofuro[3,2-*h*]quinoline (4.104o)—Synthesized according to **general procedure 2** using **4.99** and 2 equivalents of 2-vinylpyridine, and the reaction was run for 48 h. The crude product was purified via silica gel chromatography using hexanes:EtOAc (5:6 v: v + 1% NEt_3) as the mobile phase, and was isolated as a yellow oil (59.9 mg, 0.140 mmol, 70%). **^{1}H NMR** (400 MHz, $CDCl_3$) δ 8.91 (dd, *J* = 4.1, 1.6 Hz, 1H), 8.71–8.58 (m, 2H), 8.33 (d, *J* = 15.7 Hz, 1H), 7.73–7.67 (m, 2H), 7.45 (dd, *J* = 8.7, 4.1 Hz, 1H), 7.43–7.40 (m, 1H), 7.22–7.14 (m, 2H), 4.81 (d, *J* = 9.3 Hz, 1H), 4.52 (d, *J* = 9.3 Hz, 1H), 3.55 (d, *J* = 10.1 Hz, 1H), 3.51 (d, *J* = 10.1 Hz, 1H), 1.67 (s, 3H) **^{13}C NMR** (100 MHz, $CDCl_3$) δ 155.4, 155.4, 150.0, 149.8, 136.6, 136.2, 132.6, 129.5, 129.3, 128.0, 127.8, 127.6, 122.3, 122.2, 121.5, 118.6, 84.3, 47.1, 25.8, 17.6. **IR** (cm^{-1}, thin film) 2961, 1583, 1550, 1506, 1471, 1458, 1322, 1086, 961. **HRMS** (DART+) Calc'd for $C_{20}H_{18}IN_2O^+$ 429.04638, found, 429.04639.

References

1. Petrone, D.A., Le, C.M., Newman, S.G., Lautens, M.: Pd-Catalyzed carboiodination: early developments to recent advancements (Chap. 7). In: Colacot, T. (ed.) New Trends in Cross-Coupling. The Royal Society of Chemistry, Cambridge (2015)
2. Petrone, D.A., Lischka, M., Lautens, M.: Angew. Chem. Int. Ed. **10652**, 10635 (2013)
3. Schröter, S., Stock, C., Bach, T.: Tetrahedron **61**, 2245 (2005)
4. Wang, J.-R., Manabe, K.: Synthesis 1405 (2009)
5. Fairlamb, I.J.S.: Chem. Soc. Rev. **36**, 1036 (2007)
6. Minato, A., Suzuki, K., Tamao, K.: J. Am. Chem. Soc. **109**, 1257 (1987)
7. Evans, D.A., Starr, J.T.: Angew. Chem. Int. Ed. **41**, 1787 (2002)
8. Roush, W.R., Moriarty, K.J., Brown, B.B.: Tetrahedron Lett. **31**, 6509 (1990)
9. Bellina, F., Carpita, A., de Santis, M.D., Rossi, R.: Tetrahedron Lett. **35**, 6913 (1994)
10. Ratoveloamanana, V., Linstrumelle, G.: Tetrahedron Lett. **26**, 2575 (1985)
11. Handy, S.T., Zhang, Y.: Chem. Commun. 299 (2006)
12. Legault, C.Y., Garcia, Y., Merlic, C.A., Houk, K.N.: J. Am. Chem. Soc. **129**, 12664 (2007)
13. Garcia, Y., Schoenebeck, F., Legault, C.Y., Merlic, C.A., Houk, K.N.: J. Am. Chem. Soc. **131**, 6632 (2009)
14. Kada, R., Knoppova, B., Kováč, P.: Collect. Czech. Chem. Commun. **49**, 984 (1984)
15. Bach, T., Krüger, L.: Tetrahedron Lett. **39**, 1729 (1998)
16. Conreaux, D., Bossharth, E., Monteiro, N., Desbordes, P., Vors, J., Balme, G.: Org. Lett. **9**, 271 (2007)
17. Reyes, M.J., Castillo, R., Izquierdo, M.L., Alvarez-Builla, J.: Tetrahedron Lett. **47**, 6457 (2006)
18. Fahey, D.R.: J. Am. Chem. Soc. **92**, 402 (1970)
19. Just, G., Singh, R.J.: Org Chem. **54**, 4453 (1989)
20. Hayashi, T., Niizuma, S., Kamikawa, T., Suzuki, N., Uozumi, Y.: J. Am. Chem. Soc. **117**, 9101 (1995)
21. Lautens, M., Fang, Y.-Q.: Org. Lett. **5**, 3679 (2003)
22. Saget, T., Perez, D., Cramer, N.: Org. Lett. **15**, 1354 (2013)
23. Roy, A.H., Hartwig, J.F.: J. Am. Chem. Soc. **123**, 1232 (2001)
24. Roy, A.H., Hartwig, J.F.: Organometallics **23**, 1533 (2004)
25. Roy, A.H., Hartwig, J.F.: J. Am. Chem. Soc. **125**, 13944 (2003)

26. Newman, S.G., Lautens, M.: J. Am. Chem. Soc. **132**, 11416 (2010)
27. Nareddy, R., Mantilli, L., Guénée, L., Mazet, C.: Angew. Chem. Int. Ed. **51**, 3826 (2013)
28. Quesnel, J.S., Arndtsen, B.A.: J. Am. Chem. Soc. **135**, 16841 (2013)
29. Quesnel, J.S., Kayser, L.V., Fabrikant, A., Arndtsen, B.A.: Chem. Eur. J. **21**, 9550 (2015)
30. Fahey, D.R.: J. Organomet. Chem. **27**, 283 (1971)
31. Stambuli, J.P., Bühl, M., Hartwig, J.F.: J. Am. Chem. Soc. **124**, 9346 (2002)
32. Lu, Z., Hu, C., Guo, J., Li, J., Cui, Y., Jia, Y.: Org. Lett. **12**, 480 (2010)
33. Johansson Seechurn, C.C.C., Parisel, S.L., Colacot, T.J.: J. Org. Chem **76**, 7918 (2011)
34. Li, H., Johansson Seechurn, C.C.C., Colacot, T.J.: ACS Catal. **2**, 1147 (2012)
35. Pu, X., Li, H., Colacot, T.J.: J. Org. Chem. **78**, 568 (2013)
36. Edgar, K.J., Falling, S.N.: J. Org. Chem. **55**, 5287 (1990)
37. Flydare, J.A., et al.: 1-(chloromethyl)-2,3-hihydro-1H-benzo[*e*]indole dimer antibody-drug conjugate compounds, and methods of use and treatment. WO Patent **23355**, A1 (2015)
38. Lulinski, P., Kryska, A., Sasnowski, M., Skulski, L.: Synthesis **3**, 441 (2004)
39. Larock, R.C., Harrison, L.W.: J. Am. Chem. Soc. **106**, 4218 (1984)

The manufacturer's authorised representative in the EU is Springer
Nature Customer Service Centre GmbH, Europaplatz 3, 69115 Heidelberg,
Germany. If you have any concerns regarding our products, please
contact ProductSafety@springernature.com

Printed and bound by CPI Group (UK) Ltd, Croydon, CR0 4YY
30/09/2026
02225658-0013